Decomposition Techniques in Mathematical Programming

Antonio J. Conejo   Enrique Castillo
Roberto Mínguez   Raquel García-Bertrand

Decomposition Techniques in Mathematical Programming

Engineering and Science Applications

Professor Antonio J. Conejo
Universidad de Castilla – La Mancha
E.T.S. Ingenieros Industriales
Avda. Camilo José Cela s/n
13071 Ciudad Real
Spain
E-mail: antonio.conejo@uclm.es

Dr. Roberto Mínguez
Universidad de Castilla – La Mancha
E.T.S. Ingenieros de Caminos
Avda. Camilo José Cela s/n
13071 Ciudad Real
Spain
E-mail: roberto.minguez@uclm.es

Professor Enrique Castillo
Escuela de Ingenieros de Caminos
Universidad de Cantabria
Avda. de los Castros s/n
39005 Santander
Spain
E-mail: castie@unican.es

Dr. Raquel García-Bertrand
Universidad de Castilla – La Mancha
E.T.S. Ingenieros Industriales
Avda. Camilo José Cela s/n
13071 Ciudad Real
Spain
E-mail: raquel.garcia@uclm.es

Library of Congress Control Number: 2005934995

ISBN-10 3-540-27685-8 Springer Berlin Heidelberg New York
ISBN-13 978-3-540-27685-2 Springer Berlin Heidelberg New York

This work is subject to copyright. All rights are reserved, whether the whole or part of the material is concerned, specifically the rights of translation, reprinting, reuse of illustrations, recitation, broadcasting, reproduction on microfilm or in any other way, and storage in data banks. Duplication of this publication or parts thereof is permitted only under the provisions of the German Copyright Law of September 9, 1965, in its current version, and permission for use must always be obtained from Springer. Violations are liable for prosecution under the German Copyright Law.

Springer is a part of Springer Science+Business Media
springer.com
© Springer-Verlag Berlin Heidelberg 2006
Printed in The Netherlands

The use of general descriptive names, registered names, trademarks, etc. in this publication does not imply, even in the absence of a specific statement, that such names are exempt from the relevant protective laws and regulations and therefore free for general use.

Typesetting: by the authors and TechBooks using a Springer LaTeX macro package
Cover design: *design & production* GmbH, Heidelberg

Printed on acid-free paper   SPIN: 11511946   89/TechBooks   5 4 3 2 1 0

To Núria, Mireia Zhen and Olaia Xiao

To my family

To my parents, my sister, my grandma, and my aunt

To José Agustín, to my brothers Javier and Jorge Luis,
and especially to my parents

Preface

Optimization plainly dominates the design, planning, operation, and control of engineering systems. This is a book on optimization that considers particular cases of optimization problems, those with a decomposable structure that can be advantageously exploited. Those decomposable optimization problems are ubiquitous in engineering and science applications. The book considers problems with both complicating constraints and complicating variables, and analyzes linear and nonlinear problems, with and without integer variables. The decomposition techniques analyzed include Dantzig-Wolfe, Benders, Lagrangian relaxation, Augmented Lagrangian decomposition, and others. Heuristic techniques are also considered.

Additionally, a comprehensive sensitivity analysis for characterizing the solution of optimization problems is carried out. This material is particularly novel and of high practical interest.

This book is built based on many clarifying, illustrative, and computational examples, which facilitate the learning procedure. For the sake of clarity, theoretical concepts and computational algorithms are assembled based on these examples. The results are simplicity, clarity, and easy-learning.

We feel that this book is needed by the engineering community that has to tackle complex optimization problems, particularly by practitioners and researchers in Engineering, Operations Research, and Applied Economics. The descriptions of most decomposition techniques are available only in complex and specialized mathematical journals, difficult to understand by engineers. A book describing a wide range of decomposition techniques, emphasizing problem-solving, and appropriately blending theory and application, was not previously available.

The book is organized in five parts. Part I, which includes Chapter 1, provides motivating examples and illustrates how optimization problems with decomposable structure are ubiquitous. Part II describes decomposition theory, algorithms, and procedures. Particularly, Chapter 2 and 3 address solution procedures for linear programming problems with complicating constraints and complicating variables, respectively. Chapter 4 reviews and summarizes duality theory. Chapter 5 describes decomposition techniques appropriate for continuous nonlinear programming problems. Chapter 6 presents decomposition procedures relevant for mixed-integer linear and nonlinear problems. Chapter 7 considers specific decomposition techniques not analyzed in the previous chapters. Part III, which includes Chapter 8, provides a comprehensive treatment of sensitivity analysis. Part IV provides in Chapter 9 some case studies of clear interest for the engineering profession. Part V contains some of the codes in GAMS used throughout the book. Finally, Part VI contains the solutions of the even exercises proposed throughout the book.

Relevant features of this book are

1. It provides an appropriate blend of theoretical background and practical applications in engineering and science.
2. Many examples, clarifying, illustrative, and computational, are provided.
3. Applications encompass electrical, mechanical, energy, and civil engineering as well as applied mathematics and applied economics.
4. The theoretical background of the book is deep enough to be of interest to applied mathematicians.
5. Practical applications are developed up to working algorithms that can be readily used.
6. The book includes end-of-chapter exercises and the solutions of the even numbered exercises are given in a Part VI. This makes the book very practical as a textbook for graduate and postgraduate courses.
7. The book addresses decomposition in linear programming, mixed-integer linear programming, nonlinear programming, and mixed-integer nonlinear programming. It provides rigorous decomposition algorithms as well as heuristic ones.

Required background to fully understand this book is moderate and includes elementary algebra and calculus, and basic knowledge of linear and nonlinear programming.

Over the last two decades, the two senior authors of this book have been involved in research projects that required the solution of large-scale complex optimization problems. We have received advice and relevant observations from many colleagues. We would like to express our appreciation to Prof. Gerald B. Sheblé from Iowa State University, Prof. Mohammad Shahidehpour from Illinois Institute of Technology, Prof. Francisco D. Galiana from McGill University, Prof. Víctor H. Quintana from University of Waterloo, Prof. Francisco J. Prieto from Universidad Carlos III of Madrid, and Prof. Benjamin F. Hobbs from Johns Hopkins University.

We are also thankful to quite a few colleagues and former students for suggestions and insightful observations that have improved our book. Particularly, we would like to thank Prof. Steven A. Gabriel from the University of Maryland, and Prof. Bruce F. Wollenberg from the University of Minnesota.

We are deeply grateful to the University of Castilla-La Mancha, Spain, for providing us with an outstanding research environment.

Ciudad Real and Santander, Spain
June 2005

A.J. Conejo
E. Castillo
R. Mínguez
R. García-Bertrand

Contents

Part I Motivation and Introduction

1 **Motivating Examples** 3
   1.1   Motivation 3
   1.2   Introduction 7
   1.3   Linear Programming: Complicating Constraints 8
        1.3.1   Transnational Soda Company 8
        1.3.2   Stochastic Hydro Scheduling 12
        1.3.3   River Basin Operation 19
        1.3.4   Energy Production Model 23
   1.4   Linear Programming: Complicating Variables............... 28
        1.4.1   Two-Year Coal and Gas Procurement 28
        1.4.2   Capacity Expansion Planning 32
        1.4.3   The Water Supply System 36
   1.5   Nonlinear Programming: Complicating Constraints.......... 39
        1.5.1   Production Scheduling 39
        1.5.2   Operation of a Multiarea Electricity Network........ 42
        1.5.3   The Wall Design................................. 45
        1.5.4   Reliability-based Optimization
                   of a Rubblemound Breakwater 48
   1.6   Nonlinear Programming: Complicating Variables............ 53
        1.6.1   Capacity Expansion Planning: Revisited 53
   1.7   Mixed-Integer Programming: Complicating Constraints 55
        1.7.1   Unit Commitment 55
   1.8   Mixed-Integer Programming: Complicating Variables 57
        1.8.1   Capacity Expansion Planning: Revisited 2 57
        1.8.2   The Water Supply System: Revisited 60
   1.9   Concluding Remarks 61
   1.10   Exercises 62

Part II Decomposition Techniques

2 Linear Programming: Complicating Constraints 67
   2.1   Introduction 67
   2.2   Complicating Constraints: Problem Structure 70
   2.3   Decomposition 73
   2.4   The Dantzig-Wolfe Decomposition Algorithm................ 77
        2.4.1   Description 77
        2.4.2   Bounds... 87
        2.4.3   Issues Related to the Master Problem 88
        2.4.4   Alternative Formulation of the Master Problem...... 93
   2.5   Concluding Remarks 99
   2.6   Exercises 100

3 Linear Programming: Complicating Variables 107
   3.1   Introduction 107
   3.2   Complicating Variables: Problem Structure 110
   3.3   Benders Decomposition 111
        3.3.1   Description 111
        3.3.2   Bounds.. 116
        3.3.3   The Benders Decomposition Algorithm 116
        3.3.4   Subproblem Infeasibility 128
   3.4   Concluding Remarks 135
   3.5   Exercises 136

4 Duality 141
   4.1   Introduction 141
   4.2   Karush–Kuhn–Tucker First- and Second-Order Optimality
        Conditions... 142
        4.2.1   Equality Constraints and Newton Algorithm 147
   4.3   Duality in Linear Programming........................... 149
        4.3.1   Obtaining the Dual Problem from a Primal Problem
               in Standard Form.................................. 150
        4.3.2   Obtaining the Dual Problem 151
        4.3.3   Duality Theorems 154
   4.4   Duality in Nonlinear Programming....................... 161
   4.5   Illustration of Duality and Separability 176
   4.6   Concluding Remarks 181
   4.7   Exercises 181

5 Decomposition in Nonlinear Programming 187
   5.1   Introduction 187
   5.2   Complicating Constraints 187
   5.3   Lagrangian Relaxation 187

		5.3.1	Decomposition 188

         5.3.1   Decomposition 188
         5.3.2   Algorithm 194
         5.3.3   Dual Infeasibility 195
         5.3.4   Multiplier Updating............................ 195
   5.4   Augmented Lagrangian Decomposition 205
         5.4.1   Decomposition 205
         5.4.2   Algorithm 207
         5.4.3   Separability 208
         5.4.4   Multiplier Updating............................ 208
         5.4.5   Penalty Parameter Updating 208
   5.5   Optimality Condition Decomposition (OCD) 210
         5.5.1   Motivation: Modified Lagrangian Relaxation 211
         5.5.2   Decomposition Structure 213
         5.5.3   Decomposition 214
         5.5.4   Algorithm 216
         5.5.5   Convergence Properties......................... 217
   5.6   Complicating Variables 223
         5.6.1   Introduction 223
         5.6.2   Benders Decomposition......................... 223
         5.6.3   Algorithm 225
   5.7   From Lagrangian Relaxation to Dantzig-Wolfe Decomposition 233
         5.7.1   Lagrangian Relaxation in LP..................... 234
         5.7.2   Dantzig-Wolfe from Lagrangian Relaxation 236
   5.8   Concluding Remarks 238
   5.9   Exercises 239

6 Decomposition in Mixed-Integer Programming 243
   6.1   Introduction 243
   6.2   Mixed-Integer Linear Programming 244
         6.2.1   The Benders Decomposition for MILP Problems..... 245
         6.2.2   Convergence 250
   6.3   Mixed-Integer Nonlinear Programming 251
   6.4   Complicating Variables: Nonlinear Case 251
         6.4.1   The Benders Decomposition 251
         6.4.2   Subproblem Infeasibility 253
         6.4.3   Convergence 257
   6.5   Complicating Constraints: Nonlinear Case 257
         6.5.1   Outer Linearization Algorithm 258
         6.5.2   Convergence 264
   6.6   Concluding Remarks 264
   6.7   Exercises 264

7 Other Decomposition Techniques ... 271
- 7.1 Bilevel Decomposition ... 271
  - 7.1.1 A Relaxation Method ... 272
  - 7.1.2 The Cutting Hyperplane Method ... 277
- 7.2 Bilevel Programming ... 280
- 7.3 Equilibrium Problems ... 282
- 7.4 Coordinate Descent Decomposition ... 285
  - 7.4.1 Banded Matrix Structure Problems ... 287
- 7.5 Exercises ... 297

Part III Local Sensitivity Analysis

8 Local Sensitivity Analysis ... 303
- 8.1 Introduction ... 303
- 8.2 Statement of the Problem ... 304
- 8.3 Sensitivities Based on Duality Theory ... 305
  - 8.3.1 Karush–Kuhn–Tucker Conditions ... 305
  - 8.3.2 Obtaining the Set of All Dual Variable Values ... 307
  - 8.3.3 Some Sensitivities of the Objective Function ... 308
  - 8.3.4 A Practical Method for the Sensitivities of the Objective Function ... 310
  - 8.3.5 A General Formula for the Sensitivities of the Objective Function ... 310
- 8.4 A General Method for Obtaining All Sensitivities ... 315
  - 8.4.1 Determining the Set of All Feasible Perturbations ... 317
  - 8.4.2 Discussion of Directional and Partial Derivatives ... 318
  - 8.4.3 Determining Directional Derivatives if They Exist ... 320
  - 8.4.4 Partial Derivatives ... 320
  - 8.4.5 Obtaining All Sensitivities at Once ... 321
- 8.5 Particular Cases ... 321
  - 8.5.1 No Constraints ... 321
  - 8.5.2 Same Active Constraints ... 323
  - 8.5.3 The General Case ... 326
- 8.6 Sensitivities of Active Constraints ... 339
- 8.7 Exercises ... 341

Part IV Applications

9 Applications ... 349
- 9.1 The Wall Design ... 349
  - 9.1.1 Method 1: Updating Safety Factor Bounds ... 355
  - 9.1.2 Method 2: Using Cutting Planes ... 359
- 9.2 The Bridge Crane Design ... 361

		9.2.1	Obtaining Relevant Constraints 364
		9.2.2	A Numerical Example 365
	9.3	Network Constrained Unit Commitment 368
		9.3.1	Introduction 368
		9.3.2	Notation 369
		9.3.3	Problem Formulation 370
		9.3.4	Solution Approach 371
	9.4	Production Costing 374
		9.4.1	Introduction 374
		9.4.2	Notation 375
		9.4.3	Problem Formulation 376
		9.4.4	Solution Approach 377
	9.5	Hydrothermal Coordination 381
		9.5.1	Introduction 381
		9.5.2	Notation 382
		9.5.3	Problem Formulation 383
		9.5.4	Solution Approach 384
	9.6	Multiarea Optimal Power Flow 385
		9.6.1	Introduction 385
		9.6.2	Notation 386
		9.6.3	Problem Formulation 387
		9.6.4	Solution Approach 389
	9.7	Sensitivity in Regression Models 389

Part V Computer Codes

A Some GAMS Implementations 397
    A.1 Dantzig-Wolfe Algorithm 397
    A.2 Benders Decomposition Algorithm 403
    A.3 GAMS Code for the Rubblemound Breakwater Example 407
    A.4 GAMS Code for the Wall Problem 410
        A.4.1 The Relaxation Method 410
        A.4.2 The Cutting Hyperplanes Method 414

Part VI Solution to Selected Exercises

B Exercise Solutions 421
    B.1 Exercises from Chapter 1 421
    B.2 Exercises from Chapter 2 426
    B.3 Exercises from Chapter 3 435
    B.4 Exercises from Chapter 4 441
    B.5 Exercises from Chapter 5 451
    B.6 Exercises from Chapter 6 475

B.7 Exercises from Chapter 7 500
B.8 Exercises from Chapter 8 506

References 531

Index 537

Part I

Motivation and Introduction

1
Motivating Examples: Models with Decomposable Structure

1.1 Motivation

Optimization plainly dominates the design, operation and planning of engineering systems. For instance, a bridge is designed by minimizing its building costs but maintaining appropriate security standards. Similarly, railway systems are expanded and operated to minimize building and operation costs while maintaining operation and security standards. Analogously, an electric energy system is operated so that the power demanded is supplied at minimum cost while enforcing appropriate security margins.

Optimization also pertains to everyday decision making. For example, we try to buy the best house provided our budget is sufficient, and we look for the best college education for our children provided we can afford it. It also concerns minor decisions such as buying the best cup of coffee at a reasonable price, and the like. Therefore, optimization is part of our everyday activities.

Therefore, *optimization* is "the science of the best" in the sense that it helps us to make not just a reasonable decision, but the best decision *subject to* observing certain constraints describing the domain within which the decision has to be made. *Mathematical programming models* provide the appropriate framework to address these optimization decisions in a precise and formal manner.

The target or objective to be maximized (or minimized) is expressed by means of a real-valued mathematical function denominated "objective function," because this function is the objective to be maximized (or minimized). This function depends on one or several "decision" variables whose optimal values are sought.

The restrictions that have to be satisfied define what is denominated the "feasibility region" of the problem. It should be noted that this feasibility region should include many possible decisions (alternative values for the decision variables) for the optimization problem to make sense. If no decision or just one decision is possible, the optimization problem lacks practical interest because its solution is just the given unique feasible decision. The feasibility region is formally defined through equality and inequality conditions that are denominated the constraints of the problem. Each one of these conditions is mathematically expressed through one real-valued function of the decision variables. This function is equal to zero, greater than or equal to zero, lower than or equal to zero.

Therefore a mathematical programming problem, representing and optimization decision framework, presents the formal structure below:

$$\boxed{\begin{array}{ll} minimize & \text{objective function} \\ subject\ to & \text{constraints} \end{array}}$$

Depending upon the type of variables and the mathematical nature of the objective function and the functions used for the constraints, mathematical programming problems are classified in different manners. If the variables involved are continuous and both the objective function and the constraints are linear, the problem is denominated "linear programming problem." If any of the variables involved is integer or binary, while the constraints and the objective function are both linear, the problem is denominated "mixed-integer linear programming problem."

Analogously, if the objective function or any constraint is nonlinear and all variables are continuous, the problem is denominated "nonlinear programming problem." If additionally, any variable is integer, the corresponding problem is denominated "mixed-integer nonlinear programming problem."

Generally speaking, linear programming problems are routinely solved even if they involve hundred of thousands of variables and constraints. Nonlinear programming problems are easily solved provided that they meet certain regularity conditions related to the mathematical concept of convexity, which is considered throughout the following chapters of this book. Mixed-integer linear programming problems are routinely solved provided that the number of integer variables is sufficiently small, typically below one thousand. Mixed-integer nonlinear programming problems are generally hard to solve and can be numerically intractable, because: (a) a high number of integer variables, and (b) the ugly mathematical properties of the functions involved. These problems require an in-depth specific analysis before a solution procedure is tried.

From an engineering point of view, operation problems involving engineering systems are normally continuous problems, either linear or nonlinear. However, design and capacity expansion planning problems are generally mixed-integer linear or nonlinear problems. The reason is that some design or planning variables are of integer nature while most operation variables are of continuous nature.

This book considers particular cases of all these optimization problems. These cases have structural properties that can be advantageously computationally exploited. These structural properties are briefly illustrated and described below.

1.1 Motivation

Many real-world systems present certain *decentralized* structures. For instance, the highway network of two neighboring countries is in some cases highly dense in the interior of each country but rather lightly connected at the border boundary. The same fact is encountered, for instance, in telecommunication and electric energy systems.

On the other hand, investment decision of real-world systems are typically of integer nature while subsequent operation decisions are continuous. For instance, investment decisions to improve the water supply system of a city are integer because a particular facility exist or does not exist, and a facility is built instead of other possible alternatives. However, the operation decisions of the already built water supply system are continuous, e.g., pressure values to be assigned at pumping stations.

The operation problem of the highway network of the two neighboring countries naturally decomposes by country provided that we take care of the constraints related to the interconnections. Note, however, that the operation solution obtained solving both national problems independently is not the best combined solution and it might be even infeasible due to the lack of consideration of the border constraints. In this book, techniques to properly take into account border constraints while solving the problem in a decomposed manner are described.

Note that the reason to address this transnational problem in a decentralized manner is not necessarily numerical, but political; as decentralized solutions are more acceptable from a social viewpoint.

Therefore a mathematical programming problem associated with the optimization model above presents the following formal structure:

$$
\begin{array}{lll}
maximize & \text{objective}_1 & + \quad \text{objective}_2 \\
subject\ to & \text{constraints}_1 & \\
& & \text{constraints}_2 \\
& \text{common constraints} & \text{common constraints}
\end{array}
$$

where the block of constraints "constraints$_1$" is related to the first terms of the objective function "objective$_1$" and corresponds to the first country; while the second block of constraints "constraints$_2$" is related to the second part of the objective function "objective$_2$" and is related to the second country. However, the block of common constraints affects both countries (they are border constraints) and complicates the solution of the problem because it prevents the *sub-problems* associated with the two countries to be solved separately. Because of the above considerations, the common constraints are denominated *complicating constraints*.

Finally, note that it would have been possible to consider any number of countries with different border structures instead of just two.

The capacity expansion problem of a water supply system involves integer investment decision and continuous operating decisions. It is convenient to process these two sets of decisions separately, due to their different nature (integer versus continuous).

A mathematical programming problem associated with this model presents the formal structure below:

$$\begin{array}{ll} maximize & \text{objective}_1 + \text{objective}_2 \\ subject\ to & \text{constraints}_1 \qquad\qquad\quad \&\ \text{constraints}_1(\text{compl. variables}) \\ & \text{constraints}_2\ \&\ \text{constraints}_2(\text{compl. variables}) \end{array}$$

where the common part of both blocks of constraints is related to investment (integer) decisions while the noncommon parts are related to the operations (continuous) decisions. In this book, techniques to take into account separately integer and continuous decisions while achieving a solution of the whole problem are described. Treating integer variables is much more complicated than treating continuous variables, and this is why problems including such variables are denominated problems with *complicating variables*.

Additionally, note that once the integer decisions have been made, the resulting *sub-problem* decomposes by blocks, which may clearly facilitate its solution.

In dealing with nonlinear programming problems throughout this book, we assume that these problems are convex. This is a requirement for most decomposition algorithm to work. Although convexity is a strong mathematical assumption, assuming convexity is not necessarily restrictive from a practical viewpoint, as many engineering and science problems are convex in the region where the solutions of interest are located, i.e., where solution are meaningful from an engineering or science point of view.

Nowadays, people do not become totally satisfied when getting the solutions to their problems; in addition, a sensitivity analysis is asked for. Sensitivity refers to how sensitive are the solutions of a problem to the assumptions and data. A high sensitivity warns the designer about the possible consequences that can follow if the assumptions or data used are far from reality and directs him/her to the right action to prevent disastrous consequences.

In this book the problem of sensitivity analysis is analyzed in detail and several procedures are presented to derive the sensitivities of optimization problems. In particular, we study the sensitivities of the objective function and constraints, and the sensitivities of the primal and dual variables to data.

There is a close connection between decomposition and sensitivity, as it is shown in several examples used in this book. Decomposition permits the easy obtention of some sensitivities that are difficult to obtain from the initial problem.

Within the framework above, the objectives pursued in this book are the following:

1. To motivate the interest of decomposition techniques in optimization using a rich spectrum of real-world engineering examples.
2. To describe decomposition techniques related to linear and nonlinear problems that include both continuous and integer variables. These problems must present appropriate structures.
3. To illustrate these decomposition techniques with relevant engineering problems of clear practical interest.
4. To provide practical algorithms that work and can be readily used by students and practitioners.
5. To introduce the reader to the problem of sensitivity analysis and provide tools and techniques to derive the sensitivities once the optimal solution of an optimization problem is already available.

We believe that no such a book is currently available, although it is much needed for both engineering and economics graduate students and practitioners.

The organization of this book is as follows:

1. Chapter 1 provides a collection of motivating examples that illustrates how optimization problems with decomposable structure are common in the real-world.
2. Chapter 2 addresses solution procedures for linear programming problems with complicating constraints.
3. Completing the analysis of linear programming problems, Chap. 3 considers solution techniques for linear programming problems with complicating variables.
4. Chapter 4 reviews and summarizes duality theory, a requirement to develop the decomposition techniques for nonlinear problems and the sensitivity analysis presented in the following chapters.
5. Chapter 5 describes decomposition techniques appropriate for nonlinear programming (continuous) problems.
6. Chapter 6 presents decomposition procedures relevant for mixed-integer linear and nonlinear problems.
7. Chapter 7 considers specific decomposition techniques not analyzed in the previous chapters.
8. Chapter 8 provides a comprehensive treatment of sensitivity analysis for both decomposable and nondecomposable problems.
9. Finally, Chap. 9 provides some case studies of clear interest for the engineering profession.

1.2 Introduction

This chapter provides an intuitive description of different practical problems with a decomposable structure that can be exploited through a decomposition technique. These problems arise naturally in engineering and science.

To advantageously apply a decomposition technique, the problem under consideration should have the appropriate structure. Two such structures arise in practice. The first is characterized by complicating constraints, and the second by complicating variables. The complicating constraints and variables are those that complicate the solution of the problem, or prevent a straightforward solution of the problem or a solution by blocks, i.e., they make the problem more difficult to solve.

In this chapter, some practical examples are used to motivate and illustrate the problem of complicating constraints and variables. Linear problems are considered first, then nonlinear problems are dealt with, and finally, mixed-integer linear problems are analyzed.

1.3 Linear Programming: Complicating Constraints

The motivating examples in Subsects. 1.3.1 and 1.3.2 illustrate how relaxing complicating constraints make a decentralized solution of the original problem possible.

1.3.1 Transnational Soda Company

A transnational soda company manufactures soda drinks in three different countries as shown in Fig. 1.1. To produce soda, each local company needs mineral water, fruit juice, brown sugar, and the company trademark formula. It should be noted that fruit juice may vary in the composition of the soda between 20 and 30% and the formula between 2 and 4%. All components can be bought locally, at local prices, but the trademark formula is supplied from the company headquarters at a common price. The transnational company seeks to minimize its cost from operating the factories in the three countries. The total quantity of soda to be produced is fixed to 1000 m^3. The minimum cost

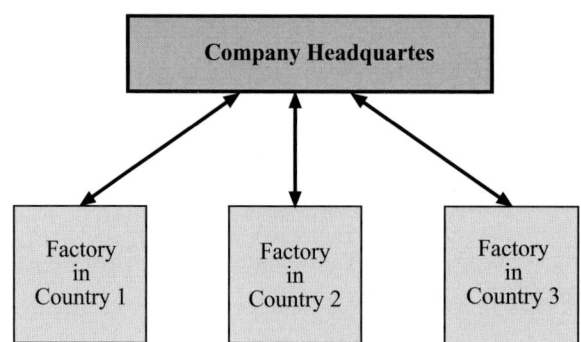

Fig. 1.1. Motivating example: transnational soda company

objective can be formulated as a linear programming problem as explained in the following paragraphs.

In the first country, soda production cost can be expressed as

$$5x_{11} + 2x_{21} + 3x_{31} + 4x_{41},$$

where x_{11}, x_{21}, x_{31}, and x_{41} are the required quantities of trademark formula, mineral water, fruit juice, and brown sugar, respectively; and \$5, \$2, \$3, and \$4, their respective market unit prices. Note that x_{ij} means the quantity of item i (trademark formula, mineral water, fruit juice, or brown sugar) from country j.

For countries 2 and 3, respectively, production costs are expressed as

$$5x_{12} + 2.2x_{22} + 3.3x_{32} + 4.4x_{42}$$

and

$$5x_{13} + 2.1x_{23} + 3.1x_{33} + 4.1x_{43}.$$

Note that prices for domestic products are different from country to country.

The Food & Drug Administration in the first country requires that any soda drink should have at most a maximum content of hydrocarbons and at least a minimum content of vitamins. These two requirements can be expressed as

$$0.1x_{11} + 0.07x_{21} + 0.08x_{31} + 0.09x_{41} \leq 24$$
$$0.1x_{11} + 0.05x_{21} + 0.07x_{31} + 0.08x_{41} \geq 19.5,$$

where 0.10, 0.07, 0.08, and 0.09 are, respectively, the per unit content of hydrocarbons of the trademark formula, the mineral water, the fruit juice, and the brown sugar; and 0.10, 0.05, 0.07, and 0.08 are respectively the per unit content of vitamins of the trademark formula, the mineral water, the fruit juice, and the brown sugar.

Food & Drug Administrations of countries 2 and 3 impose similar requirements, although minimum and maximum hydrocarbon and vitamin required quantities are slightly different than those in the first country. These requirements are formulated as

$$0.1x_{12} + 0.07x_{22} + 0.08x_{32} + 0.09x_{42} \leq 27.5$$
$$0.1x_{12} + 0.05x_{22} + 0.07x_{32} + 0.08x_{42} \geq 22$$

and

$$0.1x_{13} + 0.07x_{23} + 0.08x_{33} + 0.09x_{43} \leq 30$$
$$0.1x_{13} + 0.05x_{23} + 0.07x_{33} + 0.08x_{43} \geq 22.$$

For technical reasons, the supply of the trademark formula is limited to a maximum amount of 22 m^3. This constraint is expressed as

$$x_{11} + x_{12} + x_{13} \leq 22 \ .$$

The total amount of soda to be produced is enforced through the constraint

$$x_{11} + x_{21} + x_{31} + x_{41} + x_{12} + x_{22} + x_{32} + x_{42} + x_{13} + x_{23} + x_{33} + x_{43} = 1{,}000 \ .$$

Limits on components are enforced through the expressions below

$$0.02(x_{11} + x_{21} + x_{31} + x_{41}) \leq x_{11} \leq 0.04(x_{11} + x_{21} + x_{31} + x_{41})$$
$$0.20(x_{11} + x_{21} + x_{31} + x_{41}) \leq x_{31} \leq 0.30(x_{11} + x_{21} + x_{31} + x_{41})$$

$$0.02(x_{12} + x_{22} + x_{32} + x_{42}) \leq x_{12} \leq 0.04(x_{12} + x_{22} + x_{32} + x_{42})$$
$$0.20(x_{12} + x_{22} + x_{32} + x_{42}) \leq x_{32} \leq 0.30(x_{12} + x_{22} + x_{32} + x_{42})$$

$$0.02(x_{13} + x_{23} + x_{33} + x_{43}) \leq x_{13} \leq 0.04(x_{13} + x_{23} + x_{33} + x_{43})$$
$$0.20(x_{13} + x_{23} + x_{33} + x_{43}) \leq x_{33} \leq 0.30(x_{13} + x_{23} + x_{33} + x_{43}),$$

where coefficients $0.02, 0.20, 0.04,$ and 0.30 are used to set the limits.

Consequently, the transnational company minimum cost problem has the form

$$\underset{x_{ij};\, i=1,2,3,4;\, j=1,2,3}{\text{minimize}} \begin{pmatrix} 5x_{11} + 2x_{21} + 3x_{31} + 4x_{41} + \\ 5x_{12} + 2.2x_{22} + 3.3x_{32} + 4.4x_{42} + \\ 5x_{13} + 2.1x_{23} + 3.1x_{33} + 4.1x_{43} \end{pmatrix}$$

subject to

$$0.1x_{11} + 0.07x_{21} + 0.08x_{31} + 0.09x_{41} \leq 24$$
$$0.1x_{11} + 0.05x_{21} + 0.07x_{31} + 0.08x_{41} \geq 19.5$$
$$0.1x_{12} + 0.07x_{22} + 0.08x_{32} + 0.09x_{42} \leq 27.5$$
$$0.1x_{12} + 0.05x_{22} + 0.07x_{32} + 0.08x_{42} \geq 22$$
$$0.1x_{13} + 0.07x_{23} + 0.08x_{33} + 0.09x_{43} \leq 30$$
$$0.1x_{13} + 0.05x_{23} + 0.07x_{33} + 0.08x_{43} \geq 22$$
$$x_{11} + x_{12} + x_{13} \leq 22$$

$$x_{11} + x_{21} + x_{31} + x_{41} + x_{12} + x_{22} + x_{32} + x_{42} + x_{13} + x_{23} + x_{33} + x_{43} = 1{,}000$$

and

$$0.02(x_{11} + x_{21} + x_{31} + x_{41}) \leq x_{11} \leq 0.04(x_{11} + x_{21} + x_{31} + x_{41})$$
$$0.20(x_{11} + x_{21} + x_{31} + x_{41}) \leq x_{31} \leq 0.30(x_{11} + x_{21} + x_{31} + x_{41})$$
$$0.02(x_{12} + x_{22} + x_{32} + x_{42}) \leq x_{12} \leq 0.04(x_{12} + x_{22} + x_{32} + x_{42})$$
$$0.20(x_{12} + x_{22} + x_{32} + x_{42}) \leq x_{32} \leq 0.30(x_{12} + x_{22} + x_{32} + x_{42})$$
$$0.02(x_{13} + x_{23} + x_{33} + x_{43}) \leq x_{13} \leq 0.04(x_{13} + x_{23} + x_{33} + x_{43})$$
$$0.20(x_{13} + x_{23} + x_{33} + x_{43}) \leq x_{33} \leq 0.30(x_{13} + x_{23} + x_{33} + x_{43}).$$

The solution of this problem is shown in Table 1.1. The total production cost incurred by the company is $2,915.1.

In summary, the four main elements of the transnational company problem are:

1.3 Linear Programming: Complicating Constraints

Table 1.1. Soda company production results

Component (m³)	Country 1	Country 2	Country 3	Total
Water	147.2	198.6	118.9	464.7
Fruit juice	92.7	108.2	99.1	300.0
Brown sugar	62.9	44.7	105.7	213.3
Trademark formula	6.2	9.2	6.6	22.0
Total	309.0	360.7	330.3	1,000

Data.
m: the number of components required to make the soda drink
n: the number of countries
p_{ij}: the market price of component i in country j
h_{ij}: per unit content of hydrocarbons of component i in country j
v_{ij}: per unit content of vitamins of component i in country j
$h_j^{\max}$: the maximum allowed content of hydrocarbons in country j
$v_j^{\min}$: the minimum required content of vitamins in country j
T: the total amount of soda to be produced in all countries
t^{av}: the available amount of trademark formula
b_i^{down}: lower bound of allowed per unit content of component i in the soda drink
b_i^{up}: upper bound of allowed per unit content of component i in the soda drink.

Variables.
x_{ij}: the amount of component i to be used in country j (x_{1j} is the amount of trademark formula in country j).

It is assumed that these variables are nonnegative,

$$x_{ij} \geq 0; \quad i = 1,\ldots,m; \quad j = 1,\ldots,n . \tag{1.1}$$

Constraints. The constraints of this problem are
1. maximum hydrocarbon content constraint

$$\sum_{i=1}^{m} h_{ij} x_{ij} \leq h_j^{\max}; \quad j = 1,\ldots,n , \tag{1.2}$$

2. minimum vitamin content constraint

$$\sum_{i=1}^{m} v_{ij} x_{ij} \geq v_j^{\min}; \quad j = 1,\ldots,n , \tag{1.3}$$

3. constraints associated with the per unit content of different components

$$b_i^{\text{down}}\left(\sum_{i=1}^{m} x_{ij}\right) \leq x_{ij} \leq b_i^{\text{up}}\left(\sum_{i=1}^{m} x_{ij}\right);$$
$$i = 1, \cdots, m; \quad j = 1, \ldots, n, \tag{1.4}$$

4. total amount of available trademark formula (complicating constraint)

$$\sum_{j=1}^{n} x_{1j} \leq t^{\text{av}}, \tag{1.5}$$

5. demand constraint (complicating constraint)

$$\sum_{i=1}^{m}\sum_{j=1}^{n} x_{ij} = T. \tag{1.6}$$

The first two sets of conditions (1.2) and (1.3) state that the content of hydrocarbons and vitamins of the product are below and above the allowable limit values, respectively. The third constraint (1.4) guarantees that the amounts of the different components are between the allowable limits. The fourth constraint (1.5) states that the total amount of trademark formula is below the available amount. Finally, the fifth constraint (1.6) forces the total amount of soda produced to coincide with the desired value T.

Function to Be Optimized. We are normally interested in minimizing the total cost, i.e.,

$$\underset{x_{ij}}{\text{minimize}} \quad \sum_{i=1}^{m}\sum_{j=1}^{n} p_{ij} x_{ij}. \tag{1.7}$$

The block structure of the resulting constraint matrix is illustrated in Fig. 1.2, which shows the complicating constraints clearly.

It should be noted that the last two constraints prevent a decomposed solution of the problem above; therefore, they are complicating constraints. If this problem has to be solved in a decentralized fashion, an appropriate decomposition technique has to be applied. Such decomposition technique will be analyzed in the following chapters.

Additional examples of problems with the above decomposable structure can be found in reference [1] and in the pioneering book by Dantzig [2].

1.3.2 Stochastic Hydro Scheduling

A hydroelectric plant is associated with a hydro reservoir, as shown in Fig. 1.3. Electricity production depends on the reservoir water content and this water content depends on stochastic water inflows to the reservoir.

Consider two time periods. At the beginning of each period, the random input "amount of water inflow" (low or high) for the entire period is assumed

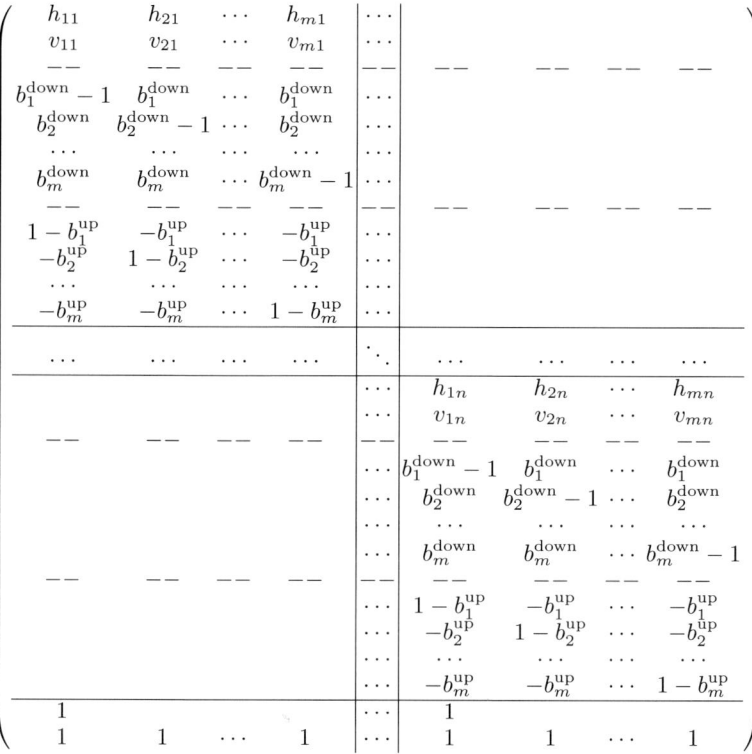

Fig. 1.2. The constraints structure of the transnational soda company problem

known, and the decision of how much water to be discharged in that period in order to produce electricity has to be made.

The objective is to maximize the profit from selling energy in the two considered periods.

Taking into account water inflow uncertainty, four scenarios are possible: low inflow in both periods (scenario 1), low inflow in the first period and high in the second (scenario 2), high inflow in the first period and low in the second (scenario 3), and high inflow in both periods (scenario 4). This implies that the inputs, O_{ts}, for scenarios 1 to 4 are

$$\{O_{11}, O_{21}\} = \{\text{low}, \text{low}\};$$
$$\{O_{12}, O_{22}\} = \{\text{low}, \text{high}\};$$
$$\{O_{13}, O_{23}\} = \{\text{high}, \text{low}\};$$
$$\{O_{14}, O_{24}\} = \{\text{high}, \text{high}\}.$$

Scenarios are illustrated in Fig. 1.4 using a scenario tree.

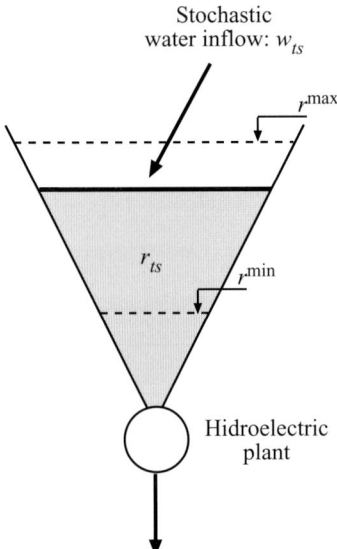

Fig. 1.3. Illustration of the reservoir data and variables in the stochastic hydro scheduling example

Considering the four scenarios, the expectation of the total profit from selling energy can be computed as

$$\sum_{s=1}^{4} p_s \left(20 \times 5 \times d_{1s} + 30 \times 5 \times d_{2s}\right),$$

where $p_1 = 0.3$ (scenario 1), $p_2 = 0.2$ (scenario 2), $p_3 = 0.2$ (scenario 3), $p_4 = 0.3$ (scenario 4) are the probabilities of occurrence of those scenarios; 20 and 30 \$/MWh are the electricity prices (considered known with certainty) for period 1 and 2, respectively; and 5 MWh/m^3 is a constant to convert water volume discharge to electric energy production; d_{1s} and d_{2s} are the water discharges during period 1 and 2, respectively, and scenario s.

The water balance in the reservoir has to be satisfied for each scenario. This is enforced through the constraints below.

$$r_{1s} = r_0 - d_{1s} + w_{1s}; \quad s = 1, 2, 3, 4$$
$$r_{2s} = r_{1s} - d_{2s} + w_{2s}; \quad s = 1, 2, 3, 4,$$

where r_{1s} and r_{2s} are the reservoir water contents at the end of periods 1 and 2, respectively, and scenario s; r_0 is the reservoir content at the beginning of period 1; and finally, w_{1s} and w_{2s} are the water inflows during periods 1 and 2, respectively, and scenario s.

For the example in question, the above equations are stated below.
Scenario 1: low and low inflows

1.3 Linear Programming: Complicating Constraints

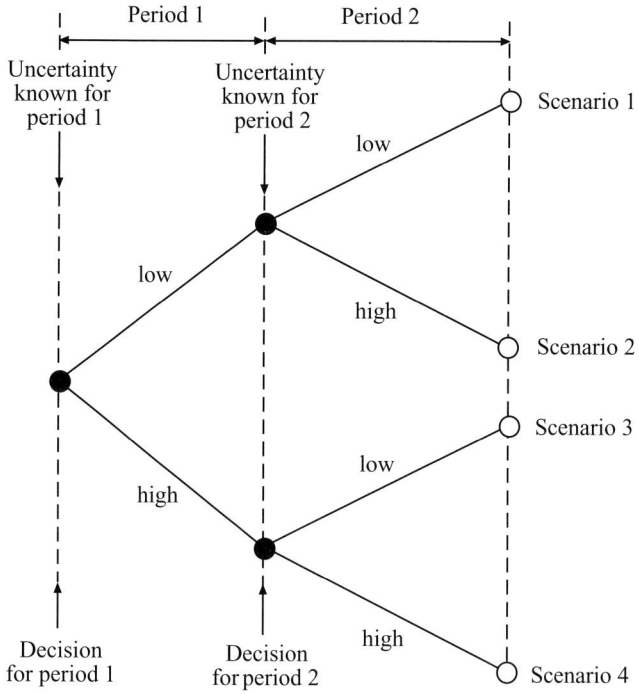

Fig. 1.4. Scenario tree for the stochastic hydro scheduling example

$$r_{11} = 50 - d_{11} + 20$$
$$r_{21} = r_{11} - d_{21} + 25 \ .$$

Scenario 2: low and high inflows

$$r_{12} = 50 - d_{12} + 20$$
$$r_{22} = r_{12} - d_{22} + 35 \ .$$

Scenario 3: high and low inflows

$$r_{13} = 50 - d_{13} + 30$$
$$r_{23} = r_{13} - d_{23} + 25 \ .$$

Scenario 4: high and high inflows

$$r_{14} = 50 - d_{14} + 30$$
$$r_{24} = r_{14} - d_{24} + 35 \ .$$

In the equations above, 50 m³ is the reservoir water content at the beginning of period 1 for all scenarios, 20 and 30 m³ the low and high water inflows during period 1, and 25 and 35 m³ the low and high water inflows in period 2.

Additional important constraints are required; those that enforce that decisions associated to two scenarios identical up to period t should be identical up to period t. These relevant constraints are denominated nonanticipativity constraints [3]. Since scenario 1 (low–low) and 2 (low–high) are equal up to the first period, it must be that

$$d_{11} = d_{12}; \qquad r_{11} = r_{12} .$$

On the other hand, since scenario 3 (high–low) and 4 (high–high) are equal up to the first period, it must be

$$d_{13} = d_{14}; \qquad r_{13} = r_{14} .$$

Reservoir content bounds are expressed as

$$r^{\min} \leq r_{ts} \leq r^{\max}; \qquad s = 1, 2, 3, 4; \quad t = 1, 2$$

and water discharge limits as

$$0 \leq d_{ts} \leq d^{\max}; \qquad s = 1, 2, 3, 4; \quad t = 1, 2 .$$

The stochastic hydro scheduling problem is finally formulated as

$$\underset{d_{ts}, r_{ts}; t = 1, 2; s = 1, 2, 3, 4}{\text{Maximize}} \quad z = k \sum_{s=1}^{4} p_s \sum_{t=1}^{2} \lambda_t d_{ts} ,$$

where k is a proportionality constant and λ_t is the benefit at time t, subject to the water balance constraints

$$r_{11} = 50 - d_{11} + 20$$
$$r_{21} = r_{11} - d_{21} + 25$$
$$r_{12} = 50 - d_{12} + 20$$
$$r_{22} = r_{12} - d_{22} + 35$$
$$r_{13} = 50 - d_{13} + 30$$
$$r_{23} = r_{13} - d_{23} + 25$$
$$r_{14} = 50 - d_{14} + 30$$
$$r_{24} = r_{14} - d_{24} + 35 ,$$

the reservoir level constraints

$$20 \leq r_{11} \leq 140; \quad 20 \leq r_{12} \leq 140; \quad 20 \leq r_{13} \leq 140; \quad 20 \leq r_{14} \leq 140$$
$$20 \leq r_{21} \leq 140; \quad 20 \leq r_{22} \leq 140; \quad 20 \leq r_{23} \leq 140; \quad 20 \leq r_{24} \leq 140 ,$$

the discharge limits constraints

1.3 Linear Programming: Complicating Constraints

$$d_{11} \leq 60; \quad d_{12} \leq 60; \quad d_{13} \leq 60; \quad d_{14} \leq 60$$
$$d_{21} \leq 60; \quad d_{22} \leq 60; \quad d_{23} \leq 60; \quad d_{24} \leq 60 \,,$$

and the nonanticipativity constraints

$$d_{11} = d_{12}; \quad r_{11} = r_{12}; \quad d_{13} = d_{14}; \quad r_{13} = r_{14} \,.$$

The solution is shown in Tables 1.2 and 1.3. Total profit from selling electricity is \$11,250.

Table 1.2. First period decisions

Inflow in period 1	Discharge in period 1 (m^3)
Low	25
High	35

Table 1.3. Second period decisions

Period 1	Inflow in period 2	Discharge in period 2 (m^3)	Scenario
Low Inflow	Low	50	1
	High	60	2
High Inflow	Low	50	3
	High	60	4

Sorting the variables in the following order $r_{11}, r_{21}, d_{11}, d_{21}, r_{12}, r_{22}, d_{12}, d_{22}, r_{13}, r_{23}, d_{13}, d_{23}, r_{14}, r_{24}, d_{14}, d_{24}$, the constraint matrix of the above problem has the form

$$\begin{pmatrix}
1 & 1 & & & & & & \\
-1 & 1 & 1 & & & & & \\
& & & 1 & 1 & & & \\
& & & -1 & 1 & 1 & & \\
& & & & & & 1 & 1 \\
& & & & & & -1 & 1 & 1 \\
& & & & & & & & 1 & 1 \\
& & & & & & & & -1 & 1 & 1 \\
& & 1 & & -1 & & & \\
1 & & & -1 & & & & \\
& & & & & 1 & & -1 \\
& & & & 1 & & -1 & \\
\end{pmatrix}$$

If the last four constraints are "relaxed," the problem decomposes by blocks. These constraints are therefore complicating constraints that prevent a distributed solution of the problem, unless an appropriate decomposition technique is used as, for instance, the Dantzig-Wolfe procedure. Such decomposition techniques will be analyzed in the following chapters.

In summary, the four main elements of the stochastic hydro scheduling problem are:

Data.
n: the number of scenarios
m: the number of periods
λ_t: the electricity price for period t
k: electric energy production to water volume discharge factor
p_s: the probability of scenario s
w_{ts}: the water inflow for period t and scenario s
r_0: initial reservoir water content
$r^{\max}$: reservoir maximum allowed water content
$r^{\min}$: reservoir minimum allowed water content
$d^{\max}$: maximum allowed discharge per time period
O_{ts}: input at time t associated with scenario s.

Variables.
d_{ts}: the water volume discharge during period t and scenario s
r_{ts}: the reservoir water content at the end of period t for scenario s.
It is assumed that these variables are nonnegative:

$$d_{ts} \geq 0; \quad t = 1, \ldots, m; \quad s = 1, \ldots, n \quad (1.8)$$

$$r_{ts} \geq 0; \quad t = 1, \ldots, m; \quad s = 1, \ldots, n. \quad (1.9)$$

Constraints. The constraints of this problem are
1. water balance constraints

$$r_{ts} = r_{t-1,s} - d_{ts} + w_{ts}; \quad t = 1, \ldots, m; \quad s = 1, \ldots, n, \quad (1.10)$$

2. allowable reservoir level constraint

$$r^{\min} \leq r_{ts} \leq r^{\max}; \quad t = 1, \ldots, m; \quad s = 1, \ldots, n, \quad (1.11)$$

3. allowable discharge constraint

$$d_{ts} \leq d^{\max}; \quad t = 1, \ldots, m; \quad s = 1, \ldots, n, \quad (1.12)$$

4. nonanticipativity constraint

$$d_{t_1 s_1} = d_{t_1 s_2} \quad \text{if } O_{ts_1} = O_{ts_2} \quad \forall t \leq t_1 \quad (1.13)$$
$$r_{t_1 s_1} = r_{t_1 s_2} \quad \text{if } O_{ts_1} = O_{ts_2} \quad \forall t \leq t_1. \quad (1.14)$$

Conditions (1.10) are the balance of water input and output for period t and scenario s. Constraints (1.11) state that the reservoir water contents at all times and scenarios are in the allowed range. Constraints (1.12) state that the water discharges are below the allowed limit. Finally, constraints (1.13) are the nonanticipativity constraints.

Function to Be Optimized. We are normally interested in maximizing the expected benefit, i.e.,

$$\underset{d_{ts}, r_{ts}; t = 1, 2, \cdots, m; s = 1, 2, \cdots, n}{\text{maximize}} \quad z = k \sum_{s=1}^{n} p_s \sum_{t=1}^{m} \lambda_t d_{ts}. \quad (1.15)$$

Further details on the above formulation can be found, for instance, in [4].

The two motivating examples below (in Subsects. 1.3.3 and 1.3.4) illustrate how relaxing complicating constraints renders a problem that can be solved in a straightforward manner.

1.3.3 River Basin Operation

Consider a river basin including two reservoirs as illustrated in Fig. 1.5. Each reservoir has associated a hydroelectric power plant that produces electricity. The natural inflows to reservoirs 1 and 2 during the period t are denoted by w_{t1} and w_{t2}, respectively. The water contents of reservoirs 1 and 2 at the end of period t are denoted, respectively, by r_{t1} and r_{t2}. The water discharged during period t by reservoir 1 and 2 are d_{t1} and d_{t2}, respectively. Reservoir contents are limited above and below by constants $r_1^{\max}$, $r_1^{\min}$, $r_2^{\max}$, and $r_2^{\min}$, respectively. Analogously, water discharge volumes are limited above by $d_1^{\max}$ and $d_2^{\max}$. It is assumed that water released in reservoir 1 reaches instantaneously reservoir 2, which is a reasonable assumption if reservoirs are not far away from each other.

The amounts of energy produced by power plants 1 and 2 during period t are proportional to the corresponding water discharges during that period t. The proportionality constants for plants 1 and 2 are k_1 and k_2, respectively.

The river system is operated to supply the local electricity demand in each period, e_t. If additional energy can be produced during period t, it is sold at market price λ_t, with the objective of maximizing profits.

Consider a time horizon of 2 h and assume that the reservoir contents at the beginning of the time horizon are r_{01} and r_{02}, for reservoirs 1 and 2, respectively.

The profit maximization problem is

$$\underset{r_{11}, r_{12}, r_{21}, r_{22}, d_{11}, d_{12}, d_{21}, d_{22}}{\text{maximize}} \quad \lambda_1(k_1 d_{11} + k_2 d_{12} - e_1) + \lambda_2(k_1 d_{21} + k_2 d_{22} - e_2)$$

subject to the water balance constraints

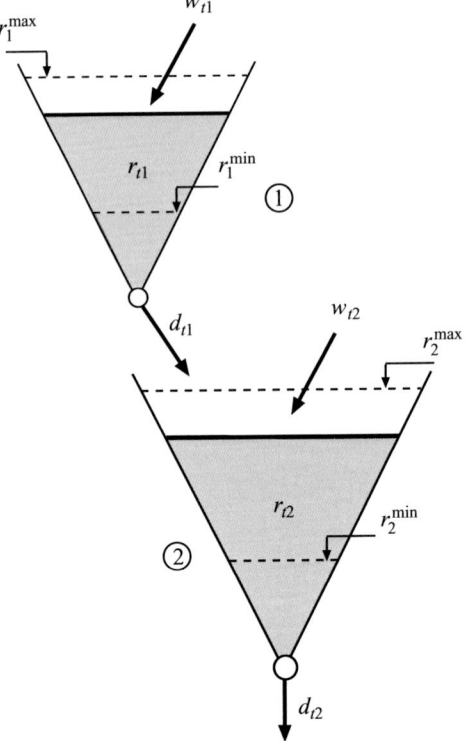

Fig. 1.5. Illustration of the river basin operation example

$$r_{11} = r_{01} + w_{11} - d_{11}$$
$$r_{21} = r_{11} + w_{21} - d_{21}$$
$$r_{12} = r_{02} + w_{12} - d_{12} + d_{11}$$
$$r_{22} = r_{12} + w_{22} - d_{22} + d_{21}$$
$$r_{21} + r_{22} = r_{01} + r_{02} + w_{11} + w_{21} + w_{12} + w_{22} - d_{12} - d_{22},$$

where the last is a redundant constraint that it is convenient to incorporate because it renders a particularly appropriate matrix form of the problem; the demand constraints

$$k_1 d_{11} + k_2 d_{12} \geq e_1 \tag{1.16}$$
$$k_1 d_{21} + k_2 d_{22} \geq e_2, \tag{1.17}$$

the reservoir level bounds

$$r_1^{\min} \leq r_{11} \leq r_1^{\max}; \ r_1^{\min} \leq r_{21} \leq r_1^{\max} \tag{1.18}$$
$$r_2^{\min} \leq r_{12} \leq r_2^{\max}; \ r_2^{\min} \leq r_{22} \leq r_2^{\max}, \tag{1.19}$$

and the allowed discharge bounds

$$0 \leq d_{11} \leq d_1^{\max}; \ 0 \leq d_{21} \leq d_1^{\max} \tag{1.20}$$
$$0 \leq d_{12} \leq d_2^{\max}; \ 0 \leq d_{22} \leq d_2^{\max}. \tag{1.21}$$

Sorting the variables in the order r_{11}, r_{21}, r_{12}, r_{22}, d_{11}, d_{21}, d_{12}, d_{22}, the matrix corresponding to constraints that are not bounds is

$$\left(\begin{array}{cccc|cccc} 1 & & & & 1 & & & \\ -1 & 1 & & & & 1 & & \\ & & 1 & & -1 & & 1 & \\ & & -1 & 1 & & -1 & & 1 \\ & -1 & & -1 & & & -1 & -1 \\ \hline & & & & k_1 & & k_2 & \\ & & & & & k_1 & & k_2 \end{array} \right).$$

The structure of this matrix reveals an exploitable structure. If the last two constraints are relaxed, the remaining matrix has a network structure which allows the use of highly efficient solution algorithms [5]. This matrix includes in each column only a 1 and a -1 (total unimodularity). However, the last two constraints prevent the use of an efficient algorithm unless an appropriate decomposition mechanism is used as, for instance, the Dantzig-Wolfe decomposition algorithm. Such decomposition technique is explained in the following chapters.

Considering for the above example the data in Tables 1.4 and 1.5, the solution of the river basin operation problem is provided in Table 1.6. Total profit from selling energy is $26,400.

Table 1.4. Reservoir data

Reservoir i	Factor k	Initial reservoir water content (m^3)	Upper limit on water discharged (m^3)	Water content Min (m^3)	Water content Max (m^3)
1	5	55	60	20	120
2	4.5	65	70	20	140

Table 1.5. Periodic data

Period t	Electricity price ($/MWh)	Electricity demand (MWh)	Water inflow in reservoir 1 (m^3)	Water inflow in reservoir 2 (m^3)
1	30	490	20	25
2	20	525	30	40

Table 1.6. Solution for the hydroelectric river basin example

Period t	Discharge plant 1 (m^3)	Discharge plant 2 (m^3)	Electricity production (MWh)	Electricity demand (MWh)	Energy sold (MWh)
1	43	70	530	490	40
2	42	70	525	525	0

In summary, the main elements of the hydroelectric profit maximization problem for a river system of n reservoirs during m time periods, are:

Data.
n: the number of reservoirs
m: the number of time periods considered
λ_t: the electricity price for period t
k_i: electric energy production to water volume discharge factor for reservoir i
w_{ti}: the water inflow in reservoir i during period t
r_{0i}: initial water content in reservoir i
$r_i^{\max}$: maximum allowed water content in reservoir i
$r_i^{\min}$: minimum allowed water content in reservoir i
$d_i^{\max}$: maximum allowed water discharge during a time period for reservoir i
e_t: electricity demand during period t
Ω_i: the set of reservoirs above reservoir i and connected to it.

Variables.
d_{ti}: the water volume discharge of reservoir i during period t
r_{ti}: the reservoir water content of reservoir i at the end of period t
It is assumed that these variables are nonnegative,

$$d_{ti} \geq 0; \quad t = 1, \ldots, m; \quad i = 1, \ldots, n \quad (1.22)$$

$$r_{ti} \geq 0; \quad t = 1, \ldots, m; \quad i = 1, \ldots, n. \quad (1.23)$$

Constraints. The constraints of this problem are
1. the water balance constraints [including constraints (1.25) which is the redundant one]

$$r_{ti} = r_{t-1,i} - d_{ti} + w_{ti} + \sum_{j \in \Omega_i} d_{tj}; \quad t = 1, \ldots, m; \quad i = 1, \ldots, n \quad (1.24)$$

$$\sum_{i=1}^{n}(r_{mi} - r_{0i}) = \sum_{t=1}^{m}\sum_{i=1}^{n} w_{ti} - \sum_{t=1}^{m} d_{tn}, \quad (1.25)$$

2. the demand constraints

$$\sum_{i=1}^{n} k_i d_{ti} \geq e_t; \quad t = 1, \ldots, m, \tag{1.26}$$

3. the reservoir level bounds

$$r_i^{\min} \leq r_{ti} \leq r_i^{\max}; \quad t = 1, \ldots, m; \; i = 1, \ldots, n, \tag{1.27}$$

4. the allowed discharge bounds

$$0 \leq d_{ti} \leq d_i^{\max}; \quad t = 1, \ldots, m; \; i = 1, \ldots, n. \tag{1.28}$$

Function to Be Optimized. We are normally interested in maximizing the expected benefit, i.e.,

$$\underset{d_{ti}, r_{ti}; t=1,2,\ldots,m; i=1,2,\ldots,n}{\text{maximize}} \quad z = \sum_{t=1}^{m} \lambda_t \left(\sum_{i=1}^{n} k_i d_{ti} - e_t \right). \tag{1.29}$$

1.3.4 Energy Production Model

Consider the triangular energy demand depicted in Fig. 1.6. In this figure, the vertical axis represents power and the horizontal axis time; therefore, the area

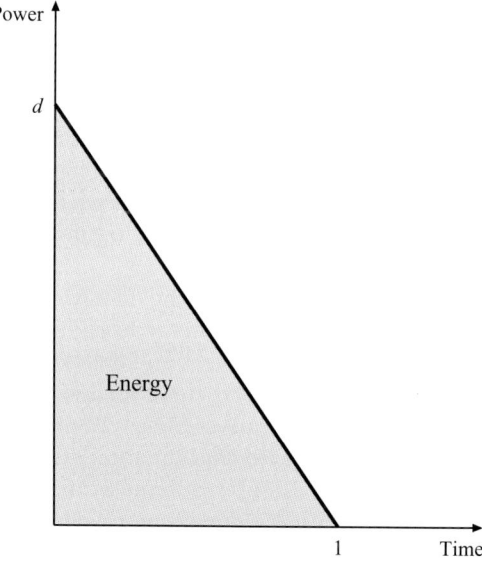

Fig. 1.6. Electricity demand curve for the energy production model

24   1 Motivating Examples

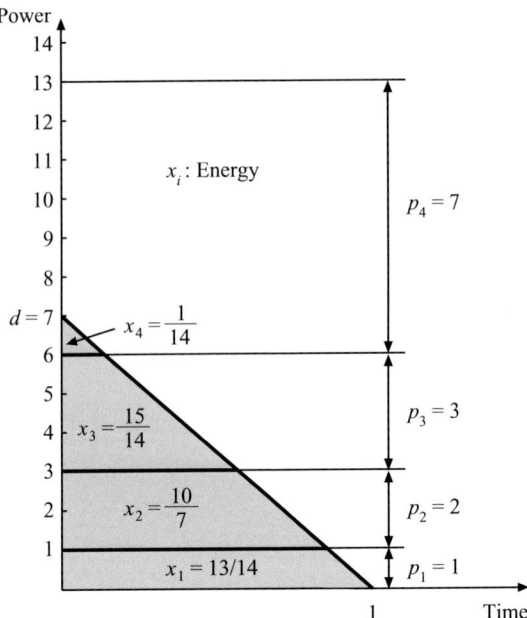

Fig. 1.7. Optimal energy production strategy

under this curve is energy. The maximum power demand is $d = 7$ MW and the considered time period is 1. A set of energy production devices are available. Their respective maximum output powers are $p_1 = 1$ MW, $p_2 = 2$ MW, $p_3 = 3$ MW and $p_4 = 7$ MW, and their production costs $c_1 = 1$ \$/MWh, $c_2 = 2$ \$/MWh, $c_3 = 3$ \$/MWh, and $c_4 = 10$ \$/MWh, respectively.

The problem to be solved consists of supplying the energy demand at minimum cost. The solution of this problem is trivial. Production devices are arranged in merit order, from the cheapest one to the most expensive one, and they are used for production in that order. This is shown in Fig. 1.7, where x_1, x_2, x_3, and x_4 are the energies produced by devices 1, 2, 3, and 4, respectively. The optimal energy production values of this minimum cost problem are $\frac{13}{14}$, $\frac{10}{7}$, $\frac{15}{14}$, and $\frac{1}{14}$ MWh.

In order to write this problem formally, the definition of the maximum energy produced by a set of production devices is introduced first. For instance, the maximum energy produced by devices 1 and 3 is denoted by $e(\{1,3\})$ and computed as the area under the demand curve if these devices are loaded together at the bottom of the demand curve. Figure 1.8 illustrates the computation of $e(\{1,3\})$. It should be observed that function $e(\{1,3\})$ is commutative, i.e., the loading order of devices 1 and 3 is immaterial.

Using the definition of the function maximum energy, the cost minimization problem previously analyzed can be stated as

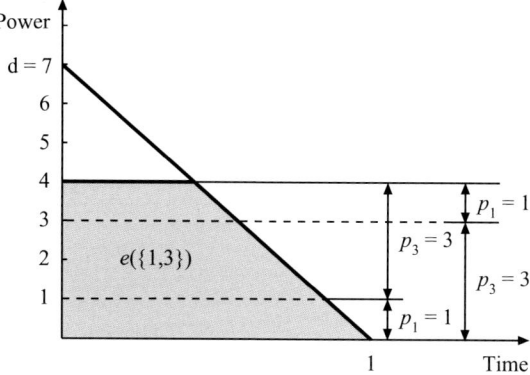

Fig. 1.8. Function $e(\{1,3\})$

$$\underset{x_1, x_2, x_3, x_4}{\text{minimize}} \quad c_1 x_1 + c_2 x_2 + c_3 x_3 + c_4 x_4$$

subject to

$$x_1 \leq e(\{1\})$$
$$x_2 \leq e(\{2\})$$
$$x_3 \leq e(\{3\})$$
$$x_4 \leq e(\{4\})$$
$$x_1 + x_2 \leq e(\{1,2\})$$
$$x_1 + x_3 \leq e(\{1,3\})$$
$$x_1 + x_4 \leq e(\{1,4\})$$
$$x_2 + x_3 \leq e(\{2,3\})$$
$$x_2 + x_4 \leq e(\{2,4\})$$
$$x_3 + x_4 \leq e(\{3,4\})$$
$$x_1 + x_2 + x_3 \leq e(\{1,2,3\})$$
$$x_1 + x_2 + x_4 \leq e(\{1,2,4\})$$
$$x_1 + x_3 + x_4 \leq e(\{1,3,4\})$$
$$x_2 + x_3 + x_4 \leq e(\{2,3,4\})$$
$$x_1 + x_2 + x_3 + x_4 = e(d) \ .$$

It should be noted that the last constraint is an equality constraint to force the devices to cover the demand. In fact, it eliminates the trivial null solution.

For the particular data considered, the above problem becomes

$$\underset{x_1, x_2, x_3, x_4}{\text{minimize}} \quad x_1 + 2x_2 + 3x_3 + 10x_4$$

subject to

$$x_1 \le 13/14$$
$$x_2 \le 12/7$$
$$x_3 \le 33/14$$
$$x_4 \le 7/2$$
$$x_1 + x_2 \le 33/14$$
$$x_1 + x_3 \le 20/7$$
$$x_1 + x_4 \le 45/14$$
$$x_2 + x_3 \le 45/14$$
$$x_2 + x_4 \le 24/7$$
$$x_3 + x_4 \le 7/2$$
$$x_1 + x_2 + x_3 \le 24/7$$
$$x_1 + x_2 + x_4 \le 7/2$$
$$x_1 + x_3 + x_4 \le 7/2$$
$$x_2 + x_3 + x_4 \le 7/2$$
$$x_1 + x_2 + x_3 + x_4 = 7/2 \,.$$

Note that the optimal solution of this problem is the solution previously obtained that can be obtained using the merit order rule, i.e., $x_1^* = \frac{13}{14}$ MWh, $x_2^* = \frac{10}{7}$ MWh, $x_3^* = \frac{15}{14}$ MWh, $x_4^* = \frac{1}{14}$ MWh.

In summary, a general statement of the problem previously analyzed must consider the following four elements:

Data.
   n: the number of production devices
   p_i: the maximum output power of device i
   c_i: the production cost of device i
   d: the power demand
$e(C)$: the maximum energy produced by the devices in the set C.
   Ω_i: the set $\{1, 2, \cdots, i\}$.

Variables.
   x_i: energy produced by device i.
   It is assumed that these variables are nonnegative,

$$x_i \ge 0; \ i = 1, \ldots, n \,. \tag{1.30}$$

Constraints. The constraints of this problem are

$$\sum_{i \in \Omega_i} x_i \le e(\Omega_i), \qquad \forall \Omega_i \tag{1.31}$$

$$\sum_{i \in \Omega_n} x_i = e(d) \,. \tag{1.32}$$

Function to Be Optimized. We are normally interested in minimizing the cost, i.e.,

$$\underset{x_1,\ldots,x_n}{\text{minimize}} \sum_{i=1}^{n} c_i x_i. \tag{1.33}$$

Each of the constraints (1.31) is called a facet [6]. The feasibility region of this linear problem is denominated a polymatroid. Further details on polymatroids and their properties can be found in references [7, 8].

A relevant observation is that the solution of this production problem becomes nontrivial if additional linear constraints are imposed. For instance, if the joint energy production of devices 1 and 2 is required to be below $\frac{25}{14}$ MWh, the merit order rule is no longer valid and the problem loses its polymatroid structure.

The example above with this additional linear constraint is

$$\underset{x_1, x_2, x_3, x_4}{\text{minimize}} \quad x_1 + 2x_2 + 3x_3 + 10x_4 \tag{1.34}$$

subject to

$$x_1 \leq 13/14$$
$$x_2 \leq 12/7$$
$$x_3 \leq 33/14$$
$$x_4 \leq 7/2$$
$$x_1 + x_2 \leq 33/14$$
$$x_1 + x_3 \leq 20/7$$
$$x_1 + x_4 \leq 45/14$$
$$x_2 + x_3 \leq 45/14$$
$$x_2 + x_4 \leq 24/7$$
$$x_3 + x_4 \leq 7/2$$
$$x_1 + x_2 + x_3 \leq 24/7$$
$$x_1 + x_2 + x_4 \leq 7/2$$
$$x_1 + x_3 + x_4 \leq 7/2$$
$$x_2 + x_3 + x_4 \leq 7/2$$
$$x_1 + x_2 + x_3 + x_4 = 7/2$$
$$x_1 + x_2 \leq 25/14, \tag{1.35}$$

whose solution is obtained using a standard solving algorithm. The solution is $x_1^* = \frac{13}{14}$ MWh, $x_2^* = \frac{6}{7}$ MWh, $x_3^* = \frac{23}{14}$ MWh and $x_4^* = \frac{1}{14}$ MWh.

It should be noted that constraint $x_1 + x_2 \leq \frac{25}{14}$ plays the role of a complicating constraint because it deprives the problem of its polymatroid structure that allows a straightforward solution.

The linear constraint (1.35) is a complicating constraint. If it is relaxed, the solution of the resulting problem is obtained in a straightforward manner using the merit order rule, i.e., loading devices from the lowest to the highest cost.

In practice, the number of production devices can be as high as 100, and the facet formulation of the minimum energy production cost problem, without additional linear constraints, requires $2^{100} - 1$ constraints, a number that prevents even writing down the problem. However, its solution is trivial using the merit order rule. Nevertheless, if additional linear constraints are included, the resulting problem becomes both unwritable and unsolvable, unless a decomposition technique is used to relax the complicating constraints [9]. Such decomposition techniques are explained in the following chapters.

1.4 Linear Programming: Complicating Variables

1.4.1 Two-Year Coal and Gas Procurement

Consider the problem of the procurement of coal and natural gas (expressed in energy units) in a factory to supply the energy demand of the present year and next year. The demand for energy this year is known with certainty and it is equal to 750 MWh. The current prices of coal and gas are 4.5 and 5.1 \$/MWh, respectively. However, the energy demand for next year is uncertain: it may be high with probability 0.3, medium with probability 0.5, and low with probability 0.2. High, medium, and low energy demands for the second year are respectively 900, 750, and 550 MWh; the corresponding prices for coal are 7.5, 6, and 3 \$/MWh, and the corresponding prices for natural gas are 8.5, 5.5, and 4 \$/MWh. Note that we have converted the units to their equivalent units in terms of electricity production. The first purchase decision of coal and natural gas is taken at the beginning of the first year, and the second one at the beginning of the second year. This second purchase decision is made once the demand for the second year is known with certainty, which allows to "correct" the purchase decision of the first year. To achieve a balanced energy supply, the total amount of either coal or natural gas, used either in the first or the second year, should be larger than one third and smaller than two thirds of the total energy demand. Figure 1.9 illustrates this decision framework. It should be noted that storage costs are considered negligible.

This 2-year coal and natural gas procurement problem can be formulated as

$$\underset{c_0, g_0, c_1, g_1, c_2, g_2, c_3, g_3}{\text{minimize}}$$

$$4.5c_0 + 5.1g_0 + 0.3(7.5c_1 + 8.5g_1) + 0.5(6c_2 + 5.5g_2) + 0.2(3c_3 + 4g_3)$$

subject to

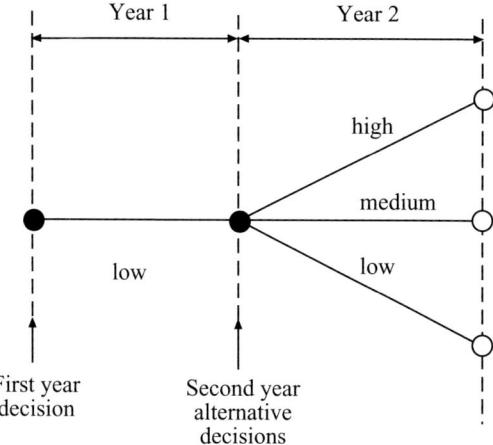

Fig. 1.9. Two-year coal and gas procurement: decision framework

$$
\begin{aligned}
c_0 + g_0 &\geq 750 \\
c_1 + g_1 \ + c_0 + g_0 &= 1650 \\
c_1 \ \ \ \ \ + c_0 &\leq 1100 \\
-c_1 \ \ \ \ \ - c_0 &\leq -550 \\
g_1 \ \ \ \ \ + g_0 &\leq 1100 \\
-g_1 \ \ \ \ \ - g_0 &\leq -550 \\
c_2 + g_2 + c_0 + g_0 &= 1500 \\
c_2 \ \ \ \ \ + c_0 &\leq 1000 \\
-c_2 \ \ \ \ \ - c_0 &\leq -500 \\
g_2 \ \ \ \ \ + g_0 &\leq 1000 \\
-g_2 \ \ \ \ \ - g_0 &\leq -500 \\
c_3 + g_3 + c_0 + g_0 &= 1300 \\
c_3 \ \ \ \ \ + c_0 &\leq 866 \\
-c_3 \ \ \ \ \ - c_0 &\leq -433 \\
g_3 \ \ \ \ \ + g_0 &\leq 866 \\
-g_3 \ \ \ \ \ - g_0 &\leq -433 \\
c_0, g_0, c_1, g_1, c_2, g_2, c_3, g_3 &\geq 0 \ ,
\end{aligned}
$$

where c_0 and g_0 are the amount of coal and natural gas purchased the first year, respectively, c_1, c_2, c_3 are the amount of coal purchased the second year for scenario 1 (high demand), 2 (medium demand), and 3 (low demand), respectively, and finally g_1, g_2, g_3 are the amount of natural gas purchased the second year for scenario 1 (high demand), 2 (medium demand), and 3 (low demand), respectively.

The first constraint is the first year demand constraint. The next constraints represent the balance energy supply during the first and the second year, the maximum and minimum coal consumption and the maximum and minimum gas consumption for each scenario, respectively.

30   1 Motivating Examples

Table 1.7. Coal and gas purchasing decisions

Year	1				2			
Demand	Unique		High		Medium		Low	
Fuel	Coal	Gas	Coal	Gas	Coal	Gas	Coal	Gas
Purchase (MW)	866	434	234	116	0	200	0	0

The solution of this problem is shown in Table 1.7. Total cost is \$7,482.7.
If the order of variables is $c_1, g_1, c_2, g_2, c_3, g_3, c_0, g_0$, the constraint matrix of the problem above is

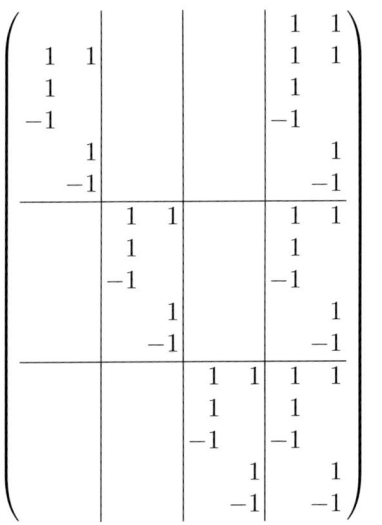

Observe that variables c_0 and g_0 are complicating variables that prevent a distributed solution of the problem. If these variables are fixed to given values, the problem decomposes by blocks. Some of the decomposition techniques explained in following chapters (such as the Benders decomposition) make use of this problem structure to allow an efficient distributed but iterative solution.

The general formulation of this problem must include the following elements:

Data.
S: the number of scenarios
p_s: the probability of scenario s in the second year
d_0: demand for the first year
d_s: demand for the second year and scenario s
a_0: coal price in the first year
a_s: coal price for scenario s in the second year

b_0: gas price in the first year
b_s: gas price for scenario s in the second year.

Variables.
c_0: amount of coal purchased the first year
c_s: amount of coal purchased the second year for scenario s
g_0: amount of gas purchased the first year
g_s: amount of gas purchased the second year for scenario s.

It is assumed that these variables are nonnegative,

$$c_0 \geq 0;$$
$$c_s \geq 0; \quad s = 1, \ldots, S$$
$$g_0 \geq 0;$$
$$g_s \geq 0; \quad s = 1, \ldots, S$$

Constraints. The constraints of this problem are
1. first year demand constraint

$$c_0 + g_0 \geq d_0 ,$$

2. supply total (first and second year) demand for all scenarios constraints

$$c_0 + g_0 + c_s + g_s = d_0 + d_s; \quad s = 1, 2, \cdots, S ,$$

3. maximum and minimum coal consumption

$$c_0 + c_s \leq \frac{2}{3}(d_0 + d_s); \quad s = 1, 2, \cdots, S$$
$$-c_0 - c_s \leq -\frac{1}{3}(d_0 + d_s); \quad s = 1, 2, \cdots, S ,$$

4. maximum and minimum gas consumption

$$g_0 + g_s \leq \frac{2}{3}(d_0 + d_s); \quad s = 1, 2, \cdots, S$$
$$-g_0 - g_s \leq -\frac{1}{3}(d_0 + d_s); \quad s = 1, 2, \cdots, S .$$

Function to Be Optimized. We are normally interested in minimizing the cost, i.e.,

$$\underset{c_0, c_s, g_0, g_s}{\text{minimize}} \quad a_0 c_0 + b_0 g_0 + \sum_{s=1}^{S} p_s \left(a_s c_s + b_s g_s \right). \tag{1.36}$$

This problem is a so-called two-stage recourse stochastic programming problem. Further details can be found in the stochastic programming books by Wallace and Kall [10], Higle and Sen [11], and Birge and Louveaux [12].

1.4.2 Capacity Expansion Planning

Consider the construction of two production facilities located at two different places to supply the demand of a large city. The locations of the two production facilities and the city are connected by roads with limited transportation capacities, as shown in Fig. 1.10. The two production facilities can be enlarged in a modular fashion, i.e., they can be expanded as needed each period of the planning horizon. Considering a 2-year analysis, the objective of the production company is to minimize investment and operation costs.

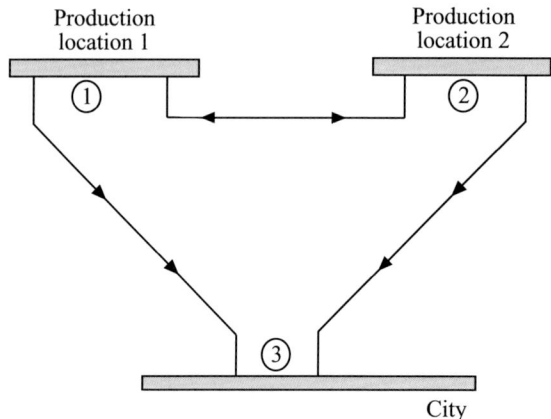

Fig. 1.10. Capacity expansion planning. Transportation network

Investment costs are expressed as

$$c_{11}x_{11} + c_{21}x_{21} + c_{12}(x_{12} - x_{11}) + c_{22}(x_{22} - x_{21}),$$

where x_{it} ($i = 1, 2; t = 1, 2$) is the production capacity already built at the beginning of period t at location i, and c_{it} ($i = 1, 2; t = 1, 2$) the building cost for period t and at location i.

The operational costs are the costs incurred in transportation. They are expressed as

$$(e_{13}f_{13,1} + e_{23}f_{23,1} + e_{12}f_{12,1} + e_{21}f_{21,1}) + (e_{13}f_{13,2} + e_{23}f_{23,2} + e_{12}f_{12,2} + e_{21}f_{21,2}),$$

where $f_{ij,t}$ ($ij = 13, 23, 12, 21; t = 1, 2$) is the quantity of product sent from location i to location j in period t, and e_{ij} ($ij = 13, 23, 12, 21$) the per unit transportation cost from location i to location j.

The constraints of this problem include the product balance at each location for each time period, as well as bounds on production and transportation capacity in each time period.

1.4 Linear Programming: Complicating Variables

Product balance equations in period 1 are

$$y_{11} = f_{13,1} + f_{12,1} - f_{21,1}$$
$$y_{21} = f_{23,1} + f_{21,1} - f_{12,1}$$
$$d_1 = f_{13,1} + f_{23,1}$$

and in period 2 are

$$y_{12} = f_{13,2} + f_{12,2} - f_{21,2}$$
$$y_{22} = f_{23,2} + f_{21,2} - f_{12,2}$$
$$d_2 = f_{13,2} + f_{23,2},$$

where the positive variables y_{it} ($i = 1, 2, t = 1, 2$) are the actual production at location i during period t; and d_t the demand of the city in period t.

Bounds on production capacity are

$$0 \leq y_{it} \leq x_{it}; \quad i = 1, 2; \quad t = 1, 2,$$

sequential bounds on available capacity are

$$x_{it} \leq x_{i,t+1}; \quad i = 1, 2; \quad t = 1,$$

absolute bounds on production capacity are

$$0 \leq x_{it} \leq x_i^{\max}; \quad i = 1, 2; \quad t = 1, 2,$$

and bounds on transportation capacity are

$$0 \leq f_{ij,t} \leq f_{ij}^{\max}; \quad ij = 12, 23, 12, 21; \quad t = 1, 2.$$

This capacity expansion planning problem can be formulated with the following elements:

Data.
d_t: demand during period t
c_{it}: building cost for location i and period t
e_{ij}: per unit transportation cost from location i to location j
$\mathcal{P}$: set of roads with transportation flow. They are ordered pairs of locations
$f_{ij}^{\max}$: maximum transportation capacity of road (i, j)
$x_i^{\max}$: maximum capacity to be built at location i.

Variables.
x_{it}: production capacity already built at the beginning of period t at location i
y_{it}: production at location i during period t

$f_{ij,t}$: production quantity sent from location i to location j during period t.
It is assumed that these variables are nonnegative,

$$x_{it} \geq 0; \quad i = 1, 2; \quad t = 1, 2$$
$$y_{it} \geq 0; \quad i = 1, 2; \quad t = 1, 2$$
$$f_{ij,t} \geq 0; \quad i = 1, 2; \quad t = 1, 2; \quad (i, j) \in \mathcal{P}.$$

Constraints. The constraints of this problem are

1. production locations balance constraints

$$y_{1t} = f_{13,t} + f_{12,t} - f_{21,t},; \quad t = 1, 2$$
$$y_{2t} = f_{23,t} + f_{21,t} - f_{12,t},; \quad t = 1, 2,$$

2. city product balance constraints

$$d_t = f_{13,t} + f_{23,t}; \quad t = 1, 2,$$

3. production bounds constraints

$$0 \leq y_{it} \leq x_{it}; \quad i = 1, 2; \quad t = 1, 2,$$

4. expansion constraints

$$x_{it} \leq x_{i,t+1}; \quad i = 1, 2; \quad t = 1,$$

5. expansion bounds

$$0 \leq x_{it} \leq x_i^{\max}; \quad i = 1, 2; \quad t = 1, 2,$$

6. transportation capacity constraints

$$0 \leq f_{ij,t} \leq f_{ij}^{\max}; \quad (i, j) \in \mathcal{P}; \quad t = 1, 2.$$

Function to Be Optimized. We are normally interested in minimizing the cost, i.e.,

$$\underset{x_{it}, y_{it}, f_{ij,t}; i = 1, 2; t = 1, 2; (i, j) \in \mathcal{P}}{\text{minimize}}$$

$$\sum_{i=1}^{2} (c_{i1} x_{i1} + c_{i2}(x_{i2} - x_{i1})) + \sum_{t=1}^{2} \sum_{(i,j) \in \mathcal{P}} e_{ij} f_{ij,t}.$$

If bounds are not taken into account, the constraint matrix of the above problem is written below

$$\begin{pmatrix}
1 & -1 & & & -1 & 1 & & & & & & & & & & \\
 & 1 & & & -1 & & 1 & -1 & & & & & & & & \\
 & & & & 1 & 1 & & & & & & & & & & \\
1 & & & & & & & & & & & & -1 & & & \\
 & 1 & & & & & & & & & & & & -1 & & \\
\hline
 & & & & & & & & 1 & -1 & & & -1 & 1 & & \\
 & & & & & & & & & 1 & & & -1 & & 1 & -1 \\
 & & & & & & & & & & & & 1 & 1 & & \\
 & & & & & & & & 1 & & & & & & -1 & \\
 & & & & & & & & & 1 & & & & & & -1 \\
\hline
 & & & & & & & & & & 1 & & -1 & & & \\
 & & & & & & & & & & & 1 & & -1 & & \\
\end{pmatrix}$$

Note that the column order is y_{11}, y_{21}, $f_{13,1}$, $f_{23,1}$, $f_{12,1}$, $f_{21,1}$; y_{12}, y_{22}, $f_{13,2}$, $f_{23,2}$, $f_{12,2}$, $f_{21,2}$; x_{11}, x_{21}, x_{12}, x_{22}.

It should be noted that variables x_{it} ($i = 1, 2; t = 1, 2$) are complicating variables. If they are fixed to given values, the problem above decomposes by time period. Complicating variables prevent a distributed solution unless a suitable decomposition technique is used. Such decomposition technique is developed in the following chapters.

For the actual values in Tables 1.8, 1.9, and 1.10, the optimal solution of this capacity expansion example is provided in Table 1.11. Total cost is $76.7, including $51.5 of investment cost and $25.2 of transportation cost.

Table 1.8. Production capacity data

Location i	Maximum production capacity
1	10
2	12

Table 1.9. Demand and building cost data

Period t	Demand	Building cost for location 1 ($)	Building cost for location 2 ($)
1	19	2.0	3.5
2	15	2.5	3.0

36   1 Motivating Examples

Table 1.10. Transportation data

Road (i,j)	Capacity	Cost ($)
(1,3)	11	0.7
(2,3)	9	0.8
(1,2)	5	0.5
(2,1)	5	0.6

Table 1.11. Solution of the capacity expansion planning example

	Location 1		Location 2		Flows			
Period t	Capacity	Production	Capacity	Production	1–3	2–3	1–2	2–1
1	10	10	9	9	10	9	0	0
2	10	10	9	5	10	5	0	0

1.4.3 The Water Supply System

Consider the water supply system in Fig. 1.11 consisting of two networks, the first containing nodes 1 to 6, and the second nodes 7 to 12, connected by a single channel.

Nodes 1 and 12 are assumed to be the water supply nodes and the rest are assumed to be consumption nodes with the flow indicated by the q variables.

Note that, in order to satisfy the balance equations, the values of the q variables must satisfy the constraint,

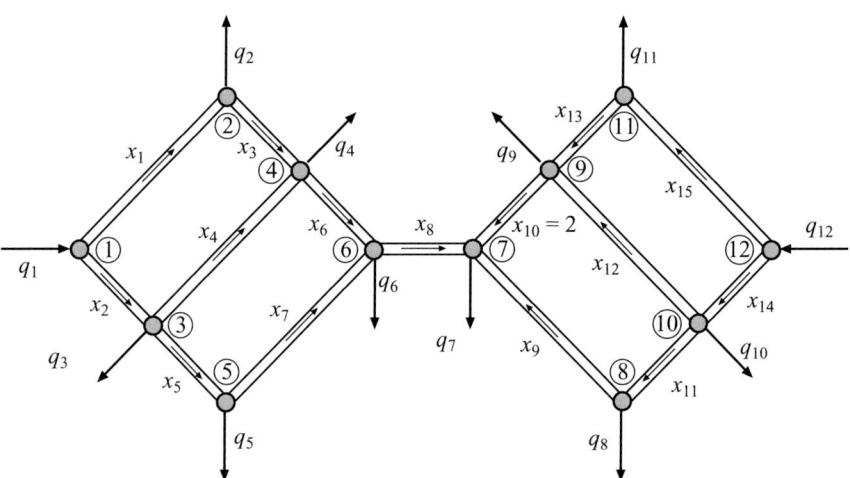

Fig. 1.11. A water supply network consisting of two networks connected by a single channel

1.4 Linear Programming: Complicating Variables

$$\sum_{i \in I} q_i = \sum_{j \in J} q_j \ ,$$

where I is the set of supply nodes, and J is the set of consumption nodes.

The flow x_i going through the connection i between nodes is considered positive ($x_i \geq 0$) because the water movement is due to the gravity (no pumps are used); it goes from higher to lower heights. To minimize the cost of operating the system, one can

$$\underset{x_i;\ i = 1, \ldots, 15}{\text{minimize}} \quad \sum_{i=1}^{15} (f_i + v_i x_i) \ ,$$

where the cost in dollars is composed by two terms, f_i is the construction cost, and v_i is the maintenance cost depending on the equilibrium flow that goes through the connection x_i, so that higher flows involve higher maintenance costs. The objective above is subject to the flow balance equations for all nodes (input amount of water equal to output amount of water including supplies and consumptions),

$$\begin{bmatrix} -1 & -1 & & & & & & & & & & & & & \\ 1 & & -1 & & & & & & & & & & & & \\ & 1 & & -1 & -1 & & & & & & & & & & \\ & & 1 & 1 & & -1 & & & & & & & & & \\ & & & & 1 & & -1 & & & & & & & & \\ & & & & & 1 & 1 & -1 & & & & & & & \\ & & & & & & & 1 & 1 & 1 & & & & & \\ & & & & & & & -1 & & & 1 & & & & \\ & & & & & & & & -1 & & & 1 & 1 & & \\ & & & & & & & & & & -1 & -1 & & 1 & \\ & & & & & & & & & & & & -1 & & 1 \\ & & & & & & & & & & & & & -1 & -1 \end{bmatrix} \begin{bmatrix} x_1 \\ x_2 \\ x_3 \\ x_4 \\ x_5 \\ x_6 \\ x_7 \\ x_8 \\ x_9 \\ x_{10} \\ x_{11} \\ x_{12} \\ x_{13} \\ x_{14} \\ x_{15} \end{bmatrix} = \begin{bmatrix} -q_1 \\ q_2 \\ q_3 \\ q_4 \\ q_5 \\ q_6 \\ q_7 \\ q_8 \\ q_9 \\ q_{10} \\ q_{11} \\ -q_{12} \end{bmatrix} .$$

Note that both supply and consumption data q_i ($i = 1, \ldots, 12$) are considered positive. The nodes have been numbered in an optimal order, so that writing the flow balance equations, the associated matrix exhibits a particular block and banded pattern, with the exception of the constraints implied by the connecting channel. In addition the width of the band is a minimum.

Any optimization problem with these constraints can be considered as a problem with a complicating constraint, or with a complicating variable. To illustrate this, we partition the above matrix in blocks of two different forms, as follows:

Formulation as a Problem with Complicating Variables (*the complicating variable is x_8*)

$$\left[\begin{array}{ccccccc|c|ccccccc}
-1 & -1 & & & & & & & & & & & & & \\
1 & & -1 & & & & & & & & & & & & \\
& 1 & & -1 & -1 & & & & & & & & & & \\
& & 1 & 1 & & -1 & & & & & & & & & \\
& & & & 1 & & -1 & & & & & & & & \\
& & & & & 1 & 1 & -1 & & & & & & & \\
\hline
& & & & & & & 1 & 1 & 1 & & & & & \\
& & & & & & & -1 & & & 1 & & & & \\
& & & & & & & & -1 & & & 1 & 1 & & \\
& & & & & & & & & -1 & -1 & & & 1 & \\
& & & & & & & & & & & -1 & & & 1 \\
& & & & & & & & & & & & -1 & -1 & \\
\end{array}\right]\left[\begin{array}{c}x_1\\x_2\\x_3\\x_4\\x_5\\x_6\\x_7\\\hline x_8\\\hline x_9\\x_{10}\\x_{11}\\x_{12}\\x_{13}\\x_{14}\\x_{15}\end{array}\right]=\left[\begin{array}{c}-q_1\\q_2\\q_3\\q_4\\q_5\\q_6\\\hline q_7\\q_8\\q_9\\q_{10}\\q_{11}\\-q_{12}\end{array}\right].$$

The matrix above has been partitioned to reveal the role of the two networks and the connection channel unknowns.

It should be noted that the network structure of the original matrix is preserved by decomposition.

Formulation as a Problem with Complicating Constraints (*the complicating constraint is constraint 6, though it could be constraint 7 too, or both*)

$$\left[\begin{array}{ccccccc|cccccccc}
-1 & -1 & & & & & & & & & & & & & \\
1 & & -1 & & & & & & & & & & & & \\
& 1 & & -1 & -1 & & & & & & & & & & \\
& & 1 & 1 & & -1 & & & & & & & & & \\
& & & & 1 & & -1 & & & & & & & & \\
\hline
& & & & & 1 & 1 & -1 & & & & & & & \\
\hline
& & & & & & & & 1 & 1 & 1 & & & & \\
& & & & & & & & -1 & & & 1 & & & \\
& & & & & & & & & -1 & & & 1 & 1 & \\
& & & & & & & & & & -1 & -1 & & & 1 \\
& & & & & & & & & & & & -1 & & 1 \\
& & & & & & & & & & & & & -1 & -1 \\
\end{array}\right]\left[\begin{array}{c}x_1\\x_2\\x_3\\x_4\\x_5\\x_6\\x_7\\\hline x_8\\x_9\\x_{10}\\x_{11}\\x_{12}\\x_{13}\\x_{14}\\x_{15}\end{array}\right]=\left[\begin{array}{c}-q_1\\q_2\\q_3\\q_4\\q_5\\\hline q_6\\\hline q_7\\q_8\\q_9\\q_{10}\\q_{11}\\-q_{12}\end{array}\right].$$

If the capacity constraints are incorporated,

$$x_i \leq x_i^{\max}; \quad i = 1, 2, \ldots, 15,$$

the banded character of the constraints is not altered. It should be noted that the network structure of the original matrix is not preserved by decomposition.

Considering the case of cost coefficients $f_i = 1$ \$/m^3; $\forall i$ and $v_i = 1$ \$/m^3; $\forall i$, capacities $x_i^{\max} = 10$ m^3; $\forall i$, and the q_i values in Fig. 1.12, the optimal cost is \$116 and the optimal solution flows are shown in Fig. 1.12.

Appropriate algorithms to take advantage of the block banded pattern of the considered problem are analyzed in this book.

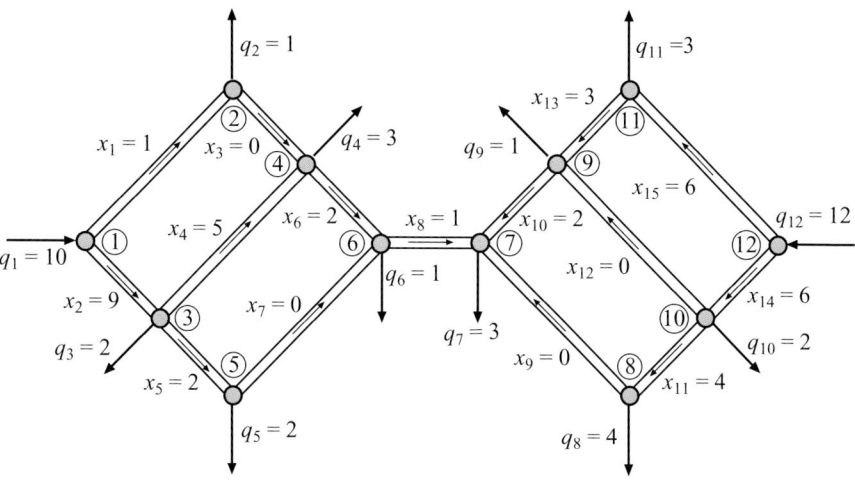

Fig. 1.12. Water supply system showing the input, output, and connection flows associated with the optimal solution

1.5 Nonlinear Programming: Complicating Constraints

1.5.1 Production Scheduling

Consider the minimum cost, multiperiod scheduling problem of several production devices serving the demand of a certain commodity. The exact demand should be served at every time period and no storage facility is available. Production devices have maximum limits on their production, and up and down ramping limits apply to the change of their respective production levels, i.e., sudden changes in production are not possible, and these limits give the maximum feasible changes.

Consider two production devices serving a 2 h demand horizon. The cost to be minimized is

$$\sum_{t=1}^{2} \sum_{i=1}^{2} \left(a_i\, x_{it} + \frac{1}{2} b_i\, x_{it}^2 \right) , \tag{1.37}$$

where a_i and b_i are the linear and quadratic cost parameters of device i, and x_{it} is its production level of device i during hour t.

The capacity limits of the production devices are

$$0 \leq x_{it} \leq x_i^{\max} ; \quad i = 1, 2; \quad t = 1, 2 , \tag{1.38}$$

where $x_i^{\max}$ is the maximum output level of device i.

Considering that the output of each device at the beginning of the planning horizon is x_i^0 ($i = 1, 2$), and that the ramping limit of each device is $r_i^{\max}$ ($i = 1, 2$), the ramping constraints are

$$x_{i1} - x_i^0 \leq r_i^{\max} ; \quad i = 1, 2 \tag{1.39}$$

1 Motivating Examples

$$x_{i2} - x_{i1} \leq r_i^{\max} \; ; \quad i = 1, 2 \tag{1.40}$$

and

$$x_i^0 - x_{i1} \leq r_i^{\max} \; ; \quad i = 1, 2 \tag{1.41}$$

$$x_{i1} - x_{i2} \leq r_i^{\max} \; ; \quad i = 1, 2 \; . \tag{1.42}$$

Finally, the demand supply constraints are

$$x_{11} + x_{21} = d_1 \tag{1.43}$$

$$x_{12} + x_{22} = d_2 \; , \tag{1.44}$$

where d_1 and d_2 are the demands in periods 1 and 2, respectively.

The last constraints couple together the decisions of the two devices serving the demands. Therefore, they are complicating constraints. If removed, the nonlinear production scheduling problem decomposes by production device.

The constraint matrix of this problem is given below. If the variable order is x_{11}, x_{12}, x_{21}, x_{22}, this matrix has the form

$$\begin{bmatrix}
1 & & & \\
 & 1 & & \\
1 & & & \\
-1 & 1 & & \\
-1 & & & \\
1 & -1 & & \\
\hline
 & & 1 & \\
 & & & 1 \\
 & & 1 & \\
 & & -1 & 1 \\
 & & -1 & \\
 & & 1 & -1 \\
\hline
1 & & 1 & \\
 & 1 & & 1
\end{bmatrix} \tag{1.45}$$

The decomposable structure of the matrix above is apparent once the last two constraints are relaxed. Note also that the nonlinear objective function is decomposable by production device.

In summary, the production scheduling problem has the following structure:

Data.
m: the number of time periods
n: the number of production devices
$x_i^{\max}$: the output capacity of device i
$r_i^{\max}$: the ramping (up and down) limit of device i
d_t: demand for period t
x_i^0: initial output level of device i
a_i, b_i: coefficients defining the nonlinear cost function of device i.

Variables.

x_{it}: the output of device i during period t.
These variables are nonnegative, i.e.,

$$x_{it} \geq 0 \; ; \quad i = 1, \ldots, n; t = 1, \ldots, m \; .$$

Constraints.

1. maximum output capacity

$$x_{it} \leq x_i^{\max} \; ; \quad i = 1, \ldots, n; t = 1, \ldots, m \; , \tag{1.46}$$

2. ramping up limits

$$x_{i1} - x_i^0 \leq r_i^{\max} \; ; \quad i = 1, \ldots, n \; , \tag{1.47}$$

$$x_{it} - x_{i,t-1} \leq r_i^{\max} \; ; \quad i = 1, \ldots, n; t = 2, \ldots, m \; , \tag{1.48}$$

3. ramping down limits

$$x_i^0 - x_{i1} \leq r_i^{\max} \; ; \quad i = 1, \ldots, n \; , \tag{1.49}$$

$$x_{i,t-1} - x_{it} \leq r_i^{\max} \; ; \quad i = 1, \ldots, n; t = 2, \ldots, m \; , \tag{1.50}$$

4. supply of demand

$$\sum_{i=1}^{n} x_{it} = d_t \; ; \quad t = 1, \ldots, m \; . \tag{1.51}$$

Function to Be Minimized.

The nonlinear function to be minimized is the production cost

$$\sum_{t=1}^{m} \sum_{i=1}^{n} \left(a_i x_{it} + \frac{1}{2} b_i x_{it}^2 \right) \; . \tag{1.52}$$

For the values provided in the Tables 1.12 and 1.13, the optimal solution is

$$x_{11}^* = 3.5; \quad x_{12}^* = 5; \quad x_{21}^* = 5.5; \quad x_{22}^* = 7 \; .$$

Finally, it should be emphasized that a decomposed solution of the nonlinear production scheduling problem is achieved by some of the decomposition techniques analyzed in the following chapters.

Table 1.12. Production device data

Device i	$x_i^{\max}$	$r_i^{\max}$	x_i^0	a_i	b_i
1	6	1.5	2.0	2.0	0.6
2	8	3.0	2.5	2.5	0.5

Table 1.13. Demand data

Period	1	2
Demand	9	12

1.5.2 Operation of a Multiarea Electricity Network

Consider the two-area electricity network of Fig. 1.13. Each area, X or Y, includes two generators and one demand interconnected through three transmission lines. Furthermore, the two areas are interconnected by one transmission line that is denominated tie-line. The maximum production capacities of the generators are $x_1^{\max}$, $x_2^{\max}$, $y_1^{\max}$, and $y_2^{\max}$, respectively; and their linear and quadratic cost coefficients are a_i^x, b_i^x, a_i^y, b_i^y ($i = 1, 2$), respectively. The hourly demands are d^x and d^y, respectively.

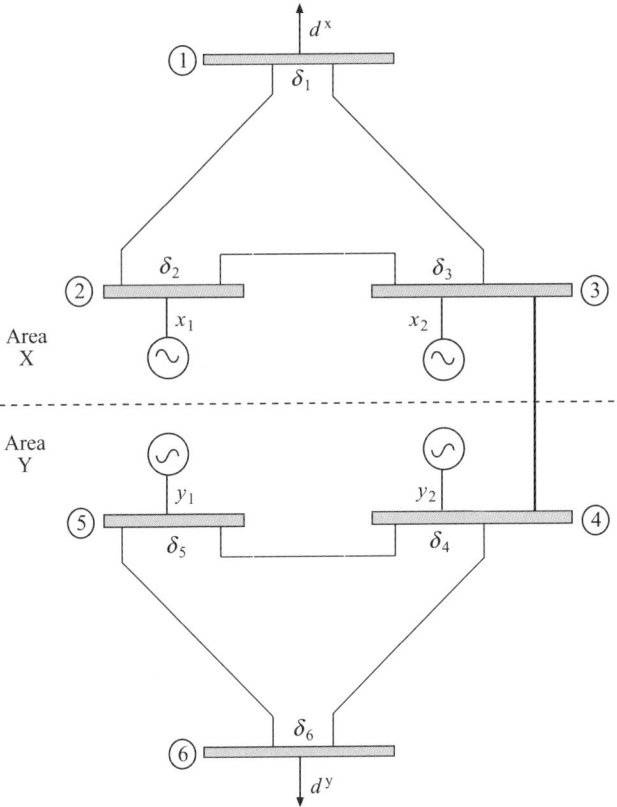

Fig. 1.13. Two-area electricity network

1.5 Nonlinear Programming: Complicating Constraints

The operation problem consists of determining the hourly energy productions of the generators x_1, x_2, y_1, and y_2, in such a manner that the production cost is minimum.

The electric energy flowing through a line between nodes i and j depends on two constructive parameters of the electricity transmission line, denominated conductance, G_{ij} and susceptance, B_{ij}, and on the relative "height" of the nodes, measured by phase variables, δ_i and δ_j, i.e.,

$$e_{ij} = e_{ij}(\delta_i, \delta_j, G_{ij}, B_{ij}) . \tag{1.53}$$

This nonlinear operation problem has the structure described below:

Data.

d^x, d^y: hourly energy demands of systems x and y, respectively
$x_1^{\max}$, $x_2^{\max}$: maximum production capacities of the two generators of area X
$y_1^{\max}$, $y_2^{\max}$: maximum production capacities of the two generators of area Y
G_{ij}, B_{ij}: conductance and susceptance (structural parameters) of line ij
$a_i^x, b_i^x, a_i^y, b_i^y$: linear and quadratic cost coefficients of generators of areas X and Y, respectively.

Variables.

x_1, x_2: energy productions of generators of area X
y_1, y_2: energy productions of generators of area Y
$\delta_1, \ldots, \delta_6$: relative "heights" or phases of nodes
e_{ij}: electric energy flowing through the line between nodes i and j.

Variables x_1, x_2, y_1, y_2 are nonnegative. Since phases always appear as differences, one phase can be set to zero, i.e., $\delta_1 = 0$.

Constraints.
1. energy balance in every node (see Fig. 1.13)

$$-d^x = e_{12} + e_{13} \tag{1.54}$$

$$x_1 = e_{21} + e_{23} \tag{1.55}$$

$$x_2 = e_{31} + e_{32} + e_{34} \tag{1.56}$$

and

$$-d^y = e_{65} + e_{64} \tag{1.57}$$

$$y_1 = e_{56} + e_{54} \tag{1.58}$$

$$y_2 = e_{45} + e_{46} + e_{43} , \tag{1.59}$$

where

$$e_{ij} = e_{ij}(\delta_i, \delta_j, G_{ij}, B_{ij}) = G_{ij}\cos(\delta_i - \delta_j) + B_{ij}\sin(\delta_i - \delta_j) - G_{ij} . \tag{1.60}$$

2. limits on production levels

$$x_1 \leq x_1^{\max} \qquad (1.61)$$
$$x_2 \leq x_2^{\max} \qquad (1.62)$$
$$y_1 \leq y_1^{\max} \qquad (1.63)$$
$$y_2 \leq y_2^{\max} . \qquad (1.64)$$

Function to Be Minimized. The nonlinear function to be minimized is the cost

$$\sum_{i=1}^{2} \left(a_i^x x_i + \frac{1}{2} b_i^x x_i^2 \right) + \sum_{i=1}^{2} \left(a_i^y y_i + \frac{1}{2} b_i^y y_i^2 \right) . \qquad (1.65)$$

The first summation corresponds to system X while the second summation corresponds to system Y.

It should be noted that if the energy balance equations at frontier nodes 3 and 4 (see Fig. 1.13) are relaxed, the problem decomposes by area. Therefore, the balance equations at the frontier nodes are complicating constraints. Some of the techniques explained in the following chapters can be used to solve the problem above in a decomposed manner.

Considering data in Tables 1.14, 1.15, and 1.16, the optimal solution of this operation problem is

$$x_1^* = 5.26 \text{ MWh}, \quad x_2^* = 4.87 \text{ MWh}, \quad y_1^* = 3.82 \text{ MWh}, \quad y_2^* = 5.65 \text{ MWh} .$$

The flow of energy in the tie line is

$$e_{34}^* = 0.89 \text{ MWh} .$$

Table 1.14. Generator data

Generator	Area X			Area Y		
	$x_i^{\max}$ (MW)	a_i^x ($/MWh)	b_i^x ($/(MW)^{2}h)	$x_i^{\max}$ (MW)	a_i^y ($/MWh)	b_i^y ($/(MW)^{2}h)
1	6	2.0	0.6	6	3.0	0.7
2	7	2.5	0.5	8	2.5	0.5

Table 1.15. Line data

Line ij	G_{ij}	B_{ij}
12	−1.0	7.0
13	−1.5	6.0
23	−0.5	7.5
45	−1.0	9.0
46	−0.9	7.0
56	−0.3	6.5
Tie-line 34	−1.3	5.5

Table 1.16. Demand data

Demand data (MWh)	d^x	d^y
	8	9.5

1.5.3 The Wall Design

Engineers in daily practice are faced with the problem of designing engineering works as bridges, dams, breakwaters, structures, for example. A good design requires minimizing the cost while satisfying some geometric and reliability constraints of the work being designed.

Consider the wall in Fig. 1.14, where a and b are the width and the height of the wall, w is its weight per unit length, t is the horizontal force acting on its right-hand side, h is the corresponding offset with respect to the soil level, and γ is the unit weight of the wall.

In this example we assume that a, b, and γ are deterministic constants, and $t \sim N(\mu_t, \sigma_t)$ and $h \sim N(\mu_h, \sigma_h)$ are independent normal random variables with the indicated mean and standard deviations.

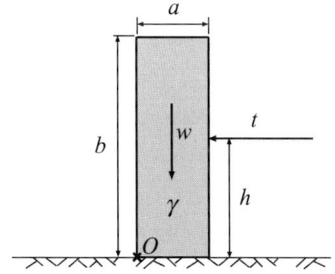

Fig. 1.14. Wall and acting forces

46   1 Motivating Examples

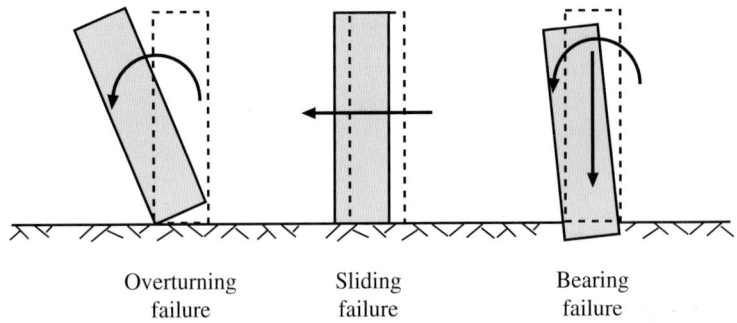

Overturning failure   Sliding failure   Bearing failure

Fig. 1.15. Illustration of the wall three modes of failure

Assume that the only the overturning failure mode as indicated in Fig. 1.15 is considered. Other failure modes (see Fig. 1.15) will be analyzed in Chap. 9, p. 349.

The overturning safety factor F_o is defined as the ratio of the stabilizing to the overturning moments with respect to some point (O in Fig. 1.14), i.e.,

$$F_\mathrm{o} = \frac{\text{Stabilizing moment}}{\text{Overturning moment}} = \frac{wa/2}{\tilde{h}\tilde{t}} = \frac{a^2b\gamma}{2\tilde{h}\tilde{t}} \geq F_\mathrm{o}^0 , \qquad (1.66)$$

where F_o^0 is the corresponding lower bound associated with the overturning failure, and the tildes refer to the characteristic values of the corresponding variables (values fixed in the engineering codes). Then, the wall is safe if and only if $F_\mathrm{o} \geq 1$.

Three different design alternatives can be used,

1. **Classical Design.** In a classical design the engineer minimizes the cost of building the engineering work subject to safety factor constraint (1.66), i.e.,

$$\underset{a,b}{\text{minimize}} \quad ab \qquad (1.67)$$

subject to

$$\frac{a^2b\gamma}{2\tilde{h}\tilde{t}} \geq F_\mathrm{o}^0 \qquad (1.68)$$

$$b = 2a , \qquad (1.69)$$

where ab is the cross section of the wall, i.e., the objective function to be optimized (cost function), and the last constraint is a geometrical constraint fixing a minimum height of the wall b_0. Characteristic values $\tilde{h}$ and $\tilde{t}$ are taken equal to their mean values μ_h and μ_t, respectively.

2. **Modern Design.** Alternatively, the modern design minimizes the cost subject to reliability constraints, i.e.,

1.5 Nonlinear Programming: Complicating Constraints

$$\underset{a,b}{\text{minimize}} \quad ab \tag{1.70}$$

subject to the reliability and geometric constraints

$$\beta(a,b) \geq \beta^0 \tag{1.71}$$
$$b = 2a , \tag{1.72}$$

where

$$\beta(a,b) = \underset{t,h}{\text{minimum}} \quad \beta(a,b,t,h) \tag{1.73}$$

subject to the overturning failure constraint

$$\frac{a^2 b \gamma}{2ht} = 1 , \tag{1.74}$$

where $\beta(a,b)$ is the reliability index associated with the overturning failure mode, and β^0 the corresponding upper bound. In this case, h and t are realizations of the corresponding random variables.

3. **Mixed Design.** There exists another design, the mixed alternative, that combines safety factors and reliability indices (see Castillo et al. 13, 15–19) and can be stated as

$$\underset{a,b}{\text{minimize}} \quad ab \tag{1.75}$$

subject to the safety factor, reliability, and geometric constraints

$$\frac{a^2 b \gamma}{2\tilde{h}\tilde{t}} \geq F_o^0 \tag{1.76}$$
$$\beta(a,b) \geq \beta^0 \tag{1.77}$$
$$b = 2a . \tag{1.78}$$

Unfortunately, the previous two alternatives cannot be solved directly, because the constraints (1.71) and (1.77) involve the additional optimization problem (1.73)–(1.74).

Therefore, constraints (1.71) and (1.77) are the complicating constraints. Consequently, these two bi-level problems cannot be solved by standard techniques and decomposition methods are required.

To perform a probabilistic design in the wall example, the joint probability density of all variables is required.

Assume that the means, characteristic values, and the standard deviations of the variables are

$$\gamma = 23 \text{ kN/m}^3, \quad \mu_h = 3 \text{ m}, \quad \mu_t = 50 \text{ kN}$$

$$\sigma_h = 0.2 \text{ m}, \quad \sigma_t = 15 \text{ kN} .$$

Assume also that the required safety factors and reliability bounds are

$$F_o^0 = 1.5, \quad \beta^0 = 3, \quad b_0 = 4 \text{ m}.$$

The random variables can be transformed to standard normal random variables (z_1, z_2) by

$$z_1 = \frac{t - \mu_t}{\sigma_t}, \quad z_2 = \frac{h - \mu_h}{\sigma_h} \tag{1.79}$$

and then, problem (1.75)–(1.78) becomes

$$\underset{a,b}{\text{minimize}} \quad ab \tag{1.80}$$

subject to the safety factor, reliability, and geometric constraints

$$\frac{a^2 b \gamma}{2 \tilde{h} \tilde{t}} \geq 1.5 \tag{1.81}$$

$$\beta(a, b) \geq 3 \tag{1.82}$$

$$b = 2a, \tag{1.83}$$

whereas it will be shown in Sect. 7.2, p. 276,

$$\beta = \underset{h,t}{\text{minimum}} \quad z_1^2 + z_2^2 \tag{1.84}$$

subject to

$$z_1 = \frac{t - \mu_t}{\sigma_t}$$

$$z_2 = \frac{h - \mu_h}{\sigma_h}$$

$$\frac{a^2 b \gamma}{2ht} = 1.$$

The solution of this problem is $a^* = 2.535$ m and $b^* = 4$ m with a cost of \$10.14 that leads to a safety factor of $F_o^* = 1.97$ and a reliability index of $\beta_o^* = 3$.

1.5.4 Reliability-based Optimization of a Rubblemound Breakwater

Consider the construction of a rubblemound breakwater (see Fig. 1.16) to protect a harbor area from high waves during storms. The breakwater must be strong enough to survive the attack of storm waves, and the crest must be high enough to prevent the intrusion of sea water into the harbor by overtopping. For simplicity, only overtopping failure is considered. Other failure modes, such as armor failure, wave transmission, are ignored.

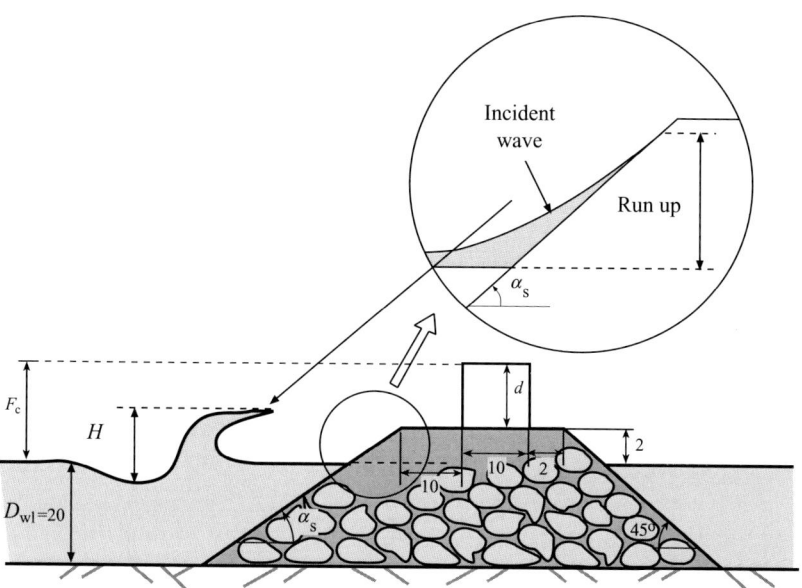

Fig. 1.16. Parameterized rubblemound breakwater used in the example

The goal is an optimal design of the breakwater based on minimizing the construction and the insurance costs against overtopping damage of the internal structures and ships.

The construction cost is

$$C_{co} = c_c v_c + c_a c_a ,$$

where v_c and c_a are the concrete and armor volumes, respectively, and c_c and c_a are the respective construction costs per unit volume.

For the sake of simplicity the insurance cost is evaluated considering the probability of overtopping failure, P_f^D, during the breakwater lifetime, D. To carry out a rigorous analysis, not only the probability of failure but how much water exceeds the freeboard level should be considered as well. Thus, the insurance cost is

$$C_{in} = 5000 + 1.25 \times 10^6 P_f^{D^2} ,$$

where the numerical constants are typical values.

The construction, insurance, and total cost versus the probability of failure are shown in Fig. 1.17. Note the decreasing and increasing character of the construction and insurance costs, respectively, as the failure probability increases, and the convex character of the total cost.

For a rubblemound breakwater of slope $\tan \alpha_s$ and freeboard F_c (see Fig. 1.16), and a given wave of height H and period T, the volume of water that overtops the structure can be estimated from the volume of water

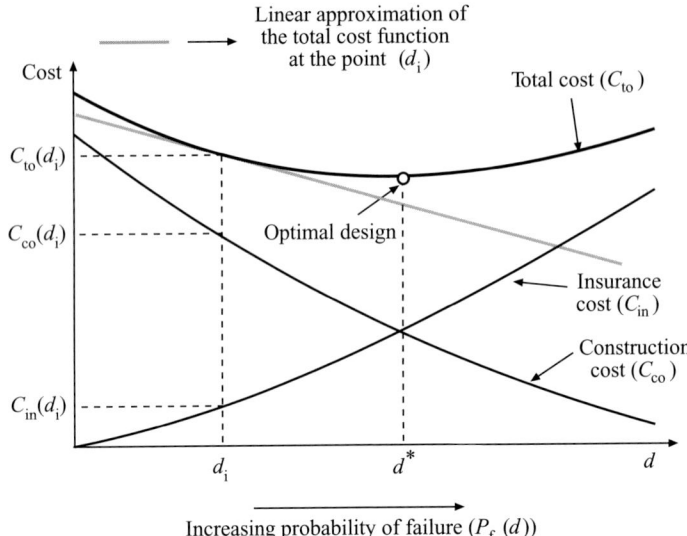

Fig. 1.17. Graphical illustration of the cost functions for the breakwater design

that would rise over the extension of the slope exceeding the freeboard level. With this approximation, overtopping (failure) occurs whenever the difference between the maximum excursion of water over the slope, R_u, called wave run-up, exceeds the freeboard F_c. Thus, overtopping failure occurs if

$$F_c - R_u < 0 \,. \tag{1.85}$$

Based on experiments, the following equation has been proposed to evaluate the dimensionless quantity R_u/H:

$$\frac{R_u}{H} = A_u \left(1 - e^{B_u I_r}\right) ,$$

where A_u and B_u are given coefficients that depend on the armor units, and I_r is the Iribarren number

$$I_r = \frac{\tan \alpha_s}{\sqrt{H/L}} ,$$

where α_s is the seaside slope angle and L is the wavelength, obtained from the dispersion equation

$$\left(\frac{2\pi}{T}\right)^2 = g \frac{2\pi}{L} \tanh \frac{2\pi D_{wl}}{L} ,$$

where D_{wl} is the design water level and g is the gravity constant.

In addition, because of construction reasons the slope α_s is limited by

$$1/5 \leq \tan \alpha_s \leq 1/2 \,.$$

The set of variables and parameters involved in this problem can be partitioned into following four subsets:

***d*: *Optimization design variables*.** The design random variables whose values are to be chosen by the optimization program to optimize the objective function (minimize the cost),
$$d = \{F_c, \tan \alpha_s\} .$$

***η*: *Nonoptimization design variables*.** The set of variables and parameters whose mean or characteristic values are fixed by the engineer or the code and must be given as data to the optimization problem,
$$\eta = \{A_u, B_u, D_{wl}, g, H, T, c_c, c_a\} .$$

***κ*: *Random model parameters*.** The set of parameters used in the probabilistic design, defining the random variability and dependence structure of the variables involved, in this case sea state descriptors,
$$\kappa = \{H_s, \bar{T}, d_{st}\} ,$$
where H_s is the significant wave height, $\bar{T}$ is the average value of the period of the seawaves, and d_{st} is the sea state duration.

***ψ*: *Auxiliary or nonbasic variables*.** The auxiliary variables whose values can be obtained from the basic variables d and η, using some formulas,
$$\psi = \{I_r, c_a, v_c, C_{co}, C_{in}, R_u, L, d\} .$$

The basic random variables in this problem are H and T, which are assumed to be independent and with cumulative distribution functions,
$$F_H(H) = 1 - e^{-2(H/H_s)^2}; \quad H \geq 0 \tag{1.86}$$
and
$$F_T(T) = 1 - e^{-0.675(T/\bar{T})^4}; \quad T \geq 0 . \tag{1.87}$$

If P_f is the probability of overtopping failure due to a single wave, the lifetime breakwater failure probability becomes
$$P_f^D(d) = 1 - (1 - P_f(d))^N , \tag{1.88}$$
where $N = d_{st}/\bar{T}$ is the mean number of waves during the design sea state for period D, and d_{st} is its duration.

Then, the design problem consists of

$$\underset{F_c, \tan \alpha_s}{\text{minimize}} \quad C_{to} = c_c v_c + c_a v_a + 5000 + 1.25 \times 10^6 P_f^{D^2}$$

subject to

$$1/5 \leq \tan \alpha_s \leq 1/2$$
$$F_c = 2 + d$$
$$v_c = 10d$$
$$v_a = \frac{1}{2}(D_{wl} + 2)\left(46 + D_{wl} + \frac{D_{wl} + 2}{\tan \alpha_s}\right)$$
$$P_f(\boldsymbol{d}) = \Phi(-\beta)$$
$$P_f^D(\boldsymbol{d}) = 1 - (1 - P_f(\boldsymbol{d}))^{(d_{st}/\bar{T})},$$

where $\Phi(\cdot)$ is the cumulative distribution function of the standard normal random variable and β is the reliability index that cannot be obtained directly and involves another optimization problem

$$\underset{H,T}{\text{minimize}} \quad \beta = \sqrt{z_1^2 + z_2^2}$$

subject to

$$\frac{R_u}{H} = A_u \left(1 - e^{B_u I_r}\right)$$
$$I_r = \frac{\tan \alpha_s}{\sqrt{H/L}}$$
$$\left(\frac{2\pi}{T}\right)^2 = g\frac{2\pi}{L} \tanh \frac{2\pi D_{wl}}{L}$$
$$\Phi(z_1) = 1 - e^{-2(H/H_s)^2}$$
$$\Phi(z_2) = 1 - e^{-0.675(T/\bar{T})^4}$$
$$F_c = R_u,$$

where z_1 and z_2 are independent standard normal random variables.

As it has been illustrated, the reliability-based optimization problems can be characterized as bi-level optimization problems (see Mínguez [13] and Mínguez et al. [14]). The upper level is the overall optimization in the design variables $\boldsymbol{d}$, in this case the minimization of the total cost function. The lower level is the reliability estimation in the $\boldsymbol{z}$ variables, necessary for evaluating the insurance cost. This bi-level structure is the reason why the decomposition procedure is needed.

Assuming the following values for the variables involved,

$$D_{wl} = 20 \text{ m}, \ A_u = 1.05, \ B_u = -0.67, \ c_c = 60 \text{ \$/m}^3,$$
$$g = 9.81 \text{ m/s}^2, \ c_a = 2.4 \text{ \$/m}^3, \ H_s = 5 \text{ m}, \ \bar{T} = 10 \text{ s}, \ d_{st} = 1 \text{ h}.$$

The solution of this problem, using the method above, is

$$F_c^* = 5.88 \text{ m}, \tan \alpha_s^* = 0.23, C_{co}^* = \$6{,}571.3, C_{in}^* = \$5{,}019.8, C_{to}^* = \$11{,}591.1.$$

Note that the method also provides the optimal reliability index and probability of run-up for a single wave and during the design sea state, respectively,

$$\beta^* = 4.738, \ P_\mathrm{f}^* = 0.00000111 \ \text{and} \ P_\mathrm{f}^{\mathrm{D}*} = 0.00039845 \ .$$

1.6 Nonlinear Programming: Complicating Variables

1.6.1 Capacity Expansion Planning: Revisited

The capacity expansion planning problem, analyzed in Sect. 1.4.2, is revisited in this section. The problem consists of constructing two electricity production facilities, located at two different places, to supply the demand of a large city. The production facilities and the city are connected by lines of limited transmission capacity, as shown in Fig. 1.10.

The production facilities can be enlarged in a modular fashion, i.e., they can be expanded as needed for each period of the planning horizon. The only differences from the original statement of the problem in Sect. 1.4.2 are that transmission costs are zero and that equations expressing the energy transmitted through lines are nonlinear.

This multiperiod capacity expansion planning problem is formulated below:

Data.
d_t: demand during time period t
c_{it}: building cost at location i for time period t
B_{ij}: susceptance (structural parameter) of line ij
G_{ij}: conductance (structural parameter) of line ij
$\mathcal{P}$: set of transmission lines; lines are represented as ordered pairs of locations, i.e., arcs
$f_{ij}^{\max}$: maximum transmission capacity of line ij
$x_i^{\max}$: maximum production capacity that can be built at location i.

Variables.
x_{it}: production capacity already built at location i at the beginning of time period t
y_{it}: actual production at location i during time period t
$f_{ij,t}$: energy sent from location i to location j during period t
δ_{it}: relative "height" at location i with respect to the reference location during period t.

It should be noted that variables x_{it} and y_{it} are nonnegative. That is

$$x_{it} \geq 0 \ ; \quad i = 1, 2; \ t = 1, 2 \tag{1.89}$$
$$y_{it} \geq 0 \ ; \quad i = 1, 2; \ t = 1, 2 \ . \tag{1.90}$$

Constraints.
The constraints of this problem are
1. energy balances at production locations 1 and 2, respectively

$$y_{1t} = f_{13,t} + f_{12,t} - f_{21,t} \; ; \quad t = 1, 2 \tag{1.91}$$
$$y_{2t} = f_{23,t} + f_{21,t} - f_{12,t} \; ; \quad t = 1, 2 \;, \tag{1.92}$$

2. energy balance in the city

$$d_t = f_{13,t} + f_{23,t} \; ; \quad t = 1, 2 \;, \tag{1.93}$$

3. production capacity limits

$$0 \leq y_{it} \leq x_{it} \; ; \quad i = 1, 2; \; t = 1, 2 \;, \tag{1.94}$$

4. expansion constraints

$$x_{it} \leq x_{i,t+1} \; ; \quad i = 1, 2; \; t = 1 \;, \tag{1.95}$$

5. expansion bounds

$$0 \leq x_{it} \leq x_i^{\max} \; ; \quad i = 1, 2; \; t = 1, 2 \;, \tag{1.96}$$

6. transmission capacity limits

$$0 \leq f_{ij,t} \leq f_{ij}^{\max} \; ; \quad (i, j) \in \mathcal{P}; \; t = 1, 2 \;, \tag{1.97}$$

7. transmitted commodity through lines

$$f_{ij,t} = G_{ij} \cos(\delta_{it} - \delta_{jt}) + B_{ij} \sin(\delta_{it} - \delta_{jt}) - G_{ij} \; ; (i, j) \in \mathcal{P}; \; t = 1, 2 \;. \tag{1.98}$$

Function to Be Minimized. The function to be minimized is cost, i.e.,

$$\sum_{i=1}^{2} [c_{i1} \, x_{i1} + c_{i2} \, (x_{i2} - x_{i1})] \;. \tag{1.99}$$

For the data provided in Tables 1.17, 1.18, 1.19, and 1.20, the optimal solution of the multiperiod capacity expansion planning problem is provided in Table 1.21. Total cost is $19.7.

It should be noted that variables x_{it} ($i = 1, 2; \; t = 1, 2$) are complicating variables. If they are fixed to given values, the nonlinear capacity expansion planning problem is decomposed by time period. Complicating variables prevent a distributed solution unless a suitable decomposition technique is used. Some of such techniques are developed in the following chapters.

Table 1.17. Transmission data

Line (i,j)	(1,2)	(1,3)	(2,1)	(2,3)
Susceptance, B_{ij}	9	15	9	18
Conductance, G_{ij}	−0.5	−0.4	−0.5	−0.7
Capacity, $f_{ij}^{\max}$	2.5	6.0	2.5	4.0

Table 1.18. Building cost data

Time period	1	2
Location 1 ($)	2.0	2.5
Location 2 ($)	3.5	3.0

Table 1.19. Demand data

Time period	1	2
Demand	7	5

Table 1.20. Expansion alternative data

Location	Maximum capacity
1	5
2	6

Table 1.21. Solution of the second version of the capacity expansion planning example

	Location 1		Location 2		Flows			
Period t	Cap.	Prod.	Cap.	Prod.	1–3	1–2	2–1	2–3
1	3.2	3.2	3.8	3.8	3.2	0.0	0.0	3.8
2	3.2	2.3	3.8	2.7	2.3	0.0	0.0	2.7

1.7 Mixed-Integer Programming: Complicating Constraints

1.7.1 Unit Commitment

Consider two electric energy consumption centers interconnected by a transmission line as illustrated in Fig. 1.18. Each consumption center has its own production facility. If economically advantageous, energy can be transmitted using the transmission line from a consumption center to the other one and vice versa. The unit commitment problem consists of determining if either one or two of the production facilities should be on line to supply the demand at minimum cost.

56   1 Motivating Examples

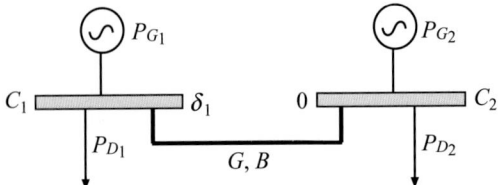

Fig. 1.18. Unit commitment of production facilities

The problem is formulated as follows:

Data.
P_{Di}: demand of consumption center i
$P_{Gi}^{\max}$: maximum capacity of production center i
$P_{Gi}^{\min}$: minimum capacity of production center i
G: conductance (structural parameter) of the transmission line
B: susceptance (structural parameter) of the transmission line
C_i: cost of producing at center i.

Variables.
P_{Gi}: production at center i
δ_1: electrical angle ("height") of production center 1 with respect to production center 2
u_i: status binary variable of production facility i; 1 if on-line and 0 if off-line.

Constraints. The constraints of this problem are
1. production capacity limits of the facilities

$$u_i P_{Gi}^{\min} \leq P_{Gi} \leq u_i P_{Gi}^{\max}; \quad i = 1, 2. \tag{1.100}$$

2. production balance at center 1

$$P_{G1} + [G(1 - \cos \delta_1) - B \sin \delta_1] = P_{D1}. \tag{1.101}$$

3. production balance at center 2

$$P_{G2} + [G(1 - \cos \delta_1) + B \sin \delta_1] = P_{D2}. \tag{1.102}$$

The two terms within square brackets are the energies transmitted through the line to centers 1 and 2, respectively.
4. security of supply

$$u_1 P_{G1}^{\max} + u_2 P_{G2}^{\max} \geq P_{D1} + P_{D2}. \tag{1.103}$$

This constraint ensures that enough capacity to supply the total demand is in operation.

Function to Be Minimized. The objective is to minimize cost, thus the objective function is

$$C_1 \, P_{G1} + C_2 \, P_{G2} \, . \tag{1.104}$$

It should be noted that the problem above is mixed-integer and nonlinear. However, nonlinear equations (1.101) and (1.102) can be linearized. This linearization renders a mixed-integer linear problem. The outer linearization procedure explained in Sect. 6.5.1, p. 258, uses an iterative linearization procedure to solve this nonlinear mixed-integer problem.

Table 1.22. Production center data

Center i	$P_{Gi}^{\max}$ (MW)	$P_{Gi}^{\min}$ (MW)	C_i ($/MWh)	P_{Di} (MWh)
1	6	1.0	25	7
2	8	1.5	15	5

Table 1.23. Line data

Line	G	B
Data	-0.5	3.5

Considering the data in Tables 1.22 and 1.23, the solution of this unit commitment problem is

$$P_{G1}^* = 4.4 \text{ MWh}; \quad P_{G2}^* = 8.0 \text{ MWh}; \quad \delta_1 = -0.927 \text{ rad} \, ,$$

with an objective function value of $230.

Note that the total production $P_{G1}^* + P_{G2}^*$ is larger than the total demand $P_{D1} + P_{D2}$ due to losses in the transmission line.

1.8 Mixed-Integer Programming: Complicating Variables

1.8.1 Capacity Expansion Planning: Revisited 2

The capacity expansion planning problem, analyzed in Sects. 1.4.2 and 1.6.1, is again revisited in this section. The problem consists of constructing two discrete-sized production facilities, located in two different places, to supply the demand of a large city. The production facilities and the city are connected by roads of limited transportation capacity, as shown in Fig. 1.10. The production facilities can be enlarged in a modular fashion, i.e., they can be expanded as needed each period of the planning horizon. However, expansion alternatives are discrete. The only differences with the original statement of the problem in Sect. 1.4.2 are that expansion alternatives are discrete and that transportation costs are quadratic.

This new discrete-sized multiperiod capacity expansion planning problem is formulated below:

Data.
d_t: demand during time period t
c_{it}: building cost at location i for time period t
e_{ij}: per unit transportation cost from location i to location j
$\mathcal{P}$: set of roads used for transportation; roads are represented as ordered pairs of locations
$f_{ij}^{\max}$: maximum transportation capacity of road (i,j)
$\{x_i^a, x_i^b, x_i^c\}$: set of possible building alternatives.

Variables.
x_{it}: discrete production capacity already built at location i at the beginning of time period t
y_{it}: actual production at location i during time period t
$f_{ij,t}$: production quantity sent from location i to location j during period t.

It should be noted that all the above variables are nonnegative. Additionally, variables x_{it} ($i = 1, 2, t = 1, 2$) are integers. That is

$$y_{it} \geq 0 ; \quad i = 1,2; \ t = 1,2 \tag{1.105}$$
$$f_{ij,t} \geq 0 ; \quad (i,j) \in \mathcal{P}; \ t = 1,2 \tag{1.106}$$
$$x_{it} \in \{x_i^a, x_i^b, x_i^c\} ; \quad i = 1,2; \ t = 1,2 \tag{1.107}$$
$$x_i^a, x_i^b, x_i^c \geq 0 ; \quad i = 1,2 . \tag{1.108}$$

Constraints. The constraints of this problem are
1. production balances at production locations 1 and 2, respectively,

$$y_{1t} = f_{13,t} + f_{12,t} - f_{21,t} ; \quad t = 1,2 \tag{1.109}$$
$$y_{2t} = f_{23,t} + f_{21,t} - f_{12,t} ; \quad t = 1,2 , \tag{1.110}$$

2. production balances in the city

$$d_t = f_{13,t} + f_{23,t} ; \quad t = 1,2 , \tag{1.111}$$

3. production capacity limits

$$0 \leq y_{it} \leq x_{it} ; \quad i = 1,2; \ t = 1,2 , \tag{1.112}$$

4. building alternatives

$$x_{it} \in \{x_i^a, x_i^b, x_i^c\} ; \quad i = 1,2; \ t = 1,2 , \tag{1.113}$$

1.8 Mixed-Integer Programming: Complicating Variables

5. transportation capacity limits

$$0 \leq f_{ij,t} \leq f_{ij}^{\max} \, ; \quad (i,j) \in \mathcal{P}; \; t = 1, 2 \, . \tag{1.114}$$

Function to Be Minimized. The function to be minimized is cost, i.e.,

$$\sum_{i=1}^{2} \left(c_{i1} \, x_{i1} + c_{i2} \left(x_{i2} - x_{i1} \right) + \sum_{(i,j) \in \mathcal{P}} e_{ij} \, f_{ij,t}^{2} \right) . \tag{1.115}$$

For the data provided in Tables 1.24, 1.25, 1.26, 1.27, the optimal solution of this discrete-sized multiperiod capacity expansion planning problem is provided in Table 1.28. Total cost is $97.91, including $22 of investment cost and $75.91 of transportation cost.

Table 1.24. Transportation data

Road (i,j)	(1,2)	(1,3)	(2,1)	(2,3)
Cost, e_{ij} ($)	0.5	0.7	0.6	0.8
Capacity, $f_{ij}^{\max}$	5	11	5	9

Table 1.25. Production cost data

Time period	1	2
Location 1	2.0	2.5
Location 2	3.5	3.0

Table 1.26. Demand data

Time period	1	2
Demand	11	9

It should be noted that integer variables x_{it} ($i = 1, 2; t = 1, 2$) are complicating variables. If they are fixed to given integer values, the mixed-integer nonlinear capacity expansion planning problem can be decomposed by time period. Complicating variables prevent a distributed solution unless a suitable decomposition technique is used. Some of such techniques are developed in the following chapters.

Table 1.27. Expansion alternative data

Expansion alternative	a	b	c
Location 1	3	5	9
Location 2	4	6	8

Table 1.28. Solution of the revisited capacity expansion planning example

Period t	Location 1		Location 2		Flows			
	Cap.	Prod.	Cap.	Prod.	1–3	1–2	2–1	2–3
1	9	7	4	4	6.15	0.85	0.00	4.85
2	5	5	4	4	4.85	0.15	0.00	4.15

1.8.2 The Water Supply System: Revisited

Consider the water supply system problem in Sect. 1.4.3. Note that the optimal solution flow in some connections is zero (no water flowing) but there is a construction cost associated with that connection. This means that the connection is built even in the case that no flow goes through. If we are interested in the minimum cost considering not only maintenance but construction as well, the following cost terms c_i in dollars associated with the connections must be considered

$$c_i = \begin{cases} 0 & \text{if } x_i = 0 \\ f_i + v_i x_i & \text{if } 0 < x_i \leq x_i^{\max} \end{cases} ; \quad i = 1, \ldots, 15 ,$$

where $x_i^{\max}$ is the maximum flow capacity for connection i, c_i is the connection cost in dollars, f_i is the construction cost, and v_i is the maintenance cost depending on the equilibrium flow that goes through the connection x_i.

This cost can be implemented using binary variables by the following set of constraints for each connection i:

$$\begin{aligned} c_i &= y_i f_i + v_i x_i, \\ x_i &\leq y_i x_i^{\max}, \\ x_i &\geq 0, \\ y_i &\in \{0, 1\} . \end{aligned} \quad (1.116)$$

Note that there are following two possibilities:

Case 1: If $y_i = 0$, then $0 \leq x_i \leq 0$, so that $x_i = 0$ and then $c_i = 0$. This means that there is no connection.

Case 2: If $y_i = 1$, then $0 \leq x_i \leq x_i^{\max}$, so that $c_i = f_i + v_i x_i$. This means that the connection exists.

One can decide which connections are necessary and what are redundant, by solving the problem,

1.9 Concluding Remarks   61

$$\begin{array}{c} \text{minimize} \\ x_i, y_i; \ i = 1, 2, \ldots, 15 \end{array} \sum_{i=1}^{15} c_i$$

subject to the flow balance equations:

$$\begin{bmatrix} -1 & -1 & & & & & & & & & & & & & \\ 1 & & -1 & & & & & & & & & & & & \\ & 1 & & -1 & -1 & & & & & & & & & & \\ & & 1 & 1 & & -1 & & & & & & & & & \\ & & & & 1 & & -1 & & & & & & & & \\ & & & & & 1 & 1 & -1 & & & & & & & \\ & & & & & & & 1 & 1 & 1 & & & & & \\ & & & & & & -1 & & 1 & & & & & & \\ & & & & & & & -1 & & & 1 & 1 & & & \\ & & & & & & & & -1 & -1 & & & 1 & & \\ & & & & & & & & & & -1 & & & 1 & \\ & & & & & & & & & & & & -1 & -1 & \end{bmatrix} \begin{bmatrix} x_1 \\ x_2 \\ x_3 \\ x_4 \\ x_5 \\ x_6 \\ x_7 \\ x_8 \\ x_9 \\ x_{10} \\ x_{11} \\ x_{12} \\ x_{13} \\ x_{14} \\ x_{15} \end{bmatrix} = \begin{bmatrix} -q_1 \\ q_2 \\ q_3 \\ q_4 \\ q_5 \\ q_6 \\ q_7 \\ q_8 \\ q_9 \\ q_{10} \\ q_{11} \\ -q_{12} \end{bmatrix}$$

and to constraints (1.116).

Using the same data as in the numerical example in page 55, one gets an optimal cost of $96 and the optimal solution flows are shown in Fig. 1.19, where the unnecessary connections have been deleted. This problem is decomposable if variable x_8 is considered a complicating variable. In that situation, the Benders decomposition algorithm can be used.

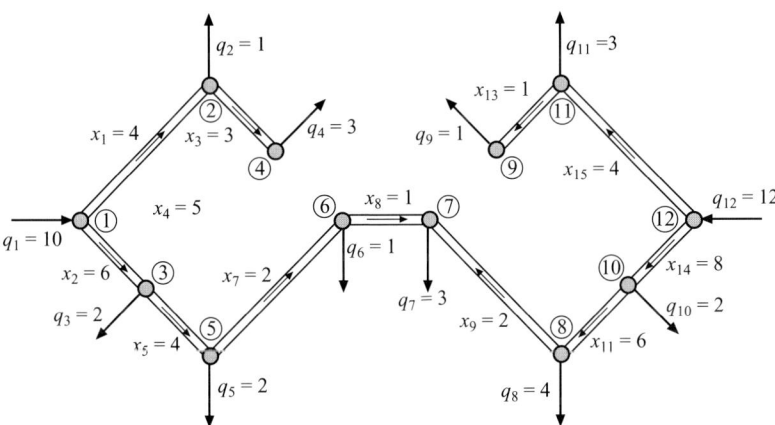

Fig. 1.19. Redesigned water supply system showing the input, output, and connection flows associated with the optimal solution

1.9 Concluding Remarks

This chapter describes a collection of motivating examples with decomposable structure. Its aim is to illustrate that the number of practical problems in engineering (Industrial, Hydroelectricity, Energy Production, Civil, etc.) and science that present decomposable structure is large and rich.

Applications range from operation to planning and design of engineering systems, and include decision making problems under uncertainty.

In the following chapters techniques to efficiently solve problems with decomposable structure are developed.

Table 1.29 shows the decomposition techniques that can be used to solve decomposable problems. The symbol ✓ means that the technique can be applied to the corresponding problem, the symbol − means that it cannot be applied, the symbol (✓) means that the technique is applicable but with a doubtful behavior (with a possible fail) and the symbol (−) means that it is not recommended to use this technique but it can be applied in some cases.

Table 1.29. Summary of decomposition techniques

Applicable Technique	Problem			
	Continuous linear		Continuous nonlinear	
	Compl. constraints	Compl. variable	Compl. constraints	Compl. variable
Dantzig-Wolfe	✓	−	✓[a]	−
Lagrangian relaxation	(−)	−	✓	−
Augmented Lagrangian	✓	−	✓	−
Optimality conditions	(−)	−	✓	−
Benders	−	✓	−	✓

	Noncontinuous linear		Noncontinuous nonlinear	
	Compl. constraints	Compl. variable	Compl. constraints	Compl. variable
Dantzig-Wolfe	✓[b]	−	✓[a]	−
Lagrangian relaxation	(✓)	✓	✓	−
Augmented Lagrangian	(✓)	✓	✓	−
Optimality conditions	−	−	✓	−
Benders	−	✓	−	✓

[a] known with names other than Dantzig-Wolfe.
[b] known as "branch and price."

1.10 Exercises

Exercise 1.1. Consider the water supply system in Subsect. 1.4.3, and answer the following questions:

1. Determine the maximum number of links (connections) that can be removed without causing service failure.
2. How this number relates to the reliability of the system?

3. Are some nodes more reliable than others? Why?
4. How can this problem be decomposed?

Exercise 1.2. Modify the wall design problem for a wall shape as indicated in Fig. 1.20, considering the additional constraints $c \geq a$ and $d \geq a$.

Exercise 1.3. Formulate a transnational manufacturing operation problem similar to the Transnational Soda Company analyzed in Sect. 1.3.1. Show its decomposable structure.

Exercise 1.4. Formulate the problem faced by the operator of a water supply system associated with a river basin. Since storage facilities are not available, hourly water demand should be met at every hour. Show the structure of this problem.

Exercise 1.5. Consider a multiarea and multiperiod production scheduling problem. Show its structure and illustrate it by means of a simple example.

Exercise 1.6. Formulate a 2-year coal oil and gas procurement problem considering that five demand scenarios are possible the last year. Show its decomposable structure.

Exercise 1.7. Formulate the multiperiod and multiarea operation problem associated with a natural gas supply network that includes storage facilities. Analyze the structure of this problem.

Exercise 1.8. Formulate a multiperiod capacity expansion planning problem including nonlinear investment and operation costs and discrete investment variables. Illustrate the structure of this problem.

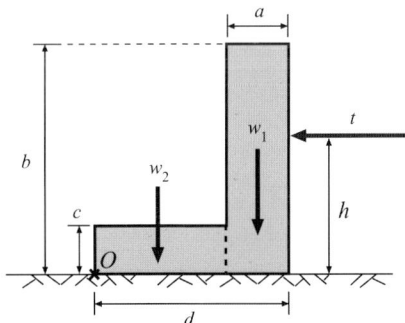

Fig. 1.20. Wall cross section

Exercise 1.9. Sketch the components of the problem associated to the engineering design of a bridge. Show and illustrate the structure of this problem.

Exercise 1.10. Formulate the 24-h unit commitment problem of production units. Consider minimum and maximum output constraints, ramping limits, and security constraint. Show the structure of this problem. Derive linear constraints to enforce minimum up time constraints.

Part II

Decomposition Techniques

2

Decomposition in Linear Programming: Complicating Constraints

2.1 Introduction

The size of a linear programming problem can be very large. One can encounter in practice problems with several hundred thousands of equations and/or unknowns. To solve these problems the use of some special techniques is either convenient or required. Alternatively, a distributed solution of large problems may be desirable for technical or practical reasons. Decomposition techniques allow certain type of problems to be solved in a decentralized or distributed fashion. Alternatively, they lead to a drastic simplification of the solution procedure of the problem under study.

For a decomposition technique to be useful, the problem at hand must have the appropriate structure. Two such cases arise in practice: the complicating constraint and the complicating variable structures. The first one is analyzed below, and the second is analyzed in Chap. 3.

In a linear programming problem, the complicating constraints involving variables from different blocks drastically complicate the solution of the problem and prevent its solution by blocks. The following example illustrates how complicating constraints impede a solution by blocks.

Illustrative Example 2.1 (Complicating constraints that prevent a distributed solution). Consider the problem

$$\underset{x_1, x_2, x_3, y_1, y_2, v_1, v_2, v_3}{\text{minimize}} \quad a_1 x_1 + a_2 x_2 + a_3 x_3 + b_1 y_1 + b_2 y_2 + c_1 v_1 + c_2 v_2 + c_3 v_3$$

subject to

$$a_{11}x_1 + a_{12}x_2 + a_{13}x_3 = e_1$$
$$a_{21}x_1 + a_{22}x_2 + a_{23}x_3 = e_2$$
$$b_{11}y_1 + b_{12}y_2 = f_1$$
$$c_{11}v_1 + c_{12}v_2 + c_{13}v_3 = g_1$$
$$c_{21}v_1 + c_{22}v_2 + c_{23}v_3 = g_2$$
$$d_{11}x_1 + d_{12}x_2 + d_{13}x_3 + d_{14}y_1 + d_{15}y_2 + d_{16}v_1 + d_{17}v_2 + d_{18}v_3 = h_1$$
$$x_1, x_2, x_3, y_1, y_2, v_1, v_2, v_3 \geq 0 \, .$$

If the last equality constraint is not enforced, i.e., it is relaxed, the above problem decomposes into the following three problems:

Subproblem 1:

$$\underset{x_1, x_2, x_3}{\text{minimize}} \quad a_1 x_1 + a_2 x_2 + a_3 x_3$$

subject to

$$a_{11}x_1 + a_{12}x_2 + a_{13}x_3 = e_1$$
$$a_{21}x_1 + a_{22}x_2 + a_{23}x_3 = e_2$$
$$x_1, x_2, x_3 \geq 0 \, .$$

Subproblem 2:

$$\underset{y_1, y_2}{\text{minimize}} \quad b_1 y_1 + b_2 y_2$$

subject to

$$b_{11}y_1 + b_{12}y_2 = f_1$$
$$y_1, y_2 \geq 0 \, .$$

Subproblem 3:

$$\underset{v_1, v_2, v_3}{\text{minimize}} \quad c_1 v_1 + c_2 v_2 + c_3 v_3$$

subject to

$$c_{11}v_1 + c_{12}v_2 + c_{13}v_3 = g_1$$
$$c_{21}v_1 + c_{22}v_2 + c_{23}v_3 = g_2$$
$$v_1, v_2, v_3 \geq 0 \, .$$

Since the last equality constraint of the original problem

$$d_{11}x_1 + d_{12}x_2 + d_{13}x_3 + d_{14}y_1 + d_{15}y_2 + d_{16}v_1 + d_{17}v_2 + d_{18}v_3 = h_1$$

involves all variables, preventing a solution by blocks, it is a complicating constraint. □

Complicating constraints may prevent a straightforward solution of the linear programming problem being considered. The next example illustrates this situation.

2.1 Introduction

Illustrative Example 2.2 (Complicating constraints that prevent an efficient solution). Consider the problem

$$\underset{x_1,\ldots,x_7}{\text{minimize}} \quad c_1 x_1 + c_2 x_2 + c_3 x_3 + c_4 x_4 + c_5 x_5 + c_6 x_6 + c_7 x_7$$

subject to

$$\begin{aligned}
a_{11}x_1 \phantom{{}+a_{42}x_2+a_{43}x_3+a_{44}x_4+a_{45}x_5+a_{46}x_6} + a_{17}x_7 &= b_1 \\
a_{22}x_2 \phantom{{}+a_{43}x_3+a_{44}x_4+a_{45}x_5+a_{46}x_6} + a_{27}x_7 &= b_2 \\
a_{37}x_7 &= b_3 \\
a_{41}x_4 + a_{42}x_2 + a_{43}x_3 + a_{44}x_4 + a_{45}x_5 + a_{46}x_6 + a_{47}x_7 &= b_4 \\
x_1, x_2, x_3, x_4, x_5, x_6, x_7 &\geq 0,
\end{aligned}$$

where

$$c_1, c_2, c_3, c_4, c_5, c_6, c_7 \geq 0.$$

If the last equality constraint is dropped, the optimal solution is trivially obtained solving the resulting system of equations, i.e.,

$$\begin{aligned}
x_7 &= \frac{b_3}{a_{37}} \\
x_1 &= \frac{b_1}{a_{11}} - \frac{a_{17}}{a_{37} a_{11}} b_3 \\
x_2 &= \frac{b_2}{a_{22}} - \frac{a_{27}}{a_{37} a_{22}} b_3 \\
x_3, x_4, x_5, x_5 &= 0.
\end{aligned}$$

Since the last constraint of the original problem

$$a_{41}x_1 + a_{42}x_2 + a_{43}x_3 + a_{44}x_4 + a_{45}x_5 + a_{46}x_6 + a_{47}x_7 = b_4$$

prevents a straightforward solution of the problem, it is a complicating constraint. □

Decomposition procedures are computational techniques that indirectly consider the complicating constraints. The price that has to be paid for such a simplification is repetition. That is, instead of solving the original problem with complicating constraints, two problems are solved iteratively (i.e., repetitively): a simple so-called master problem and a problem similar to the original one but without complicating constraints. In this manner, complicating constraints are progressively taken into account. The decomposition techniques analyzed in this chapter attain optimality within a finite number of iterations. This robust behavior is of particular interest in many practical applications. Decomposition techniques for problems with complicating constraints are developed in the following sections.

2.2 Complicating Constraints: Problem Structure

Consider the linear programming problem

$$\underset{x_1, x_2, \ldots, x_n}{\text{minimize}} \quad \sum_{j=1}^{n} c_j x_j \qquad (2.1)$$

subject to

$$\sum_{j=1}^{n} e_{ij} x_j = f_i; \qquad i = 1, \ldots, q \qquad (2.2)$$

$$\sum_{j=1}^{n} a_{ij} x_j = b_i; \qquad i = 1, \ldots, m \qquad (2.3)$$

$$0 \leq x_j \leq x_j^{\text{up}}; \quad j = 1, \ldots, n, \qquad (2.4)$$

where constraints (2.2) have a decomposable structure in r blocks, each of size n_k ($k = 1, 2, \ldots, r$), i.e., they can be written as

$$\sum_{j=n_{k-1}+1}^{n_k} e_{ij} x_j = f_i; \qquad i = q_{k-1}+1, \ldots, q_k; \qquad k = 1, 2, \ldots, r. \qquad (2.5)$$

Note that $n_0 = q_0 = 0$, $q_r = q$ and $n_r = n$.

On the other hand, since constraints (2.3) have no decomposable structure, they are the complicating constraints.

Note that upper bounds x_j^{up} ($j = 1, \ldots, n$) are considered for all optimization variables x_j ($j = 1, \ldots, n$). This assumption allows dealing with a compact (finite) feasible region, leading to a simpler theoretical analysis of the problem (2.1)–(2.4). This assumption is justified by the bounded nature of most engineering variables.

Figure 2.1 shows the structure of the above problem for the case $r = 3$. This particular problem can be written as

$$\underset{\boldsymbol{x}^{[1]}, \boldsymbol{x}^{[2]}, \boldsymbol{x}^{[3]}}{\text{minimize}} \quad \left((\boldsymbol{c}^{[1]})^T \mid (\boldsymbol{c}^{[2]})^T \mid (\boldsymbol{c}^{[3]})^T \right) \begin{pmatrix} \boldsymbol{x}^{[1]} \\ - \\ \boldsymbol{x}^{[2]} \\ - \\ \boldsymbol{x}^{[3]} \end{pmatrix},$$

where the superindices in brackets refer to partitions, subject to

2.2 Complicating Constraints: Problem Structure

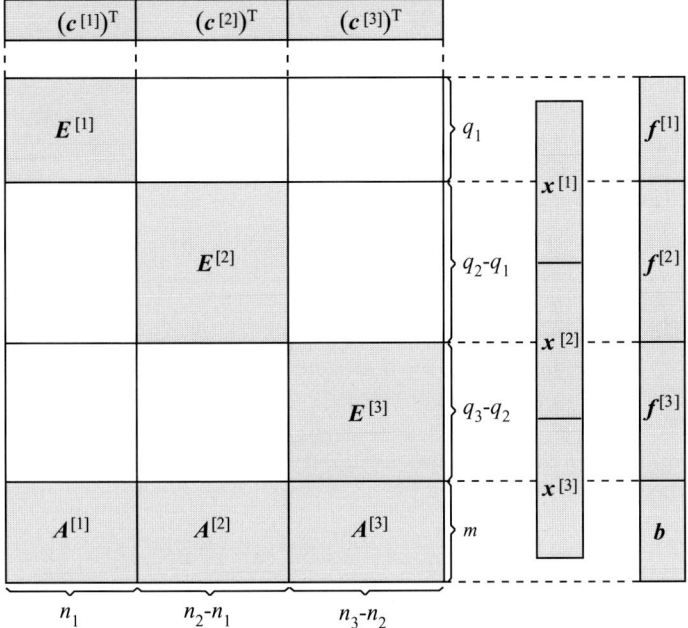

Fig. 2.1. Decomposable matrix with complicating constraints

$$\begin{pmatrix} E^{[1]} & | & & | & \\ -- & -- & -- & -- & -- \\ & | & E^{[2]} & | & \\ -- & -- & -- & -- & -- \\ & | & & | & E^{[3]} \\ -- & -- & -- & -- & -- \\ A^{[1]} & | & A^{[2]} & | & A^{[3]} \end{pmatrix} \begin{pmatrix} x^{[1]} \\ -- \\ x^{[2]} \\ -- \\ x^{[3]} \end{pmatrix} = \begin{pmatrix} f^{[1]} \\ -- \\ f^{[2]} \\ -- \\ f^{[3]} \\ -- \\ b \end{pmatrix}$$

$$0 \leq x^{[1]} \leq x^{[1]\mathrm{up}}$$
$$0 \leq x^{[2]} \leq x^{[2]\mathrm{up}}$$
$$0 \leq x^{[3]} \leq x^{[3]\mathrm{up}}.$$

In general, the initial problem can be written as

$$\underset{x^{[1]}, x^{[2]}, \ldots, x^{[r]}}{\text{minimize}} \quad \sum_{k=1}^{r} \left(c^{[k]}\right)^T x^{[k]} \tag{2.6}$$

subject to

$$E^{[k]} x^{[k]} = f^{[k]}; \; k = 1, \ldots, r \tag{2.7}$$

$$\sum_{k=1}^{r} A^{[k]} x^{[k]} = b \tag{2.8}$$

$$0 \leq x^{[k]} \leq x^{[k]\mathrm{up}}; \; k = 1, \ldots, r, \tag{2.9}$$

where for each $k = 1, 2, \ldots, r$, we have

$$\left(c^{[k]}\right)^T = \begin{pmatrix} c_{n_{k-1}+1} & \cdots & c_{n_k} \end{pmatrix} \tag{2.10}$$

$$\left(x^{[k]}\right)^T = \begin{pmatrix} x_{n_{k-1}+1} & \cdots & x_{n_k} \end{pmatrix} \tag{2.11}$$

$$\left(x^{[k]\text{up}}\right)^T = \begin{pmatrix} x^{\text{up}}_{n_{k-1}+1} & \cdots & x^{\text{up}}_{n_k} \end{pmatrix} \tag{2.12}$$

$$E^{[k]} = (e_{ij}); \quad i = q_{k-1}+1, \ldots, q_k; \quad j = n_{k-1}+1, \ldots, n_k \tag{2.13}$$

$$\left(f^{[k]}\right)^T = \begin{pmatrix} f_{q_{k-1}+1} & \cdots & f_{q_k} \end{pmatrix} \tag{2.14}$$

$$A^{[k]} = (a_{ij}); \quad i = 1, \ldots, m; \quad j = n_{k-1}+1, \ldots, n_k \tag{2.15}$$

$$b^T = \begin{pmatrix} b_1 & \cdots & b_m \end{pmatrix}. \tag{2.16}$$

If complicating constraints are ignored, i.e., they are relaxed, the original problem becomes

$$\underset{x^{[k]}, x^{[2]}, \ldots, x^{[r]}}{\text{minimize}} \quad \sum_{k=1}^{r} \left(c^{[k]}\right)^T x^{[k]} \tag{2.17}$$

subject to

$$E^{[k]} x^{[k]} = f^{[k]}; \quad k = 1, \ldots, r \tag{2.18}$$

$$0 \leq x^{[k]} \leq x^{[k]\text{up}}; \quad k = 1, \ldots, r. \tag{2.19}$$

This problem (2.17)–(2.19) is called the relaxed version of the problem. The decomposed kth-subproblem is therefore

$$\underset{x^{[k]}}{\text{minimize}} \quad \left(c^{[k]}\right)^T x^{[k]} \tag{2.20}$$

subject to

$$E^{[k]} x^{[k]} = f^{[k]} \tag{2.21}$$

$$0 \leq x^{[k]} \leq x^{[k]\text{up}} \tag{2.22}$$

or

$$\underset{x_{n_{k-1}+1}, x_{n_{k-1}+2}, \ldots, x_{n_k}}{\text{minimize}} \quad \sum_{j=n_{k-1}+1}^{n_k} c_j x_j \tag{2.23}$$

subject to

$$\sum_{j=n_{k-1}+1}^{n_k} e_{ij} x_j = f_i; \quad i = q_{k-1}+1, \ldots, q_k \tag{2.24}$$

$$0 \leq x_j \leq x_j^{\text{up}}; \quad j = n_{k-1}+1, \ldots, n_k. \tag{2.25}$$

The following example shows a decomposable linear programming problem with complicating constraints.

Illustrative Example 2.3 (Problem with decomposable structure).
The problem

$$\underset{x_1,x_2,x_3,y_1,y_2,y_3,z_1,z_2,z_3,w_1}{\text{minimize}} \quad -4x_1 - y_1 - 6z_1$$

subject to

$$\begin{aligned}
x_1 - x_2 &= 1 \\
x_1 + x_3 &= 2 \\
y_1 - y_2 &= 1 \\
y_1 + y_3 &= 2 \\
z_1 - z_2 &= 1 \\
z_1 + z_3 &= 2 \\
3x_1 + 2y_1 + 4z_1 + w_1 &= 17 \\
x_1, x_2, x_3, y_1, y_2, y_3, z_1, z_2, z_3, w_1 &\geq 0
\end{aligned}$$

has a decomposable structure in three blocks, where the last equality

$$3x_1 + 2y_1 + 4z_1 + w_1 = 17$$

is the complicating constraint. □

2.3 Decomposition

This section motivates the decomposition algorithm developed in the following sections.

Suppose that each of the subproblems (relaxed problem) is solved p times with different and arbitrary objective functions, and assume that the p basic feasible solutions of the relaxed problems are

$$\begin{matrix} x_1^{(1)}, & x_2^{(1)} & \ldots & x_n^{(1)} \\ \vdots & \vdots & \vdots & \vdots \\ x_1^{(p)}, & x_2^{(p)} & \ldots & x_n^{(p)} \end{matrix} \quad (2.26)$$

where $x_j^{(s)}$ is the jth component of solution s, where all variables from all the subproblems are considered, and the corresponding p optimal objective function values are

$$\begin{matrix} z^{(1)} \\ z^{(2)} \\ \vdots \\ z^{(p)} \end{matrix} \quad (2.27)$$

where $z^{(s)}$ is the objective function value of solution s. Note that we use superindices of the form (s) to refer to the sth solution.

The values of the m complicating constraints for the above p solutions are

$$r_1^{(1)}, r_2^{(1)} \ldots r_m^{(1)}$$
$$\vdots \quad \vdots \quad \vdots \quad \vdots \quad \quad (2.28)$$
$$r_1^{(p)}, r_2^{(p)} \ldots r_m^{(p)},$$

where $r_i^{(s)}$ is the value of the ith complicating constraint for the sth solution, i.e.,

$$r_i^{(s)} = \sum_{j=1}^{n} a_{ij} x_j^{(s)} .$$

For the derivations below, it should be emphasized that a linear convex combination of basic feasible solutions of a linear programming problem is a feasible solution of that problem.

The above p basic feasible solutions of the relaxed problem can be used to produce a feasible solution of the original (nonrelaxed) problem. The case in which a feasible solution cannot be generated is treated below. This is done by solving the weighting problem below, which is the so-called master problem

$$\underset{u_1,\ldots,u_p}{\text{minimize}} \quad \sum_{s=1}^{p} z^{(s)} u_s \qquad (2.29)$$

subject to

$$\sum_{s=1}^{p} r_i^{(s)} u_s = b_i : \lambda_i; \quad i = 1, \ldots, m \qquad (2.30)$$

$$\sum_{s=1}^{p} u_s = 1 : \sigma \qquad (2.31)$$

$$u_s \geq 0; \; s = 1, \ldots, p , \qquad (2.32)$$

where the corresponding dual variables λ_i and σ are indicated.

The following observations are in order:

1. Any solution of the above problem is a convex combination of basic feasible solutions of the relaxed problem; therefore, it is itself a basic feasible solution of the relaxed problem.
2. Complicating constraints are enforced; therefore, the solution of the above problem is a basic feasible solution for the original (nonrelaxed) problem.

Consider that a prospective new basic feasible solution is added to the problem above. The objective function value of this solution is z and its complicating constraints values are $r_1, \ldots, r_m$.

The new weighting problem becomes

$$\underset{u_1,\ldots,u_p,u}{\text{minimize}} \quad \sum_{s=1}^{p} z^{(s)} u_s + zu \qquad (2.33)$$

subject to

$$\sum_{s=1}^{p} \left(r_i^{(s)} u_s + r_i u\right) = b_i : \lambda_i; \qquad i = 1, \ldots, m \qquad (2.34)$$

$$\sum_{s=1}^{p} u_s + u = 1 : \sigma \qquad (2.35)$$

$$u, u_s \geq 0; \ s = 1, \ldots, p. \qquad (2.36)$$

The reduced cost (see Bazaraa et al. [20], Castillo et al. [21]) of the new weighting variable u, associated with the tentative new basic feasible solution, can be computed as

$$d = z - \sum_{i=1}^{m} \lambda_i r_i - \sigma. \qquad (2.37)$$

Taking into account that

$$z = \sum_{j=1}^{n} c_j x_j, \qquad (2.38)$$

where $x_j (j = 1, \ldots, n)$ is the new prospective basic feasible solution, and that

$$r_i = \sum_{j=1}^{n} a_{ij} x_j, \qquad (2.39)$$

the reduced cost of weighting variable u becomes

$$d = \sum_{j=1}^{n} c_j x_j - \sum_{i=1}^{m} \lambda_i \left(\sum_{j=1}^{n} a_{ij} x_j\right) - \sigma, \qquad (2.40)$$

which reduces to

$$d = \sum_{j=1}^{n} \left(c_j - \sum_{i=1}^{m} \lambda_i a_{ij}\right) x_j - \sigma. \qquad (2.41)$$

If the tentative basic feasible solution is to be added to the set of previous ones, the reduced cost associated with its weighting variable should be negative and preferably a minimum. To find the minimum reduced cost associated with the weighting variable u and corresponding to a basic feasible solution of the relaxed problem, the following problem is solved.

$$\underset{x_1, x_2, \ldots, x_n}{\text{minimize}} \quad \sum_{j=1}^{n} \left(c_j - \sum_{i=1}^{m} \lambda_i a_{ij}\right) x_j - \sigma$$

subject to

$$\sum_{j=1}^{n} e_{ij} x_j = f_i ; \qquad i = 1, \ldots, q$$

$$0 \leq x_j \leq x_j^{\text{up}} ; \qquad j = 1, \ldots, n .$$

Note that the constraints of the relaxed problem must be added.

If constant σ is dropped from the objective function, the problem above becomes

$$\underset{x_1, x_2, \ldots, x_n}{\text{minimize}} \quad \sum_{j=1}^{n} \left(c_j - \sum_{i=1}^{m} \lambda_i a_{ij} \right) x_j \tag{2.42}$$

subject to

$$\sum_{j=1}^{n} e_{ij} x_j = f_i ; \qquad i = 1, \ldots, q \tag{2.43}$$

$$0 \leq x_j \leq x_j^{\text{up}} ; \qquad j = 1, \ldots, n . \tag{2.44}$$

It should be noted that the above problem has the same structure as the relaxed problem but with different objective functions. Therefore, it can be solved by blocks. The subproblem associated with block k is

$$\underset{x_{n_{k-1}+1}, x_{n_{k-1}+2}, \ldots, x_{n_k}}{\text{minimize}} \quad \sum_{j=n_{k-1}+1}^{n_k} \left(c_j - \sum_{i=1}^{m} \lambda_i a_{ij} \right) x_j \tag{2.45}$$

subject to

$$\sum_{j=n_{k-1}+1}^{n_k} e_{ij} x_j = f_i ; \qquad i = q_{k-1}+1, \ldots, q_k \tag{2.46}$$

$$0 \leq x_j \leq x_j^{\text{up}} ; \qquad j = n_{k-1}+1, \ldots, n_k . \tag{2.47}$$

From the analysis carried out, the following conclusions can be drawn:

1. To determine whether or not a given basic feasible solution of the relaxed problem should be added to the weighting problem, the subproblems associated with the relaxed problem should be solved with modified objective functions.
2. The modified objective function of subproblem k is

$$\sum_{j=n_{k-1}+1}^{n_k} \left(c_j - \sum_{i=1}^{m} \lambda_i a_{ij} \right) x_j . \tag{2.48}$$

In the above objective function, each cost coefficient has the form

$$\bar{c}_j = c_j - \sum_{i=1}^{m} \lambda_i a_{ij} , \tag{2.49}$$

that is, the cost coefficient $\bar{c}_j$ depends on the dual variable of every complicating constraint and on the column j of the corresponding complicating constraint matrix $\boldsymbol{A}$.

3. Once the subproblems are solved, the optimal value v^o of the objective function of the relaxed problem is computed as

$$v^o = \sum_{j=1}^{n} \left(c_j - \sum_{i=1}^{m} \lambda_i^o a_{ij} \right) x_j^o, \qquad (2.50)$$

where x_j^o ($j = 1, \ldots, n$) is the solution of the relaxed problem, and λ_i^o ($i = 1, \ldots, m$) are the optimal values of the dual variables associated with complicating constraints.

The minimum reduced cost is then

$$d^o = v^o - \sigma^o, \qquad (2.51)$$

where σ^o is the optimal value of the dual variable associated with the convex combination constraint (2.35).

4. Based on the minimum reduced cost value, it can be decided whether or not to include the tentative basic feasible solution associated with the weighting variable u. This is done as follows:

   a) If $v^o \geq \sigma^o$, the tentative basic feasible solution cannot improve the current solution of the weighting problem because its reduced cost is nonnegative.

   b) If, on the other hand, $v^o < \sigma^o$, the tentative basic feasible solution should be included in the weighting problem because its reduced cost is negative and this can be used to attain a basic feasible solution with smaller objective function value than the current one.

The above remarks allow us to propose the solution algorithm described in the next section.

Because of its importance, we dedicate a section to the Dantzig-Wolfe decomposition method.

2.4 The Dantzig-Wolfe Decomposition Algorithm

In this section the Dantzig-Wolfe decomposition algorithm is described in detail.

2.4.1 Description

The Dantzig-Wolfe decomposition algorithm works as follows.

Algorithm 2.1 (The Dantzig-Wolfe decomposition algorithm).

Input. A linear programming problem with complicating constraints.
Output. The solution of the linear programming problem obtained after using the Dantzig-Wolfe decomposition algorithm.

Step 0: Initialization. Initialize the iteration counter, $\nu = 1$. Obtain $p^{(\nu)}$ distinct solutions of the relaxed problem by solving $p^{(\nu)}$ times ($\ell = 1, \ldots, p^{(\nu)}$) each of the r subproblems ($k = 1, \ldots, r$) below

$$\underset{x_{n_{k-1}+1}, \ldots, x_{n_k}}{\text{minimize}} \quad \sum_{j=n_{k-1}+1}^{n_k} \hat{c}_j^{(\ell)} x_j \tag{2.52}$$

subject to

$$\sum_{j=n_{k-1}+1}^{n_k} e_{ij} x_j = f_i \,; \quad i = q_{k-1}+1, \ldots, q_k \tag{2.53}$$

$$0 \leq x_j \leq x_j^{\text{up}} \,; \quad j = n_{k-1}+1, \ldots, n_k \,, \tag{2.54}$$

where $\hat{c}_j^{(\ell)}$ ($j = n_{k-1}+1, \ldots, n_k; k = 1, \ldots, r; \ell = 1, \ldots, p^{(\nu)}$) are arbitrary cost coefficients to attain the $p^{(\nu)}$ initial solutions of the r subproblems.

Step 1: Master problem solution. Solve the master problem

$$\underset{u_1, \ldots, u_{p^{(\nu)}}}{\text{minimize}} \quad \sum_{s=1}^{p^{(\nu)}} z^{(s)} u_s \tag{2.55}$$

subject to

$$\sum_{s=1}^{p^{(\nu)}} r_i^{(s)} u_s = b_i : \lambda_i; \quad i = 1, \ldots, m \tag{2.56}$$

$$\sum_{s=1}^{p^{(\nu)}} u_s = 1 : \sigma \tag{2.57}$$

$$u_s \geq 0; \; s = 1, \ldots, p^{(\nu)} \tag{2.58}$$

to obtain the solution $u_1^{(\nu)}, \ldots, u_{p^{(\nu)}}^{(\nu)}$, and the dual variable values $\lambda_1^{(\nu)}, \ldots, \lambda_m^{(\nu)}$ and $\sigma^{(\nu)}$.

Step 2: Relaxed problem solution. Generate a solution of the relaxed problem by solving the r subproblems ($k = 1, \ldots, r$) below.

$$\underset{x_{n_{k-1}+1}, x_{n_{k-1}+2}, \ldots, x_{n_k}}{\text{minimize}} \quad \sum_{j=n_{k-1}+1}^{n_k} \left(c_j - \sum_{i=1}^{m} \lambda_i^{(\nu)} a_{ij} \right) x_j \tag{2.59}$$

subject to

$$\sum_{j=n_{k-1}+1}^{n_k} e_{ij}x_j = f_i \; ; \; i = q_{k-1}+1,\ldots,q_k \tag{2.60}$$

$$0 \leq x_j \leq x_j^{\text{up}} \; ; \; j = n_{k-1}+1,\ldots,n_k \tag{2.61}$$

to obtain a solution of the relaxed problem, i.e., $x_1^{(p^{(\nu)}+1)},\ldots,x_n^{(p^{(\nu)}+1)}$, and its objective function value $v^{(\nu)}$, i.e.,

$$v^{(\nu)} = \sum_{j=1}^{n} \left(c_j - \sum_{i=1}^{m} \lambda_i^{(\nu)} a_{ij}\right) x_j^{(p^{(\nu)}+1)}. \tag{2.62}$$

The objective function value of the original problem is

$$z^{(p^{(\nu)}+1)} = \sum_{j=1}^{n} c_j x_j^{(p^{(\nu)}+1)} \tag{2.63}$$

and the value of every complicating constraint is

$$r_i^{(p^{(\nu)}+1)} = \sum_{j=1}^{n} a_{ij} x_j^{(p^{(\nu)}+1)}; \quad i = 1,\ldots,m. \tag{2.64}$$

Step 3: Convergence checking. If $v^{(\nu)} \geq \sigma^{(\nu)}$, the optimal solution of the original problem has been achieved. It is computed as

$$x_j^* = \sum_{s=1}^{p^{(\nu)}} u_s^{(\nu)} x_j^{(s)}; \quad j = 1,\ldots,n \tag{2.65}$$

and the algorithm concludes.

Else if $v^{(\nu)} < \sigma^{(\nu)}$, the relaxed problem current solution can be used to improve the solution of the master problem. Update the iteration counter, $\nu \leftarrow \nu + 1$, and the number of available solutions of the relaxed problem, $p^{(\nu+1)} = p^{(\nu)} + 1$. Go to Step 1. □

A GAMS implementation of the Dantzig-Wolfe decomposition algorithm is given in the Appendix A, p. 397.

Computational Example 2.1 (The Dantzig-Wolfe decomposition). Consider the problem below

$$\underset{x_1,x_2,x_3}{\text{minimize}} \quad z = -4x_1 - x_2 - 6x_3$$

subject to

2 Linear Programming: Complicating Constraints

$$\begin{aligned}
-x_1 &\leq -1 \\
x_1 &\leq 2 \\
-x_2 &\leq -1 \\
x_2 &\leq 2 \\
-x_3 &\leq -1 \\
x_3 &\leq 2 \\
3x_1 + 2x_2 + 4x_3 &\leq 17 \\
x_1, x_2, x_3 &\geq 0 \ .
\end{aligned}$$

Note that this problem has a decomposable structure and one complicating constraint. Its optimal solution is

$$x_1^* = 2, \ x_2^* = 3/2, \ x_3^* = 2 \ .$$

This problem is solved in the following steps using the Dantzig-Wolfe decomposition algorithm as previously stated in Subsect. 2.4.1.

Step 0: Initialization. The iteration counter is initialized, $\nu = 1$. Two ($p^{(1)} = 2$) solutions for the relaxed problem are obtained solving the three subproblems twice.

First, cost coefficients $\hat{c}_1^{(1)} = -1$, $\hat{c}_2^{(1)} = -1$, and $\hat{c}_3^{(1)} = -1$ are used. The subproblems for the first solution are

$$\underset{x_1}{\text{minimize}} \quad -x_1$$

subject to

$$1 \leq x_1 \leq 2 \ ,$$

whose solution is $x_1^{(1)} = 2$, and

$$\underset{x_2}{\text{minimize}} \quad -x_2$$

subject to

$$1 \leq x_2 \leq 2 \ ,$$

whose solution is $x_2^{(1)} = 2$, and

$$\underset{x_3}{\text{minimize}} \quad -x_3$$

subject to

$$1 \leq x_3 \leq 2$$

whose solution is $x_3^{(1)} = 2$.

The objective function of the relaxed problem is $z^{(1)} = -22$ and the complicating constraint value is $r_1^{(1)} = 18$.

Using cost coefficients $\hat{c}_1^{(2)} = 1$, $\hat{c}_2^{(2)} = 1$, and $\hat{c}_3^{(2)} = -1$, the subproblems are solved again to derive the second solution for the relaxed problem. This solution is $x_1^{(2)} = 1$, $x_2^{(2)} = 1$, and $x_3^{(2)} = 2$, leading to an objective function value $z^{(2)} = -17$ and a complicating constraint value $r_1^{(2)} = 13$.

Step 1: Master problem solution. The master problem below is solved.

$$\underset{u_1, u_2}{\text{minimize}} \quad -22u_1 - 17u_2$$

subject to

$$18u_1 + 13u_2 \leq 17 : \lambda_1$$
$$u_1 \phantom{{}+13u_2} + u_2 = 1 : \sigma$$
$$u_1, u_2 \geq 0 .$$

Its solution is $u_1^{(1)} = \frac{4}{5}$ and $u_2^{(1)} = \frac{1}{5}$ with dual variable values $\lambda_1^{(1)} = -1$ and $\sigma^{(1)} = -4$.

Step 2: Relaxed problem solution. The subproblems are solved below to obtain a solution for the current relaxed problem.
The objective function of the first subproblem is

$$\left(c_1 - \lambda_1^{(1)} a_{11} \right) x_1 = (-4 + 3)x_1 = -x_1$$

and its solution, obtained by inspection, is $x_1^{(3)} = 2$.
The objective function of the second subproblem is

$$\left(c_2 - \lambda_1^{(1)} a_{12} \right) x_2 = (-1 + 2)x_2 = x_2$$

and its solution is $x_2^{(3)} = 1$.
Finally, the objective function of the third subproblem is

$$\left(c_3 - \lambda_1^{(1)} a_{13} \right) x_3 = (6 + 1)x_3 - -2x_3$$

and its solution is $x_3^{(3)} = 2$.
For this relaxed problem solution ($x_1^{(3)} = 2$, $x_2^{(3)} = 1$, $x_3^{(3)} = 2$), the objective function value of the original problem is $z^{(3)} = -21$ and the value of the complicating constraint $r_1^{(3)} = 16$.

Step 3: Convergence checking. The objective function value of the current relaxed problem is

$$v^{(1)} = -x_1^{(3)} + x_2^{(3)} - 2x_3^{(3)} = -5 .$$

Note that $v^{(1)} < \sigma^{(1)}$ ($-5 < -4$) and therefore the current solution of the relaxed problem can be used to improve the solution of the master problem.

The iteration counter is updated, $\nu = 1 + 1 = 2$, and the number of available solutions of the relaxed problem is also updated, $p^{(2)} = 2 + 1 = 3$. The algorithm continues in Step 1.

Step 1: Master problem solution. The master problem below is solved.

$$\underset{u_1, u_2, u_3}{\text{minimize}} \quad -22u_1 - 17u_2 - 21u_3$$

subject to

$$18u_1 + 13u_2 + 16u_3 \leq 17 : \lambda_1$$
$$u_1 + u_2 + u_3 = 1 : \sigma$$
$$u_1, u_2, u_3 \geq 0 .$$

Its solution is $u_1^{(2)} = \frac{1}{2}$, $u_2^{(2)} = 0$, and $u_3^{(2)} = \frac{1}{2}$ with dual variable values $\lambda_1^{(2)} = -\frac{1}{2}$ and $\sigma^{(2)} = -13$.

Step 2: Relaxed problem solution. The subproblems are solved below to obtain a solution for the current relaxed problem.

The objective function of the first subproblem is

$$\left(c_1 - \lambda_1^{(2)} a_{11}\right) x_1 = \left(-4 + \frac{3}{2}\right) x_1 = -\frac{5}{2} x_1$$

that renders $x_1^{(4)} = 2$.

The objective function of the second subproblem is

$$\left(c_2 - \lambda_1^{(2)} a_{12}\right) x_2 = (-1 + 1) x_2 = 0$$

that renders $x_2^{(4)} = 1$.

Finally, the objective function of the third subproblem is

$$\left(c_3 - \lambda_1^{(2)} a_{13}\right) x_3 = (-6 + 2) x_3 = -4 x_3$$

and its solution is $x_3^{(4)} = 2$.

For this relaxed problem solution ($x_1^{(4)} = 2$, $x_2^{(4)} = 1$, $x_3^{(4)} = 2$), the objective function of the original problem is $z^{(4)} = -21$ and the value of the complicating constraint $r_1^{(4)} = 16$.

Step 3: Convergence checking. The objective function value of the current relaxed problem is

$$v^{(2)} = -\frac{5}{2} x_1^{(4)} - 4 x_3^{(4)} = -13 .$$

Note that $v^{(2)} \geq \sigma^{(2)}$ ($-13 \geq -13$) and therefore the optimal solution of the original problem has been attained, i.e.,

2.4 The Dantzig-Wolfe Decomposition Algorithm

$$\begin{pmatrix} x_1^* \\ x_2^* \\ x_3^* \end{pmatrix} = u_1^{(2)} \begin{pmatrix} x_1^{(1)} \\ x_2^{(1)} \\ x_3^{(1)} \end{pmatrix} + u_2^{(2)} \begin{pmatrix} x_1^{(2)} \\ x_2^{(2)} \\ x_3^{(2)} \end{pmatrix} + u_3^{(2)} \begin{pmatrix} x_1^{(3)} \\ x_2^{(3)} \\ x_3^{(3)} \end{pmatrix}$$

and

$$\begin{pmatrix} x_1^* \\ x_2^* \\ x_3^* \end{pmatrix} = \frac{1}{2}\begin{pmatrix} 2 \\ 2 \\ 2 \end{pmatrix} + 0\begin{pmatrix} 1 \\ 1 \\ 2 \end{pmatrix} + \frac{1}{2}\begin{pmatrix} 2 \\ 1 \\ 2 \end{pmatrix} = \begin{pmatrix} 2 \\ \frac{3}{2} \\ 2 \end{pmatrix}.$$

□

The following example has been designed to illustrate a geometric interpretation of the Dantzig-Wolfe decomposition technique.

Computational Example 2.2 (The Dantzig-Wolfe algorithm). Consider the problem

$$\underset{x_1, x_2}{\text{minimize}} \quad z = 2x_1 + x_2 \tag{2.66}$$

subject to

$$\begin{aligned} x_1 &\leq 5 \\ x_2 &\leq 5 \\ x_1 + x_2 &\leq 9 \\ x_1 - x_2 &\leq 4 \\ -x_1 - x_2 &\leq -2 \\ -3x_1 - x_2 &\leq -3 \\ x_1, x_2 &\geq 0, \end{aligned} \tag{2.67}$$

where the last four constraints in (2.67) are the complicating constraints.

As shown in Fig. 2.2a, where the original feasible region has been shaded, and the objective function contours drawn, it is clear that the global solution of this problem is

$$z^* = 2.5, \quad x_1^* = 0.5, \quad x_2^* = 1.5 .$$

Next, we use the Dantzig-Wolfe algorithm, which is illustrated in Table 2.1, where the solutions of the master problems and subproblems are shown for each iteration, until convergence.

Step 0: Initialization. As it has been explained, the initialization part of the algorithm requires a minimum of $p^{(\nu)} \geq 1$ solutions of the relaxed problem with arbitrary objective functions, to obtain a set of extreme points of the relaxed problem. We consider the following two arbitrary optimization problems:

$$\underset{x_1, x_2}{\text{minimize}} \quad z = -x_1 - x_2$$

and

$$\underset{x_1, x_2}{\text{minimize}} \quad z = -2x_1 + x_2$$

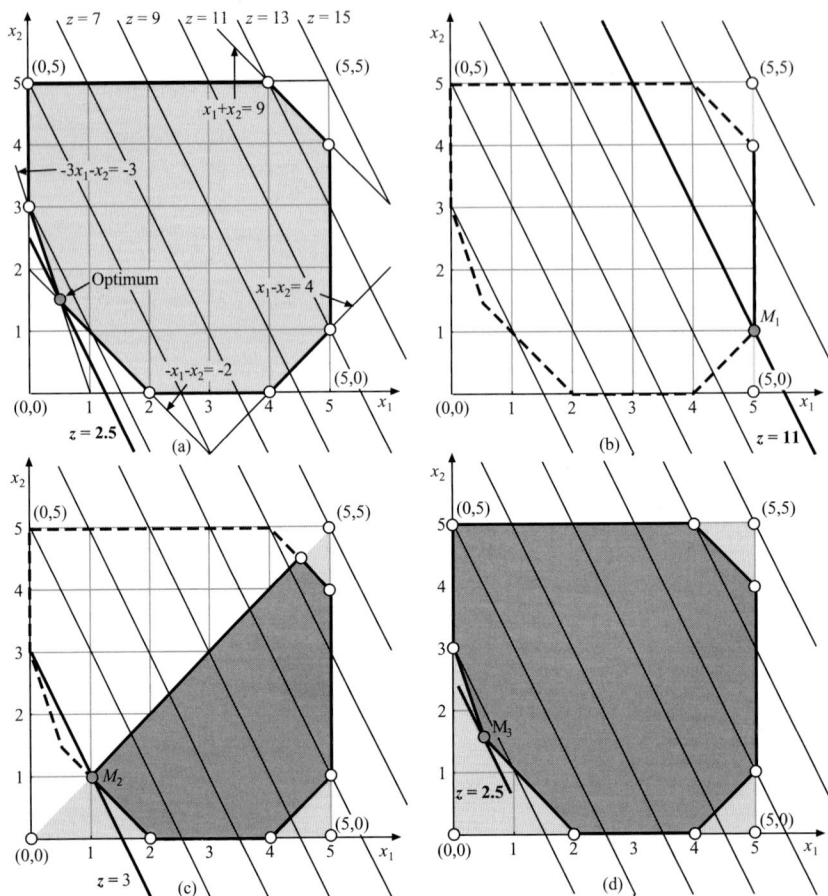

Fig. 2.2. Graphical illustration of the Dantzig-Wolfe decomposition algorithm

subject to

$$x_1 \leq 5$$
$$x_2 \leq 5$$
$$x_1, x_2 \geq 0 \, ,$$

whose solutions are the extreme points $(5,5)$ and $(5,0)$, respectively. Next, we evaluate the complicating constraints [the last four constraints in (2.67)], not including the non-negativity constraints, and the target objective function (2.66), and obtain the values of r_1, r_2, r_3, r_4, and z in Table 2.1 (iteration 0). This ends the initialization step.

Step 1: Master problem solution. The master problem finds the point that minimizes the original objective function in the intersection of the set of linear convex combinations of the extreme points $(5,5)$ and $(5,0)$ [this is the

2.4 The Dantzig-Wolfe Decomposition Algorithm

Table 2.1. Solutions of the master problem and the subproblems in Example (option 1)

Iteration	Initial solutions for the subproblems							
ν	$x_1^{(\nu)}$	$x_2^{(\nu)}$	$r_1^{(\nu)}$	$r_2^{(\nu)}$	$r_3^{(\nu)}$	$r_4^{(\nu)}$	$z^{(\nu)}$	$v^{(\nu)}$
0–1	5.0	5.0	10.0	0.0	−10.0	−20.0	15.0	–
0–2	5.0	0.0	5.0	5.0	-5.0	−15.0	10.0	–
	Solutions for the subproblem							
1	0.0	0.0	0.0	0.0	0.0	0.0	0.0	0.0
2	0.0	5.0	5.0	−5.0	−5.0	−5.0	5.0	−2.5
3	–	–	–	–	–	–	–	0.0

Iteration	Master problem solutions									
ν	$u_1^{(\nu)}$	$u_2^{(\nu)}$	$u_3^{(\nu)}$	$u_4^{(\nu)}$	$\lambda_1^{(\nu)}$	$\lambda_2^{(\nu)}$	$\lambda_3^{(\nu)}$	$\lambda_4^{(\nu)}$	$\sigma^{(\nu)}$	Feasible
1	0.2	0.8	–	–	0.0	−1.0	0.0	0.0	15.0	Yes
2	0.2	0.0	0.8	–	0.0	0.0	−1.5	0.0	0.0	Yes
3	0.1	0.0	0.7	0.2	0.0	0.0	−0.5	−0.5	0.0	Yes

segment $(5,5) - (5,0)$], and the original feasible region, i.e., in the segment $(5,4) - (5,1)$ [see Fig. 2.2b]. Since the master problem looks for the optimal solution in this intersection region, the point denoted by M_1 in Fig. 2.2b is obtained, with associated values of the primal variables, u_1 and u_2, and the dual variables $\lambda_1, \lambda_2, \lambda_3, \lambda_4$, and σ, as shown in Table 2.1 (master problem iteration 1). Note that only the second complicating constraint is active; this implies that the only λ-value different from zero is λ_2. Similarly, since the solution of the master problem is on the boundary of the feasible region of the relaxed problem, the value of σ is different from zero.

Step 2: Relaxed problem solution. The subproblems look for the extreme point in the relaxed feasible region to be added to the master problem so that the largest improvement in the objective function value is obtained. We have two options for selecting the extreme point to be incorporated (see Fig. 2.2c): $(0,0)$ and $(0,5)$. The extreme point $(0,5)$ would allow us to move to point $P : (0,5)$, and the extreme point $(0,0)$, to the point $M_2 : (1,1)$.

Clearly, the optimum is obtained by adding the point $(0,0)$, and this is the point obtained after solving the relaxed problem. Next, the complicating constraints and the target objective function are evaluated at this point, and the values of r_1, r_2, r_3, r_4, and z are obtained (see subproblem iteration 1 in Table 2.1)

Step 3: Convergence checking. Since $v^{(1)} = 0 < \sigma^{(1)} = 15$, we go to Step 1.

2 Linear Programming: Complicating Constraints

Step 1: Master problem solution. Now the intersection of the set of linear convex combinations of the points in the set $\{(5,5),(5,0),(0,0)\}$ with the original feasible region, that is also indicated in Fig. 2.2c, is obtained. From this, the solution of the master problem can be easily obtained (point M_2 in the figure). The associated values of the primal, u_1, u_2, and u_3, and the dual variables $\lambda_1, \lambda_2, \lambda_3, \lambda_4$, and σ, are shown in Table 2.1 (master problem iteration 2). Note that only the third complicating constraint is active; this implies that the only λ-value different from zero is λ_3. Similarly, since the solution of the master problem is not in the boundary of the feasible region of relaxed problem, the value of σ is zero.

Step 2: Relaxed problem solution. Then, the subproblem step looks for the extreme point in the relaxed feasible region to be added, which is the point $(0, 5)$ because it is the only one remaining. Next, the complicating constraints and the original objective function are evaluated at this point, and the values of r_1, r_2, r_3, r_4, and z are obtained (see subproblem iteration 2 in Table 2.1).

Step 3: Convergence checking. Since $v^{(2)} = -2.5 < \sigma^{(2)} = 0$, we go to Step 1.

Step 1: Master problem solution. The new relaxed feasible region (gray and shaded region), becomes the linear convex combination of the points in the set $\{(5,5),(5,0),(0,0),(0,5)\}$. The intersection of this region with the initial feasible region is also indicated in the Fig. 2.2d, from which the solution of the master problem can be easily obtained (point M_3 in the figure). The associated values for the primal u_1, u_2, u_3, and u_4 and dual variables $\lambda_1, \lambda_2, \lambda_3, \lambda_4$, and σ, are shown in Table 2.1 (master problem iteration 3). Note that only the third and the fourth complicating constraints are active; this implies that the only λ-value different from zero are λ_3 and λ_4. Similarly, since the solution of the master problem is not at the boundary of the feasible region of the relaxed problem, the value of σ is zero.

Step 2: Relaxed problem solution. Since no extreme point can be added, the algorithm continues in Step 3.

Step 3: Convergence checking. Since $\nu^{(3)} = 0 \geq \sigma^{(3)} = 0$, the optimal solution has been obtained and the algorithm returns the optimal solution using the following expressions:

$$z^* = u_1^{(3)} z^{(0-1)} + u_2^{(3)} z^{(0-2)} + u_3^{(3)} z^{(1)} + u_4^{(3)} z^{(2)}$$
$$= 0.1 \times 15 + 0.0 \times 10 + 0.7 \times 0 + 0.2 \times 5 = 2.5$$

and

$$\begin{aligned}(x_1^*, x_2^*) &= u_1^{(3)}(x_1^{(0-1)}, x_2^{(0-1)}) + u_2^{(3)}(x_1^{(0-2)}, x_2^{(0-2)}) + u_3^{(3)}(x_1^{(1)}, x_2^{(1)}) \\ &\quad + u_4^{(3)}(x_1^{(2)}, x_2^{(2)}) \\ &= 0.1 \times (5,5) + 0.0 \times (5,0) + 0.7 \times (0,0) + 0.2 \times (0,5) \\ &= (0.5, 1.5) ,\end{aligned}$$

where superscript $(0 - \nu)$ indicates initial solution ν. □

2.4.2 Bounds

An upper and a lower bound of the objective function value that are obtained as the Dantzig-Wolfe algorithm progresses, are derived below. These bounds are of interest to stop the solution procedure once a prespecified error tolerance is satisfied.

The upper bound is readily available once the master problem objective function value is available. The master problem is, in fact, a restricted version of the original problem and, therefore, its objective function value is an upper bound of the optimal objective function value of the original problem. At iteration ν, the objective function of the master problem is

$$\sum_{i=1}^{p^{(\nu)}} z^{(i)} u_i^{(\nu)} .$$

An upper bound of the optimal objective function value of the original problem is therefore

$$z_{\text{up}}^{(\nu)} = \sum_{i=1}^{p^{(\nu)}} z^{(i)} u_i^{(\nu)} . \quad (2.68)$$

A lower bound is easily obtained from the solutions of the subproblems. The relaxed problem at iteration ν is

$$\underset{x_j; j=1,\ldots,n}{\text{minimize}} \quad \sum_{j=1}^{n} \left(c_j - \sum_{i=1}^{m} \lambda_i^{(\nu)} a_{ij} \right) x_j \quad (2.69)$$

subject to

$$\sum_{j=1}^{n} e_{ij} x_j = f_i ; \quad i = 1,\ldots,q \quad (2.70)$$

$$0 \leq x_j \leq x_j^{\text{up}} ; \quad j = 1,\ldots,n \quad (2.71)$$

and its optimal objective function value is $v^{(\nu)}$.

Consider a feasible solution x_j $(j = 1,\ldots,n)$ of the original problem. Therefore, this solution meets the complicating constraints, i.e.,

$$\sum_{j=1}^{n} a_{ij} x_j = b_i; \quad i = 1,\ldots,m. \quad (2.72)$$

Since $v^{(\nu)}$ is the optimal objective function value of problem (2.69)–(2.71) above, substituting the arbitrary feasible solution into the objective function of that problem renders

$$\sum_{j=1}^{n}\left(c_j - \sum_{i=1}^{m}\lambda_i^{(\nu)}a_{ij}\right)x_j \geq v^{(\nu)} \tag{2.73}$$

and

$$\sum_{j=1}^{n} c_j x_j \geq v^{(\nu)} + \sum_{i=1}^{m}\lambda_i^{(\nu)}\sum_{j=1}^{n} a_{ij}x_j \tag{2.74}$$

so that using (2.72) yields

$$\sum_{j=1}^{n} c_j x_j \geq v^{(\nu)} + \sum_{i=1}^{m}\lambda_i^{(\nu)}b_i \ . \tag{2.75}$$

Due to the fact that x_j ($j = 1, \ldots, n$) is an arbitrary feasible solution of the original problem, the inequality above allows writing

$$z_{\text{down}}^{(\nu)} = v^{(\nu)} + \sum_{i=1}^{m}\lambda_i^{(\nu)}b_i \ , \tag{2.76}$$

where $z_{\text{down}}^{(\nu)}$ is a lower bound of the optimal objective function value of the original problem.

2.4.3 Issues Related to the Master Problem

The possible infeasibility of the master problem is considered in this subsection and an alternative always-feasible master problem is formulated. The price paid is a larger number of variables and dissimilar cost coefficients in the objective function.

The selection of p solutions of the relaxed problem are usually motivated by engineering considerations and it allows us the formulation of a feasible master problem. However, in some instances, the formulation of a feasible master problem might not be simple. In such a situation, an always feasible master problem that includes artificial variables can be formulated. This problem has the form

$$\underset{u_1, \ldots, u_p; v_1, \ldots, v_m, w}{\text{minimize}} \quad \sum_{i=1}^{p} z^{(i)} u_i + M\left(\sum_{j=1}^{m} v_j + w\right) \tag{2.77}$$

subject to

$$\sum_{i=1}^{p} r_j^{(i)} u_i + v_j - w = b_j : \lambda_j; \ j = 1, \ldots, m \qquad (2.78)$$

$$\sum_{i=1}^{p} u_i = 1 : \sigma \qquad (2.79)$$

$$u_i \geq 0; \ i = 1, \ldots, p \qquad (2.80)$$

$$0 \leq v_j \leq v_j^{\text{up}}; \ j = 1, \ldots, m \qquad (2.81)$$

$$0 \leq w \leq w^{\text{up}}, \qquad (2.82)$$

where v_j ($j = 1, \ldots, m$) and w are the artificial variables, and v_j^{up} ($j = 1, \ldots, m$) and w^{up} are their respective upper bounds.

Note that the artificial variable makes problem (2.77)–(2.82) always feasible.

To illustrate how this master problem behaves, and its geometrical interpretation, we include the following example.

Computational Example 2.3 (The Dantzig-Wolfe example revisited). Consider the same problem as in Example 2.2, but assume that only the first arbitrary objective function is used in the initialization Step 0.

Step 0: Initialization. Then, we consider the following arbitrary optimization problems

$$\underset{x_1, x_2}{\text{minimize}} \quad z = -x_1 - x_2$$

subject to

$$\begin{aligned} x_1 &\leq 5 \\ x_2 &\leq 5, \end{aligned}$$

whose solution is the extreme point $(5, 5)$. Then, we evaluate the complicating constraints [the last four constraints in (2.67)] and the original objective function (2.66), and obtain the values of r_1, r_2, r_3, r_4, and z in Table 2.2 (iteration 0).

Step 1: Master problem solution. The master problem is infeasible because the intersection of the region generated by all linear convex combinations of the point $(5, 5)$ reduces to this point, which is not in the original feasible region.

In the modified master problem, we obtain feasibility by modifying the hyperplane boundaries, translating them by the minimum amount required to attain feasibility. This means that those constraints that lead to feasibility are kept ($v_j = w$), and those that lead to infeasibility are modified the minimum amount to get feasibility.

In our example, only the constraint $x_1 + x_2 = 9$ is not satisfied by the point $(5, 5)$, so it is replaced by $x_1 + x_2 = 10$, as illustrated in Fig. 2.3a.

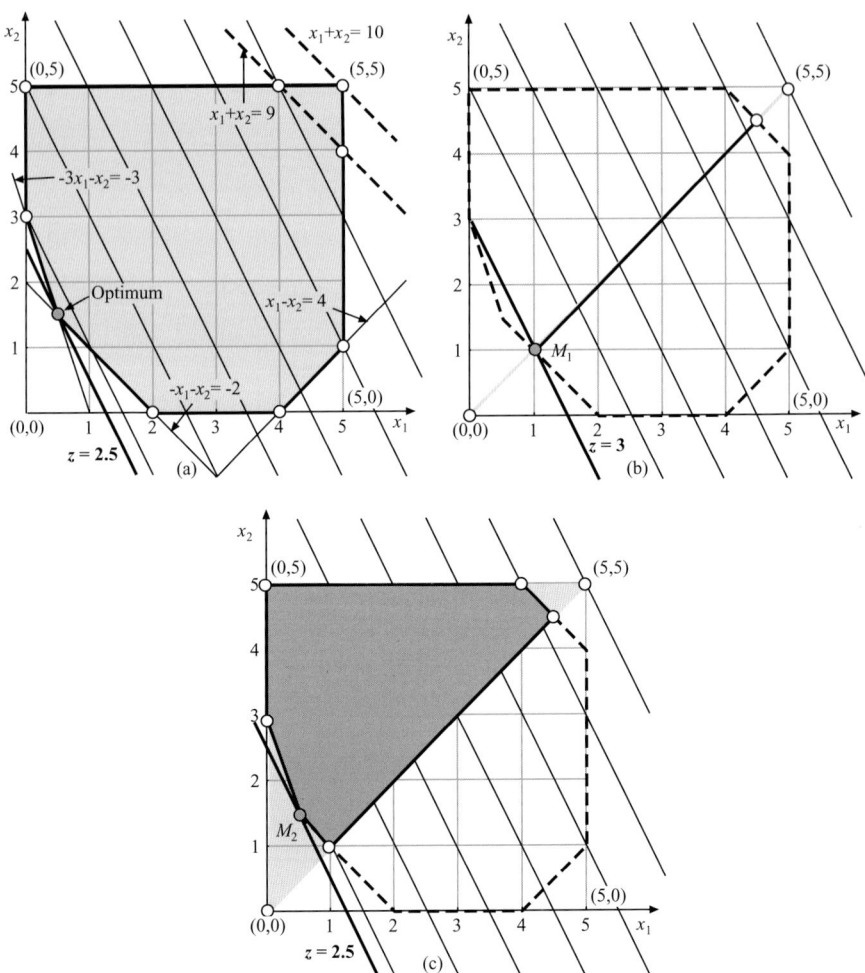

Fig. 2.3. Graphical illustration of how the feasible region is modified to make the master problem feasible

Since the master problem looks for the optimal solution in this modified intersection region, the point $(5,5)$ is obtained, whose associated values of the primal, u_1, and the dual variables $\lambda_1, \lambda_2, \lambda_3, \lambda_4$, and σ, are shown in Table 2.2 (master problem iteration 1).

Step 2: Relaxed problem solution. The subproblems look for the extreme point in the relaxed feasible region to be added to the master problem so that the largest improvement in the objective function value is obtained. We have three options for selecting the extreme point to be incorporated (see Fig. 2.3b): $(0,0)$, $(0,5)$, and $(5,0)$. The extreme point $(0,5)$ would allow us to move to point $P : (0,5)$, the extreme point $(0,0)$, to the point $M_1 : (1,1)$, and the extreme point $(5,0)$, to the point $(5,1)$.

Clearly, the optimum is obtained by adding the point $(0,0)$, and this is the point obtained after solving the relaxed problem. Next, the complicating constraints and the original objective function are evaluated at this point, and the values of r_1, r_2, r_3, r_4, and z are obtained (see subproblem iteration 1 in Table 2.2).

Step 3: Convergence checking. Since $v^{(1)} = 0 < \sigma^{(1)} = 215$, the procedure continues in Step 1.

Step 1: Master problem solution. Now the intersection of the set of linear convex combinations of the points in the set $\{(5,5),(0,0)\}$ with the target feasible region that is the segment $(1,1)$–$(4.5,4.5)$, is obtained. From it, the solution of the master problem can be easily obtained (point M_1 in Fig. 2.3b). The associated values of the primal, u_1 and u_2, and the dual variables $\lambda_1, \lambda_2, \lambda_3, \lambda_4$, and σ, are shown in Table 2.2 (master problem iteration 2).

Step 2: Relaxed problem solution. Next, the subproblem step looks for the extreme point in the relaxed feasible region to be added. We have two options for selecting the extreme point to be incorporated (see Fig. 2.3c): $(0,5)$ and $(5,0)$. The extreme point $(0,5)$ would allow us to move to point $M_2 : (0.5, 1.5)$, and the extreme point $(5,0)$ would not allow us further improvement. So, we add the point $(0,5)$. Next, the complicating constraints and the original objective function are evaluated at this point, and the values of r_1, r_2, r_3, r_4, and z are obtained (see subproblem iteration 2 in Table 2.2).

Step 3: Convergence checking. Since $v^{(2)} = -2.5 < \sigma^{(2)} = 0$, the algorithm continues in Step 1.

Step 1: Master problem solution. The new relaxed feasible region (gray and shaded region), becomes the linear convex combination of the points in the set $\{(5,5),(0,0),(0,5)\}$. The intersection of this region with the initial feasible region is also indicated in the Fig. 2.3c, from which the solution of the master problem can be easily obtained (point M_2 in Fig. 2.3c). The associated values for the primal u_1, u_2, and u_3 and dual variables $\lambda_1, \lambda_2, \lambda_3, \lambda_4$, and σ, are shown in Table 2.2 (master problem iteration 3).

Step 2: Relaxed problem solution. Since no extreme point can be added, the algorithm continues in Step 3.

Step 3: Convergence checking. Since $v^{(3)} = 0 \geq \sigma^{(3)} = 0$, the optimal solution has been obtained and the algorithm returns the optimal solution using the following expressions:

Table 2.2. Solutions of the master problems and subproblems in Example 2.3

Iteration	Bounds		Initial solutions for the subproblems							
ν	Lower	Upper	$x_1^{(\nu)}$	$x_2^{(\nu)}$	$r_1^{(\nu)}$	$r_2^{(\nu)}$	$r_3^{(\nu)}$	$r_4^{(\nu)}$	$z^{(\nu)}$	$v^{(\nu)}$
0	$-\infty$	∞	5.0	5.0	10.0	0.0	-10.0	-20.0	15.0	$-$
			Solutions for the subproblem							
1	-180.0	35.0	0.0	0.0	0.0	0.0	0.0	0.0	0.0	0.0
2	0.50	3.00	0.0	5.0	5.0	-5.0	-5.0	-5.0	5.0	-2.5
3	2.50	2.50	$-$	$-$	$-$	$-$	$-$	$-$	$-$	0.0

Iteration	Bounds		Master solutions								
ν	Lower	Upper	$u_1^{(\nu)}$	$u_2^{(\nu)}$	$u_3^{(\nu)}$	$\lambda_1^{(\nu)}$	$\lambda_2^{(\nu)}$	$\lambda_3^{(\nu)}$	$\lambda_4^{(\nu)}$	$\sigma^{(\nu)}$	Feasible
1	$-\infty$	35.0	1.0	$-$	$-$	-20.0	0.0	0.0	0.0	215	No
2	-180.0	3.00	0.2	0.8	$-$	0.0	0.0	-1.5	0.0	0.0	Yes
3	0.50	2.50	0.1	0.7	0.2	0.0	0.0	-0.5	-0.5	0.0	Yes

$$z^* = u_1^{(3)} z^{(0)} + u_2^{(3)} z^{(1)} + u_3^{(3)} z^{(2)} = 0.1 \times 15 + 0.7 \times 0 + 0.2 \times 5 = 2.5$$

and

$$(x_1^*, x_2^*) = u_1^{(3)}(x_1^{(0)}, x_2^{(0)}) + u_2^{(3)}(x_1^{(1)}, x_2^{(1)}) + u_3^{(3)}(x_1^{(2)}, x_2^{(2)})$$
$$= 0.1 \times (5,5) + 0.7 \times (0,0) + 0.2 \times (0,5) = (0.5, 1.5) \ .$$

Using expressions (2.68)–(2.76) lower and upper bounds on the solution of Example 2.3 are computed and plotted in Fig. 2.4. Observe how bounds approach each other until the optimal solution is attained. □

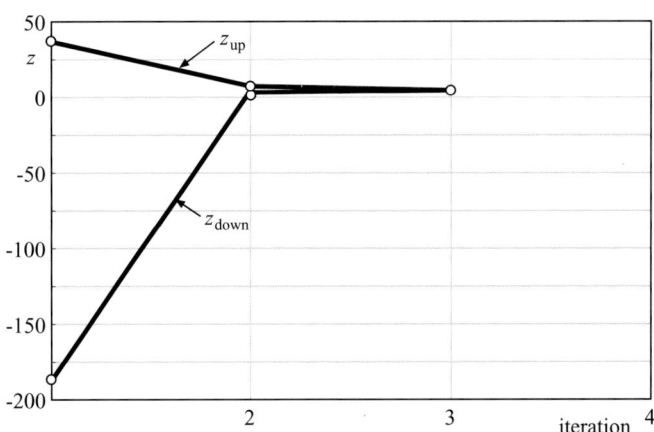

Fig. 2.4. Evolution of the upper and lower bounds of the objective function in Example 2.3

2.4 The Dantzig-Wolfe Decomposition Algorithm

To comprehend the Dantzig-Wolfe decomposition, we encourage the reader to determine the solutions of the master and subproblems corresponding to the following cases:

1. The initialization step leads to the extreme point: $(5,0)$.
2. The initialization step leads to the extreme point: $(0,5)$.
3. The initialization step leads to the extreme point: $(0,0)$.
4. The initialization step leads to the extreme points: $(5,0)$ and $(0,5)$.

2.4.4 Alternative Formulation of the Master Problem

In some practical applications, it is convenient to formulate the master problem (2.29)–(2.32) in the alternative format stated below.

$$\underset{u_{ij};\, i=1,\ldots,p;\, j=1,\ldots,r}{\text{minimize}} \quad \sum_{i=1}^{p}\sum_{j=1}^{r} z^{(ij)} u_{ij} \tag{2.83}$$

subject to

$$\sum_{i=1}^{p}\sum_{j=1}^{r} r_{ij}^{(\ell)} u_{ij} = b^{(\ell)} \,:\, \lambda^{(\ell)} \,;\, \ell = 1,\ldots,m \tag{2.84}$$

$$\sum_{i=1}^{p} u_{ij} = 1 \,:\, \sigma_j \,;\, j = 1,\ldots,r \tag{2.85}$$

$$u_{ij} \geq 0 \,;\, i = 1,\ldots,p;\, j = 1,\ldots,r \,, \tag{2.86}$$

where $z^{(ij)}$ is the objective function value for the solution i of the subproblem j and $r_{ij}^{(\ell)}$ is the contribution to the right-hand-side values of the complicating constraint ℓ of the solution i of the subproblem j. In Fig. 2.5 a graphical comparison between the master problem formulation presented in this section and the one presented in Sect. 2.4.3 is shown.

Note that the above formulation relies on the fact that a convex combination of any number of basic feasible solutions of any subproblem is a feasible solution of this subproblem. Note also that $u_{1j},\ldots,u_{pj}$ are the convex combination variables corresponding to subproblem j ($j = 1,\ldots,r$). The r constraints (2.85) in the problem above are equivalent to the single constraint (2.31) in the original formulation of the master problem.

The following observations are in order:

1. Any solution of problem (2.83)–(2.86) is a convex linear combination of basic feasible solutions of the subproblems and therefore of the relaxed problem; thus, it is itself a basic feasible solution of the relaxed problem.
2. Complicating constraints are enforced; therefore, the solution of the above problem is a basic feasible solution for the original (nonrelaxed) problem.

A valuable geometrical interpretation associated with this alternative approach is illustrated in the example below.

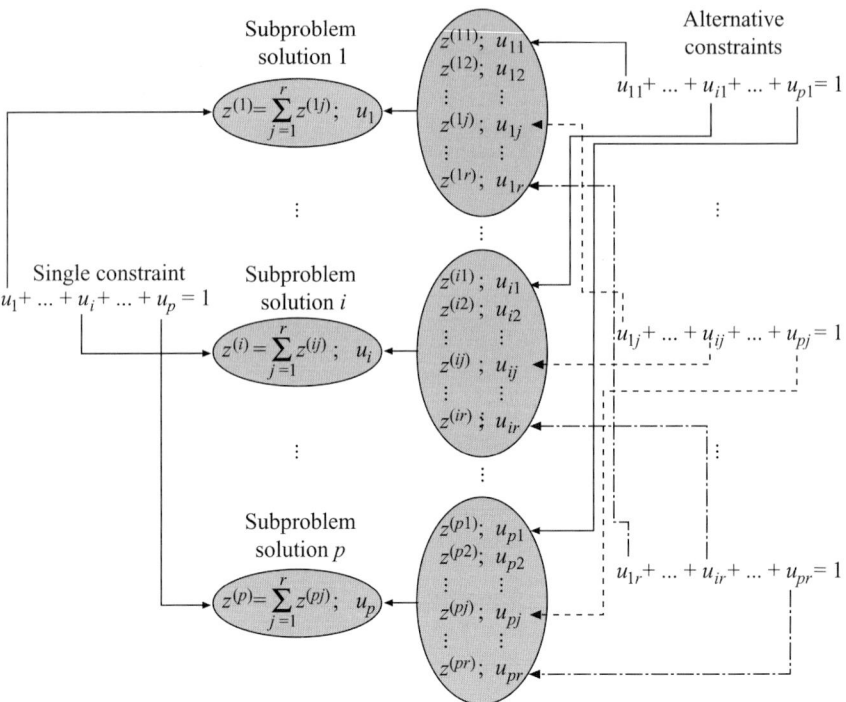

Fig. 2.5. Relationship of the two alternative master problem formulations

Computational Example 2.4 (Alternative master problem). Consider the problem

$$\underset{x_1, x_2, x_3}{\text{minimize}} \quad z = x_1 + x_2 + x_3 \qquad (2.87)$$

subject to

$$\begin{array}{rl}
x_1 & \leq 5 \\
x_2 & \leq 5 \\
1/2 x_1 - x_2 & \leq -1/2 \\
x_3 & \leq 4 \\
-x_1 + x_2 + x_3 & \leq 0 \\
x_1, x_2, x_3 & \geq 0 \,.
\end{array} \qquad (2.88)$$

Note that this problem has a decomposable structure in two blocks and one complicating constraint. Its optimal solution is

$$z^* = 2, \; x_1^* = 1, \; x_2^* = 1, \; x_3^* = 0 \,.$$

Figure 2.6 illustrates the feasibility region of problem (2.87)–(2.88), where Fig. 2.6a represents the feasible region of that problem not considering the single complicating constraint (relaxed problem), Fig. 2.6b shows the feasible

2.4 The Dantzig-Wolfe Decomposition Algorithm

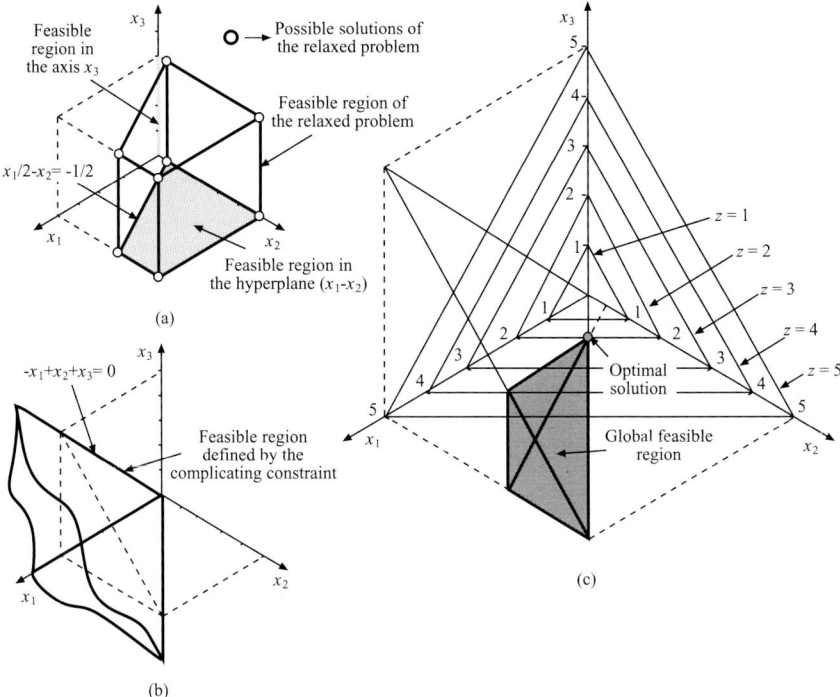

Fig. 2.6. Graphical illustration of the feasibility region of the Computational Example 2.4

region defined by the complicating constraint, and Fig. 2.6c shows the feasible region of the original problem, and the objective function contours.

The problem above is solved using the Dantzig-Wolfe algorithm and the two alternative master problem definitions.

Step 0: Initialization. Consider the objective function

$$z = 5x_1 + 2x_2 - x_3,$$

to obtain an initial feasible solution of the relaxed problem for both decomposition algorithms. This solution, represented in Fig. 2.7, is $S_1 = (0, 0.5, 4)$. The complicating constraint [last constraint in (2.88)] and the objective function (2.87), are then evaluated to obtain $r^{(0)}$ and $z^{(0)}$ that are shown in Table 2.3 and $r_1^{(0)}, r_2^{(0)}, z_1^{(0)}$, and $z_2^{(0)}$ that are shown in Table 2.4.

Step 1: Master problem solutions. The master problems are solved obtaining the solution $M_1 = (0, 0.5, 4)$ shown in Fig. 2.7. Values for the primal variables $u_1^{(1)}, u_{11}^{(1)}, u_{12}^{(1)}$ and the dual variables $\lambda^{(1)}, \sigma^{(1)}, \sigma_1^{(1)}, \sigma_2^{(1)}$ are shown in Tables 2.3 and 2.4, respectively.

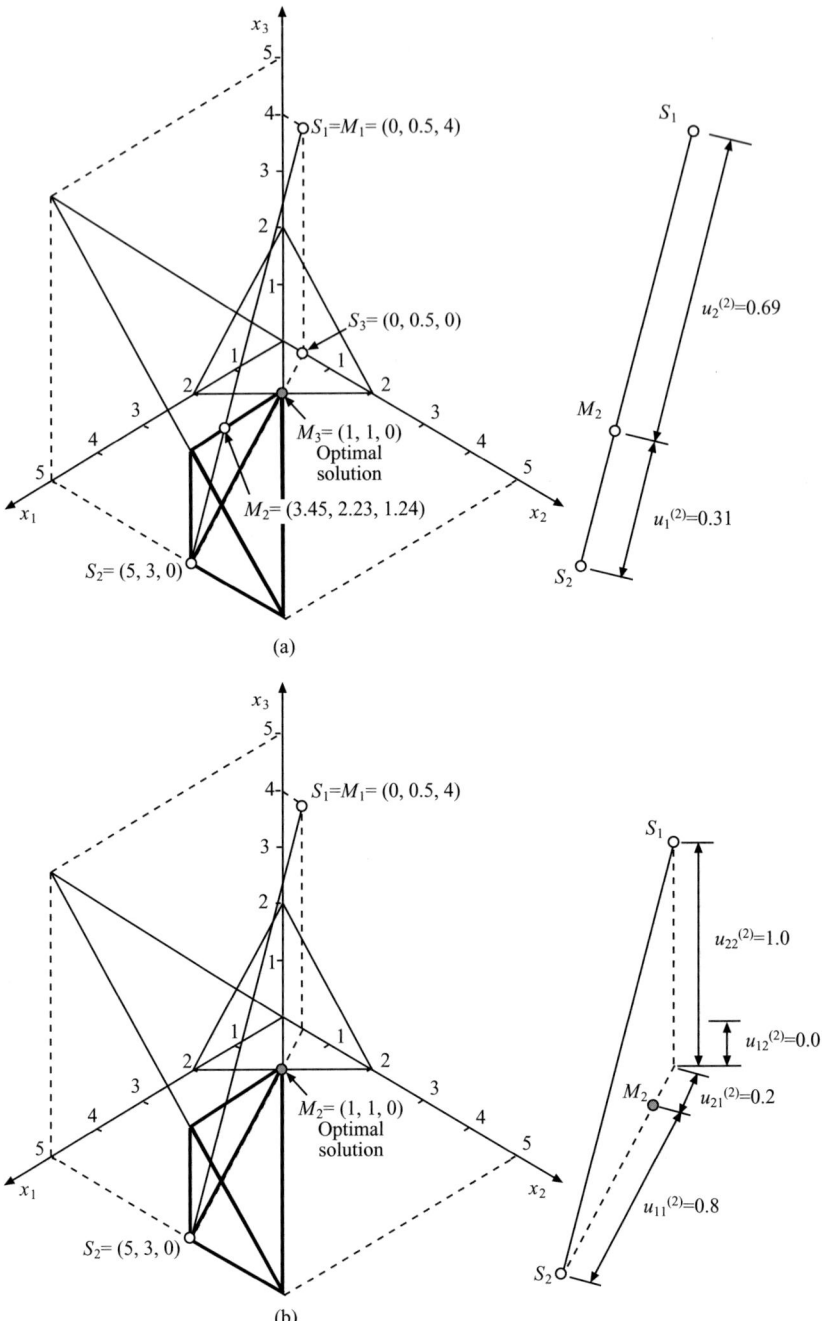

Fig. 2.7. Graphical illustration of the functioning of the two alternative algorithms used in Computational Example 2.4

2.4 The Dantzig-Wolfe Decomposition Algorithm

Table 2.3. Solutions of the master problem and the subproblems in Computational Example 2.4 using the first master problem formulation

Iteration	Bounds		Initial solutions for the subproblem				
ν	Lower	Upper	$x_1^{(\nu)}$	$x_2^{(\nu)}$	$x_3^{(\nu)}$	$r^{(\nu)}$	$z^{(\nu)}$
0	−1.0E+8	1.0E+8	0.0	0.5	4.0	4.5	4.5
		Solutions for the subproblem					
1	−32.00	94.50	5.0	3.0	0.0	−2.0	8.0
2	0.77	6.92	0.0	0.5	0.0	0.5	0.5
3	2.00	2.00	−	−	−	−	−

Iteration	Master solutions					
ν	$u_1^{(\nu)}$	$u_2^{(\nu)}$	$u_3^{(\nu)}$	$\lambda^{(\nu)}$	$\sigma^{(\nu)}$	Feasible
1	1.0	0.0	0.0	−20.0	94.5	No
2	0.31	0.69	0.0	−0.54	6.92	Yes
3	0.0	0.2	0.8	−3.0	2.0	Yes

Table 2.4. Solutions of the master problem and the subproblems in Computational Example 2.4 using the alternative master problem formulation

Iteration	Bounds		Initial solutions for the subproblem							
ν	Lower	Upper	$x_1^{(\nu)}$	$x_2^{(\nu)}$	$x_3^{(\nu)}$	$r_1^{(\nu)}$	$r_2^{(\nu)}$	$r^{(\nu)}$	$z_1^{(\nu)}$	$z_2^{(\nu)}$	$z^{(\nu)}$
0	−1.0E+8	1.0E+8	0.0	0.5	4.0	0.5	4.0	4.5	0.5	4.0	4.5
	Solutions for the subproblem										
1	−32.00	94.50	5.0	3.0	0.0	−2.0	0.0	−2.0	8.0	0.0	8.0
2	2.00	2.00	−	−	−	−	−	−	−	−	−

Iteration	Master solutions								
ν	$u_{11}^{(\nu)}$	$u_{12}^{(\nu)}$	$u_{21}^{(\nu)}$	$u_{22}^{(\nu)}$	$\lambda^{(\nu)}$	$\sigma_1^{(\nu)}$	$\sigma_2^{(\nu)}$	$\sigma^{(\nu)}$	Feasible
1	1.0	1.0	0.0	0.0	−20.0	10.5	84.0	94.5	No
2	0.8	0.0	0.2	1.0	−3.0	2.0	0.0	2.0	Yes

Step 2: Relaxed problem solution. The objective of the subproblems is to attain the most convenient solution inside the relaxed feasible region to be added to the master problem. This results in solution $S_2 = (5, 3, 0)$ shown in Fig. 2.7. The complicating constraint and the objective function are then evaluated to obtain $r^{(1)}$ and $z^{(1)}$ that are shown in Table 2.3 and $r_1^{(1)}, r_2^{(1)}, z_1^{(1)},$ and $z_2^{(1)}$ that are shown in Table 2.4.

Step 3: Convergence checking. Since

$$v^{(1)} = z^{(1)} = 8 < \sigma^{(1)} = 94.5$$

and

$$v^{(1)} = z_1^{(1)} + z_2^{(1)} = 8 < \sigma_1^{(1)} + \sigma_2^{(1)} = 94.5 ,$$

the current solution of the relaxed problem can be used to improve the solution of the master problem.

Step 1: Master problem solutions. Up to this point, both decomposition algorithms work identically, as shown in Fig. 2.7. In this step, they follow different paths.

- The master problem (2.77)–(2.82) is solved finding the solution M_2, shown in Fig. 2.7a. This solution minimizes the original objective function in the intersection of (i) the set of linear convex combinations of solutions S_1 and S_2, and (ii) the original feasible region. The values of the primal variables $u_1^{(2)}, u_2^{(2)}$ and dual variables $\lambda^{(2)}$, and $\sigma^{(2)}$ are shown in Table 2.3.

$$M_2 = u_1^{(2)} S_1 + u_2^{(2)} S_2 = 0.31 \begin{pmatrix} 0 \\ 0.5 \\ 4 \end{pmatrix} + 0.69 \begin{pmatrix} 5 \\ 3 \\ 0 \end{pmatrix} = \begin{pmatrix} 3.45 \\ 2.23 \\ 1.24 \end{pmatrix}.$$

- The alternative master problem (2.83)–(2.86) is solved finding the solution M_2 shown in Fig. 2.7b. Note that this solution minimizes the original objective function in the intersection of (i) the set of linear convex combinations decomposed by blocks of the solutions S_1 and S_2, and (ii) the global feasible region. The associated values of the primal variables $u_{11}^{(2)}, u_{12}^{(2)}, u_{21}^{(2)}, u_{22}^{(2)}$ and dual variables $\lambda^{(2)}, \sigma_1^{(2)}$, and $\sigma_2^{(2)}$ are shown in Table 2.4.

$$\begin{pmatrix} x_1 \\ x_2 \end{pmatrix} = u_{11}^{(2)} \begin{pmatrix} x_1 \\ x_2 \end{pmatrix}_{S_1} + u_{21}^{(2)} \begin{pmatrix} x_1 \\ x_2 \end{pmatrix}_{S_2}$$

$$= 0.8 \begin{pmatrix} 0 \\ 0.5 \end{pmatrix} + 0.2 \begin{pmatrix} 5 \\ 3 \end{pmatrix} = \begin{pmatrix} 1 \\ 1 \end{pmatrix}$$

$$(x_3) = u_{12}^{(2)} (x_3)_{S_1} + u_{22}^{(2)} (x_3)_{S_2}$$
$$= 0.0 (4) + 1.0 (0) = (0)$$

$$M_2 = \begin{pmatrix} x_1 \\ x_2 \\ x_3 \end{pmatrix} = \begin{pmatrix} 1 \\ 1 \\ 0 \end{pmatrix}.$$

Step 2: Relaxed problem solution. The subproblem to be solved is the same for both decomposition approaches. Its target is to find the most convenient solution inside the relaxed feasible region to be added to the master problem. This solution is $S_3 = (0, 0.5, 0)$ shown in Fig. 2.7. The complicating constraint and the target objective function are then evaluated to obtain $r^{(1)}$ and $z^{(1)}$, which are shown in Table 2.3.

Step 3: Convergence checking. Note that this step is different for both decompositions algorithms.

- As
$$v^{(2)} = z^{(2)} = 0.5 < \sigma^{(2)} = 6.92 ,$$
the current solution of the relaxed problem can be used to improve the solution of the master problem in the first approach.
- Since
$$v^{(2)} = z_1^{(1)} + z_2^{(1)} = \sigma_1^{(2)} + \sigma_2^{(2)} = 2 ,$$
the optimal solution has been obtained using the second alternative. This optimal solution is M_2, shown in Fig. 2.7b. Therefore, this algorithm concludes.

Step 1: Master problem solution. The master problem solution using the first approach is $M_3 = (1, 1, 0)$, shown in Fig. 2.7a. The values of the primal variables $u_1^{(3)}, u_2^{(3)}, u_3^{(3)}$ and the dual variables $\lambda^{(3)}$, and $\sigma^{(3)}$ are shown in Table 2.3. Note that these values are optimal. Therefore the algorithm concludes. □

Concerning the alternative formulation (2.83)–(2.86) of the master problem, the following observations are in order:

1. Convex combination of basic feasible solutions of any subproblem are treated independently from other subproblems. This may provide the master problem with more flexibility to attain the optimal solution of the original problem.
2. Despite of a larger number of variables in the master problem, it has been observed in practical applications (see [9]) that this alternative master problem performs usually better than the initial one.

2.5 Concluding Remarks

This chapter analyzes linear problems that include complicating constraints. If these constraints are relaxed, the original problem decomposes by blocks or attains such a structure that its solution is straightforward. This circumstance occurs often in engineering and science problems. The Dantzig-Wolfe decomposition algorithm is motivated, derived, and illustrated in this chapter. Alternative formulations of the master problem are considered, and bounds on the optimal value of the objective function are provided. Diverse geometrical interpretations enrich the algebra-oriented algorithms and quite a few illustrative examples are analyzed in detail.

The decomposition technique analyzed in this chapter to solve the original linear problem is the so-called Dantzig-Wolfe decomposition procedure. This decomposition is also analyzed in the excellent references by Bazaraa et al.[5], Chvatal [22], and Luenberger [23], in the application-oriented manual by Bradley et al. [1], and in the historical book by Dantzig [2].

2.6 Exercises

Exercise 2.1. The problem faced by a multinational company that manufactures one product in different countries is analyzed in Sect. 1.3.1, p. 8. This problem is formulated as a linear programming problem that includes complicating constraints.

Solve the numerical example presented in that section using Dantzig-Wolfe decomposition. Analyze the numerical behavior of the decomposition algorithm and show that the result obtained are identical to those provided in Sect. 1.3.1.

Exercise 2.2. Given the problem

$$\underset{x_1, x_2, x_3, x_4}{\text{minimize}} \quad z = -2x_1 - x_2 - x_3 + x_4$$

subject to

$$\begin{aligned}
x_1 + 2x_2 &\leq 5 \\
-x_3 + x_4 &\leq 2 \\
2x_3 + x_4 &\leq 6 \\
x_1 + x_3 &\leq 2 \\
x_1 + x_2 + 2x_4 &\leq 3 \\
x_1, x_2, x_3, x_4 &\geq 0 .
\end{aligned}$$

1. Check that the following vector (x_1, x_2, x_3, x_4) is a solution:

$$x_1 = 1, \quad x_2 = 2, \quad x_3 = 1, \quad x_4 = 0, \quad z = -5.$$

2. Using the Dantzig-Wolfe decomposition algorithm and by minimizing the objective functions,

$$\begin{aligned}
z_1 &= -x_1 - x_2 + x_4 \\
z_2 &= x_1 + x_2 - x_3 \\
z_3 &= x_1 - x_3 + x_4 \\
z_4 &= 2x_1 + x_2 + 3x_4 ,
\end{aligned}$$

obtain the two different feasible solutions (x_1, x_2, x_3, x_4) of the relaxed problem and the associated values of r_i and z, shown in Table 2.5.

Table 2.5. Initial solutions for the subproblems and new added solutions using the Dantzig-Wolfe decomposition algorithm for Exercise 2.2

Iteration ν	Bounds		Initial solutions for the subproblems						
	Lower	Upper	$x_1^{(\nu)}$	$x_2^{(\nu)}$	$x_3^{(\nu)}$	$x_4^{(\nu)}$	$r_1^{(\nu)}$	$r_2^{(\nu)}$	$z^{(\nu)}$
0–1	$-\infty$	∞	5.00	0.00	0.00	0.00	5.00	5.00	-10.00
0–2	$-\infty$	∞	0.00	0.00	3.00	0.00	3.00	0.00	-3.00
			Subproblem solutions						
1	-42.50	17.00	0.00	2.50	0.00	0.00	0.00	2.50	-2.50
2	-5.00	-5.00	–	–	–	–	–	–	–

Iteration ν	Bounds		Master solutions						
	Lower	Upper	$u_1^{(\nu)}$	$u_2^{(\nu)}$	$u_3^{(\nu)}$	$\lambda_1^{(\nu)}$	$\lambda_2^{(\nu)}$	$\sigma^{(\nu)}$	Feasible
1	$-\infty$	17.00	0.00	1.00	0.00	-20.00	0.00	57.00	No
2	-42.50	-5.00	0.30	0.10	0.50	-1.00	-1.00	0.00	Yes

3. Show that using the Dantzig-Wolfe decomposition algorithm the following solution is obtained

$$x_1 = 1.6, \quad x_2 = 1.4, \quad x_3 = 0.4, \quad x_4 = 0, \quad z = -5 .$$

4. Compare and discuss the resulting solution and that given in item 1 above.

Exercise 2.3. David builds electrical cable using 2 type of alloys, A and B. Alloy A contains 80% of copper and 20% of aluminum, whereas alloy B contains 68% of copper and 32% of aluminum. Costs of alloys A and B are $80 and $60, respectively. In order to produce 1 unit of cable and ensuring that the cable manufactured does not contain more that 25% of aluminum, which are the quantities of alloys A and B that David should use to minimize his manufacturing cost?

Consider the constraint limiting the amount of aluminum a complicating constraint and solve the problem using Dantzig-Wolfe decomposition.

Exercise 2.4. Given the problem

$$\underset{x_1, x_2, \ldots, x_{10}}{\text{minimize}} \quad z = -4x_1 - x_4 - 6x_7$$

subject to

$$\begin{aligned}
x_1 - x_2 &= 1 \\
x_1 + x_3 &= 2 \\
x_4 - x_5 &= 1 \\
x_4 + x_6 &= 2 \\
x_7 - x_8 &= 1 \\
x_7 + x_9 &= 2 \\
3x_1 + 2x_4 + 4x_7 + x_{10} &= 17 \\
x_1, x_2, x_3, x_4, x_5, x_6, x_7, x_8, x_9, x_{10} &\geq 0 .
\end{aligned}$$

Show, using the Dantzig-Wolfe decomposition algorithm, that its solution is

$$x_1 = 2, \; x_2 = 1, \; x_3 = 0, \; x_4 = 1.5, \; x_5 = 0.5,$$
$$x_6 = 0.5, \; x_7 = 2, \; x_8 = 1, \; x_9 = 0, \; x_{10} = 0, \; z = -21.5 .$$

Exercise 2.5. The multireservoir hydroelectric operating planning problem stated in Sect. 1.3.3, p. 19, of Chap. 1 is linear and includes complicating constraints.

Solve the numerical example described in Sect. 1.3.3 using Dantzig-Wolfe decomposition and compare the results obtained with those provided in that section.

Exercise 2.6. Given the problem

$$\underset{x_1, x_2, x_3, x_4}{\text{minimize}} \quad z = -2x_1 - x_2 - x_3 + x_4$$

subject to

$$\begin{aligned}
x_1 - x_2 &\leq 0 \\
x_1 + 2x_2 &\leq 3 \\
-x_3 + x_4 &\leq 0 \\
3x_3 + x_4 &\leq 4 \\
x_1 + x_3 &\leq 2 \\
x_1 + 4x_2 + 2x_4 &\leq 7 \\
x_1, x_2, x_3, x_4 &\geq 0 ,
\end{aligned}$$

show, using the Dantzig-Wolfe decomposition algorithm, that its solution is

$$x_1 = 1, \; x_2 = 1, \; x_3 = 1, \; x_4 = 0, \; z = -4 .$$

Exercise 2.7. Consider the hydroelectric river system depicted in Fig. 2.8. The system should be operated so that the demand for electricity is served in every time period of the planning horizon in such a way that total cost is minimum. Data is provided in the Tables 2.6, 2.7, and 2.8. The conversion factor is used to convert water discharge volume to energy.

Solve this multiperiod operation planning problem using the Dantzig-Wolfe decomposition so that the problem decomposes by reservoir.

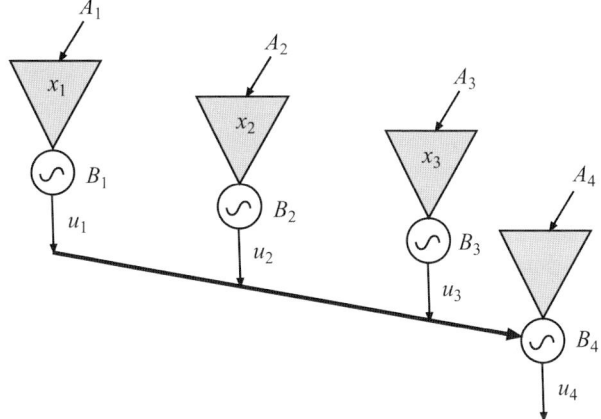

Fig. 2.8. Hydroelectric river system for Exercise 2.7

Table 2.6. Hydroelectric plant data for Exercise 2.7

Hydro plant data				
Unit	1	2	3	4
Initial Volume (hm^3)	104	205	55	0
Maximum Volume (hm^3)	1,000	1,000	1,000	0
Minimum Volume (hm^3)	0	0	0	0
Maximum Discharge (hm^3/h)	30	30	30	80
Minimum Discharge (hm^3/h)	0	0	0	0
Cost ($)	20	10	5	0
Conversion Factor (MWh/hm^3)	10	10	10	10

Table 2.7. Inflow to reservoirs for Exercise 2.7

Inflow to reservoirs (hm^3)				
Reservoir	1	2	3	4
Period 1	35	25	20	10
Period 2	36	26	21	9
Period 3	37	27	22	8
Period 4	36	26	21	7
Period 5	35	25	20	6

Table 2.8. Demand data for Exercise 2.7

Demand data					
Hour	1	2	3	4	5
Demand (MWh)	1,000	1,200	1,300	1,400	1,200

Exercise 2.8. The stochastic programming linear problem formulated in Sect. 1.3.2, p. 12, consists in determining the production policy of a hydroelectric system under water inflow uncertainty. The target is to achieve maximum expected benefit from selling energy. This problem includes complicating constraints and therefore can be conveniently solved using the Dantzig-Wolfe decomposition.

Use the Dantzig-Wolfe decomposition to solve the hydro scheduling numerical example stated in Sect. 1.3.2, and compare the results obtained with those stated in that section.

Exercise 2.9. Electric Power Alpha serves a system including four nodes and four lines. The generating nodes are 1 and 2 and the consumption nodes 3 and 4. Similarly, electric Power Beta serves other system that also includes four nodes and four lines. Nodes 5 and 6 are generating nodes and nodes 7 and 8 are demand nodes. Both companies have agreed in interconnecting their system to minimize cost and to improve security. The interconnection line connects nodes 4 and 8.

1. Compute the saving resulting from using the interconnection. To do this, solve the operation problem of the interconnected system, and compare its optimal solution with the optimal solutions obtained if the two systems are operated independently and without taking into account the interconnection.
2. Solve the operation problem of the interconnected system using the Dantzig-Wolfe decomposition.
3. Give an economic interpretation of the coordinating parameters of the decomposition.

Electric Power Alpha data are provided in the Tables 2.9 and 2.10. Electric Power Beta data are given in the Tables 2.11 and 2.12. Interconnection line data are provided in the Table 2.13.

Note that equations describing how electricity is transmitted through a transmission line are explained in Sect. 1.5.2, p. 42.

Table 2.9. Line data for system served by Electric Power Alpha

From/to	Conductance	Susceptance	Maximum capacity (MW)
1–2	−0.0064	0.4000	0.3
1–3	−0.0016	0.2857	0.5
2–4	−0.0033	0.3333	0.4
3–4	−0.0016	0.2500	0.6

Table 2.10. Node data system served by Electric Power Alpha

Node	Generating cost ($/MWh)	Demand (MW)
1	6	0.00
2	7	0.00
3	0	0.35
4	0	0.45

Table 2.11. Line data system served by Electric Power Beta

From/to	Conductance	Susceptance	Maximum capacity (MW)
5–6	−0.0056	0.3845	0.3
5–7	−0.0021	0.2878	0.5
6–8	−0.0033	0.3225	0.8
7–8	−0.0014	0.2439	0.6

Table 2.12. Node data system served by Electric Power Beta

Node	Generating cost ($/MWh)	Demand (MW)
5	8	0.00
6	9	0.00
7	0	0.35
8	0	0.45

Table 2.13. Interconnection line data

From/to	Conductance	Susceptance	Maximum Capacity (MW)
4–8	−0.0033	0.3333	0.6

Exercise 2.10. To supply the energy demand depicted in Fig. 2.9, five production devices are available. Their respective production costs ($/MWh) and powers (MW) are 1, 2, 3, 4, 5 and 1, 2, 3, 3, 5. Knowing that the joint production of devices 1 and 3 should be below 3 and that the joint production devices 4 and 5 should be above 4, find the optimal schedule of the production devices. In order to do so, analyze first the structure of the problem and then solve it using the Dantzig-Wolfe decomposition procedure. Provide an economical interpretation of the equivalent costs of the subproblems.

Exercise 2.11. The multi-year energy model studied in Sect. 1.3.4, p. 23, is a large-scale linear programming problem that includes complicating constraints. These complicating constraints are few while the noncomplicating ones are many. If the complicating constraints are ignored, the resulting problem attains polymatroid structure and its solution is straightforwardly obtained using a greedy algorithm.

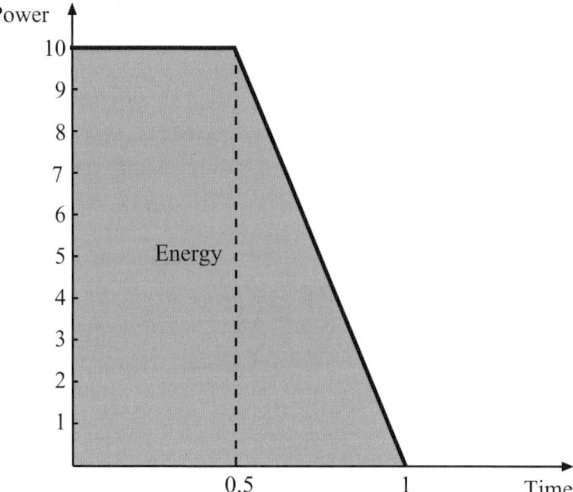

Fig. 2.9. Energy demand for Exercise 2.10

Use the Dantzig-Wolfe decomposition technique to solve the numerical example stated in Sect. 1.3.4, solving subproblems through a greedy algorithm. Analyze the numerical behavior of the Dantzig-Wolfe algorithm for this particular problem.

Exercise 2.12. A transnational plane maker manufactures engines centrally but fuselages locally in three different locations where planes are built and sold. In the first location available labor time and fuselage material are respectively 100 and 55. In the second location 120 and 40, and in the third are 60 and 60.

The manufacture of a plane requires 10 labor units and 15 fuselage material unit plus one engine.

1. Write an optimization problem to determine the maximum number of planes that can be manufactured.
2. Consider the number of planes a real variable and solve the problem using the Dantzig-Wolfe decomposition algorithm.
3. How to solve this problem if the number of planes is considered an integer variable?

3
Decomposition in Linear Programming: Complicating Variables

3.1 Introduction

In this chapter we address linear programming problems with complicating variables. In a linear programming problem, the complicating variables are those variables preventing a solution of the problem by blocks, i.e., a straightforward solution of the problem.

The following example illustrates the way complicating variables make it impossible a distributed solution of a linear programming problem.

Illustrative Example 3.1 (Complicating variables that prevent a distributed solution). Consider the problem

$$\underset{\alpha_1,\alpha_2,\beta_1,\gamma_1,\gamma_2,\lambda_1}{\text{maximize}} \quad e_1\alpha_1 + e_2\alpha_2 + f_1\beta_1 + g_1\gamma_1 + g_2\gamma_2 + h_1\lambda_1$$

subject to

$$\begin{array}{rcl}
a_{11}\alpha_1 + a_{12}\alpha_2 & + d_{11}\lambda_1 & \leq a_1 \\
a_{21}\alpha_1 + a_{22}\alpha_2 & + d_{21}\lambda_1 & \leq a_2 \\
a_{31}\alpha_1 + a_{32}\alpha_2 & + d_{31}\lambda_1 & \leq a_3 \\
b_{11}\beta_1 & + d_{41}\lambda_1 & \leq b_1 \\
b_{21}\beta_1 & + d_{51}\lambda_1 & \leq b_2 \\
c_{11}\gamma_1 + c_{12}\gamma_2 & + d_{61}\lambda_1 & \leq c_1 \\
c_{21}\gamma_1 + c_{22}\gamma_2 & + d_{71}\lambda_1 & \leq c_2 \\
c_{31}\gamma_1 + c_{32}\gamma_2 & + d_{81}\lambda_1 & \leq c_3 \, .
\end{array}$$

If variable λ_1 is given the fixed value λ_1^{fixed}, the problem decomposes into the three subproblems:

Subproblem 1:

$$\underset{\alpha_1,\alpha_2}{\text{maximize}} \quad e_1\alpha_1 + e_2\alpha_2$$

subject to

$$a_{11}\alpha_1 + a_{12}\alpha_2 \leq a_1 - d_{11}\lambda_1^{\text{fixed}}$$
$$a_{21}\alpha_1 + a_{22}\alpha_2 \leq a_2 - d_{21}\lambda_1^{\text{fixed}}$$
$$a_{31}\alpha_1 + a_{32}\alpha_2 \leq a_3 - d_{31}\lambda_1^{\text{fixed}}.$$

Subproblem 2:

$$\underset{\beta_1}{\text{maximize}} \quad f_1\beta_1$$

subject to

$$b_{11}\beta_1 \leq b_1 - d_{41}\lambda_1^{\text{fixed}}$$
$$b_{21}\beta_1 \leq b_2 - d_{51}\lambda_1^{\text{fixed}}.$$

Subproblem 3:

$$\underset{\gamma_1, \gamma_2}{\text{maximize}} \quad g_1\gamma_1 + g_2\gamma_2$$

subject to

$$c_{11}\gamma_1 + c_{12}\gamma_2 \leq c_1 - d_{61}\lambda_1^{\text{fixed}}$$
$$c_{21}\gamma_1 + c_{22}\gamma_2 \leq c_2 - d_{71}\lambda_1^{\text{fixed}}$$
$$c_{31}\gamma_1 + c_{32}\gamma_2 \leq c_3 - d_{81}\lambda_1^{\text{fixed}}.$$

$\square$

The example below illustrates how complicating variables prevent a straightforward solution of a linear programming problem.

Illustrative Example 3.2 (Complicating variables preventing a straightforward solution). Consider the problem

$$\underset{\alpha_1, \alpha_2, \alpha_3, \beta_1}{\text{maximize}} \quad b_1\alpha_1 + b_2\alpha_2 + b_3\alpha_3 + b_4\beta_1$$

subject to

$$a_{11}\alpha_1 \qquad\qquad\qquad + a_{14}\beta_1 \leq c_1$$
$$\qquad a_{22}\alpha_2 \qquad\qquad + a_{24}\beta_1 \leq c_2$$
$$\qquad\qquad\qquad\qquad a_{34}\beta_1 \leq c_3$$
$$\qquad\qquad\qquad\qquad a_{44}\beta_1 \leq c_4$$
$$\qquad\qquad\qquad\qquad a_{54}\beta_1 \leq c_5$$
$$\qquad\qquad\qquad\qquad a_{64}\beta_1 \leq c_6$$
$$a_{71}\alpha_1 + a_{72}\alpha_2 + a_{73}\alpha_3 + a_{74}\beta_1 \leq c_7,$$

where

$$b_1, b_2, b_3, b_4 > 0.$$

If variable β_1 is fixed to the value β_1^{fixed}, the problem above has a straightforward solution. Note that the selection of the fixed value of β_1 should meet the constraints below

$$\beta_1^{\text{fixed}} \leq \frac{c_3}{a_{34}}, \quad \beta_1^{\text{fixed}} \leq \frac{c_4}{a_{44}}, \quad \beta_1^{\text{fixed}} \leq \frac{c_5}{a_{54}}, \quad \beta_1^{\text{fixed}} \leq \frac{c_6}{a_{64}}.$$

The trivial solution is computed below. Variable α_1 is computed first.

$$\alpha_1 \leq \frac{c_1}{a_{11}} - \frac{a_{14}}{a_{11}}\beta_1^{\text{fixed}}$$

and since $b_1 > 0$,

$$\alpha_1 = \frac{c_1}{a_{11}} - \frac{a_{14}}{a_{11}}\beta_1^{\text{fixed}}.$$

Then, variable α_2 is computed as

$$\alpha_2 \leq \frac{c_2}{a_{22}} - \frac{a_{24}}{a_{22}}\beta_1^{\text{fixed}}$$

and since $b_2 > 0$,

$$\alpha_2 = \frac{c_2}{a_{22}} - \frac{a_{24}}{a_{22}}\beta_1^{\text{fixed}}.$$

Finally, once α_1 and α_2 are known, α_3 is computed from

$$\alpha_3 \leq \frac{c_7}{a_{73}} - \frac{a_{74}}{a_{73}}\beta_1^{\text{fixed}} - \frac{a_{71}}{a_{73}}\alpha_1 - \frac{a_{72}}{a_{73}}\alpha_2$$

and since $b_3 > 0$, the above inequality becomes binding, and

$$\alpha_3 = \frac{c_7}{a_{73}} - \frac{a_{74}}{a_{73}}\beta_1^{\text{fixed}} - \frac{a_{71}}{a_{73}}\alpha_1 - \frac{a_{72}}{a_{73}}\alpha_2.$$

□

These examples and those analyzed in the preceding chapter suggest the following theorem.

Theorem 3.1 (Primal and dual decomposability). *If a linear programming problem has a decomposable structure with complicating constraints, its dual linear programming problem has a decomposable structure with complicating variables. And conversely, if a linear programming problem has a decomposable structure with complicating variables, its dual linear programming problem has a decomposable structure with complicating constraints.* □

Proof. The proof of Theorem 3.1 follows in a straightforward manner from the definition of the dual problem of a linear programming problem. □

3.2 Complicating Variables: Problem Structure

Consider the initial problem

$$\underset{x_1,\ldots,x_n;\, y_1,\ldots,y_m}{\text{minimize}} \quad \sum_{i=1}^{n} c_i\, x_i + \sum_{j=1}^{m} d_j\, y_j \tag{3.1}$$

subject to

$$\sum_{i=1}^{n} a_{\ell i}\, x_i + \sum_{j=1}^{m} e_{\ell j}\, y_j \;\leq\; b^{(\ell)}; \quad \ell = 1, \ldots, q \tag{3.2}$$

$$0 \leq x_i \;\leq\; x_i^{\text{up}}; \quad i = 1, \ldots, n \tag{3.3}$$
$$0 \leq y_j \;\leq\; y_j^{\text{up}}; \quad j = 1, \ldots, m, \tag{3.4}$$

where x_i ($i = 1, \ldots, n$) are the complicating variables.

Complicating variables make the solution of problem (3.1)–(3.4) difficult. If they are fixed to given values, problem (3.1)–(3.4) becomes substantially simpler. This is so because either it decomposes in subproblems or it attains such a structure that its solution is straightforward.

A problem with complicating variables can be transformed, using the dual problem, into a problem with complicating constraints. The example below illustrates how the dual approach can be used.

Illustrative Example 3.3 (Complicating variables: Dual problem).
The problem

$$\underset{y_1, y_2, y_3, y_4, y_5, x_1}{\text{maximize}} \quad 4y_1 + 3y_2 + 2y_3 + 3y_4 + 2y_5 + 3x_1$$

subject to

$$\begin{aligned}
y_1 + 2y_2 \quad\quad\quad\quad\quad\quad\quad + 2x_1 &\leq 3 \\
2y_1 + y_2 \quad\quad\quad\quad\quad\quad\quad + x_1 &\leq 3 \\
-2y_1 + 3y_2 \quad\quad\quad\quad\quad\quad + x_1 &\leq 7 \\
y_3 \quad\quad\quad\quad + 3x_1 &\leq 4 \\
2y_3 \quad\quad\quad\quad - x_1 &\leq 3 \\
y_4 \quad\quad\quad\quad\quad\quad &\leq 1 \\
2y_4 + 4y_5 + 3x_1 &\leq 5 \\
3y_4 + y_5 \quad - x_1 &\leq 4
\end{aligned}$$

has a decomposable structure and the complicating variable is x_1.
Its dual problem is

$$\underset{u_1, u_2, \ldots, u_8}{\text{minimize}} \quad 3u_1 + 3u_2 + 7u_3 + 4u_4 + 3u_5 + u_6 + 5u_7 + 4u_8$$

subject to

$$
\begin{aligned}
u_1 + 2u_2 - 2u_3 &= 4 \\
2u_1 + u_2 + 3u_3 &= 3 \\
u_4 + 2u_5 &= 2 \\
u_6 + 2u_7 + 3u_8 &= 3 \\
4u_7 + u_8 &= 2 \\
2u_1 + u_2 + u_3 + 3u_4 - u_5 + 3u_7 - u_8 &= 3 \\
u_1, u_2, u_3, u_4, u_5, u_6, u_7, u_8 &\leq 0,
\end{aligned}
$$

which is a problem with complicating constraints. □

3.3 Benders Decomposition

The Benders decomposition algorithm allows us to solve a linear programming problem with complicating variables in a distributed manner at the cost of iteration. Benders decomposition is described in this section.

3.3.1 Description

The solution of problem (3.1)–(3.4) can be obtained parameterizing this problem as a function of the complicating variables $x_1, \ldots, x_n$. This is done as follows.

Alternative Formulation of the Initial Problem (3.1)–(3.4):

$$\underset{x_1,\ldots,x_n}{\text{minimize}} \quad \sum_{i=1}^{n} c_i\, x_i + \alpha(x_1, \ldots, x_n) \tag{3.5}$$

subject to

$$0 \leq x_i \leq x_i^{\text{up}}; \quad i = 1, \ldots, n, \tag{3.6}$$

where

$$\alpha(x_1, \ldots, x_n) = \underset{y_1,\ldots,y_m}{\text{minimum}} \sum_{j=1}^{m} d_j\, y_j \tag{3.7}$$

subject to

$$\sum_{j=1}^{m} e_{\ell j}\, y_j \leq b^{(\ell)} - \sum_{i=1}^{n} a_{\ell i}\, x_i; \quad \ell = 1, 2, \ldots, q \tag{3.8}$$

$$0 \leq y_j \leq y_j^{\text{up}}; \quad j = 1, \ldots, m, \tag{3.9}$$

where $\alpha(x_1,\ldots,x_n)$ is the function that provides the optimal objective function value of problem (3.7)–(3.9) for given values of the complicating variables $x_1,\ldots,x_n$. It should be noted that function $\alpha(x_1,\ldots,x_n)$ is convex by construction, as shown below.

Theorem 3.2 (Convexity of $\alpha(x_1,\ldots,x_n)$). *The function $\alpha(x_1,\ldots,x_n)$ defined by (3.7)–(3.9) is convex.*

□

Proof. Taking into account that the feasible region associated with the linear programming problem (3.1)–(3.4) has the structure of a convex polytope, it is clear that the feasible region associated with the function $\alpha(x_1,\ldots,x_n)$ defined by (3.7)–(3.9) is a subset of the initial one, restricted to the fixed values of the complicating variables $x_1,\ldots,x_n$, and considering that these variables are inside the feasible region defined by constraints (3.2)–(3.4). Then, consider two feasible solutions of problem (3.1)–(3.4) $\boldsymbol{v}^{(1)} = (x^{(1)}, y^{(1)})^T$ and $\boldsymbol{v}^{(2)} = (x^{(2)}, y^{(2)})^T$ in such a way that for the complicating variables $\boldsymbol{x}^{(1)} = (x_1^{(1)},\ldots,x_n^{(1)})^T$ and $\boldsymbol{x}^{(2)} = (x_1^{(2)},\ldots,x_n^{(2)})^T$ the associated solutions from problem (3.7)–(3.9) are $\boldsymbol{y}^{(1)}$ and $\boldsymbol{y}^{(2)}$, respectively, with associated objective function values,

$$\alpha\left(\boldsymbol{x}^{(1)}\right) = \sum_{j=1}^{m} d_j\, y_j^{(1)}, \tag{3.10}$$

$$\alpha\left(\boldsymbol{x}^{(2)}\right) = \sum_{j=1}^{m} d_j\, y_j^{(2)}. \tag{3.11}$$

Consider a linear convex combination of $\boldsymbol{v}^{(1)}$ and $\boldsymbol{v}^{(2)}$. By definition of convex set, inside the feasible region defined by constraints (3.2)–(3.4), one gets

$$\boldsymbol{v}^{(3)} = \lambda \boldsymbol{v}^{(1)} + (1-\lambda)\boldsymbol{v}^{(2)} = \lambda \begin{pmatrix} x^{(1)} \\ y^{(1)} \end{pmatrix} + (1-\lambda) \begin{pmatrix} x^{(2)} \\ y^{(2)} \end{pmatrix} = \begin{pmatrix} x^{(3)} \\ y^{(3)} \end{pmatrix}.$$

The objective function of problem (3.7)–(3.9) can be evaluated at $\boldsymbol{y}^{(3)}$, resulting in

$$\begin{aligned}
\sum_{j=1}^{m} d_j\, y_j^{(3)} &= \sum_{j=1}^{m} d_j \left(\lambda y_j^{(1)} + (1-\lambda) y_j^{(2)}\right) \\
&= \lambda \sum_{j=1}^{m} d_j\, y_j^{(1)} + (1-\lambda)\sum_{j=1}^{m} d_j\, y_j^{(2)} \qquad (3.12)\\
&= \lambda\, \alpha\left(\boldsymbol{x}^{(1)}\right) + (1-\lambda)\, \alpha\left(\boldsymbol{x}^{(2)}\right).
\end{aligned}$$

On the other hand, problem (3.7)–(3.9) can be solved for $\boldsymbol{x}^{(3)} = \lambda \boldsymbol{x}^{(1)} + (1-\lambda)\boldsymbol{x}^{(2)}$ resulting in $\boldsymbol{y}^*$ with objective function value $\alpha\left(\boldsymbol{x}^{(3)}\right) = \sum_{j=1}^{m} d_j\, y_j^*$.

Note that

$$\sum_{j=1}^{m} d_j\, y_j^* \le \sum_{j=1}^{m} d_j\, y_j^{(3)} \qquad (3.13)$$

results in

$$\alpha\left(x^{(3)}\right) \le \lambda\, \alpha\left(x^{(1)}\right) + (1-\lambda)\, \alpha\left(x^{(2)}\right), \qquad (3.14)$$

which shows the convexity of the function $\alpha(x)$. □

In Fig. 3.1, a 3-D graphical interpretation of the function $\alpha(x_1,\ldots,x_n)$ is shown. Note that, since $\alpha(x_1,\ldots,x_n)$ is the minimum of $z = d^T y$ in the feasible region, the piece-wise linear structure of $\alpha(x_1,\ldots,x_n)$ is due to the linear character of both the objective function and the feasible region boundaries. Figure 3.1 shows the feasible region (in grey) associated with the set of constraints (3.8) and (3.9), the points where the minima are attained for each value of x, and the corresponding points on the hyperplane $z = d^T y$, which once projected on the X–Z plane leads to the $\alpha(x_1,\ldots,x_n)$ function.

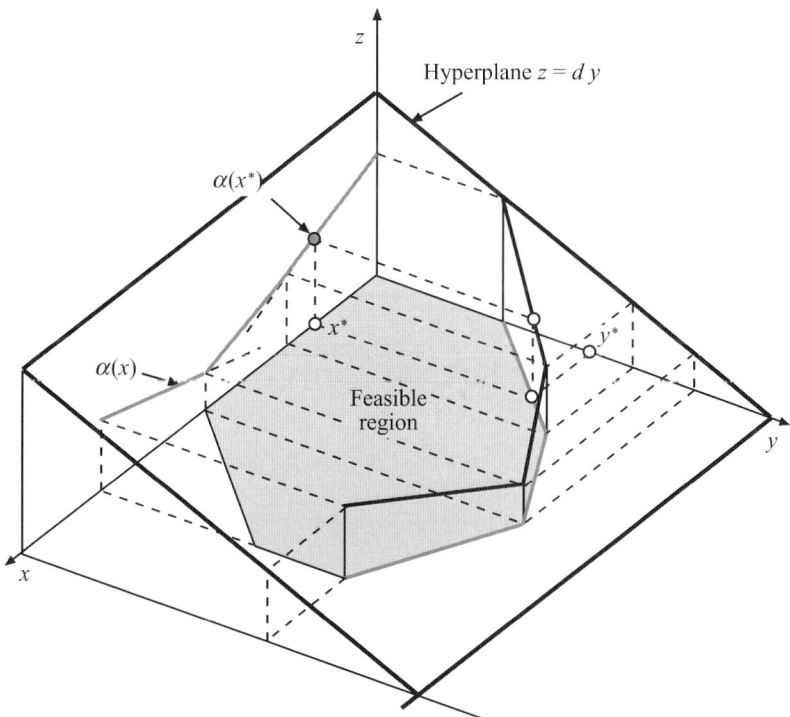

Fig. 3.1. 3-D graphical illustration of the α function

Problem (3.5)–(3.6), which is equivalent to the original one, has the advantage of depending only on the complicating variables, and this can be exploited computationally. However, an exact formulation is harder to obtain than solving the original problem (3.1)–(3.4) because function $\alpha(x_1, \ldots, x_n)$ has to be determined exactly, and this is computationally intensive. Nevertheless, function $\alpha(x_1, \ldots, x_n)$, being convex, can be easily approximated from below using hyperplanes. In practice, this approximation has to be iterative, so that it is improved at every iteration. This iterative approximation renders the Benders decomposition algorithm sketched below:

1. Approximate $\alpha(x_1, \ldots, x_n)$ from below using hyperplanes.
2. Solve problem (3.5)–(3.6) using the approximation of $\alpha(x_1, \ldots, x_n)$ instead of $\alpha(x_1, \ldots, x_n)$.
3. Improve the approximation of $\alpha(x_1, \ldots, x_n)$ using additional hyperplanes. These additional hyperplanes are obtained solving the so-called subproblem or subproblems.
4. If the approximation of $\alpha(x_1, \ldots, x_n)$ is good enough, stop; otherwise, continue with Step 2.

Note that once the approximation of $\alpha(x_1, \ldots, x_n)$ is sufficiently good, the solution of the original problem (3.1)–(3.4) is obtained through the solution of problem (3.5)–(3.6) that depends only on the complicating variables.

In what follows the Benders decomposition algorithm is developed rigorously, a relaxed version of problem (3.5)–(3.6), denominated the master problem, is formulated in the following.

Consider given values for the complicating variables $x_1^{(k)}, \ldots, x_n^{(k)}$, so that $0 \leq x_i^{(k)} \leq x_i^{\text{up}}$ ($i = 1, \ldots, n$), and consider the following:

Subproblem:

$$\underset{y_1, \ldots, y_m}{\text{minimize}} \quad \sum_{j=1}^{m} d_j\, y_j \tag{3.15}$$

subject to

$$\sum_{i=1}^{n} a_{\ell i}\, x_i + \sum_{j=1}^{m} e_{\ell j}\, y_j \leq b^{(\ell)} \quad ; \quad \ell = 1, \ldots, q \tag{3.16}$$

$$0 \leq y_j \leq y_j^{\text{up}} \quad ; \quad j = 1, \ldots, m \tag{3.17}$$

$$x_i = x_i^{(k)} : \lambda_i \quad ; \quad i = 1, \ldots, n. \tag{3.18}$$

This problem, which is denominated the subproblem, typically, either it decomposes by blocks in subproblems or it attains such a structure such that it is much easier to solve than the original problem.

Problem (3.15)–(3.18) is a particular instance of the original problem, i.e., a problem more restricted than the original one. Its solution is denoted by

$y_1^{(k)}, \ldots, y_m^{(k)}$. The optimal values of the dual variables associated with the constraints that fix the values of the complicating variables are $\lambda_1^{(k)}, \ldots, \lambda_n^{(k)}$.

This solution allows us formulating a relaxed version of problem (3.5)–(3.6), i.e., the master problem formulated below.

Master problem 1:

$$\underset{x_1, \ldots, x_n, \alpha}{\text{minimize}} \quad \sum_{i=1}^{n} c_i\, x_i + \alpha \tag{3.19}$$

subject to

$$\alpha \geq \sum_{j=1}^{m} d_j\, y_j^{(k)} + \sum_{i=1}^{n} \lambda_i^{(k)}(x_i - x_i^{(k)}) \tag{3.20}$$

$$0 \leq x_i \leq x_i^{\text{up}}; \quad i = 1, \ldots, n \tag{3.21}$$

$$\alpha \geq \alpha^{\text{down}}, \tag{3.22}$$

where α is a scalar and α^{down} is a bound that can be determined from physical or economical considerations pertaining to the problem under study.

It should be noted that problem (3.19)–(3.22) is a relaxed version of problem (3.5)–(3.6) because it approximates from below problem (3.5)–(3.6).

Considering ν solutions for the complicating variables, $x_i^{(k)}$ ($i = 1, \ldots, n$; $k = 1, \ldots, \nu$), the corresponding number of solutions of subproblem (3.15)–(3.18) is ν, i.e., $y_j^{(k)}$ ($j = 1, \ldots, m$; $k = 1, \ldots, \nu$).

In this situation, a better relaxed master problem can be formulated. This is so because ν hyperplanes reconstruct the function $\alpha(x_1, \ldots, x_n)$, instead of just 1, as in problem (3.19)–(3.22). This problem is also denominated master problem and has the following form:

Master problem 2:

$$\underset{x_1, \ldots, x_n, \alpha}{\text{minimize}} \quad \sum_{i=1}^{n} c_i\, x_i + \alpha \tag{3.23}$$

subject to

$$\sum_{j=1}^{m} d_j\, y_j^{(k)} + \sum_{i=1}^{n} \lambda_i^{(k)}(x_i - x_i^{(k)}) \leq \alpha \quad ; \quad k = 1, \ldots, \nu \tag{3.24}$$

$$0 \leq x_i \leq x_i^{\text{up}} \quad ; \quad i = 1, \ldots, n \tag{3.25}$$

$$\alpha \geq \alpha^{\text{down}}. \tag{3.26}$$

Constraints (3.24) are denominated Benders cuts. The solution of the above problem is denoted by $x_1^{(k+1)}, \ldots, x_n^{(k+1)}$; $\alpha^{(k+1)}$.

The above considerations allow us building the Benders algorithm to be described in Subsect. 3.3.3, p. 116. Before constructing this algorithms, appropriate bounds of the objective function are derived.

3.3.2 Bounds

It should be noted that the problem (3.23)–(3.26) is a relaxed version of the original problem and its objective function approximates from below the objective function of the original problem. Therefore, for iteration $k+1$, the optimal value of the objective function of problem (3.23)–(3.26) is a lower bound of the optimal value of the objective function of the original problem, i.e.,

$$z_{\text{down}}^{(k+1)} = \sum_{i=1}^{n} c_i \, x_i^{(k+1)} + \alpha^{(k+1)} . \qquad (3.27)$$

On the other hand, problem (3.15)–(3.18), the subproblem, is a further restricted version of the original problem. Therefore, its optimal objective function value is an upper bound of the optimal value of the objective function of the original problem, i.e.,

$$z_{\text{up}}^{(k+1)} = \sum_{i=1}^{n} c_i \, x_i^{(k+1)} + \sum_{j=1}^{m} d_j \, y_j^{(k+1)} . \qquad (3.28)$$

Remark 3.1. As the lower bound (3.27) is the solution of a relaxed problem that becomes progressively less and less relaxed as iterations increase, this lower bound is monotonously increasing with the iterations.

3.3.3 The Benders Decomposition Algorithm

The Benders decomposition algorithm works as follows.

Algorithm 3.1 (The Benders decomposition algorithm).

Input. An LP problem with complicating variables, and a small tolerance value ε to control convergence.

Output. The solution of the LP problem obtained after using the Benders decomposition algorithm.

Step 0: Initialization. Initialize the iteration counter, $\nu = 1$. Solve the initial master problem below

$$\underset{x_1,\ldots,x_n,\alpha}{\text{minimize}} \quad \sum_{i=1}^{n} c_i x_i + \alpha \qquad (3.29)$$

subject to

$$0 \leq x_i \leq x_i^{\text{up}}; \quad i = 1,\ldots,n \qquad (3.30)$$

$$\alpha \geq \alpha^{\text{down}} . \qquad (3.31)$$

This problem has the trivial solution: $\alpha^{(1)} = \alpha^{\text{down}}$, and $x_i^{(1)} = 0$ if $c_i \geq 0$, and $x_i^{(1)} = x_i^{\text{up}}$ if $c_i < 0$.

3.3 Benders Decomposition

Step 1: Subproblem solution. The subproblem below is solved

$$\underset{y_1,\ldots,y_m}{\text{minimize}} \quad \sum_{j=1}^{m} d_j\, y_j \tag{3.32}$$

subject to

$$\sum_{i=1}^{n} a_{\ell i}\, x_i + \sum_{j=1}^{m} e_{\ell j}\, y_j \leq b^{(\ell)}; \quad \ell = 1,\ldots,q \tag{3.33}$$

$$0 \leq y_j \leq y_j^{\text{up}}; \quad j = 1,\ldots,m \tag{3.34}$$

$$x_i = x_i^{(\nu)} : \lambda_i; \quad i = 1,\ldots,n. \tag{3.35}$$

The solution of this problem is $y_1^{(\nu)},\ldots,y_m^{(\nu)}$, with dual variable values $\lambda_1^{(\nu)},\ldots,\lambda_n^{(\nu)}$.

Step 2: Convergence checking. Compute an upper bound of the optimal value of the objective function of the original problem:

$$z_{\text{up}}^{(\nu)} = \sum_{i=1}^{n} c_i\, x_i^{(\nu)} + \sum_{j=1}^{m} d_j\, y_j^{(\nu)} \tag{3.36}$$

and compute a lower bound of the optimal value of the objective function of the original problem:

$$z_{\text{down}}^{(\nu)} = \sum_{i=1}^{n} c_i\, x_i^{(\nu)} + \alpha^{(\nu)}. \tag{3.37}$$

If $z_{\text{up}}^{(\nu)} - z_{\text{down}}^{(\nu)} < \varepsilon$, stop, the optimal solution is $x_1^{(\nu)},\ldots,x_n^{(\nu)}$ and $y_1^{(\nu)},\ldots,y_m^{(\nu)}$. Otherwise, the algorithm continues with the next step.

It should be noted that bounds do not have to be computed, i.e., if $\sum_{j=1}^{m} d_j\, y_j^{(\nu)} - \alpha^{(\nu)} < \varepsilon$, stop, the optimal solution is $x_1^{(\nu)},\ldots,x_n^{(\nu)}$ and $y_1^{(\nu)},\ldots,y_m^{(\nu)}$.

Step 3: Master problem solution. Update the iteration counter, $\nu \leftarrow \nu+1$. Solve the master problem

$$\underset{x_1,\ldots,x_n,\alpha}{\text{minimize}} \quad \sum_{i=1}^{n} c_i x_i + \alpha \tag{3.38}$$

subject to

$$\sum_{j=1}^{m} d_j y_j^{(k)} + \sum_{i=1}^{n} \lambda_i^{(k)}(x_i - x_i^{(k)}) \leq \alpha; \quad k = 1,\ldots,\nu-1 \tag{3.39}$$

$$0 \leq x_i \leq x_i^{\text{up}}; \quad i = 1,\ldots,n \tag{3.40}$$

$$\alpha \geq \alpha^{\text{down}}. \tag{3.41}$$

Once the solution of this problem, $x_1^{(\nu)}, \ldots, x_n^{(\nu)}$ and $\alpha^{(\nu)}$, has been obtained, the algorithm continues in Step 1.

In the previous derivations and for the sake of simplicity, it has been considered the particular case in which the decomposition leads to a single subproblem.

For the general case of multiple subproblems, note that the values of the dual variables obtained in each subproblem, after fixing complicating variables to given values, are different. They are denoted by $\lambda_{i,s}^{(k)}$, where k is the iteration counter, i is the index corresponding to the complicating variable i, and s is the subproblem index.

The master problem constraint (3.39) for this general case is formulated as

$$\sum_{j=1}^{m} d_j y_j^{(k)} + \sum_{s=1}^{r} \sum_{i=1}^{n} \lambda_{i,s}^{(k)} (x_i - x_i^{(k)}) \leq \alpha; \quad k = 1, \ldots, \nu - 1 . \quad (3.42)$$

If dual variables are obtained for all subproblems in Step 1, and the above master problem is solved in Step 3 instead of the originally formulated one, the decomposition algorithm remains unchanged.

A problem with decomposable structure in three subproblems and complicating variables is illustrated in Fig. 3.2. In this figure, a superscript in square brackets indicates the corresponding subproblem. □

A GAMS implementation of the Benders decomposition algorithm is given in the Appendix A, p. 403.

Computational Example 3.1 (The Benders decomposition algorithm). Consider the problem

$$\underset{x,y}{\text{minimize}} \quad z = -y - x/4$$

subject to

$$\begin{aligned} y - x &\leq 5 \\ y - \tfrac{1}{2}x &\leq \tfrac{15}{2} \\ y + \tfrac{1}{2}x &\leq \tfrac{35}{2} \\ -y + x &\leq 10 \\ 0 \leq x &\leq 16 \\ y &\geq 0 . \end{aligned}$$

The solution of this example is illustrated in Fig. 3.3, where the feasible region is shown and the optimal values are $x^* = 10$ and $y^* = \tfrac{25}{2}$ with an objective function value equal to $z^* = -15$.

If variable x is considered to be a complicating variable, the above problem can be solved using the Benders decomposition (Algorithm 3.1). The process

3.3 Benders Decomposition   119

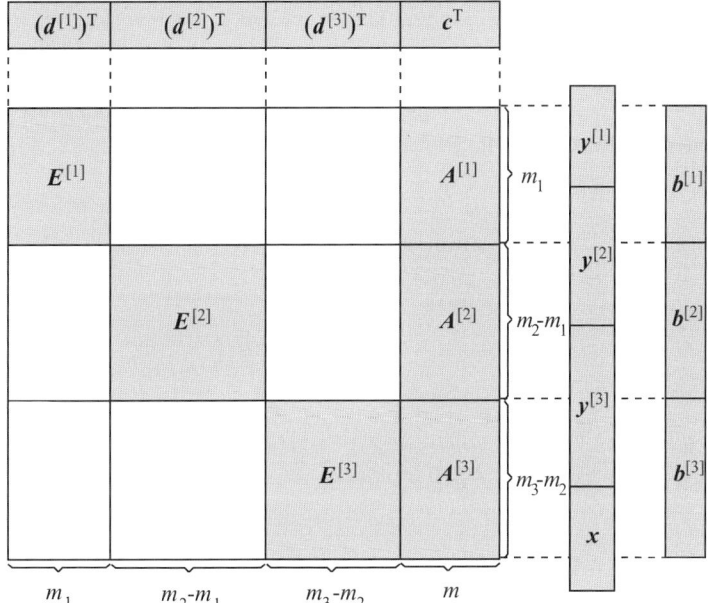

Fig. 3.2. Decomposable matrix with complicating variables

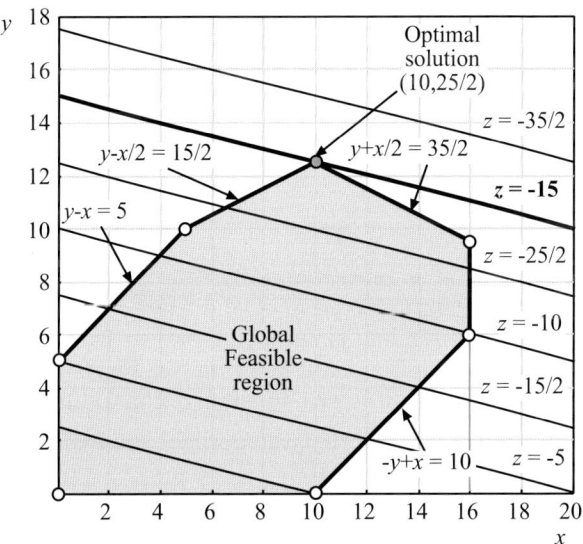

Fig. 3.3. Graphical illustration of Computational Example 3.1

is illustrated in Figs. 3.4–3.7, where the graphical interpretations in the (x, y) and (x, z) subspaces are given. Both illustrations allow a three dimensional (3-D) understanding of the example.

Step 0: Initialization. The iteration counter is initialized, $\nu = 1$. The initial master problem is solved

$$\underset{x, \alpha}{\text{minimize}} \quad -\frac{1}{4}x + \alpha$$

subject to

$$0 \leq x \leq 16$$
$$-25 \leq \alpha.$$

The solution of this problem is $x^{(1)} = 16$, point $M^{(1)}$ in Fig. 3.4b, and $\alpha^{(1)} = -25$ (see Figs. 3.4a,b). The value for variable α is equal to its lower bound, -25.

Step 1: Subproblem solution. The subproblem below is solved

$$\underset{y}{\text{minimize}} \quad z = -y$$

subject to

$$y - x \leq 5$$
$$y - \tfrac{1}{2}x \leq \tfrac{15}{2}$$
$$y + \tfrac{1}{2}x \leq \tfrac{35}{2}$$
$$-y + x \leq 10$$
$$y \geq 0$$
$$x = 16 : \lambda.$$

The solution of this problem is the point $S^{(1)} = \left(x^{(1)}, y^{(1)}\right) = \left(16, \tfrac{19}{2}\right)$, which is located inside the initial feasible region (see Fig. 3.4a), and minimizes the objective function z. The objective function optimal value is $z = -\tfrac{27}{2}$, i.e., the $z_{\text{up}}^{(1)}$ value in Fig. 3.4b. Note that the optimal value of the dual variable associated with the constraint $x = 16$ is $\lambda^{(1)} = \tfrac{1}{2}$.

Step 2: Convergence checking. An upper bound of the objective function optimal value is computed as

$$z_{\text{up}}^{(1)} = -\frac{1}{4}x^{(1)} - y^{(1)} = -\frac{1}{4} \times 16 - \frac{19}{2} = -\frac{27}{2},$$

where the first term is associated with the complicating variable, $x^{(1)}$, and the second term is associated with the $y^{(1)}$ variable (see Fig. 3.4b).

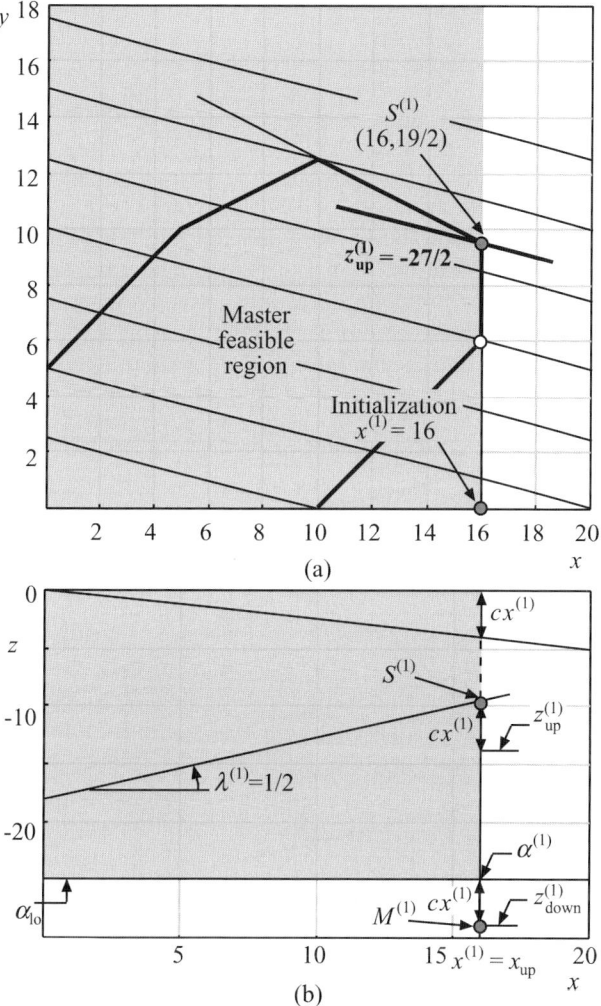

Fig. 3.4. Graphical illustration of the decomposition algorithm for iteration $\nu = 1$

A lower bound of the objective function optimal value is computed as

$$z_{\text{down}}^{(1)} = -\frac{1}{4}x^{(1)} + \alpha^{(1)} = -\frac{1}{4} \times 16 - 25 = -29 \ .$$

Since the difference $z_{\text{up}}^{(1)} - z_{\text{down}}^{(1)} = \frac{31}{2} > \varepsilon$, the procedure continues.

Step 3: Master problem solution. The iteration counter is updated, $\nu = 1 + 1 = 2$. The master problem below is solved

$$\underset{x,\alpha}{\text{minimize}} \quad -\frac{1}{4}x + \alpha$$

subject to
$$-\frac{19}{2} + \frac{1}{2}(x-16) \leq \alpha$$
$$0 \leq x \leq 16$$
$$-25 \leq \alpha,$$

where the first constraint is the Benders cut 1 (see Figs. 3.5a and 3.5b) associated with the previous iteration.

The solution of this problem is the point $M^{(2)}$ in both Figs. 3.5a and 3.5b and $\alpha^{(2)} = -\frac{35}{2}$. The point $M^{(2)}$ in Fig. 3.5a is the intersection of the Benders cut obtained in the previous iteration (active constraint for subproblem 1) and the line $x = 0$ (lower bound of the complicating variable x). Similarly, the point $M^{(2)}$ in Fig. 3.5b is the intersection of the tangent hyperplane and

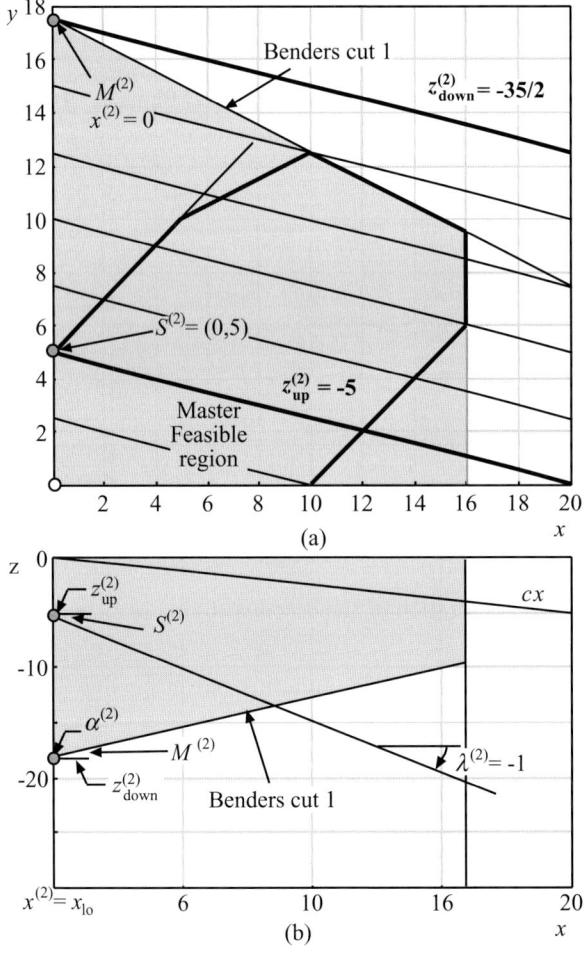

Fig. 3.5. Graphical illustration of the decomposition algorithm for iteration $\nu = 2$

the α function at the point $x^{(1)} = 16$ (lower bound of the shadow region in Fig. 3.5b) and the line $x = 0$.

Step 1: Subproblem solution. The subproblem below is solved

$$\underset{y}{\text{minimize}} \quad z = -y$$

subject to

$$\begin{aligned}
y - x &\leq 5 \\
y - \tfrac{1}{2}x &\leq \tfrac{15}{2} \\
y + \tfrac{1}{2}x &\leq \tfrac{35}{2} \\
-y + x &\leq 10 \\
y &\geq 0 \\
x &= 0 : \lambda.
\end{aligned}$$

The solution of this problem is $y^{(2)} = 5$, point $S^{(2)}$ in Fig. 3.5a that minimizes the objective function for the fixed value of $x^{(2)} = 0$. Its z-value is $z = -5$, as can be seen in Fig. 3.5b. The optimal dual variable associated with the constraint $x = 0$, $\lambda^{(2)} = -1$, is the slope of the α function at the point $x^{(2)} = 0$ (see Fig. 3.5b).

Step 2: Convergence checking. Upper and lower bounds of the objective function optimal value are computed as

$$z_{\text{up}}^{(2)} = -y^{(2)} - \frac{1}{4}x^{(2)} = -5 - \frac{1}{4} \times 0 = -5$$

and

$$z_{\text{down}}^{(1)} = -\frac{1}{4}x^{(2)} + \alpha^{(2)} = -\frac{1}{4} \times 0 - \frac{35}{2} = -\frac{35}{2}.$$

Since the difference $z_{\text{up}}^{(2)} - z_{\text{down}}^{(2)} = -5 - \frac{35}{2} = \frac{25}{2} > \varepsilon$, the procedure continues.

Step 3: Master problem solution. The iteration counter is updated, $\nu = 2 + 1 = 3$, and the master problem below is solved

$$\underset{x, \alpha}{\text{minimize}} \quad -\frac{1}{4}x + \alpha$$

subject to

$$\begin{aligned}
-\frac{19}{2} + \frac{1}{2}(x - 16) &\leq \alpha \\
-5 - (x - 0) &\leq \alpha \\
0 \leq x &\leq 16 \\
-25 &\leq \alpha,
\end{aligned}$$

124   3 Linear Programming: Complicating Variables

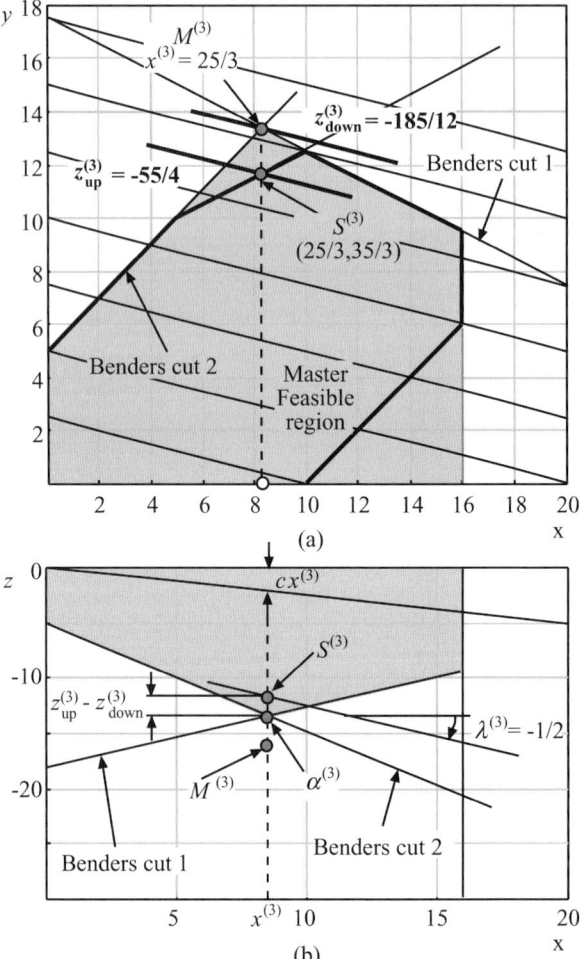

Fig. 3.6. Graphical illustration of the decomposition algorithm for iteration $\nu = 3$

where the first two restrictions are the Benders cuts 1 and 2 (see Figs. 3.6a and 3.6b) associated with the previous iterations 1 and 2.

The solution of this problem is $x^{(3)} = \frac{25}{3}$, point $M^{(3)}$ in Fig. 3.6a, and $\alpha^{(3)} = -\frac{40}{3}$ in Fig. 3.6b, resulting from the intersection of the Benders cuts 1 and 2 in both Figs. 3.6a and 3.6b. The objective function value is the z-value associated with the point $M^{(3)}$ (see Fig. 3.6b) and is equal to $-\frac{185}{12} = -15.416$.

Step 1: Subproblem solution. The subproblem below is solved

$$\underset{y}{\text{minimize}} \quad z = -y$$

subject to

$$\begin{aligned} y - x &\le 5 \\ y - \tfrac{1}{2}x &\le \tfrac{15}{2} \\ y + \tfrac{1}{2}x &\le \tfrac{35}{2} \\ -y + x &\le 10 \\ y &\ge 0 \\ x &= \tfrac{25}{3} : \lambda \, . \end{aligned}$$

The solution of the problem above is the point $S^{(3)} = \left(x^{(3)}, y^{(3)}\right) = \left(\tfrac{25}{3}, \tfrac{35}{3}\right)$ in Fig. 3.6a with an objective function value of $-\tfrac{35}{3}$ (see the z-value of point $S^{(3)}$ in Fig. 3.6b). The optimal dual variable associated with the constraint $x = \tfrac{25}{3}$, $\lambda^{(3)} = -0.5$, is the slope of the α function at the point $x^{(3)} = \tfrac{25}{3}$ (see Fig. 3.6b).

Step 2: Convergence checking. Upper and lower bounds of the objective function optimal value are computed as

$$z_{\text{up}}^{(3)} = -y^{(3)} - \tfrac{1}{4}x^{(3)} = -\tfrac{35}{3} - \tfrac{1}{4} \times \tfrac{25}{3} = -\tfrac{55}{4}$$

and

$$z_{\text{down}}^{(3)} = -\tfrac{1}{4}x^{(3)} + \alpha^{(3)} = -\tfrac{1}{4} \times \tfrac{25}{3} - \tfrac{40}{3} = -\tfrac{185}{12} \, .$$

Since the difference $z_{\text{up}}^{(3)} - z_{\text{down}}^{(3)} = \tfrac{5}{3} > \varepsilon$, the procedure continues.

Step 3: Master problem solution. The iteration counter is updated, $\nu = 3 + 1 = 4$, and the master problem below is solved

$$\underset{x,\alpha}{\text{minimize}} \quad -\tfrac{1}{4}x + \alpha$$

subject to

$$\begin{aligned} -\tfrac{19}{2} + \tfrac{1}{2}(x - 16) &\le \alpha \\ -5 - (x - 0) &\le \alpha \\ -\tfrac{35}{3} - \tfrac{1}{2}\left(x - \tfrac{25}{3}\right) &\le \alpha \\ 0 &\le x \le 16 \\ -25 &\le \alpha \, , \end{aligned}$$

where the first three restrictions are the Benders cuts 1, 2, and 3 (see Figs. 3.7a and 3.7b) associated with the previous iterations.

The solution of this problem is $x^{(4)} = 10$, point $M^{(4)}$ in Fig. 3.7a, and $\alpha^{(4)} = -\tfrac{25}{2}$ in Fig. 3.7b, resulting from the intersection of the Benders cuts 1 and 3 in both Figs. 3.7a and 3.7b. The objective function z-value associated with the point $M^{(4)}$ is $z = -15$.

126   3 Linear Programming: Complicating Variables

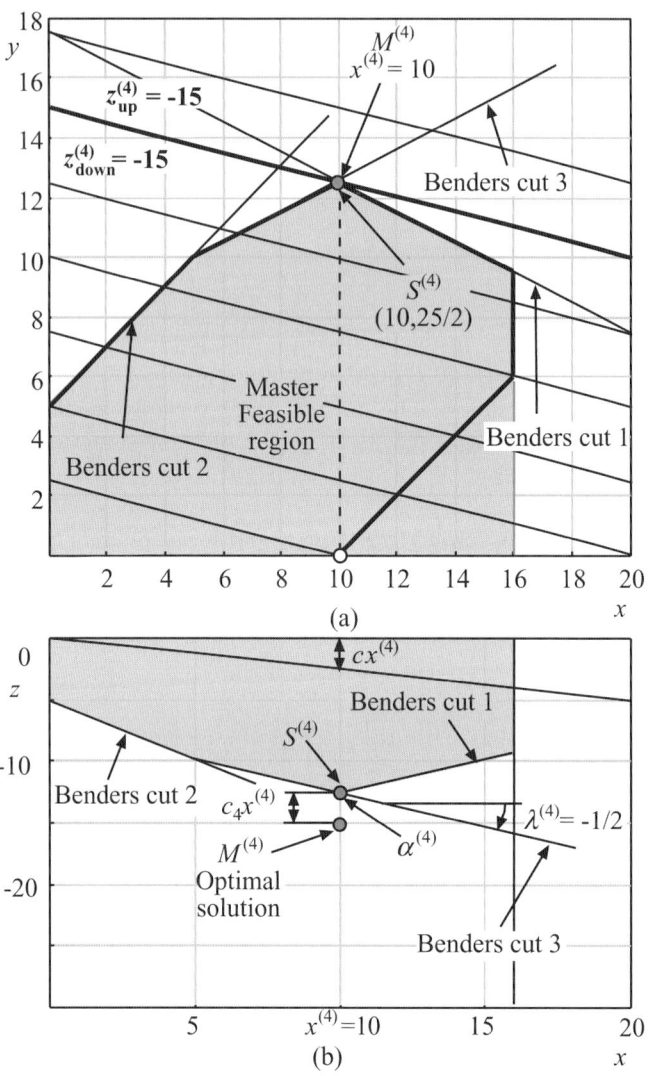

Fig. 3.7. Graphical illustration of the decomposition algorithm for iteration $\nu = 4$

Step 1: Subproblem solution. The subproblem below is solved

$$\underset{y}{\text{minimize}} \quad z = -y$$

3.3 Benders Decomposition

subject to

$$\begin{aligned}
y - x &\leq 5 \\
y - \tfrac{1}{2}x &\leq \tfrac{15}{2} \\
y + \tfrac{1}{2}x &\leq \tfrac{35}{2} \\
-y + x &\leq 10 \\
y &\geq 0 \\
x &= 10 \ : \ \lambda \ .
\end{aligned}$$

The solution of the problem above is the point $S^{(4)} = \left(x^{(4)}, y^{(4)}\right) = \left(10, \tfrac{25}{2}\right)$ in Fig. 3.7a with an objective function value of $-\tfrac{25}{2}$ (see the z-value of point $S^{(4)}$ in Fig. 3.7b). The optimal dual variable associated with the constraint $x = 10$, $\lambda^{(4)} = -\tfrac{1}{2}$, is the slope of the α function at the point $x^{(4)} = 10$ (see Fig. 3.7b).

Step 2: Convergence checking. Upper and lower bounds of the objective function optimal value are computed as

$$z_{\text{up}}^{(4)} = -y^{(4)} - \tfrac{1}{4} x^{(4)} = -12.5 - \tfrac{1}{4} \times 10 = -15$$

and

$$z_{\text{down}}^{(4)} = -\tfrac{1}{4} x^{(4)} + \alpha^{(4)} = -\tfrac{1}{4} \times 10 - 12.5 = -15 \ .$$

Because both bounds are equal, the difference $z_{\text{up}}^{(4)} - z_{\text{down}}^{(4)} = -15 + 15 = 0 < \varepsilon$, therefore, the optimal solution has been found

$$x^* = 10; \ y^* = \tfrac{25}{2}; \ z = -15 \ .$$

The values of the objective function bounds and master and subproblem variables are shown in Table 3.1. In Fig. 3.8 the evolution of the bounds $z_{\text{up}}^{(\nu)}$ and $z_{\text{down}}^{(\nu)}$ is shown.   □

Table 3.1. Evolution of the values of the master and subproblem variables as the decomposition algorithm, progresses

ν	$x^{(\nu)}$	$y^{(\nu)}$	$\alpha^{(\nu)}$	$\lambda^{(\nu)}$	$z_{\text{up}}^{(\nu)}$	$z_{\text{down}}^{(\nu)}$
1	16	$\tfrac{19}{2}$	-25	$\tfrac{1}{2}$	$\tfrac{-27}{2}$	-29
2	0	5	$\tfrac{-35}{2}$	-1	-5	$\tfrac{-35}{2}$
3	$\tfrac{25}{3}$	$\tfrac{35}{3}$	$\tfrac{-40}{3}$	$\tfrac{-1}{2}$	$\tfrac{-55}{4}$	$\tfrac{-185}{12}$
4	10	$\tfrac{25}{2}$	$\tfrac{-25}{2}$	$\tfrac{-1}{2}$	-15	-15

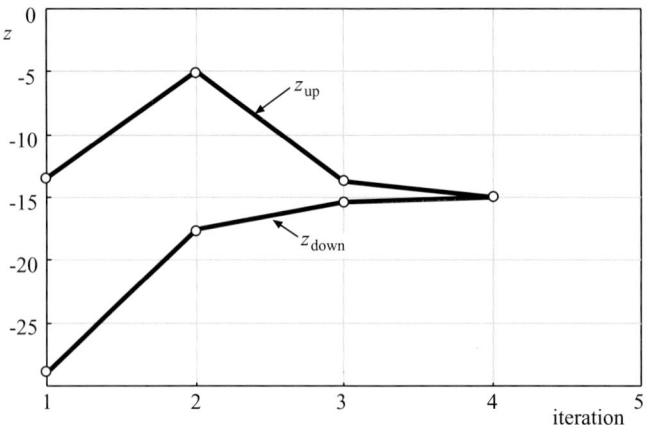

Fig. 3.8. Evolution of the upper and lower bounds of the objective function in Example 3.1

3.3.4 Subproblem Infeasibility

The possible infeasibility of the subproblem is considered in this subsection and an alternative always-feasible subproblem is formulated. The price paid is a larger number of variables and dissimilar cost coefficients in the objective function. This always-feasible problem has the following form:

$$\underset{y_1,\ldots,y_m;\,v_1,\ldots,v_q;\,w}{\text{minimize}} \quad \sum_{j=1}^{m} d_j\, y_j + M \sum_{\ell=1}^{q} (v_\ell + w) \qquad (3.43)$$

subject to

$$\sum_{i=1}^{n} a_{\ell i}\, x_i + \sum_{j=1}^{m} e_{\ell j}\, y_j + v_\ell - w = b^{(\ell)}; \quad \ell = 1,\ldots,q \qquad (3.44)$$

$$0 \le y_j \le y_j^{\text{up}}; \quad j = 1,\ldots,m \qquad (3.45)$$

$$x_i = x_i^{(k)} : \lambda_i; \quad i = 1,\ldots,n \qquad (3.46)$$

$$0 \le v_\ell \le v_\ell^{\text{up}}; \quad \ell = 1,\ldots,q \qquad (3.47)$$

$$0 \le w \le w^{\text{up}}, \qquad (3.48)$$

where M is a large enough positive constant, v_ℓ ($\ell = 1,\ldots,q$) and w are the artificial variables, and v_ℓ^{up} ($\ell = 1,\ldots,q$) and w^{up} are their respective upper bounds.

To illustrate how this method works we solve Example 3.2.

Additionally, this problem decomposes in three subproblems and therefore it illustrates the Benders decomposition algorithm using constraint (3.42).

3.3 Benders Decomposition

Computational Example 3.2 (The Benders decomposition algorithm). Consider the problem

$$\underset{x_1, x_2, y_1, y_2, y_3}{\text{minimize}} \quad z = -2y_1 - y_2 + y_3 + 3x_1 - 3x_2$$

subject to

$$\begin{aligned}
y_1 + x_1 + x_2 &\leq 3 \\
2y_2 + 3x_1 &\leq 12 \\
y_3 - 7x_2 &\leq -16 \\
-x_1 + x_2 &\leq 2 \\
x_1, \, x_2, y_1, \, y_2, y_3 &\geq 0 \, .
\end{aligned}$$

The optimal values of this problem are $x_1^* = 0.3$, $x_2^* = 2.3$, $y_1^* = 0.4$, $y_2^* = 5.6$, and $y_3^* = 0.0$ with an objective function value equal to $z^* = -12.43$.

If variables x_1 and x_2 are considered to be complicating variables, the above problem is solved using the Benders Decomposition Algorithm 3.1.

Step 0: Initialization. The iteration counter is initialized, $\nu = 1$. The initial master problem is solved

$$\underset{x_1, x_2, \alpha}{\text{minimize}} \quad 3x_1 - 3x_2 + \alpha$$

subject to

$$\begin{aligned}
-x_1 + x_2 &\leq 2 \\
\alpha &\geq -100 \, .
\end{aligned}$$

The solution of this problem is $x_1^{(1)} = 0.0$, $x_2^{(1)} = 2.0$, and $\alpha^{(1)} = -100$.

Step 1: Subproblem solution. The subproblems below are solved.
The first subproblem is

$$\underset{y_1}{\text{minimize}} \quad z_1 = -2y_1$$

subject to

$$\begin{aligned}
y_1 + x_1 + x_2 &\leq 3 \\
x_1 &= 0 \quad : \lambda_{1,1} \\
x_2 &= 2 \quad : \lambda_{2,1} \\
y_1 &\geq 0,
\end{aligned}$$

whose solution is $y_1^{(1)} = 1.0$, $\lambda_{1,1}^{(1)} = 2.0$, and $\lambda_{2,1}^{(1)} = 2.0$ with an objective function value $z_1^{(1)} = -2.0$.

The second subproblem is

$$\underset{y_2}{\text{minimize}} \quad z_2 = -y_2$$

subject to

$$\begin{array}{rcll} y_2 + 3x_1 & \leq & 12 & \\ x_1 & = & 0 & : \lambda_{1,2} \\ x_2 & = & 2 & : \lambda_{2,2} \\ y_2 & \geq & 0, & \end{array}$$

whose solution is $y_2^{(1)} = 6.0$, $\lambda_{1,2}^{(1)} = 1.5$, and $\lambda_{2,2}^{(1)} = 0.0$ with an objective function value $z_2^{(1)} = -6.0$.

The third subproblem is infeasible, so artificial variables v_3 and w are included in this subproblem:

$$\underset{y_3, v_3, w}{\text{minimize}} \quad z_3 = y_3 + 20(v_3 + w)$$

subject to

$$\begin{array}{rcll} y_3 - 7x_2 + v_3 - w & \leq & -16 & \\ x_1 & = & 0 & : \lambda_{1,3} \\ x_2 & = & 2 & : \lambda_{2,3} \\ y_3 & \geq & 0, & \end{array}$$

whose solution is $y_3^{(1)} = 0$, $v_3 = 0$, $w = 2$, $\lambda_{1,3}^{(1)} = 0$, and $\lambda_{2,3}^{(1)} = -140$ with an objective function value $z_3^{(1)} = 40$.

The complete solution of all subproblems is $y_1^{(1)} = 1$, $y_2^{(1)} = 6$, $y_3^{(1)} = 0$, $\lambda_1^{(1)} = \lambda_{1,1}^{(1)} + \lambda_{1,2}^{(1)} + \lambda_{1,3}^{(1)} = 3.5$, and $\lambda_2^{(1)} = \lambda_{2,1}^{(1)} + \lambda_{2,2}^{(1)} + \lambda_{2,3}^{(1)} = -138$ with an objective function value $z^{(1)} = z_1^{(1)} + z_2^{(1)} + z_3^{(1)} = 32$.

Step 2: Convergence checking. An upper bound of the objective function optimal value is computed as

$$z_{\text{up}}^{(1)} = -2y_1^{(1)} - y_2^{(1)} + y_3^{(1)} + 3x_1^{(1)} - 3x_2^{(1)} = 26 \ .$$

A lower bound of the objective function optimal value is computed as

$$z_{\text{down}}^{(1)} = 3x_1^{(1)} - 3x_2^{(1)} + \alpha^{(1)} = -106 \ .$$

The difference $z_{\text{up}}^{(1)} - z_{\text{down}}^{(1)} = 132$ is not equal to 0, therefore, the procedure continues.

Step 3: Master problem solution. The iteration counter is updated, $\nu = 1 + 1 = 2$. The master problem below is solved

3.3 Benders Decomposition

$$\text{minimize} \quad 3x_1 - 3x_2 + \alpha$$
$$x_1, x_2, \alpha$$

subject to
$$32 + 3.5 \times x_1 - 138 \times (x_2 - 2) \leq \alpha$$
$$-x_1 + x_2 \leq 2$$
$$\alpha \geq -100 .$$

The solution of this problem is $x_1^{(2)} = 1$, $x_2^{(2)} = 3$, and $\alpha^{(2)} = -100$. The objective function value is $z^{(2)} = -106$.

The algorithm continues in Step 1.

Step 1: Subproblem solution. The subproblems below are solved. The first subproblem is infeasible, so artificial variables v_1 and w are included in the subproblem:

$$\text{minimize} \quad z_1 = -2y_1 + 20(v_1 + w)$$
$$y_1, v_1, w$$

subject to

$$y_1 + x_1 + x_2 + v_1 - w \leq 3$$
$$x_1 = 1 \; : \; \lambda_{1,1}$$
$$x_2 = 3 \; : \; \lambda_{2,1}$$
$$y_1 \geq 0,$$

whose solution is $y_1^{(2)} = 0$, $v_1 = 0$, $w = 1$, $\lambda_{1,1}^{(2)} = 20$, and $\lambda_{2,1}^{(2)} = 20$ with an objective function value $z_1^{(2)} = -19.3$.

The second subproblem is

$$\text{minimize} \quad z_2 = -y_2$$
$$y_2$$

subject to

$$y_2 + 3x_1 \leq 12$$
$$x_1 = 1 \; : \; \lambda_{1,2}$$
$$x_2 = 3 \; : \; \lambda_{2,2}$$
$$y_2 \geq 0,$$

whose solution is $y_2^{(2)} = 4.5$, $\lambda_{1,2}^{(2)} = 1.5$, and $\lambda_{2,2}^{(2)} = 0$ with an objective function value $z_2^{(2)} = -4.5$.

The third subproblem is

$$\text{minimize} \quad z_3 = y_3$$
$$y_3$$

subject to

$$\begin{aligned} y_3 \quad -7x_2 &\leq -16 \\ x_1 &= 1 \quad : \quad \lambda_{1,3} \\ x_2 &= 3 \quad : \quad \lambda_{2,3} \\ y_3 &\geq 0, \end{aligned}$$

whose solution is $y_3^{(2)} = 0$, $\lambda_{1,3}^{(2)} = 0$, and $\lambda_{2,3}^{(2)} = 0$ with an objective function value $z_3^{(2)} = 0$.

The complete solution of all subproblems is $y_1^{(2)} = 0$, $y_2^{(2)} = 4.5$, $y_3^{(2)} = 0$, $\lambda_1^{(2)} = 21.5$, and $\lambda_2^{(2)} = 20$ with an objective function value $z^{(2)} = 14.8$.

Step 2: Convergence checking. An upper bound of the objective function optimal value is computed as

$$z_{\text{up}}^{(2)} = -2y_1^{(2)} - y_2^{(2)} + y_3^{(2)} + 3x_1^{(2)} - 3x_2^{(2)} = 8.7 \ .$$

A lower bound of the objective function optimal value is computed as

$$z_{\text{down}}^{(2)} = 3x_1^{(2)} - 3x_2^{(2)} + \alpha^{(2)} = -106 \ .$$

The difference $z_{\text{up}}^{(2)} - z_{\text{down}}^{(2)} = 114.7$ is not equal to 0, therefore, the procedure continues.

Step 3: Master problem solution. The iteration counter is updated, $\nu = 2 + 1 = 3$.

The master problem below is solved:

$$\begin{aligned} \underset{x_1, x_2, \alpha}{\text{minimize}} \quad & 3x_1 - 3x_2 + \alpha \end{aligned}$$

subject to

$$\begin{aligned} 32 + 3.5 \times x_1 - 138 \times (x_2 - 2) &\leq \alpha \\ 14.8 + 21.5 \times (x_1 - 1) + 20 \times (x_2 - 3) &\leq \alpha \\ -x_1 + x_2 &\leq 2 \\ \alpha &\geq -100 \ . \end{aligned}$$

The solution of this problem is $x_1^{(3)} = \frac{1}{3}$, $x_2^{(3)} = \frac{7}{3}$, and $\alpha^{(3)} = -12.3$. The objective function value is $z^{(3)} = -18.3$.

The algorithm continues in Step 1.

Step 1: Subproblem solution. The subproblems below are solved. The first subproblem is

$$\underset{y_1}{\text{minimize}} \quad z_1 = -2y_1$$

3.3 Benders Decomposition

subject to

$$y_1 + x_1 + x_2 \leq 3$$
$$x_1 = \tfrac{1}{3} \quad : \quad \lambda_{1,1}$$
$$x_2 = \tfrac{7}{3} \quad : \quad \lambda_{2,1}$$
$$y_1 \geq 0,$$

whose solution is $y_1^{(3)} = 0.3$, $\lambda_{1,1}^{(3)} = 2$, and $\lambda_{2,1}^{(3)} = 2$ with an objective function value $z_1^{(3)} = -0.7$.

The second subproblem is

$$\text{minimize} \quad z_2 = -y_2$$
$$y_2$$

subject to

$$y_2 + 3x_1 \leq 12$$
$$x_1 = \tfrac{1}{3} \quad : \quad \lambda_{1,2}$$
$$x_2 = \tfrac{7}{3} \quad : \quad \lambda_{2,2}$$
$$y_2 \geq 0,$$

whose solution is $y_2^{(3)} = 5.5$, $\lambda_{1,2}^{(3)} = 1.5$, and $\lambda_{2,2}^{(3)} = 0$ with an objective function value $z_2^{(3)} = -5.5$.

The third subproblem is

$$\text{minimize} \quad z_3 = y_3$$
$$y_3$$

subject to

$$y_3 - 7x_2 \leq -16$$
$$x_1 = \tfrac{1}{3} \quad : \quad \lambda_{1,3}$$
$$x_2 = \tfrac{7}{3} \quad : \quad \lambda_{2,3}$$
$$y_3 \geq 0,$$

whose solution is $y_3^{(3)} = 0$, $\lambda_{1,3}^{(3)} = 0$, and $\lambda_{2,3}^{(3)} = 0$ with an objective function value $z_3^{(3)} = 0$.

The complete solution of all subproblems is $y_1^{(3)} = 0.3$, $y_2^{(3)} = 5.5$, $y_3^{(3)} = 0$, $\lambda_1^{(3)} = 3.5$, and $\lambda_2^{(3)} = 2$ with an objective function value $z^{(3)} = -6.2$.

Step 2: Convergence checking. An upper bound of the objective function optimal value is computed as

$$z_{\text{up}}^{(3)} = -2y_1^{(3)} - y_2^{(3)} + y_3^{(3)} + 3x_1^{(3)} - 3x_2^{(3)} = -12.2 \; .$$

A lower bound of the objective function optimal value is computed as

$$z_{\text{down}}^{(3)} = 3x_1^{(3)} - 3x_2^{(3)} + \alpha^{(3)} = -18.3 \ .$$

The difference $z_{\text{up}}^{(3)} - z_{\text{down}}^{(3)} = 6.1$ is not equal to 0, therefore, the procedure continues.

Step 3: Master problem solution. The iteration counter is updated, $\nu = 3 + 1 = 4$.

The master problem below is solved

$$\underset{x_1, x_2, \alpha}{\text{minimize}} \quad 3x_1 - 3x_2 + \alpha$$

subject to

$$\begin{aligned} 32 + 3.5 \times x_1 - 138 \times (x_2 - 2) &\le \alpha \\ 14.8 + 21.5 \times (x_1 - 1) + 20 \times (x_2 - 3) &\le \alpha \\ -6.2 + 3.5 \times (x_1 - \tfrac{1}{3}) + 2 \times (x_2 - \tfrac{7}{3}) &\le \alpha \\ -x_1 + x_2 &\le 2 \\ \alpha &\ge -100 \ . \end{aligned}$$

The solution of this problem is $x_1^{(4)} = 0.28$, $x_2^{(4)} = 2.28$, and $\alpha^{(4)} = -6.4$. The objective function value is $z^{(4)} = -12.4$.

The algorithm continues in Step 1.

Step 1: Subproblem solution. The subproblems below are solved.
The first subproblem is

$$\underset{y_1}{\text{minimize}} \quad z_1 = -2y_1$$

subject to

$$\begin{aligned} y_1 + x_1 + x_2 &\le 3 \\ x_1 &= 0.28 \ : \ \lambda_{1,1} \\ x_2 &= 2.28 \ : \ \lambda_{2,1} \\ y_1 &\ge 0, \end{aligned}$$

whose solution is $y_1^{(4)} = 0.4$, $\lambda_{1,1}^{(4)} = 2$, and $\lambda_{2,1}^{(4)} = 2$ with an objective function value $z_1^{(4)} = -0.8$.

The second subproblem is

$$\underset{y_2}{\text{minimize}} \quad z_2 = -y_2$$

subject to
$$y_2 + 3x_1 \leq 12$$
$$x_1 = 0.28 : \lambda_{1,2}$$
$$x_2 = 2.28 : \lambda_{2,2}$$
$$y_2 \geq 0,$$

whose solution is $y_2^{(4)} = 5.6$, $\lambda_{1,2}^{(4)} = 1.5$, and $\lambda_{2,2}^{(4)} = 0$ with an objective function value $z_2^{(4)} = -5.6$.

The third subproblem is

$$\underset{y_3}{\text{minimize}} \quad z_3 = y_3$$

subject to
$$y_3 - 7x_2 \leq -16$$
$$x_1 = 0.28 : \lambda_{1,3}$$
$$x_2 = 2.28 : \lambda_{2,3}$$
$$y_3 \geq 0,$$

whose solution is $y_3^{(4)} = 0$, $\lambda_{1,3}^{(4)} = 0$, and $\lambda_{2,3}^{(4)} = 0$ with an objective function value $z_3^{(4)} = 0$.

The complete solution of all subproblems is $y_1^{(4)} = 0.4$, $y_2^{(4)} = 5.6$, $y_3^{(4)} = 0$, $\lambda_1^{(4)} = 3.5$, and $\lambda_2^{(4)} = 2$ with an objective function value $z^{(4)} = -6.5$.

Step 2: Convergence checking. An upper bound of the objective function optimal value is computed as

$$z_{up}^{(4)} = -2y_1^{(4)} - y_2^{(4)} + y_3^{(4)} + 3x_1^{(4)} - 3x_2^{(4)} = -12.4 .$$

A lower bound of the objective function optimal value is computed as

$$z_{down}^{(4)} = 3x_1^{(4)} - 3x_2^{(4)} + \alpha^{(4)} = -12.4 .$$

The difference $z_{up}^{(4)} - z_{down}^{(4)}$ is equal to 0, therefore, the optimal solution has been found.

It is $x_1^* = 0.3$, $x_2^* = 2.3$, $y_1^* = 0.4$, $y_2^* = 5.6$, and $y_3^* = 0$, and the optimal objective function value is -12.4. □

3.4 Concluding Remarks

This chapter analyzes linear programming problems with complicating variables. If these variables are fixed to given values, the original problem either decomposes by blocks or attains such a structure that its solution is straightforward. This situation often occurs in practical engineering and science problems.

The decomposition explained in this chapter was originally proposed by Benders [24] and it is named Benders' decomposition on his behalf. This decomposition allows us the distributed solution of linear programming problems including complicating variables at the cost of repetition. Benders decomposition was later extended to nonlinear problems and made known to a broader technical community by Geoffrion [25]. Relevant insights and details are presented in the manual by Lasdon [26] and in the text by Floudas [27].

Benders Decomposition is precisely derived and then illustrated algebraic and geometrically through the detailed solution of diverse examples throughout the chapter.

Finally, it should be noted that the linear programming problem considered may have both complicating variables and constraints. In such situation, a nested decomposition can be used. For instance, an outer Benders Decomposition algorithm may deal with complicating variables while an inner Dantzig–Wolfe procedure may deal with complicating constraints.

3.5 Exercises

Exercise 3.1. The stochastic linear programming problem addressed in Sect. 1.4.1, p. 28, consists in procuring coal and gas for a 2-year time horizon, so that heating needs are properly covered at minimum expected cost. This problem has a complicating variable structure and therefore it can be solved using Benders decomposition.

Solve the numerical example given in Sect. 1.4.1 through the Benders decomposition algorithm and verify that the solution coincides with that provided in that section. Analyze the convergence behavior of the Benders algorithm for this particular problem.

Exercise 3.2. Consider the following problem

$$\underset{x_1, x_2, x_3, x_4, x_5}{\text{minimize}} \quad 2x_1 + 2.5x_2 + 0.5x_3 + 4x_4 + 3x_5$$

subject to

$$
\begin{aligned}
-2x_1 + 3x_2 -4x_5 &\leq -4 \\
2x_1 + 4x_2 + x_5 &\leq 2.5 \\
2x_3 - x_4 - x_5 &\leq 0.5 \\
-0.5x_3 - x_4 + 3x_5 &\leq -3 \\
x_1, x_2, x_3, x_4, x_5 &\geq 0 \, .
\end{aligned}
$$

1. Check that the following vector $(x_1, x_2, x_3, x_4, x_5)$ is a solution:

$$x_1 = 1, \ x_2 = 0, \ x_3 = 2.2, \ x_4 = 3.4, \ x_5 = 0.5, \ z = 18.2 \ .$$

2. Using the Benders decomposition algorithm obtain the final solution.

Exercise 3.3. Peter builds two types of transformers and he has available 6 units of ferromagnetic material and 28 h of time. Type I transformer manufacturing requires 2 unit of ferromagnetic material and 7 h of work, whereas type II transformer manufacturing requires 1 unit of ferromagnetic material and 8 h of work. Selling prices for transformers I and II are $120 and $80, respectively. How many transformers of each type should Peter build to maximize his profits?

Considering the number of type I transformers a complicating variable, solve the problem using Benders decomposition.

Exercise 3.4. Given the problem

$$\begin{array}{cc} \text{maximize} & y_1 + 3y_2 + y_3 + 4x_1 \\ y_1, y_2, y_3, x_1 & \end{array}$$

subject to

$$\begin{aligned}
-y_1 + x_1 &\leq 1 \\
2y_2 + 2x_1 &\leq 4 \\
x_1 &\leq 4 \\
2x_1 &\leq 6 \\
-x_1 &\leq -1 \\
2y_1 + y_2 + 2y_3 + 2x_1 &\leq 9 \\
y_1, y_2, y_3, x_1 &\geq 0 \ .
\end{aligned}$$

Find the optimal solution considering x_1 as a complicating variable.

Exercise 3.5. Consider a 2-year coal, gas, and oil procurement problem to supply the energy demand of a factory. The demand for energy the first year is 900 and the prices of coal, gas, and oil are 4.5, 4.9, and 5.1, respectively. Demand for energy the second year can be 900 with probability 0.3, 750 with probability 0.4, and 55 with probability 0.3. The corresponding prices for coal are 7.5, 6, and 3; for gas 9, 5, and 4; and for oil 8.5, 5.5, and 3.5. Formulate a linear problem whose solution determines the optimal purchase policy for the first and the second year. Show that the problem has a complicating variable structure and solve it using Benders decomposition.

Exercise 3.6. The multiperiod investment problem formulated and studied in Sect. 1.4.2, p. 32, is linear and exhibits complicating variable structure. Therefore, it can be solved using Bender decomposition.

138   3 Linear Programming: Complicating Variables

Solve the numerical case study presented in Sect. 1.4.2 using Benders decomposition. Check the results with those provided in that section. Analyze the numerical behavior of Benders decomposition in this particular example.

Exercise 3.7. The city water supply problem stated in Sect. 1.4.3, p. 36, is linear and includes complicating variables.

Solve this problem using Benders decomposition and verify that the results obtained are the same that those provided in Sect. 1.4.3.

Exercise 3.8. Consider two production devices serving a 3-period demand of 100, 140, and 200 units. The operating range of devices 1 and 2 are within 10 and 150, and 50 and 180 units, respectively. The production of each device cannot change above 60 units from one period to the next one. Production costs of devices 1 and 2 are, respectively, $10 and $12 per unit, and start-up costs are $4 and $2 per unit, respectively. The devices are not working before the considered time horizon.

1. Formulate the optimal scheduling problem that allows determining the start-up and shut-down sequence of the production devices that minimize production cost while serving the demand.
2. Considering that the binary variables are complicating ones and use Benders decomposition to solve the problem.

Exercise 3.9. Consider the capacity expansion planning of two production facilities to supply the demand of two cities during a 2-year time horizon. The interconnection of production and demand locations are specified in Fig. 3.9. Demands at location 1 for years 1 and 2 are 8 and 6, respectively, and for location 2, 11, and 9 respectively. The maximum capacity to be built at locations

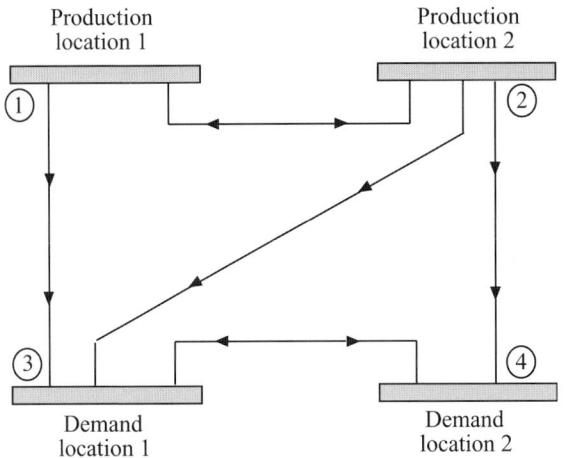

Fig. 3.9. Transportation network for Exercise 3.9

Table 3.2. Building cost for the problem of Exercise 3.9

Period t	Building Cost ($)	
	Location 1	Location 2
1	2.0	3.5
2	2.5	3.0

Table 3.3. Transportation cost and capacities for Exercise 3.9

Road (i–j)	Capacity	Cost ($)
1–2	11	0.5
1–3	9	0.6
2–3	5	0.7
2–4	5	0.8
3–4	4	0.4

1 and 2 are 10 and 12, respectively. Building cost are provided in Table 3.2. Transportation costs and capacities are provided in Table 3.3.

Formulate a problem to determine the production capacity to be built at each location each year. Show that this problem has a complicating variable decomposable structure. Solve it using Benders decomposition.

4
Duality

4.1 Introduction

This chapter deals with duality in mathematical programming. Given a mathematical programming problem, called the primal problem, there exists another associated mathematical programming problem, called the dual problem, closely related with it. Since duality is symmetric, i.e., the dual of the dual problem is the primal, both problems are said to be dual to each other. In the following, we provide simple examples that lead to a better understanding of this important concept.

Under certain assumptions, the primal and dual problems have the same optimal objective function value; hence it is possible to solve the primal problem indirectly by solving its corresponding dual problem. This can lead to important computational advantages. In some other cases, the primal and dual problems have different optimal objective function values. The difference between the optimal values of the primal and dual objective functions is denominated duality gap.

As it will be shown, the values of the dual variables give the sensitivities of the objective function with respect to changes to discuss in the constraints.

Furthermore, dual variables constitute the information interchanged between master problems and subproblems in most decomposition schemes; therefore, duality is important to discuss in this book.

In Sect. 4.2 the important Karush–Kuhn–Tucker (KKT) optimality conditions are described. Since the mathematical statement of dual problems have substantial differences in linear and nonlinear programming, we deal with them in Sects. 4.3 and 4.4, respectively. However, their equivalence is shown. Finally, Sect. 4.5 illustrates duality and separability.

4.2 Karush–Kuhn–Tucker First- and Second-Order Optimality Conditions

The general problem of mathematical programming, also referred to as the nonlinear programming problem (NLPP), can be stated as follows:

$$\underset{x_1,\ldots,x_n}{\text{minimize}} \quad z = f(x_1,\ldots,x_n) \tag{4.1}$$

subject to

$$\begin{aligned} h_1(x_1,\ldots,x_n) &= 0 \\ &\vdots \\ h_\ell(x_1,\ldots,x_n) &= 0 \\ g_1(x_1,\ldots,x_n) &\leq 0 \\ &\vdots \\ g_m(x_1,\ldots,x_n) &\leq 0. \end{aligned}$$

In compact form the previous model can be stated as follows:

$$\underset{\boldsymbol{x}}{\text{minimize}} \quad z = f(\boldsymbol{x}) \tag{4.2}$$

subject to

$$\boldsymbol{h}(\boldsymbol{x}) = \boldsymbol{0} \tag{4.3}$$
$$\boldsymbol{g}(\boldsymbol{x}) \leq \boldsymbol{0}, \tag{4.4}$$

where $\boldsymbol{x} = (x_1,\ldots,x_n)^T$ is the vector of the decision variables, $f : \mathbb{R}^n \to \mathbb{R}$ is the objective function, and $\boldsymbol{h} : \mathbb{R}^n \to \mathbb{R}^\ell$ and $\boldsymbol{r} : \mathbb{R}^n \to \mathbb{R}^m$. Note that $\boldsymbol{h}(\boldsymbol{x}) = (h_1(\boldsymbol{x}),\ldots,h_\ell(\boldsymbol{x}))^T$ and $\boldsymbol{r}(\boldsymbol{x}) = (g_1(\boldsymbol{x}),\ldots,g_m(\boldsymbol{x}))^T$ are the equality and the inequality constraints, respectively. For this problem to be nonlinear, at least one of the functions involved in its formulation must be nonlinear. Any vector $\boldsymbol{x} \in \mathbb{R}^n$ that satisfies the constraints is said to be a feasible solution, and the set of all feasible solutions is referred to as the feasible region.

One of the most important theoretical results in the field of nonlinear programming are the conditions of Karush, Kuhn, and Tucker. They must be satisfied at any constrained optimum, local, or global, of any linear and most nonlinear programming problems. Moreover, they form the basis for the development of many computational algorithms. In addition, the criteria for stopping many algorithms, specifically, for recognizing if a local constrained optimum has been achieved, are derived directly from them.

In unconstrained differentiable problems the gradient is equal to zero at the minimum. In constrained differentiable problems the gradient is not necessarily equal to zero. This is due to the constraints of the problem. Karush–Kuhn–Tucker conditions (KKTC) generalize the necessary conditions for unconstrained problems to constrained problems.

4.2 Karush–Kuhn–Tucker First- and Second-Order Optimality Conditions

Definition 4.1 (Karush–Kuhn–Tucker first-order conditions). *The vector $x \in \mathbb{R}^n$ satisfies the KKTCs for the NLPP (4.2)–(4.4) if there exists vectors $\boldsymbol{\mu} \in \mathbb{R}^m$ and $\boldsymbol{\lambda} \in \mathbb{R}^\ell$ such that*

$$\nabla f(\boldsymbol{x}) + \sum_{k=1}^{\ell} \lambda_k \nabla h_k(\boldsymbol{x}) + \sum_{j=1}^{m} \mu_j \nabla g_j(\boldsymbol{x}) = \boldsymbol{0} \tag{4.5}$$

$$h_k(\boldsymbol{x}) = 0, \quad k = 1, \ldots, \ell \tag{4.6}$$

$$g_j(\boldsymbol{x}) \leq 0, \quad j = 1, \ldots, m \tag{4.7}$$

$$\mu_j \, g_j(\boldsymbol{x}) = 0, \quad j = 1, \ldots, m \tag{4.8}$$

$$\mu_j \geq 0, \quad j = 1, \ldots, m. \tag{4.9}$$

□

The vectors $\boldsymbol{\mu}$ and $\boldsymbol{\lambda}$ are called the Kuhn–Tucker multipliers. Condition (4.8) is known as the complementary slackness condition and requires the non-negativity of the multipliers of the inequality constraints (4.9) and is referred to as the dual feasibility condition, and constraints (4.6)–(4.7) are called the primal feasibility conditions.

The genesis of these first-order optimality conditions (KKTC) can be motivated in the case of two independent variables for the case of (i) one equality or inequality constraint, (ii) two inequality constraints, and (iii) one equality and one inequality constraint, as shown in Figs. 4.1, 4.2, and 4.3, respectively.

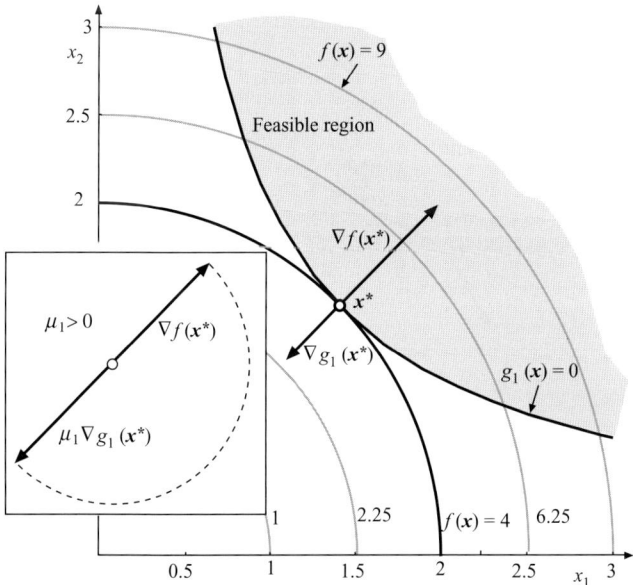

Fig. 4.1. Illustration of the Karush–Kuhn–Tucker conditions (KKTCs) for the case of one inequality constraint in the bidimensional case

144   4 Duality

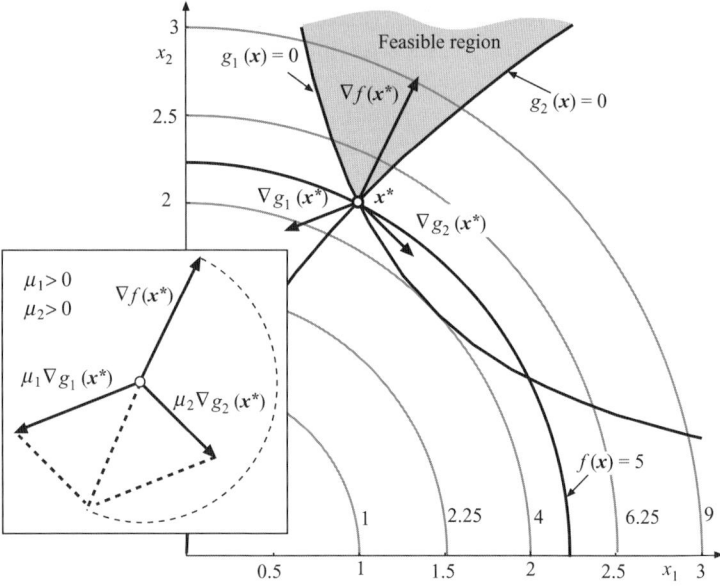

Fig. 4.2. Illustration of the KKTCs for the case of two inequality constraints in the bidimensional case

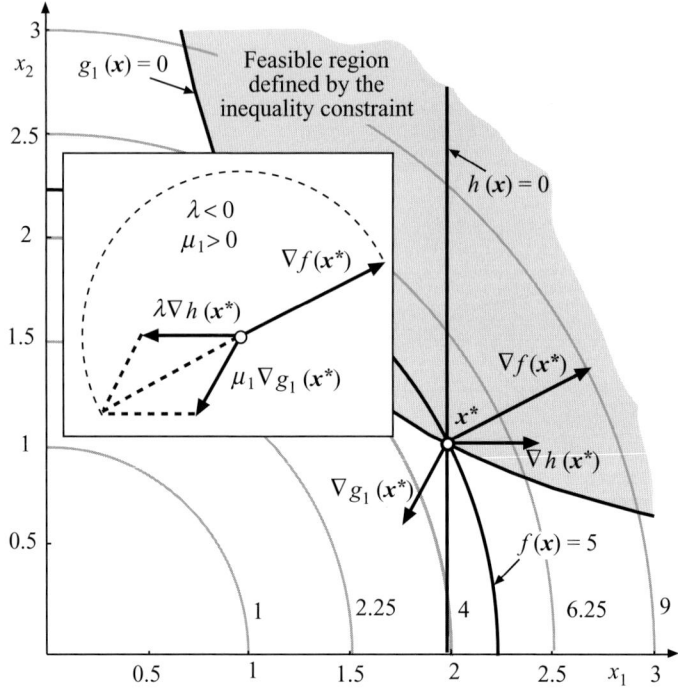

Fig. 4.3. Illustration of the KKTCs for the case of one inequality constraint and one equality constraint in the bidimensional case

4.2 Karush–Kuhn–Tucker First- and Second-Order Optimality Conditions

Consider the case of an inequality constraint (see Fig. 4.1), which separates the plane $\mathbb{R}^2$ into two regions. In one of them the constraint is satisfied, and in the other one it is not. Feasible points are those in the feasible region including its border curve. If the minimum of the objective function is attained at the interior of the feasible region, the constraint is not binding and the associated multiplier vanishes. If, on the contrary, the minimum is attained at the boundary, the constraint is binding. The problem is then equivalent to one with the corresponding equality constraint, and at the minimum, the gradients of the objective function and the constraint must be parallel (see Fig. 4.1). They must be pointing in opposite directions because the objective function increases as we move toward the interior of the feasible region, while the constraint function becomes negative and thus it decreases. The multiplier is then positive. This is what optimality conditions say in this situation.

If we add a new inequality constraint (see Fig. 4.2), (4.5) requires that, if we multiply the gradient of each binding constraint $[g_1(\boldsymbol{x}^*)$ and $g_2(\boldsymbol{x}^*)]$ by its corresponding Lagrange multiplier, the vector sum of these two vectors must equal the negative of the gradient of the objective function, as shown in Fig. 4.2.

Next, consider the case of one equality constraint and one inequality constraint (see Fig. 4.3). Satisfying the constraints is equivalent to moving along the curve that represents the equality constraint inside the feasible region defined by the inequality constraint. While moving along this curve, the objective function contour curves must intersect in such a way that the objective function value decreases. When the inequality constraint is strictly satisfied, the gradients of the objective function and the constraints are linearly dependent. This is what the first-order optimality conditions state (see Fig. 4.3).

Remark 4.1 (Special cases). If a type of constraint is not present in an NLPP, then the multiplier associated with the "absent" constraint is equal to zero, and the constraint is dropped from the formulation of the KKTCs. The form of the KKTCs for these cases are:

1. Unconstrained problems. In this case we have only the condition

$$\nabla f(\boldsymbol{x}) = \boldsymbol{0} \, .$$

2. Problems with only equality constraints. The KKTCs are an extension of the classical principle of the multiplier method. This method appears when NLPP have only equality constraints, and the KKTCs has the form

$$\nabla f(\boldsymbol{x}) + \sum_{k=1}^{\ell} \lambda_k \nabla h_k(\boldsymbol{x}) = \boldsymbol{0} \qquad (4.10)$$

$$h_k(\boldsymbol{x}) = 0; \qquad k = 1, \ldots, \ell \, .$$

3. Problems with only inequality constraints. KKT conditions read

$$\nabla f(\boldsymbol{x}) + \sum_{j=1}^{m} \mu_j \nabla g_j(\boldsymbol{x}) = \boldsymbol{0}$$

$$\begin{aligned} g_j(\boldsymbol{x}) &\leq 0; & j &= 1,\ldots,m \\ \mu_j\, g_j(\boldsymbol{x}) &= 0; & j &= 1,\ldots,m \\ \mu_j &\geq 0; & j &= 1,\ldots,m\ . \end{aligned} \quad (4.11)$$

If we define the Lagrangian function by

$$\mathcal{L}(\boldsymbol{x},\boldsymbol{\mu},\boldsymbol{\lambda}) = f(\boldsymbol{x}) + \boldsymbol{\lambda}^T \boldsymbol{h}(\boldsymbol{x}) + \boldsymbol{\mu}^T \boldsymbol{g}(\boldsymbol{x})\ ,$$

we can write the KKTCs in compact form as

$$\begin{aligned} \nabla_{\boldsymbol{x}} \mathcal{L}(\boldsymbol{x},\boldsymbol{\mu},\boldsymbol{\lambda}) &= \boldsymbol{0} \\ \nabla_{\boldsymbol{\lambda}} \mathcal{L}(\boldsymbol{x},\boldsymbol{\mu},\boldsymbol{\lambda}) &= \boldsymbol{0} \\ \nabla_{\boldsymbol{\mu}} \mathcal{L}(\boldsymbol{x},\boldsymbol{\mu},\boldsymbol{\lambda}) &\leq \boldsymbol{0} \\ \boldsymbol{\mu}^T \nabla_{\boldsymbol{\mu}} \mathcal{L}(\boldsymbol{x},\boldsymbol{\mu},\boldsymbol{\lambda}) &= 0 \\ \boldsymbol{\mu} &\geq \boldsymbol{0}\ . \end{aligned}$$

Note that $\boldsymbol{\mu}^T \nabla_{\boldsymbol{\mu}} \mathcal{L}(\boldsymbol{x},\boldsymbol{\mu},\boldsymbol{\lambda}) = 0$ is equivalent to $\mu_j\, g_j(\boldsymbol{x}) = 0$ ($j = 1,\ldots,m$) only because $\nabla_{\boldsymbol{\mu}} \mathcal{L}(\boldsymbol{x},\boldsymbol{\mu},\boldsymbol{\lambda}) \leq 0$ and $\boldsymbol{\mu} \geq \boldsymbol{0}$.

Illustrative Example 4.1 (The KKT conditions). Consider the following optimization problem:

$$\underset{x_1,x_2}{\text{minimize}} \quad -x_1^2 - x_2^2 \quad (4.12)$$

subject to

$$\begin{aligned} x_1 + x_2 &\leq 1 \\ x_1 &\geq 0 \\ x_2 &\geq 0\ . \end{aligned} \quad (4.13)$$

The Lagrangian function is

$$\mathcal{L}(x_1, x_2, \mu_1, \mu_2, \mu_3) = -x_1^2 - x_2^2 + \mu_1(x_1 + x_2 - 1) + \mu_2(-x_1) + \mu_3(-x_2)$$

and the KKT conditions become

$$\frac{\partial \mathcal{L}(x_1, x_2, \mu_1, \mu_2, \mu_3)}{\partial x_1} = -2x_1 + \mu_1 - \mu_2 = 0 \Rightarrow x_1 = \frac{\mu_1 - \mu_2}{2} \quad (4.14)$$

$$\frac{\partial \mathcal{L}(x_1, x_2, \mu_1, \mu_2, \mu_3)}{\partial x_2} = -2x_2 + \mu_1 - \mu_3 = 0 \Rightarrow x_2 = \frac{\mu_1 - \mu_3}{2} \quad (4.15)$$

$$x_1 + x_2 \leq 1 \quad (4.16)$$
$$-x_1 \leq 0 \quad (4.17)$$
$$-x_2 \leq 0 \quad (4.18)$$

4.2 Karush–Kuhn–Tucker First- and Second-Order Optimality Conditions

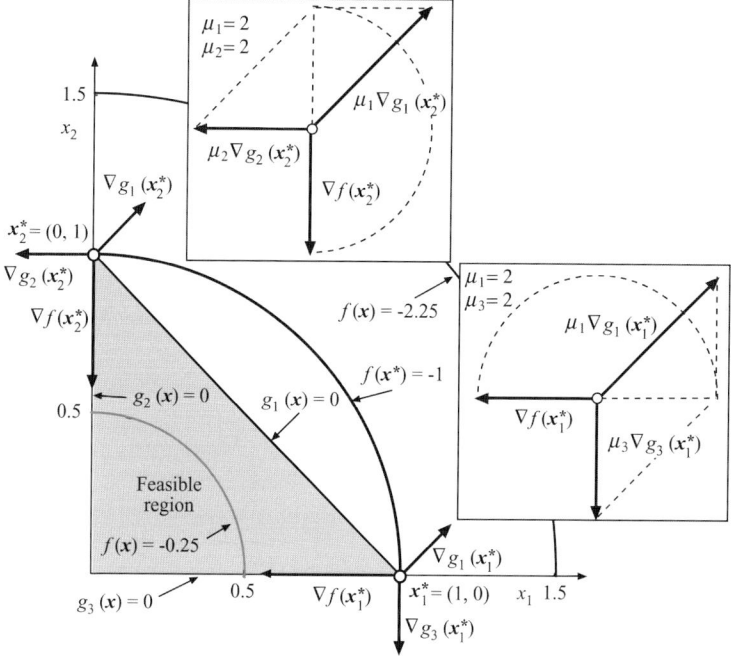

Fig. 4.4. Illustration of the minimization problem in Example 4.1

$$\mu_1(x_1 + x_2 - 1) = 0 \tag{4.19}$$
$$\mu_2(-x_1) = 0 \tag{4.20}$$
$$\mu_3(-x_2) = 0 \tag{4.21}$$
$$\mu_1, \mu_2, \mu_3 \geq 0, \tag{4.22}$$

which has following two solutions:

$$x_1 = 1, \ x_2 = 0, \ \mu_1 = 2, \ \mu_2 = 0, \ \mu_3 = 2, \ z = -1$$

and

$$x_1 = 0, \ x_2 = 1, \ \mu_1 = 2, \ \mu_2 = 2, \ \mu_3 = 0, \ z = -1 .$$

These solutions are illustrated in Fig. 4.4, where the two optimal points are shown together with the gradients of the objective function and the active constraints as the KKT condition state. Note that they are linearly dependent. □

4.2.1 Equality Constraints and Newton Algorithm

The KKTCs for the problem

$$\underset{x}{\text{minimize}} \quad z = f(x) \tag{4.23}$$

subject to
$$h(x) = 0 \tag{4.24}$$

constitute the system of nonlinear equations
$$\begin{aligned}\nabla_x \mathcal{L}(x, \lambda) &= 0 \\ h(x) &= 0, \end{aligned} \tag{4.25}$$

where $\mathcal{L}(x, \lambda) = f(x) + \lambda^T h(x)$.

The system of $n + \ell$ nonlinear equations presented above can be solved by the Newton algorithm. If z denotes (x, λ) and $F(z)$ denotes system (4.25), the Taylor expansion of this system for $\|\Delta z\|$ sufficiently small is

$$F(z + \Delta z) \approx F(z) + \nabla_z F(z) \, \Delta z.$$

Note that the double bar notation denotes a vector norm (2-*norm*) defined as [28]:
$$\|x\| = (|x_1|^2 + |x_2|^2 + \cdots + |x_n|^2)^{\frac{1}{2}} = (x^T x)^{\frac{1}{2}}. \tag{4.26}$$

To achieve $F(z) = 0$, it is convenient to find a direction Δz so that $F(z + \Delta z) = 0$. This direction can be computed from

$$\nabla_z F(z) \Delta z = -F(z),$$

where $\nabla_z F(z)$ can be expressed as

$$\nabla_z F(z) = \nabla_{(x,\lambda)} F(x, \lambda) = \begin{pmatrix} \nabla_{xx} \mathcal{L}(x, \lambda) & \nabla_x^T h(x) \\ \nabla_x h(x) & 0 \end{pmatrix}$$

and where $\nabla_x h(x)$ is the Jacobian of $h(x)$.

The matrix above is denominated the KKT matrix of problem (4.23)–(4.24), and the system

$$\begin{pmatrix} \nabla_{xx} \mathcal{L}(x, \lambda) & \nabla_x^T h(x) \\ \nabla_x h(x) & 0 \end{pmatrix} \begin{pmatrix} \Delta x \\ \Delta \lambda \end{pmatrix} = - \begin{pmatrix} \nabla_x \mathcal{L}(x, \lambda) \\ h(x) \end{pmatrix}$$

is denominated the KKT system of problem (4.23)–(4.24). It constitutes a Newton iteration to solve system (4.25).

Definition 4.2 (Regular point). *The solution x^* of the LP problem (4.2)–(4.4) is said to be a regular point of the constraints if the gradient vectors of the active constraints are linearly independent.* □

Definition 4.3 (Degenerate inequality constraint). *An inequality constraint is said to be degenerate if it is active and the associated μ-multiplier is null.* □

Definition 4.4 (Second-order sufficient conditions). *Assume that $f, h, g \in C^2$. The following conditions are sufficient for a point x^* satisfying (4.3)–(4.4) to be a strict relative minimum of the problem (4.2)–(4.4):*

(a) *Constraints (4.5), (4.8), and (4.9) hold.*
(b) *The Hessian matrix $L(x^*) = F(x^*) + \lambda^T H(x^*) + \mu^T G(x^*)$ is positive definite on the subspace*

$$\{y : \nabla h(x^*)^T y = 0, \nabla g_j(x^*)^T y = 0;\ \forall j \in J\},$$

where $J = \{j : g_j(x^*) = 0,\ \mu_j > 0\}$.

□

Definition 4.5 (Second-order necessary conditions). *Assume that $f, h, g \in C^2$. If x^* is a relative minimum regular point of the problem (4.2)–(4.4), then there exists vectors λ and $\mu \geq 0$ such that constraints (4.5) and (4.9) hold and the Hessian matrix $L(x^*) = F(x^*) + \lambda^T H(x^*) + \mu^T G(x^*)$ is positive semidefinite on the tangent subspace of the active constraints at x^*.*

□

4.3 Duality in Linear Programming

In this section we deal with duality in linear programming. We start by giving the definition of a dual problem.

Definition 4.6 (Dual problem). *Given the linear programming problem*

$$\underset{x}{\text{minimize}} \quad z = c^T x \qquad (4.27)$$

subject to

$$\begin{aligned} Ax &\geq b \\ x &\geq 0, \end{aligned} \qquad (4.28)$$

its dual problem is

$$\underset{y}{\text{maximize}} \quad z = b^T y \qquad (4.29)$$

subject to

$$\begin{aligned} A^T y &\leq c \\ y &\geq 0, \end{aligned} \qquad (4.30)$$

where $y = (y_1, \ldots, y_m)^T$ are called dual variables. □

We identify the first problem as the primal problem, and the second one as its dual counterpart. Note how the same elements (the matrix A, and the vectors b and c) determine both problems. Although the primal problem is not in its standard form, this format enables us to see more clearly the symmetry between the primal and the corresponding dual problem. The following theorem shows that the dual of the dual is the primal.

Theorem 4.1 (Symmetry of the duality relationship). *Duality is a symmetric relationship; i.e., if problem D is the dual of problem P, then P is also the dual of D.* □

Proof. To see this, we rewrite the dual problem (4.29)–(4.30) as a minimization problem with constraints of the form $\geq$, as in constraints (4.27)–(4.28)

$$\underset{y}{\text{minimize}} \quad z = -b^T y$$

subject to

$$\begin{aligned} -A^T y &\geq -c \\ y &\geq 0 \, . \end{aligned} \tag{4.31}$$

Then, according to Definition 4.6, its dual is

$$\underset{x}{\text{maximize}} \quad z = -c^T x$$

subject to

$$\begin{aligned} -Ax &\leq -b \\ x &\geq 0 \, , \end{aligned} \tag{4.32}$$

which is equivalent to the primal problem (4.27)–(4.28). □

Remark 4.2. As it can be observed, every constraint of the primal problem has a dual variable associated with it, the coefficients of the objective function of the primal problem are the right-hand side terms of the constraints of the dual problem and vice versa, and the coefficient matrix of the constraints of the dual problem is the transpose of the constraint matrix of the primal one. The primal is a minimization problem but its dual is a maximization problem.

4.3.1 Obtaining the Dual Problem from a Primal Problem in Standard Form

In this section we address the problem of finding the dual problem when the primal one is given in its standard form. To answer this question all we have to do is to use Definition 4.6 of the dual problem.

Theorem 4.2 (Dual problem in standard form). *The dual problem of the problem in standard form*

$$\underset{x}{\text{minimize}} \quad z = c^T x \tag{4.33}$$

subject to

$$\begin{aligned} Ax &= b \\ x &\geq 0 \end{aligned} \tag{4.34}$$

is

$$\text{maximize} \quad z = b^T y \qquad (4.35)$$
$$y$$

subject to
$$A^T y \leq c \,. \qquad (4.36)$$

□

Proof. The equality $Ax = b$ in constraint (4.34) can be replaced by the two equivalent inequalities $Ax \geq b$ and $-Ax \geq -b$. Then, we can write the problem (4.33)–(4.34) as

$$\text{minimize} \quad z = c^T x$$
$$x$$

subject to
$$\begin{aligned} Ax &\geq b \\ -Ax &\geq -b \\ x &\geq 0 \,. \end{aligned}$$

Using Definition 4.6, the dual of this problem is

$$\text{maximize} \quad z = b^T y^{(1)} - b^T y^{(2)}$$
$$y^{(1)}, y^{(2)}$$

that is equivalent to
$$\text{maximize} \quad z = b^T y \,,$$
$$y$$

where $y = y^{(1)} - y^{(2)}$ is not restricted in sign, subject to

$$A^T y^{(1)} - A^T y^{(2)} \leq c \,,$$

which is equivalent to
$$A^T y \leq c \,.$$

□

Remark 4.3. If the relationship between the primal and dual problems (Theorem 4.2) would have been taken as the definition of dual problem, then the initially given definition would have resulted as a theorem.

4.3.2 Obtaining the Dual Problem

A linear programming problem whose formulation is not in the form (4.27)–(4.28) also has a dual problem. To facilitate the obtention of the dual of a given problem in any form or the primal problem from the dual one, one can use the following sets of rules:

Rules to Be Used if the Primal Problem Is a Minimization Problem:

Rule 1. A minimization primal problem leads to a maximization dual problem, and a maximization dual problem leads to a minimization primal problem.

Rule 2. An equality constraint in the primal (dual) problem implies that the corresponding dual (primal) variable is unrestricted.

Rule 3. A $\geq$ ($\leq$) inequality constraint in the primal (dual) problem implies that its dual (primal) variable is nonnegative.

Rule 4. A $\leq$ ($\geq$) inequality constraint in the primal (dual) problem implies that its dual (primal) variable is nonpositive.

Rule 5. A primal (dual) nonnegative variable leads to a $\leq$ ($\geq$) inequality constraint in the dual (primal) problem.

Rule 6. A primal (dual) nonpositive variable leads to a $\geq$ ($\leq$) inequality constraint in the dual (primal) problem.

Rule 7. A primal (dual) unrestricted variable leads to an equality constraint in the dual (primal) problem.

Rules to Be Used if the Primal Problem Is a Maximization Problem:

Rule 1. A maximization primal problem leads to a minimization dual problem, and a minimization dual problem leads to a maximization primal problem.

Rule 2. An equality constraint in the primal (dual) problem implies that the corresponding dual (primal) variable is unrestricted.

Rule 3. A $\geq$ ($\leq$) inequality constraint in the primal (dual) problem implies that its dual (primal) variable is nonpositive.

Rule 4. A $\leq$ ($\geq$) inequality constraint in the primal (dual) problem implies that its dual (primal) variable is nonnegative.

Rule 5. A primal (dual) nonnegative variable leads to a $\geq$ ($\leq$) inequality constraint in the dual (primal) problem.

Rule 6. A primal (dual) nonpositive variable leads to a $\leq$ ($\geq$) inequality constraint in the dual (primal) problem.

Rule 7. A primal (dual) unrestricted variable leads to an equality constraint in the dual (primal) problem.

The above rules are summarized in Fig. 4.5.

Illustrative Example 4.2 (Dual problem). The dual of the linear programming problem

$$\underset{x_1, x_2, x_3}{\text{minimize}} \quad z = x_1 + x_2 - x_3$$

4.3 Duality in Linear Programming

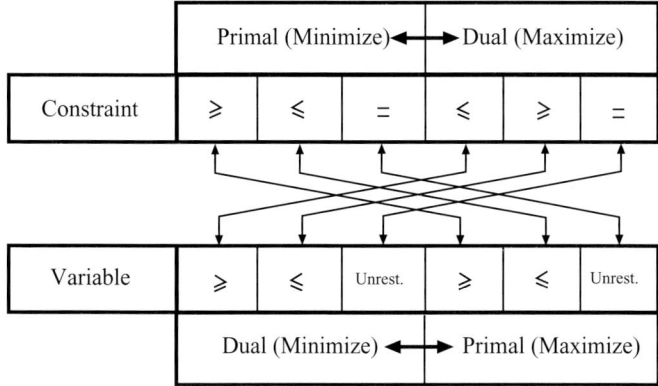

Fig. 4.5. Schematic representation of rules for obtaining dual problems from primal ones and vice versa

subject to
$$\begin{aligned}
2x_1 +x_2 & \geq 3 \\
x_1 -x_3 & = 0 \\
x_2 +x_3 & \leq 2 \\
x_3 & \geq 0 \\
x_2 & \leq 0
\end{aligned} \qquad (4.37)$$

is

$$\text{maximize} \quad z = 3y_1 + 2y_3$$
$$y_1, y_2$$

subject to
$$\begin{aligned}
2y_1 +y_2 & = 1 \\
y_1 +y_3 & \geq 1 \\
-y_2 +y_3 & \leq -1 \\
y_1 & \geq 0 \\
y_3 & \leq 0.
\end{aligned} \qquad (4.38)$$

To see this, we apply the rules as follows:

Rule 1. Since the primal problem is a minimization problem, the dual one is a maximization problem.

Rule 2. Since the second constraint in the primal problem is an equality constraint, the second dual variable y_2 is unrestricted.

Rule 3. Since the first constraint in the primal problem is $\geq$, the first dual variable y_1 is nonnegative.

Rule 4. Since the third constraint in the primal problem is $\leq$, the third dual variable y_3 is nonpositive.

Rule 5. Since the third primal variable x_3 is nonnegative, the third dual constraint is $\leq$.

Rule 6. Since the second primal variable x_2 is nonpositive, the second dual constraint is $\geq$.

Rule 7. Since the first primal variable x_1 is unrestricted, the first dual constraint is an equality.

Using the same rules to the dual problem, we can recover the primal one as follows:

Rule 1. Since the dual is a maximization problem, the primal is a minimization problem.

Rule 2. Since the first constraint in the dual problem is an equality constraint, the first primal variable x_1 is unrestricted.

Rule 3. Since the third constraint in the dual problem is $\leq$, the third primal variable x_3 is nonnegative.

Rule 4. Since the second constraint in the dual problem is $\geq$, the second primal variable x_2 is nonpositive.

Rule 5. Since the first dual variable y_1 is nonnegative, the first primal constraint is $\geq$.

Rule 6. Since the third dual variable y_3 is nonpositive, the third primal constraint is $\leq$.

Rule 7. Since the second dual variable y_2 is unrestricted, the second primal constraint is an equality.

□

4.3.3 Duality Theorems

The importance of the primal–dual relationship is established in the following theorems.

Lemma 4.1 (Weak duality lemma). *Let P be the LPP (4.27)–(4.28), and D its dual (4.29)–(4.30). Let x be a feasible solution of P and y a feasible solution of D. Then*

$$b^T y \leq c^T x . \qquad (4.39)$$

Proof. The proof is simple. If x and y are feasible for P and D, respectively, then

$$Ax \geq b, \quad y \geq 0, \quad A^T y \leq c, \quad x \geq 0 .$$

Note, by the nonnegativity of x and y, that

$$b^T y \leq x^T A^T y \leq x^T c = c^T x .$$

□

4.3 Duality in Linear Programming

Corollary 4.1. *If $b^T \bar{y} = c^T \bar{x}$ for some particular vectors $\bar{x}$ and $\bar{y}$, feasible for P and D, respectively, then the optimal solutions of the dual and primal problems are $\bar{y}$ and $\bar{x}$, respectively. Furthermore, if $\bar{x}$ is an optimal solution for the primal problem, then, there is an optimal solution $\bar{y}$ for the dual problem, and the optimal objective function value of both problems coincides with the common value $b^T \bar{y} = c^T \bar{x}$. Otherwise, one or both of the two sets of feasible solutions is empty.*

Proof. Proving this corollary is straightforward if we realize that, using constraint (4.39), we obtain

$$c^T \bar{x} = b^T \bar{y} \leq \max_{y}\{b^T y | A^T y \leq c\}$$
$$\leq \min_{x}\{c^T x | Ax \geq b, \ x \geq 0\}$$
$$\leq c^T \bar{x} = b^T \bar{y}.$$

Hence, all inequalities are in reality equalities and $\bar{x}$ and $\bar{y}$ must be optimal solutions of P and D, respectively, as claimed. $\square$

The Corollary 4.1 asserts that both problems, the primal P and the dual D, admit optimal solutions simultaneously.

In summary, one of the following two statements holds:

1. Both problems have optimal solution and their optimal values coincide.
2. One problem has unbounded optimum and the other one has an empty feasible region.

Illustrative Example 4.3 (Primal and dual of a mill problem and sensitivities). A small mill manufactures two types of wood tables. Each table of type 1 requires 4 machine-hours for cutting from the wood stock to size and 4 machine-hours finishing time (assembly, painting, etc.). Similarly, each table of type 2 requires 3 machine-hours of cutting time and 7 machine-hours of finishing time. The cutting and finishing equipment capacities available per day are 40 and 56 machine-hours, respectively. Finally, each type 1 table provides $70 profit, whereas each type 2 table provides $90 profit. These data are summarized in Table 4.1.

Table 4.1. Data of the mill planning problem

	Table type		Machine-hours available per day
	1	2	
Cutting time (h)	4	3	40
Finishing time (h)	4	7	56
Profit ($)	70	90	

Primal and dual problems are analyzed below.

Primal

In order to maximize the daily total profit, we wish to know the quantity of each type of tables produced per day. The linear programming model of the problem is

$$\underset{x_1, x_2}{\text{maximize}} \quad z = 70x_1 + 90x_2$$

subject to

$$\begin{array}{rcl} 4x_1 + 3x_2 & \leq & 40 \\ 4x_1 + 7x_2 & \leq & 56 \\ x_1, x_2 & \geq & 0 \,, \end{array} \quad (4.40)$$

where x_1 and x_2 are the quantities of table types 1 and 2, respectively, to be produced per day.

The optimal solution of this model, as can be seen in Fig. 4.6, indicates a daily production of seven type 1 tables and four type 2 tables, and a daily maximum profit of \$850. This result indicates that both cutting and finishing resources are fully utilized, because both constraints are binding.

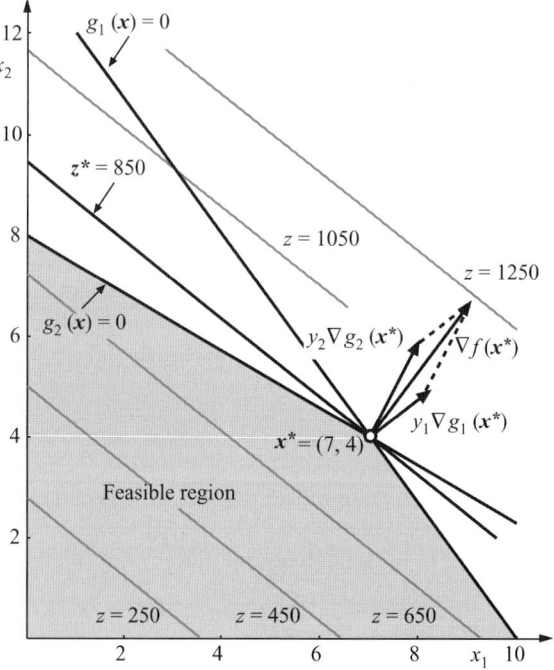

Fig. 4.6. Graphical analysis of the mill planning problem

Now, suppose that we wish to obtain higher daily profits. Then, an expansion of the available capacities is necessary. Suppose that the available finishing capacity can be expanded from 56 machine-hours to 72 machine-hours per day. How does this expansion affect the total daily profit? The answer is given by the solution of the following linear programming problem, where we have

$$\text{maximize} \quad z = 70x_1 + 90x_2$$
$$x_1, x_2$$

subject to
$$\begin{array}{rcl} 4x_1 + 3x_2 & \leq & 40 \\ 4x_1 + 7x_2 & \leq & 72 \\ x_1, x_2 & \geq & 0 \,. \end{array} \quad (4.41)$$

In this case the optimal solution is $x_1 = 4$ and $x_2 = 8$ with a maximum daily profit of \$1,000 (see Fig. 4.7). This solution indicates that we can increase the daily profit by \$150 if we increase the daily finishing capacity by $72-56 = 16$ machine-hours. The rate \$$(1{,}000-850)/16 = \$150/16 = \$75/8$, at which the objective function value will increase as the finishing capacity is increased by 1 h, is called the sensitivity or shadow price (also dual price) of the finishing resource constraint.

A formal treatment of sensitivities is provided in Theorem 4.5 in p. 175.

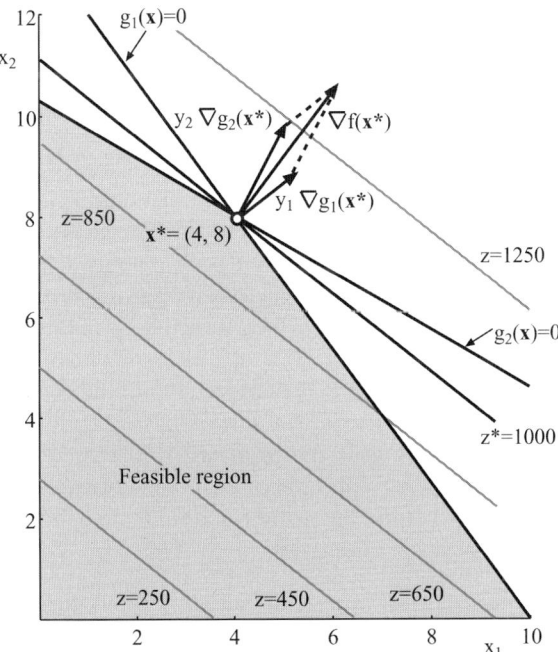

Fig. 4.7. Graphical analysis of the expanded mill planning problem

158   4 Duality

In general, the shadow price of a constraint is the change in the optimal value of the objective function per unit change in the right-hand side value of the constraint, assuming that all other parameter values of the problem remain unchanged. For many linear programming applications the shadow prices are at least as important as the solution of the problem itself; they indicate whether certain changes in the use of productive capacities affect the objective function value. Shadow prices can be obtained directly from the solution of the primal problem or determined by formulating and solving the dual problem.

Note that for both problems (4.40) and (4.41), as the same constraints remain active, the dual variable values are identical.

Dual

The dual of the planning mill problem (4.40) is stated as follows:

$$\underset{y_1, y_2}{\text{minimize}} \quad z = 40y_1 + 56y_2$$

subject to

$$\begin{array}{rcl} 4y_1 + 4y_2 & \geq & 70 \\ 3y_1 + 7y_2 & \geq & 90 \\ y_1, y_2 & \geq & 0 \,. \end{array} \qquad (4.42)$$

The optimal solution of this problem is $y_1 = 65/8$, $y_2 = 75/8$ and the optimal objective function value is 850. Note that y_1 and y_2 are the shadow prices of cutting and finishing resources, respectively, and optimal objective function values of problems (4.40) and (4.42) coincide.

The dual problem (4.42) can be interpreted as follows. Assume that we are willing to sell machine time (rather than produce and sell tables), provided that we can obtain at least the same level of total profit. In this case, producing tables and selling machine time would be two equivalent alternatives; the y_1 and y_2 variables would represent the sale prices of 1 h of work on the cutting and finishing machines, respectively. In order to maintain a competitive business we should set the total daily profit as low as possible, i.e, we should minimize the function $40y_1 + 56y_2$, where the 40 and 56 coefficients represent the machine-hours available per day for sale of cutting and finishing machines. The constraints (4.42) state that the total prices of the cutting and finishing hours required for producing one unit of each type of table must not be less than the profit expected from the sale of one unit of each type of table, and that prices must be positive quantities. □

The primal dual relationship is further illustrated through the example below.

Illustrative Example 4.4 (Communication net). We are interested in sending a message through a communication net from one node A to some

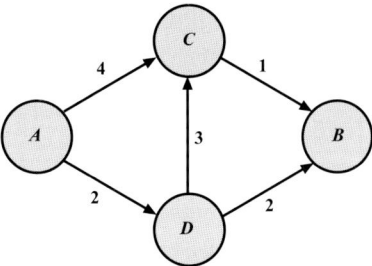

Fig. 4.8. Communication net in Example 4.4

other node B with minimal cost. The network consists of four nodes, A, B, C, and D, and there are channels connecting A and C, A and D, D and C, C and B, and D and B, with associated costs per unit of message 4, 2, 3, 1, and 2, respectively (see Fig. 4.8).

Primal

If variable x_{PQ} denotes the fraction of the message going through the channel connecting any node P with any other node Q, the following constraints must be satisfied:

1. No fraction of the total message is lost in node C

$$x_{AC} + x_{DC} = x_{CB}. \tag{4.43}$$

2. No part of the message is lost in node D

$$x_{AD} = x_{DC} + x_{DB}. \tag{4.44}$$

3. The complete message arrives at B

$$1 = x_{CB} + x_{DB}. \tag{4.45}$$

If we write the above constraints in matrix form, we get

$$\begin{array}{rrrrrl}
x_{AC} & +x_{DC} & -x_{CB} & & = & 0 \\
& x_{AD} & -x_{DC} & & -x_{DB} & = & 0 \\
& & & x_{CB} & +x_{DB} & = & 1 \\
x_{AC}, x_{AD}, x_{DC}, x_{CB}, x_{DB} & & & & \geq & 0.
\end{array} \tag{4.46}$$

Note that if we add the last two constraints and subtract the first one, we get $x_{AC} + x_{AD} = 1$. This means that the whole message departs from A. Since it can be obtained as a linear combination of the other three, it must not be included to avoid redundancy.

Consequently, the primal approach to this problem consists of minimizing the total cost of sending the messages,

$$\underset{x_{AC}, x_{AD}, x_{DC}, x_{CB}, x_{DB}}{\text{minimize}} \quad z = 4x_{AC} + 2x_{AD} + 3x_{DC} + x_{CB} + 2x_{DB}$$

subject to
$$\begin{array}{rcl}
x_{AC} + x_{DC} - x_{CB} & = & 0 \\
x_{AD} - x_{DC} \quad\quad -x_{DB} & = & 0 \\
x_{CB} + x_{DB} & = & 1 \\
x_{AC}, x_{AD}, x_{DC}, x_{CB}, x_{DB} & \geq & 0.
\end{array} \quad (4.47)$$

Dual

Alternatively, we can assume that we can buy the information of the messages at each node and sell them at a different node. Thus, the dual approach can be interpreted as finding the prices y_A, y_B, y_C, y_D of the fractions of the total message in each node, assuming that fractions of messages are bought in one node and sold in another. The net benefit obtained in such a transaction would be the difference of prices between the arriving and departing nodes. If we normalize these conditions by setting $y_A = 0$, the benefits are

$$\begin{array}{l}
y_C - y_A = y_C \\
y_D - y_A = y_D \\
y_C - y_D \\
y_B - y_C \\
y_B - y_D.
\end{array}$$

We must ensure that these benefits are not greater than the prices we already know,

$$\begin{array}{rcl}
y_C & \leq & 4 \\
y_D & \leq & 2 \\
y_C - y_D & \leq & 3 \\
y_B - y_C & \leq & 1 \\
y_B - y_D & \leq & 2.
\end{array}$$

The objective is now to maximize the revenue when taking the total message from node A to node B, i.e., the dual problem becomes

$$\underset{y_C, y_D, y_B}{\text{maximize}} \quad y_B$$

subject to
$$\begin{array}{rcl}
y_C & \leq & 4 \\
y_D & \leq & 2 \\
y_C - y_D & \leq & 3 \\
-y_C + y_B & \leq & 1 \\
-y_D + y_B & \leq & 2.
\end{array} \quad (4.48)$$

□

4.4 Duality in Nonlinear Programming

In this section the results obtained for linear programming problems are generalized to nonlinear programming problems. Consider the following general nonlinear primal problem:

$$\minimize_{x} \quad z_P = f(x) \tag{4.49}$$

subject to

$$\begin{aligned} h(x) &= 0 \\ g(x) &\leq 0, \end{aligned} \tag{4.50}$$

where $f : \mathbb{R}^n \to \mathbb{R}$, $h : \mathbb{R}^n \to \mathbb{R}^\ell$, $g : \mathbb{R}^n \to \mathbb{R}^m$.

The dual problem requires the introduction of the dual function defined as

$$\phi(\lambda, \mu) = \Infimum_{x} \quad \{f(x) + \lambda^T h(x) + \mu^T g(x)\}, \tag{4.51}$$

where λ^*, μ^* are the multipliers associated with the constraints (4.50) for the optimal solution x^* of problem (4.49)–(4.50).

The dual problem of the primal problem (4.49)–(4.50) is then defined as follows:

$$\maximize_{\lambda, \mu} \quad z_D = \phi(\lambda, \mu) \tag{4.52}$$

subject to

$$\mu \geq 0. \tag{4.53}$$

Using the Lagrangian

$$\mathcal{L}(x, \lambda, \mu) = f(x) + \lambda^T h(x) + \mu^T g(x),$$

we can rewrite the dual problem as

$$\maximize_{\lambda, \mu; \mu \geq 0} \left[\Infimum_{x} \quad \mathcal{L}(x, \lambda, \mu) \right]. \tag{4.54}$$

Illustrative Example 4.5 (Dual function).

$$\minimize_{x_1, x_2} \quad z = x_1^2 + x_2 \tag{4.55}$$

subject to

$$\begin{aligned} -x_1 + 2x_2 &= 0 \\ x_1 x_2 &\geq 2. \end{aligned} \tag{4.56}$$

The Lagrangian function is

$$\mathcal{L}(x_1, x_2, \lambda, \mu) = x_1^2 + x_2 + \lambda(-x_1 + 2x_2) + \mu_2(-x_1 x_2 + 2)$$

and the KKT conditions become

$$\frac{\partial \mathcal{L}(x_1, x_2, \lambda, \mu)}{\partial x_1} = 2x_1 - \lambda - \mu x_2 = 0 \quad (4.57)$$

$$\frac{\partial \mathcal{L}(x_1, x_2, \lambda, \mu)}{\partial x_2} = 1 + 2\lambda - \mu x_1 = 0 \quad (4.58)$$

$$-x_1 + 2x_2 = 0 \quad (4.59)$$

$$-x_1 x_2 + 2 \leq 0 \quad (4.60)$$

$$\mu(-x_1 x_2 + 2) = 0 \quad (4.61)$$

$$\mu \geq 0, \quad (4.62)$$

which lead to the following primal and dual solutions:

$$z^* = 5, \quad x_1^* = 2, \quad x_2^* = 1, \quad \lambda^* = 7/4, \quad \mu^* = 9/4 \ .$$

For obtaining the dual problem we first calculate the dual function

$$\phi(\lambda, \mu) = \inf_{x_1, x_2} \mathcal{L}(x, y, \lambda, \mu) = \inf_{x_1, x_2} \left[x_1^2 + x_2 + \lambda(-x_1 + 2x_2) + \mu_2(-x_1 x_2 + 2) \right] \ ,$$

and since

$$\frac{\partial \mathcal{L}(x_1, x_2, \lambda, \mu)}{\partial x_1} = 2x_1 - \lambda - \mu x_2 = 0$$

$$\frac{\partial \mathcal{L}(x_1, x_2, \lambda, \mu)}{\partial x_2} = 1 + 2\lambda - \mu x_1 = 0 \ ,$$

which leads to

$$x_1 = \frac{1 + 2\lambda}{\mu}, \quad y = \frac{2 + 4\lambda - \lambda\mu}{\mu^2} \ ,$$

the dual function becomes

$$\phi(\lambda, \mu) = \frac{(1 + 2\lambda)^2 - \lambda\mu(1 + 2\lambda) + 2\mu^3}{\mu^2}$$

and then the dual problem is

$$\underset{\lambda, \mu}{\text{maximize}} \quad \phi(\lambda, \mu) = \frac{(1 + 2\lambda)^2 - \lambda\mu(1 + 2\lambda) + 2\mu^3}{\mu^2}$$

subject to $\mu \geq 0$.

Finally, since

$$\frac{\partial \phi(\lambda, \mu)}{\partial \lambda} = \frac{4\lambda(2-\mu) - \mu + 4}{\mu^2} = 0$$

$$\frac{\partial \phi(\lambda, \mu)}{\partial \mu} = 2 + \frac{(1+2\lambda)[\lambda(\mu-4) - 2]}{\mu^3} = 0$$

leads to

$$\lambda^* = 7/4, \quad \mu^* = 9/4 \,,$$

which is the solution of the dual problem, and obviously coincides with that obtained from the KKT conditions above. □

Illustrative Example 4.6 (Dual function). Consider the same problem as the Illustrative Example 4.1. To obtain the dual function $\phi(\mu_1, \mu_2, \mu_3)$ one needs to obtain the Infimum,

$$\phi(\mu_1, \mu_2, \mu_3) = \underset{x_1, x_2}{\text{Infimum}} \quad \mathcal{L}(x_1, x_2, \mu_1, \mu_2, \mu_3) \,.$$

From conditions (4.14) and (4.15), and since

$$\frac{\partial \mathcal{L}(x_1, x_2, \mu_1, \mu_2, \mu_3)}{\partial x_1} = -2x_1 + \mu_1 - \mu_2 = 0$$

$$\frac{\partial \mathcal{L}(x_1, x_2, \mu_1, \mu_2, \mu_3)}{\partial x_2} = -2x_2 + \mu_1 - \mu_3 = 0$$

leads to

$$x_1 = \frac{\mu_1 - \mu_2}{2}, \quad x_2 = \frac{\mu_1 - \mu_3}{2} \,,$$

the dual function becomes

$$\phi(\mu_1, \mu_2, \mu_3) = \frac{1}{4} \left[(\mu_1 - \mu_2)^2 + (\mu_1 - \mu_3)^2 - 4\mu_1 \right] \,.$$

Then, the dual problem becomes

$$\underset{\mu_1, \mu_2, \mu_3}{\text{maximize}} \quad \phi(\mu_1, \mu_2, \mu_3)$$

subject to

$$\mu_1, \mu_2, \mu_3 \geq 0 \,,$$

which has following two possible solutions:

$$\mu_1 = 2, \quad \mu_2 = 0, \quad \mu_3 = 2, \quad \phi(\mu_1, \mu_2, \mu_3) = -1$$

and

$$\mu_1 = 2, \quad \mu_2 = 2, \quad \mu_3 = 0, \quad \phi(\mu_1, \mu_2, \mu_3) = -1 \,.$$

□

Remark 4.4. We assume that f, h, and g are such that the infimum of the Lagrangian function is always attained at some x, so that the "infimum" operation in problems (4.51) and (4.54) can be replaced by the "minimum" operation. Then, problem (4.54) is referred to as the *max–min dual problem*.

In this situation, if we denote by $x(\lambda, \mu)$ a point where the minimum of the Lagrangian is attained (considered as a function of the multipliers), we can write

$$\phi(\lambda, \mu) = f(x(\lambda, \mu)) + \lambda^T h(x(\lambda, \mu)) + \mu^T g(x(\lambda, \mu)) .$$

On the other hand, under the assumption of convexity of f and g, the Lagrangian is convex in x, and therefore at the minimum its gradient should vanish

$$\nabla \phi(\lambda, \mu) = \nabla f(x(\lambda, \mu)) + \lambda^T \nabla h(x(\lambda, \mu)) + \mu^T \nabla g(x(\lambda, \mu)) = 0 . \quad (4.63)$$

This last identity can be utilized to derive an expression for the gradient of the dual function as well as its Hessian. Indeed, according to the chain rule

$$\nabla_\mu \phi(\lambda, \mu) = \left[\nabla f(x(\lambda, \mu)) + \lambda^T \nabla h(x(\lambda, \mu)) + \mu^T \nabla g(x(\lambda, \mu)) \right]^T$$
$$\times \nabla_\mu x(\lambda, \mu) + g(x(\lambda, \mu)) , \quad (4.64)$$

and by condition (4.63)

$$\nabla_\mu \phi(\lambda, \mu) = g(x(\lambda, \mu)) . \quad (4.65)$$

One must be careful in using expresion (4.65) because of the discontinuity of expression (4.64) with respect to μ. Nevertheless, this is not a problem in practice.

Similarly, we obtain

$$\nabla_\lambda \phi(\lambda, \mu) = h(x(\lambda, \mu)) . \quad (4.66)$$

Note that the gradient of the dual function is the vector of mismatches of the corresponding constraints.

Remark 4.5. For nondifferentiable functions, the subgradients play the same as the gradients play for differentiable functions. Therefore, the important concept of subgradient is defined below.

Definition 4.7 (Subgradient and subdifferential). *Let C be a convex nonempty set in $\Re^n$ and let $\phi : C \to \Re$ be convex. Then, α is called a subgradient of $\phi(\lambda)$ at $\tilde{\lambda} \in C$ if*

$$\phi(\lambda) \geq \phi(\tilde{\lambda}) + \alpha^T (\lambda - \tilde{\lambda}); \quad \forall \lambda \in C . \quad (4.67)$$

Analogously, let $\phi : C \to \Re$ be concave. Then, α is called a subgradient of $\phi(\lambda)$ at $\tilde{\lambda} \in C$ if

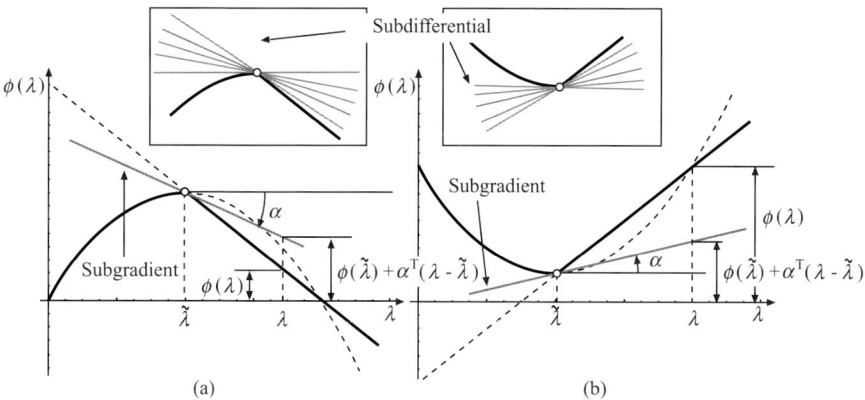

Fig. 4.9. Subgradient and subdifferential graphical illustration: (**a**) concave function and (**b**) convex function

$$\phi(\boldsymbol{\lambda}) \leq \phi(\tilde{\boldsymbol{\lambda}}) + \boldsymbol{\alpha}^T(\boldsymbol{\lambda} - \tilde{\boldsymbol{\lambda}}); \quad \forall \boldsymbol{\lambda} \in \mathcal{C} . \tag{4.68}$$

The set of all subgradients of $\phi(\boldsymbol{\lambda})$ in $\tilde{\boldsymbol{\lambda}}$ is a convex set known as the subdifferential of $\phi(\boldsymbol{\lambda})$ at $\tilde{\boldsymbol{\lambda}}$.

Figure 4.9 illustrates the concepts of subgradient and subdifferential. □

Illustrative Example 4.7 (Subgradient). Consider the following problem:

$$\underset{x_1, x_2}{\text{minimize}} \quad -2x_1 + x_2$$

subject to

$$x_1 + x_2 = \frac{5}{2}$$
$$(x_1, x_2) \in \boldsymbol{X},$$

where $\boldsymbol{X} = \{(0,0), (0,2), (2,0), (2,2), (5/4, 5/4)\}$.

The dual function $\phi(\lambda)$ is obtained solving the following problem:

$$\phi(\lambda) = \underset{x_1, x_2}{\text{minimum}} \quad -2\,x_1 + x_2 + \lambda \left(x_1 + x_2 - \frac{5}{2} \right)$$

subject to

$$(x_1, x_2) \in \boldsymbol{X} .$$

166   4 Duality

The explicit expression of the dual function for this example is given by

$$\phi(\lambda) = \begin{cases} -2 + \dfrac{3\lambda}{2} & \text{if} \quad \lambda \leq -1 \\ -4 - \dfrac{\lambda}{2} & \text{if} \quad -1 \leq \lambda \leq 2 \\ -\dfrac{5\lambda}{2} & \text{if} \quad \lambda \geq 2 \,. \end{cases}$$

Differentiating with respect to λ we obtain

$$\alpha(\lambda) = \dfrac{d\phi(\lambda)}{d\lambda} = \begin{cases} \dfrac{3}{2} & \text{if} \quad \lambda \leq -1 \\ -\dfrac{1}{2} & \text{if} \quad -1 \leq \lambda \leq 2 \\ -\dfrac{5}{2} & \text{if} \quad \lambda \geq 2 \,, \end{cases}$$

which is the subgradient of the dual function. Note that at the values $\lambda = -1$ and $\lambda = 2$ there are many possible subgradients, thus their corresponding subdifferentials are

$$\begin{aligned} \tilde{\lambda} &= -1; \quad \alpha \in [-1/2, 3/2] \\ \tilde{\lambda} &= 2; \quad \alpha \in [-5/2, -1/2] \,, \end{aligned}$$

as shown in Fig. 4.10. □

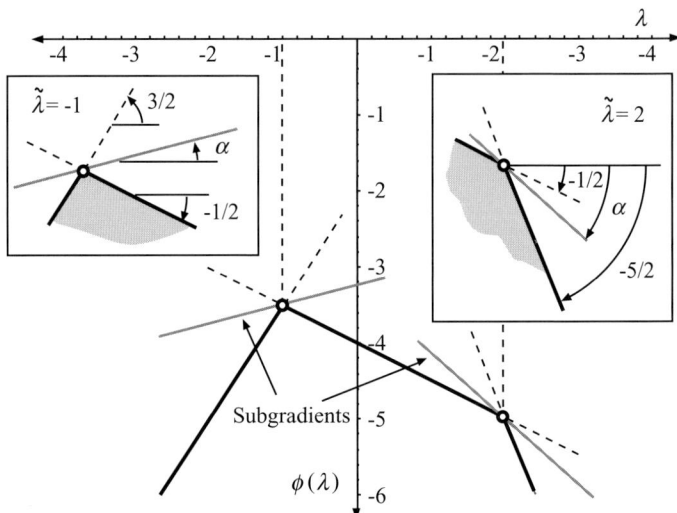

Fig. 4.10. Graphical illustration of the dual function, subgradient, and subdifferential for the Illustrative Example 4.7

Concerning the Hessian, note that by differentiating conditions (4.65) and (4.66) with respect to $\boldsymbol{\mu}$ and $\boldsymbol{\lambda}$, respectively, we obtain

$$\nabla^2_{\boldsymbol{\mu}}\phi(\boldsymbol{\lambda},\boldsymbol{\mu}) = \nabla_{\boldsymbol{x}}g(\boldsymbol{x}(\boldsymbol{\lambda},\boldsymbol{\mu}))\ \nabla_{\boldsymbol{\mu}}\boldsymbol{x}(\boldsymbol{\lambda},\boldsymbol{\mu}) \tag{4.69}$$

and

$$\nabla^2_{\boldsymbol{\lambda}}\phi(\boldsymbol{\lambda},\boldsymbol{\mu}) = \nabla_{\boldsymbol{x}}h(\boldsymbol{x}(\boldsymbol{\lambda},\boldsymbol{\mu}))\ \nabla_{\boldsymbol{\lambda}}\boldsymbol{x}(\boldsymbol{\lambda},\boldsymbol{\mu})\ . \tag{4.70}$$

To obtain an expression for $\nabla_{\boldsymbol{\mu}}\boldsymbol{x}(\boldsymbol{\lambda},\boldsymbol{\mu})$ and $\nabla_{\boldsymbol{\lambda}}\boldsymbol{x}(\boldsymbol{\lambda},\boldsymbol{\mu})$, we differentiate expression (4.63) with respect to $\boldsymbol{\mu}$ and $\boldsymbol{\lambda}$, respectively. The results are

$$\nabla^2_{\boldsymbol{x}}\mathcal{L}(\boldsymbol{x}(\boldsymbol{\lambda},\boldsymbol{\mu}),\boldsymbol{\lambda},\boldsymbol{\mu})\ \nabla_{\boldsymbol{\mu}}\boldsymbol{x}(\boldsymbol{\lambda},\boldsymbol{\mu}) + \nabla_{\boldsymbol{x}}g(\boldsymbol{x}(\boldsymbol{\lambda},\boldsymbol{\mu})) = \mathbf{0}\ , \tag{4.71}$$

and a similar equation exchanging $\boldsymbol{\lambda}$ by $\boldsymbol{\mu}$ and h by g, i.e.,

$$\nabla^2_{\boldsymbol{x}}\mathcal{L}(\boldsymbol{x}(\boldsymbol{\lambda},\boldsymbol{\mu}),\boldsymbol{\lambda},\boldsymbol{\mu})\ \nabla_{\boldsymbol{\lambda}}\boldsymbol{x}(\boldsymbol{\lambda},\boldsymbol{\mu}) + \nabla_{\boldsymbol{x}}h(\boldsymbol{x}(\boldsymbol{\lambda},\boldsymbol{\mu})) = \mathbf{0}\ . \tag{4.72}$$

Substituting these equations into (4.69) and (4.70), respectively, we obtain

$$\nabla^2_{\boldsymbol{\mu}}\phi(\boldsymbol{\lambda},\boldsymbol{\mu}) = -\nabla_{\boldsymbol{x}}g(\boldsymbol{x}(\boldsymbol{\lambda},\boldsymbol{\mu}))[\nabla^2_{\boldsymbol{x}}\mathcal{L}(\boldsymbol{x}(\boldsymbol{\lambda},\boldsymbol{\mu}),\boldsymbol{\lambda},\boldsymbol{\mu})]^{-1}\ \nabla_{\boldsymbol{x}}g(\boldsymbol{x}(\boldsymbol{\lambda},\boldsymbol{\mu}))\ , \tag{4.73}$$

and the parallel formula for $\nabla^2_{\boldsymbol{\lambda}}\phi(\boldsymbol{\lambda},\boldsymbol{\mu})$ is

$$\nabla^2_{\boldsymbol{\lambda}}\phi(\boldsymbol{\lambda},\boldsymbol{\mu}) = -\nabla_{\boldsymbol{x}}h(\boldsymbol{x}(\boldsymbol{\lambda},\boldsymbol{\mu}))[\nabla^2_{\boldsymbol{x}}\mathcal{L}(\boldsymbol{x}(\boldsymbol{\lambda},\boldsymbol{\mu}),\boldsymbol{\lambda},\boldsymbol{\mu})]^{-1}\ \nabla_{\boldsymbol{x}}h(\boldsymbol{x}(\boldsymbol{\lambda},\boldsymbol{\mu}))\ . \tag{4.74}$$

Finally, if we are interested in the mixed second partial derivatives

$$\nabla^2_{\boldsymbol{\lambda}\boldsymbol{\mu}}\phi(\boldsymbol{\lambda},\boldsymbol{\mu})\ ,$$

similar computations lead to

$$\nabla^2_{\boldsymbol{\lambda}\boldsymbol{\mu}}\phi(\boldsymbol{\lambda},\boldsymbol{\mu}) = -\nabla_{\boldsymbol{x}}g(\boldsymbol{x}(\boldsymbol{\lambda},\boldsymbol{\mu}))[\nabla^2_{\boldsymbol{x}}\mathcal{L}(\boldsymbol{x}(\boldsymbol{\lambda},\boldsymbol{\mu}),\boldsymbol{\lambda},\boldsymbol{\mu})]^{-1}\ \nabla_{\boldsymbol{x}}h(\boldsymbol{x}(\boldsymbol{\lambda},\boldsymbol{\mu}))\ , \tag{4.75}$$

and similarly

$$\nabla^2_{\boldsymbol{\lambda}\boldsymbol{\mu}}\phi(\boldsymbol{\lambda},\boldsymbol{\mu}) = -\nabla_{\boldsymbol{x}}h(\boldsymbol{x}(\boldsymbol{\lambda},\boldsymbol{\mu}))[\nabla^2_{\boldsymbol{x}}\mathcal{L}(\boldsymbol{x}(\boldsymbol{\lambda},\boldsymbol{\mu}),\boldsymbol{\lambda},\boldsymbol{\mu})]^{-1}\ \nabla_{\boldsymbol{x}}g(\boldsymbol{x}(\boldsymbol{\lambda},\boldsymbol{\mu}))\ . \tag{4.76}$$

To show that the definition of a dual problem in nonlinear programming is equivalent to Definition 4.6 for linear programming problems, we provide the following example.

Illustrative Example 4.8 (Linear programming). Consider the following dual problem of the linear programming primal problem:

$$\text{minimize } \boldsymbol{c}^T\boldsymbol{x} \tag{4.77}$$

subject to

$$Ax = a \quad (4.78)$$
$$Bx \leq b. \quad (4.79)$$

Using the rules in Sect. 4.3.2, p. 151 the dual problem becomes:

$$\underset{\lambda,\mu}{\text{maximize}} \quad \lambda^T a + \mu^T b \quad (4.80)$$

subject to

$$A^T \lambda + B^T \mu = c \quad (4.81)$$

and

$$\mu \leq 0. \quad (4.82)$$

On the other hand, the Lagrangian function of (4.77)–(4.79) becomes

$$\begin{aligned} \mathcal{L}(x, \lambda, \mu) &= c^T x + \lambda^T (Ax - a) + \mu^T (Bx - b) \\ &= (c^T + \lambda^T A + \mu^T B)x - \lambda^T a - \mu^T b. \end{aligned} \quad (4.83)$$

For this function to have a minimum we must have

$$(c^T + \lambda^T A + \mu^T B) = 0, \quad (4.84)$$

which leads to

$$\lambda^T A + \mu^T B = -c^T, \quad (4.85)$$

and the minimum is

$$\phi(\lambda, \mu) = -\lambda^T a - \mu^T b. \quad (4.86)$$

Thus, the dual problem becomes

$$\underset{\lambda,\mu}{\text{maximize}} \quad -\lambda^T a - \mu^T b \quad (4.87)$$

subject to

$$\lambda^T A + \mu^T B = -c^T \quad (4.88)$$

and

$$\mu \geq 0, \quad (4.89)$$

which, changing signs to λ and μ, is equivalent to (4.80)–(4.82).
Note that the dual variables are λ and μ. □

4.4 Duality in Nonlinear Programming

Illustrative Example 4.9 (Numerical example). Consider the following problem:

$$\text{minimize}_{x_1, x_2} \quad z_P = 40x_1 + 56x_2$$

subject to

$$\begin{aligned} 4x_1 + 4x_2 &\geq 70 \\ 3x_1 + 7x_2 &\geq 90 \\ x_1, x_2 &\geq 0. \end{aligned}$$

The optimal solution of this problem is $x_1^* = 65/8$, $x_2^* = 75/8$, and the optimal objective function value is 850.

The above problem can be written as

$$\text{minimize}_{x_1, x_2} \quad z = 40x_1 + 56x_2$$

subject to

$$\begin{aligned} -4x_1 - 4x_2 &\leq -70 \\ -3x_1 - 7x_2 &\leq -90 \\ -x_1 &\leq 0 \\ -x_2 &\leq 0. \end{aligned}$$

The Lagrangian function for this problem is

$$\begin{aligned} \mathcal{L}(x, \mu) &= 40x_1 + 56x_2 + \mu_1(-4x_1 - 4x_2 + 70) + \mu_2(-3x_1 - 7x_2 + 90) \\ &\quad - \mu_3 x_1 - \mu_4 x_2 \\ &= (40 - 4\mu_1 - 3\mu_2 - \mu_3)x_1 + (56 - 4\mu_1 - 7\mu_2 - \mu_4)x_2 \\ &\quad + 70\mu_1 + 90\mu_2 , \end{aligned}$$

which has a minimum only if (note the non-negativity of x_1, x_2)

$$\begin{aligned} 4\mu_1 + 3\mu_2 + \mu_3 &= 40 \\ 4\mu_1 + 7\mu_2 + \mu_4 &= 56. \end{aligned}$$

The minimum is

$$\phi(\mu) = 70\mu_1 + 90\mu_2 .$$

Then, we conclude that the dual problem is

$$\text{maximize}_{\mu_1, \mu_2} \quad \phi(\mu) = 70\mu_1 + 90\mu_2$$

subject to

$$\begin{aligned} 4\mu_1 + 3\mu_2 + \mu_3 &= 40 \\ 4\mu_1 + 7\mu_2 + \mu_4 &= 56 \\ \mu_1 &\geq 0 \\ \mu_2 &\geq 0 \\ \mu_3 &\geq 0 \\ \mu_4 &\geq 0, \end{aligned}$$

which, considering μ_3 and μ_4 as slack variables, is equivalent to

$$\begin{array}{ll} \text{maximize} & z_D = 70\mu_1 + 90\mu_2 \\ \mu_1, \mu_2 & \end{array}$$

$$\begin{array}{rcl} 4\mu_1 + 3\mu_2 & \leq & 40 \\ 4\mu_1 + 7\mu_2 & \leq & 56 \\ \mu_1 & \geq & 0 \\ \mu_2 & \geq & 0. \end{array}$$

The optimal solution of this problem is $\mu_1^* = 7$, $\mu_2^* = 4$, and the optimal objective function value is 850.

This means (see Table 4.2) that the primal objective function z_P will change in 7 units per each unit of change of the right-hand side whose value is 70, and 4 units per each unit of change of the right-hand side whose value is 90, and that dual objective function z_D will change in 65/8 units per each unit of change of the right-hand side 40, and 75/8 units per each unit of change of the right-hand side 56. □

The following theorem shows that any value of the objective function of the dual problem is a lower bound of the optimal value of the objective function of the primal problem. This may be used to terminate computations in an iterative algorithm where such values are available.

Theorem 4.3 (Weak duality). *For any feasible solution x of the primal problem (4.49)–(4.50) and for any feasible solution, λ, μ, of the dual problem (4.52)–(4.53), the following holds*

$$f(x) \geq \phi(\lambda, \mu) . \tag{4.90}$$

□

Table 4.2. Numerical example: primal and dual problems with their solutions

	Primal			Dual	
Minimize x_1, x_2	$z_P = 40x_1 + 56x_2$		Maximize μ_1, μ_2	$z_D = 70\mu_1 + 90\mu_2$	
	subject to			subject to	
	$4x_1 + 4x_2$	≥ 70		$4\mu_1 + 3\mu_2$	≤ 40
	$3x_1 + 7x_2$	≥ 90		$4\mu_1 + 7\mu_2$	≤ 56
	x_1	≥ 0		μ_1	≥ 0
	x_2	≥ 0		μ_2	≥ 0
	Optimal solution				
	$x_1^* = 65/8, \; x_2^* = 75/8$			$\mu_1^* = 7, \; \mu_2^* = 4$	

4.4 Duality in Nonlinear Programming

Proof. From the definition of λ, μ, for every feasible x and for every $\mu \geq 0, \mu \in \mathbb{R}^m$ and $\lambda \in \mathbb{R}^\ell$, we have

$$\phi(\lambda, \mu) = \underset{y}{\text{Infimum}} \left[f(y) + \lambda^T h(y) + \mu^T g(y) \right] \leq f(x) + \lambda^T h(x) + \mu^T g(x) \tag{4.91}$$

and

$$f(x) + \lambda^T h(x) + \mu^T g(x) \leq f(x) \tag{4.92}$$

because $h(x) = 0$ and $\mu^T g(x) \leq 0$ for being feasible solutions, therefore

$$\phi(\lambda, \mu) \leq f(x) . \tag{4.93}$$

$\square$

If the feasible regions of both primal and dual problems are nonempty, then expression (4.90) implies that $\sup(\phi)$ and $\inf(f)$, both taken over their respective constraint sets, are finite. Certainly, neither the supremum nor the infimum need to be attained at any feasible point. If they are, then expression (4.90) yields the result that if both primal an dual are feasible, both have optimal solutions. A corollary of high practical interest is stated below.

Corollary 4.2 (Primal and dual optimality).
1. The following relation always holds

$$\sup\{\phi(\lambda, \mu) | \mu \geq 0\} \leq \inf\{f(x) | h(x) = 0, g(x) \leq 0\} . \tag{4.94}$$

2. If $f(x^*) = \phi(\lambda^*, \mu^*)$ for some feasible solution x^* of the primal problem (4.49)–(4.50) and for some feasible solution (λ^*, μ^*) of the dual problem (4.52)–(4.53), then x^* and (λ^*, μ^*) are optimal solutions of the primal and dual problems, respectively. This solution defines a saddle point of the Lagrangian corresponding to a maximum with respect to λ and μ and a minimum with respect to x (see Fig. 4.11).

3. If $\sup\{\phi(\lambda, \mu) : \mu \geq 0\} = +\infty$, then the primal problem has no feasible solution.

4. If $\inf\{f(x) : h(x) = 0, g(x) \leq 0\} = -\infty$, then $\phi(\lambda, \mu) = -\infty$ for every $\lambda \in \mathbb{R}^\ell$ and $\mu \geq 0$.

The proof of this corollary can be found for instance in [20].

If (4.94) does not hold with equality, the definition of duality gap, stated below, is of interest.

Definition 4.8 (Duality gap). *The difference*

$$\sup\{\phi(\lambda, \mu) | \mu \geq 0\} - \inf\{f(x) | h(x) = 0, g(x) \leq 0\}$$

is called the duality gap of the dual problems (4.49)–(4.50) and (4.52)–(4.53).

$\square$

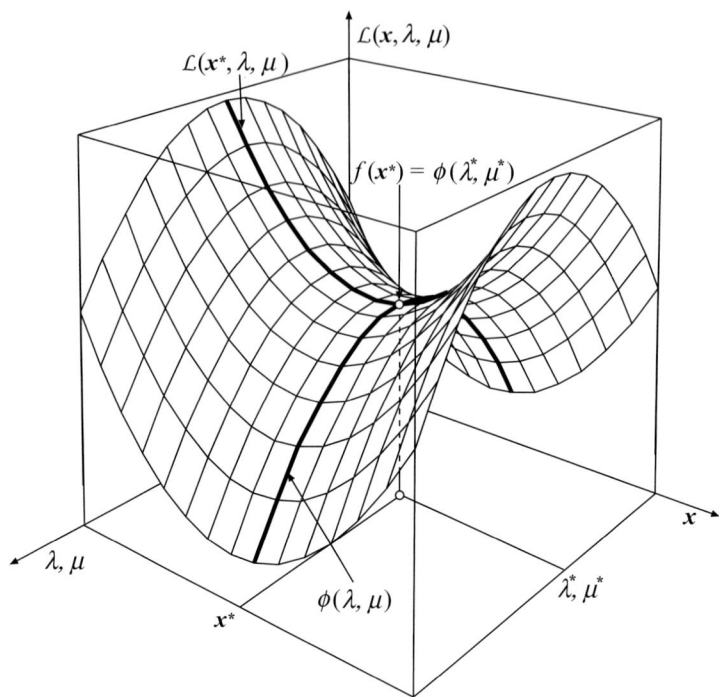

Fig. 4.11. Graphical illustration of saddle point $(\boldsymbol{x}^*, \boldsymbol{\lambda}^*, \boldsymbol{\mu}^*)$ of the Lagrangian

Illustrative Example 4.10 (Duality gap). Consider the same problem as the one considered in the Illustrative Example 4.7,

$$\underset{x_1, x_2}{\text{minimize}} \quad -2x_1 + x_2$$

subject to

$$x_1 + x_2 = \frac{5}{2}$$
$$(x_1, x_2) \in \boldsymbol{X},$$

where $\boldsymbol{X} = \{(0,0), (0,2), (2,0), (2,2), (5/4, 5/4)\}$. Its dual function is given by the explicit expression

$$\phi(\lambda) = \begin{cases} -2 + \dfrac{3\lambda}{2} & \text{if} \quad \lambda \leq -1 \\ -4 - \dfrac{\lambda}{2} & \text{if} \quad -1 \leq \lambda \leq 2 \\ -\dfrac{5\lambda}{2} & \text{if} \quad \lambda \geq 2, \end{cases}$$

shown in Fig. 4.12. The optimal solution of the dual problem is $\lambda^* = -1$ with objective function value $\phi(\lambda^*) = -7/2$.

4.4 Duality in Nonlinear Programming

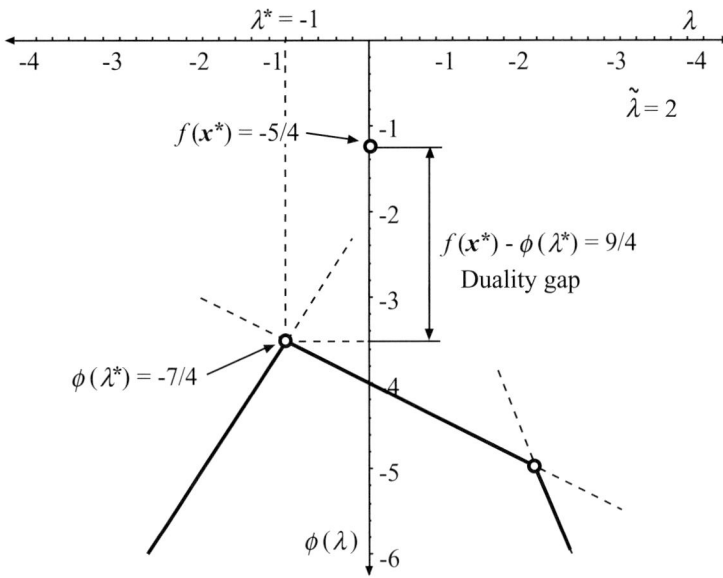

Fig. 4.12. Graphical illustration of the duality gap for the Illustrative Example 4.10

It is easy to verify that $x_1^* = 5/4$ and $x_2^* = 5/4$ is the optimal solution of the primal problem with an objective function value equal to $f(\boldsymbol{x}^*) = -5/4$.

Therefore, the difference $f(\boldsymbol{x}^*) - \phi(\boldsymbol{\lambda}^*) = -5/4 + 7/2 = 9/4$ is the duality gap shown in Fig. 4.12. □

If a multiplier vector solves the dual problem and there is no duality gap, i.e., if the relationship (4.94) is strictly satisfied, then the solutions of the Lagrangian associated with this multiplier vector are optimal solutions of the primal problem. This result introduces a way to solve the primal problem by means of solving the dual problem. The main result consists of obtaining conditions that guarantee the nonexistence of the duality gap. This is the situation of problems in which convexity conditions hold.

Theorem 4.4 (Duality theorem for convex problems). *Consider a convex mathematical programming problem such as (4.49)–(4.50). If $\boldsymbol{x}^*$ solves the primal problem, then its associated vector of multipliers $(\boldsymbol{\lambda}^*, \boldsymbol{\mu}^*)$ solves the dual problem, and $(\boldsymbol{\mu}^*)^T \boldsymbol{g}(\boldsymbol{x}^*) = 0$. Conversely, if $(\boldsymbol{\lambda}^*, \boldsymbol{\mu}^*)$ solves the dual problem, and there exists a solution, $\boldsymbol{x}^*$, of the Lagrangian associated with this vector of multipliers such that $(\boldsymbol{\mu}^*)^T \boldsymbol{g}(\boldsymbol{x}^*) = 0$, then $\boldsymbol{x}^*$ is an optimal solution of the primal problem.* □

Proof. If $\boldsymbol{x}^*$ solves the primal problem and $(\boldsymbol{\mu}^*, \boldsymbol{\lambda}^*)$ is its associated vector of multipliers, we know that they should be a solution of the KKT conditions

$$\nabla f(\boldsymbol{x}^*) + \boldsymbol{\lambda}^* \nabla h(\boldsymbol{x}^*) + \boldsymbol{\mu}^* \nabla g(\boldsymbol{x}^*) = \mathbf{0}$$
$$\boldsymbol{\mu}^* g(\boldsymbol{x}^*) = \mathbf{0}$$
$$\boldsymbol{\mu}^* \geq \mathbf{0}$$
$$g(\boldsymbol{x}^*) \leq \mathbf{0}$$
$$h(\boldsymbol{x}^*) = \mathbf{0}.$$

The minimization problem defining $\phi(\boldsymbol{\lambda}^*, \boldsymbol{\mu}^*)$ is

$$\phi(\boldsymbol{\lambda}^*, \boldsymbol{\mu}^*) = \text{Inf}_{\boldsymbol{x}} \ \{f(\boldsymbol{x}) + \boldsymbol{\lambda}^{*T} h(\boldsymbol{x}) + \boldsymbol{\mu}^{*T} g(\boldsymbol{x})\} \ .$$

The condition coming from the above KKT system

$$\nabla f(\boldsymbol{x}^*) + \boldsymbol{\lambda}^* \nabla h(\boldsymbol{x}^*) + \boldsymbol{\mu}^* \nabla g(\boldsymbol{x}^*) = \mathbf{0} \ ,$$

together with the convexity of the objective function (note that $\boldsymbol{\mu}^* \geq \mathbf{0}$)

$$f(\boldsymbol{x}) + \boldsymbol{\lambda}^{*T} h(\boldsymbol{x}) + \boldsymbol{\mu}^{*T} g(\boldsymbol{x}) \ ,$$

imply that $\boldsymbol{x}^*$ is a local minimizer for the problem to determine $\phi(\boldsymbol{\lambda}^*, \boldsymbol{\mu}^*)$ and therefore

$$\phi(\boldsymbol{\lambda}^*, \boldsymbol{\mu}^*) = f(\boldsymbol{x}^*) \ ,$$

and by Corollary 4.2 the pair $(\boldsymbol{\lambda}^*, \boldsymbol{\mu}^*)$ is an optimal solution of the dual problem. Note that the condition

$$\boldsymbol{\mu}^* g(\boldsymbol{x}^*) = \mathbf{0}$$

also holds.

Conversely, assume that the pair $(\boldsymbol{\lambda}^*, \boldsymbol{\mu}^*)$ is an optimal solution for the dual problem, and that $\boldsymbol{x}^*$ is a corresponding solution of the associated Lagrangian such that $(\boldsymbol{\mu}^*)^T g(\boldsymbol{x}^*) = 0$. Applying the KKT conditions to the dual, it can be concluded that

$$\nabla_{\boldsymbol{\mu}} \phi(\boldsymbol{\lambda}^*, \boldsymbol{\mu}^*) \leq \mathbf{0}, \quad \nabla_{\boldsymbol{\lambda}} \phi(\boldsymbol{\lambda}^*, \boldsymbol{\mu}^*) = \mathbf{0} \ .$$

But taking into account expressions (4.65) and (4.66), we obtain

$$g(\boldsymbol{x}^*) \leq \mathbf{0}; \ h(\boldsymbol{x}^*) = \mathbf{0} \ .$$

This implies that $\boldsymbol{x}^*$ is feasible for the primal. Then

$$\phi(\boldsymbol{\lambda}^*, \boldsymbol{\mu}^*) = f(\boldsymbol{x}^*) + \boldsymbol{\lambda}^{*T} h(\boldsymbol{x}^*) + \boldsymbol{\mu}^{*T} g(\boldsymbol{x}^*) = f(\boldsymbol{x}^*) \ ,$$

and by Corollary 4.2 we conclude the proof. □

Illustrative Example 4.11 (Dual problem). Consider the following (primal) problem:

$$\text{minimize}_{x_1, x_2} \quad z_P = x_1^2 + x_2^2$$

subject to

$$x_1 + x_2 \leq 4$$
$$-x_1 \leq 0$$
$$-x_2 \leq 0 .$$

The unique optimal solution to this problem is $(x_1^*, x_2^*) = (0,0)$ with the optimal objective function value equal to 0. Let us formulate the dual problem. We need first to find the dual function

$$\begin{aligned}\phi(\mu_1, \mu_2, \mu_3) &= \min_{\boldsymbol{x} \in \mathbb{R}^2} \{x_1^2 + x_2^2 + \mu_1(x_1 + x_2 - 4) + \mu_2(-x_1) + \mu_3(-x_2)\} \\ &= \min_{x_1}\{x_1^2 + (\mu_1 - \mu_2)x_1\} + \min_{x_2}\{x_2^2 + (\mu_1 - \mu_3)x_2\} - 4\mu_1 .\end{aligned}$$

After algebra manipulation we explicitly find the dual function

$$\phi(\boldsymbol{\mu}) = -\frac{1}{4}[(\mu_1 - \mu_2)^2 + (\mu_1 - \mu_3)^2] - 4\mu_1 ,$$

an the dual problem becomes

$$\text{supremum}_{\boldsymbol{\mu} \geq 0} \quad \phi(\boldsymbol{\mu}) .$$

Note that the dual function is concave. The dual problem optimum is attained at $\mu_1^* = \mu_2^* = \mu_3^* = 0$ and $z_D^* = \phi(\boldsymbol{\mu}^*) = 0$. □

In the following, relevant sensitivity results are derived.

Theorem 4.5 (Sensitivities are given by dual variables). *Consider the following general nonlinear primal problem:*

$$\text{minimize}_{\boldsymbol{x}} \quad z_P = f(\boldsymbol{x}) \tag{4.95}$$

subject to

$$\begin{aligned}\boldsymbol{h}(\boldsymbol{x}) &= \boldsymbol{a} : \boldsymbol{\lambda} \\ \boldsymbol{g}(\boldsymbol{x}) &\leq \boldsymbol{b} : \boldsymbol{\mu} ,\end{aligned} \tag{4.96}$$

where $\boldsymbol{\lambda}$ and $\boldsymbol{\mu}$ are the dual variables and assume that its solution is a regular point and that no degenerate inequality constraints exist. Assume also that sufficient conditions (4.4) for a minimum hold. Then, the values of the dual variables, also called shadow prices, give the sensitivity of the objective function optimal value to changes in the primal constraints, i.e.,

$$\left.\frac{\partial f(\boldsymbol{x})}{\partial a_k}\right|_{\boldsymbol{x}^*} = -\lambda_k, \quad \left.\frac{\partial f(\boldsymbol{x})}{\partial b_j}\right|_{\boldsymbol{x}^*} = -\mu_j . \tag{4.97}$$

□

Proof. If the optimal values for the primal and dual exist, they satisfy

$$f(x) = \phi(\lambda, \mu) = f(x) + \lambda^T h(x) + \mu^T g(x) - \lambda^T a - \mu^T b$$
$$= f(x) + \lambda^T h(x) + \mu^T g(x) - \sum_{k=1}^{\ell} \lambda_k a_k - \sum_{j=1}^{m} \mu_j b_j ,$$

then, we have (4.97), which proves that the values of the dual variables are the sensitivities of the objective function to changes in the right-hand sides of (4.96). □

It is relevant at this point to stress that in some instances it may be important to look at duality partially, in the sense that some but not all constraints are dualized. Those restrictions not dualized are incorporated as such in the definition of the dual function. This is in fact the underlying driving motivation for the Lagrangian decomposition techniques.

4.5 Illustration of Duality and Separability

To illustrate the relationship between primal and dual problems and separability, consider a single commodity that is produced and supplied under the following conditions:

1. Different producers compete in a single market.
2. Demand in every time period of a study horizon has to be strictly satisfied.
3. No storage capacity is available.
4. Producers have different production cost functions and production constraints.

We consider two different approaches to this problem (i) the centralized or primal approach and (ii) the competitive or dual approach.

A centralized (primal) approach to this problem is as follows. Producers elect a system operator which is in charge of guaranteeing the supply of demand for every time period. To this end, the system operator:

1. is entitled to know the actual production cost function of every producer and its production constraints,
2. knows the actual commodity demand in every time period,
3. has the power to dictate the actual production of every producer.

The main elements of this problem are:

Data.
$c_{ti}(p_{ti})$: production cost of producer i in period t to produce p_{ti}
Π_i: producer i feasible operating region
d_t: demand in period t

4.5 Illustration of Duality and Separability

T: number of time periods.
I: number of producers.

Variables.
p_{ti}: amount of commodity produced by producer i during period t.

Constraints.

$$\sum_{i=1}^{I} p_{ti} = d_t; \quad t = 1, \ldots, T \quad (4.98)$$

$$p_{ti} \in \Pi_i; \quad t = 1, \ldots, T; \; i = 1, \ldots, I. \quad (4.99)$$

Constraints (4.98) enforce for every time period the strict supply of demand, and constraints (4.99) enforce the operational limits of every producer.

Function to Be Optimized. Because the operator has the objective of minimizing total production cost, it has to solve the following problem:

$$\underset{p_{ti}}{\text{minimize}} \quad \sum_{t=1}^{T} \sum_{i=1}^{I} c_{ti}(p_{ti}), \quad (4.100)$$

which is the sum of the production costs of all producers over all time periods.

Thus, the primal problem is as follows:

$$\underset{p_{ti}}{\text{minimize}} \quad \sum_{t=1}^{T} \sum_{i=1}^{I} c_{ti}(p_{ti}) \quad (4.101)$$

subject to

$$\sum_{i=1}^{I} p_{ti} = d_t; \quad t = 1, \ldots, T \quad (4.102)$$

$$p_{ti} \in \Pi_i; \quad t = 1, \ldots, T; \; i = 1, \ldots, I. \quad (4.103)$$

The solution of this problem provides the optimal level of production of every producer. Then the system operator communicates its optimal production to every producer, which actually produces this optimal value.

Note that the approach explained is a centralized one. The system operator has knowledge of the production cost functions and the production constraints of every producer can determine the actual (optimal) production of every producer, and has the power to force every producer to produce its centrally determined optimal value.

We analyze then the structure of the dual problem of the cost minimization primal problem (4.101)–(4.103).

If partial duality over constraints (4.102) is carried out, the Lagrangian function has the form

$$\mathcal{L}(p_{ti}, \lambda_t) = \sum_{t=1}^{T} \sum_{i=1}^{I} c_{ti}(p_{ti}) + \sum_{t=1}^{T} \lambda_t \left(d_t - \sum_{i=1}^{I} p_{ti} \right). \qquad (4.104)$$

Arranging terms, this Lagrangian function can be expressed as

$$\mathcal{L}(p_{ti}, \lambda_t) = \sum_{i=1}^{I} \left[\sum_{t=1}^{T} \left(c_{ti}(p_{ti}) - \lambda_t \, p_{ti} \right) \right] + \sum_{t=1}^{T} \lambda_t \, d_t. \qquad (4.105)$$

The evaluation of the dual function for a given value of the Lagrange multipliers $\lambda_t = \widehat{\lambda}_t$ $(t = 1, \ldots, T)$ is achieved by solving the problem

$$\underset{p_{ti}}{\text{minimize}} \quad \mathcal{L}(p_{ti}, \widehat{\lambda}_t) \qquad (4.106)$$

subject to

$$p_{ti} \in \Pi_i; \quad t = 1, \ldots, T; \; i = 1, \ldots, I. \qquad (4.107)$$

Using the explicit expression (4.105) for the Lagrangian function, the problem presented above entails

$$\underset{p_{ti}}{\text{minimize}} \quad \sum_{i=1}^{I} \left[\sum_{t=1}^{T} \left(c_{ti}(p_{ti}) - \widehat{\lambda}_t \, p_{ti} \right) \right] + \sum_{t=1}^{T} \widehat{\lambda}_t \, d_t \qquad (4.108)$$

subject to

$$p_{ti} \in \Pi_i; \quad t = 1, \ldots, T; \; i = 1, \ldots, I. \qquad (4.109)$$

Taking into account that the expression $\sum_{t=1}^{T} \widehat{\lambda}_t \, d_t$ is constant, the previous problem reduces to

$$\underset{p_{ti}}{\text{minimize}} \quad \sum_{i=1}^{I} \left[\sum_{t=1}^{T} \left(c_{ti}(p_{ti}) - \widehat{\lambda}_t \, p_{ti} \right) \right] \qquad (4.110)$$

subject to

$$p_{ti} \in \Pi_i; \quad t = 1; \ldots, T; \; i = 1, \ldots, I. \qquad (4.111)$$

This problem has a very remarkable property of being separable, i.e., it decomposes by producer. The individual problem of producer j is therefore (note the sign change of the objective function)

$$\underset{p_{tj}}{\text{maximize}} \quad \sum_{t=1}^{T} \left[\widehat{\lambda}_t p_{tj} - c_{tj}(p_{tj}) \right] \qquad (4.112)$$

subject to

$$p_{tj} \in \Pi_j; \quad t = 1, \ldots, T. \qquad (4.113)$$

Therefore, the evaluation of the dual function can be achieved by solving problem (4.112)–(4.113) for every producer, namely, I times.

What is again particularly remarkable is the interpretation of problem (4.112)–(4.113); it represents the maximization of the profit of producer j subject to its production constraints. The Lagrange multipliers λ_t ($t = 1, \ldots, T$) are naturally interpreted as the selling prices of the commodity in periods $t = 1, \ldots, T$.

It can be concluded that to evaluate the dual function, every producer has to solve its own profit maximization problem, and then the operator adds the maximum profit of every producer and the constant term $\sum_{t=1}^{T} \widehat{\lambda}_t d_t$.

The duality theorem states that, under certain convexity assumptions, the solutions of the primal and the dual problems are the same in terms of the objective function value. It also states that the evaluation of the dual function for the optimal values of Lagrangian multipliers λ_t^* ($t = 1, \ldots, T$) results in optimal values for productions p_{ti}^* ($t = 1, \ldots, T; i = 1, \ldots, I$). As a consequence, the solution of the primal problem can be obtained solving the dual one.

Duality theory also states that under very general conditions in the structure of the primal problem, the dual function is concave, and therefore, it can be maximized using a simple steepest-ascent procedure.

To use a steepest-ascent procedure, it is necessary to know the gradient of the dual function. Duality theory provides us with a simple expression for the gradient of the dual function [see expressions (4.65) and (4.66)]. For productions p_{ti} ($t = 1, \ldots, T; i = 1, \ldots, I$) the components of the gradient of the dual function are

$$d_t - \sum_{i=1}^{I} p_{ti}; \quad t = 1, \ldots, T. \qquad (4.114)$$

That is, the component t of the gradient of the dual function is the actual demand imbalance in period t.

A steepest-ascent procedure to solve the dual problem (and therefore the primal one) includes the following steps:

1. The system operator sets up initial values of Lagrange multipliers (selling prices).
2. Using the Lagrangian multipliers presented above, every producer maximizes its own profit meeting its own production constraints.

180   4 Duality

3. The operator calculates the demand imbalance in every time period (components of the gradient of the dual function).
4. If the norm of the gradient of the dual function is not sufficiently small, a steepest-ascent iteration is performed and the procedure continues in 2. Otherwise the procedure concludes, the dual solution has been attained.

The preceding four-steps algorithm can be interpreted in the framework of a competitive market. This is done below:

1. The market operator broadcasts to producers the initial ($k = 1$) value of selling price for every period, $\lambda_t^{(k)}$ ($t = 1, \ldots, T$).
2. Every producer j solves its profit maximization problem,

$$\underset{p_{tj}}{\text{maximize}} \quad \sum_{t=1}^{T} \left(\lambda_t^{(k)} p_{tj} - c_{tj}(p_{tj}) \right) \quad (4.115)$$

subject to

$$p_{tj} \in \Pi_j; \quad \forall t, \quad (4.116)$$

and sends to the market operator its optimal production schedule, $p_{tj}^{(k)}$ ($t = 1, \ldots, T$ ($j = 1, \ldots, I$).
3. The market operator computes the demand imbalance in every time period

$$d_t - \sum_{i=1}^{I} p_{ti}^{(k)}; \quad t = 1, \ldots, T, \quad (4.117)$$

and updates prices proportionally to imbalances

$$\lambda_t^{(k+1)} = \lambda_t^{(k)} + K \left(d_t - \sum_{i=1}^{I} p_{ti}^{(k)} \right); \quad t = 1, \ldots, T, \quad (4.118)$$

where K is a proportionality constant.
4. If prices remain constant in two consecutive rounds, stop; the optimal solution (market equilibrium) has been attained. Otherwise, continue with Step 2.

This dual solution algorithm can be interpreted as the functioning of a multiround auction. In fact, economists call this iterative procedure a Walrasian auction [29].

A Walrasian auction preserves the privacy of the corporate information (cost function and production constraints) of every producer and allows them to maximize their respective own profits.

Some comments are in order. The primal approach is centralized, and the system operator has complete information of the production cost function and operation constraints of every producer. Moreover, it is given the power of deciding how much every producer should produce.

As opposed to the primal approach, the dual approach is decentralized in nature and every producer maintains the privacy of its production cost function and operation constraints. Furthermore, every producer decides how much to produce to maximize its own benefits.

Under some assumptions, duality theory guarantees that both approaches yield identical results.

4.6 Concluding Remarks

This chapter addresses the classical topic of duality in both linear and nonlinear programming. After stating the Karush–Kuhn–Tucker optimality conditions, the dual problem is derived for both linear and nonlinear problems. The chapter concludes illustrating how duality can be used to attain a separable dual problem.

Dual variables are the crucial communication instruments used by the subproblems that result from applying any decomposition procedure to a decomposable problem. Moreover, dual variables are sensitivity parameters that inform on the objective function changes due to changes on the right-hand sides of the constraints.

This chapter provides background material of high interest to address the following chapters. Additional material on duality can be found in the classical manuals by Bazaraa et al. [20] and by Luenberger [23].

4.7 Exercises

Exercise 4.1. Solve the following problems by applying KKTCs:

1.
$$\underset{x_1, x_2}{\text{maximize}} \quad z = x_1 - \exp(-x_2)$$

subject to

$$-\sin x_1 + x_2 \leq 0$$
$$x_1 \leq 3.$$

2.
$$\underset{x_1, x_2}{\text{minimize}} \quad z = (x_1 - 4)^2 + (x_2 - 3)^2$$

subject to

$$x_1^2 - x_2 \leq 0$$
$$x_2 \leq 4.$$

3.
$$\text{maximize}_{x_1, x_2} \quad z = x_2$$

subject to
$$\begin{aligned} x_1^2 - x_2^2 &\leq 4 \\ -x_1^2 + x_2 &\leq 0 \\ x_1, x_2 &\geq 0 \,. \end{aligned}$$

Exercise 4.2. Consider the following problem:
$$\text{minimize}_{x_1, x_2, x_3, x_4} \quad z = 3x_1 + x_2 - x_4$$

subject to
$$\begin{aligned} x_1 + x_2 - x_3 - x_4 &= 4 \\ 2x_1 - x_2 + x_4 &\leq 0 \\ 3x_2 + x_3 - 2x_4 &\geq 1 \\ x_1, x_2 &\geq 0 \\ x_4 &\leq 0 \,. \end{aligned}$$

Build the dual problem, solve both problems, and check that the optimal values coincide.

Exercise 4.3. Consider the following problem:
$$\text{minimize}_{x_1, x_2} \quad z = x_1$$

subject to
$$x_1^2 + x_2^2 = 1 \,.$$

Compute the dual function and show that it is a concave function. Draw it. Find the optimal solutions for the dual and primal problems by means of comparing their objective functions.

Exercise 4.4. Consider the following problem:
$$\text{minimize}_{x, y} \quad z = x^2 + y^2$$

subject to
$$\begin{aligned} x &= 5 \\ xy &\geq 3 \,. \end{aligned}$$

Solve the primal problem. Compute explicitly the dual function and solve the dual problem.

Exercise 4.5. Consider the following problem:

$$\begin{array}{c} \text{minimize} \\ x_1, x_2 \end{array} \quad z = (x_1 - 4)^2 + (x_2 - 1)^2$$

subject to

$$2x_1 + x_2 \geq 6$$
$$-x_1^2 + x_2 \geq -2 .$$

1. Find the optimal solution for this primal problem geometrically.
2. Obtain graphically the optimal multipliers values associated with both constraints, i.e., the optimal solution of the dual problem.
3. Verify that KKT conditions hold.

Exercise 4.6. Consider the following problem:

$$\begin{array}{c} \text{minimize} \\ x_1, x_2 \end{array} \quad z = 2x_1^2 + x_2^2 - 2x_1 x_2 - 6x_2 - 4x_1$$

subject to

$$x_1^2 + x_2^2 = 1$$
$$-x_1 + 2x_2 \leq 0$$
$$x_1 + x_2 \leq 8$$
$$x_1, x_2 \geq 0 .$$

1. Solve the problem using the KKTCs.
2. Compute explicitly the dual function.
3. Does a duality gap exist?

Exercise 4.7. Consider the same problem as the one in Illustrative Example 4.7, p. 165. Obtain the explicit expression of the dual function.

Exercise 4.8. Find the minimum length of a ladder that must lean against a wall if a box of dimensions a and b is placed right at the corner of that same wall (Fig. 4.13). Formulate and solve the primal and dual problems.

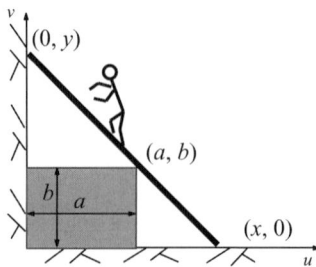

Fig. 4.13. Illustration of the ladder problem in Exercise 4.8

Exercise 4.9. Consider the following problem:

$$\underset{x_1,x_2}{\text{minimize}} \quad z = x_1 + x_2$$

subject to

$$x_1 - x_2 = 0$$
$$(x_1, x_2) \in X,$$

where $X = \{(-4,0), (4,0), (0,-4), (0,4), (-1,-1), (1,1)\}$.

1. Compute explicitly the dual function.
2. Obtain graphically the solution of the dual problem.
3. Solve the primal problem. Does a duality gap exist?

Exercise 4.10. Find the closest point of the surface $xyz = 1$ to the origin for $x, y, z \geq 0$. Solve the primal and dual problems.

Exercise 4.11. Write and solve the dual problem of the following problem

$$\underset{x_1,x_2,x_3,x_4}{\text{maximize}} \quad z = 3x_1 - 4x_2 + 9x_3 + x_4$$

subject to

$$\begin{aligned} x_1 - 5x_2 + x_3 &\geq 0 \\ 3x_1 - 5x_2 + x_3 &\geq 10 \\ x_2, x_3 &\geq 0 \\ x_4 &\leq 0. \end{aligned}$$

Exercise 4.12. Using the KKT conditions find the possible candidates (x_1, x_2) and $(\lambda, \mu_1, \mu_2, \mu_3)$ for solving the following problem:

$$\underset{x_1,x_2}{\text{minimize}} \quad Z = -x_1 + x_2$$

subject to

$$-x_1^2 + x_2 = 0 \ : \ \lambda$$
$$x_1^2 + x_2^2 - 4 \leq 0 \ : \ \mu_1$$
$$-x_1 \leq 0 \ : \ \mu_2$$
$$-x_2 \leq 0 \ : \ \mu_3 \ ,$$

and its dual, where $\lambda, \mu_1, \mu_2,$ and μ_3 are the dual variables.

Exercise 4.13. Consider the case of three producers of a good competing in a single market during a time horizon of 4 months with demands $5, 6, 7,$ and 4, respectively, and assume that no storage capacity is available. Suppose that the producers have the following production cost functions:

$$c_{t1}(p_{t1}) = 1 + 2p_{t1}^2, \quad c_{t2}(p_{t2}) = 2 + 3p_{t2}^2, \quad \text{and} \quad c_{t3}(p_{t3}) = 3 + p_{t3}^2,$$

where the subindex ti refers to producer i and period t, and it is assumed that the production costs are independent of the period t. Assume also that the maximum productions of the three producers are respectively,

$$C_1 = 2, \quad C_2 = 3, \quad \text{and} \quad C_3 = 3 \ .$$

1. Solve the problem using the centralized approach described in Sect. 4.5 and show that the optimal solution gives a minimum cost of $95.1636 for the following productions:

Producer	Time period			
	1	2	3	4
1	1.36	1.80	2.00	1.09
2	0.91	1.20	2.00	0.73
3	2.73	3.00	3.00	2.18

2. Solve the problem using the decentralized approach and two different values of the K constant to update the prices. Compare the number of iterations required for a given error in the prices of the goods.
3. Use a large value of K and check that the process blows up.

5
Decomposition in Nonlinear Programming

5.1 Introduction

This chapter analyzes nonlinear problems with decomposable structure. The complicating constraint and the complicating variable cases are both considered, and three procedures are analyzed for dealing with the complicating constraint case: the Lagrangian relaxation (LR), the augmented Lagrangian decomposition (ALD), and the optimality condition decomposition (OCD). The last procedure presents the most efficient computational behavior in most cases. Finally, for the complicating variable case, the Benders decomposition (BD) procedure is reviewed.

5.2 Complicating Constraints

As previously stated in the linear case, complicating constraints are constraints that if relaxed, the resulting problem decomposes in several simpler problems. Alternatively, the resulting problem attains such a structure that its solution is simple. The decomposable case is the most interesting one in practice and the one mostly considered in this chapter. In Sects. 5.3, 5.4, and 5.5, three different decomposition algorithms are studied. Sections 5.3 and 5.4 present the basic theory of the Lagrangian relaxation (LR) and the augmented Lagrangian (AL) decomposition procedures, respectively. Section 5.5 presents a partitioning technique based on decomposing the optimality conditions of the original problem, which is denominated optimality condition decomposition (OCD).

5.3 Lagrangian Relaxation

In this section, the Lagrangian relaxation (LR) decomposition procedure is explained.

5.3.1 Decomposition

For the LR technique to be applied advantageously to a mathematical programming problem, the problem should have the following structure:

$$\underset{x}{\text{minimize}} \quad f(x) \tag{5.1}$$

subject to

$$a(x) = 0 \tag{5.2}$$
$$b(x) \leq 0 \tag{5.3}$$
$$c(x) = 0 \tag{5.4}$$
$$d(x) \leq 0, \tag{5.5}$$

where $f(x) : \mathbb{R}^n \to \mathbb{R}$, $a(x) : \mathbb{R}^n \to \mathbb{R}^{n_a}$, $b(x) : \mathbb{R}^n \to \mathbb{R}^{n_b}$, $c(x) : \mathbb{R}^n \to \mathbb{R}^{n_c}$, $d(x) : \mathbb{R}^n \to \mathbb{R}^{n_d}$, and n_a, n_b, n_c, and n_d are scalars. Constraints $c(x) = 0$ and $d(x) \leq 0$ are the complicating constraints, i.e., constraints that if relaxed, problem (5.1)–(5.5) becomes drastically simplified. The Lagrangian function (LF) is defined as [20, 23]

$$\mathcal{L}(x, \lambda, \mu) = f(x) + \lambda^T c(x) + \mu^T d(x), \tag{5.6}$$

where λ and μ are the Lagrange multiplier vectors. Under regularity and convexity assumptions (Sect. 4.2 in Chap. 4) the dual function (DF) is defined as

$$\phi(\lambda, \mu) = \underset{x}{\text{minimum}} \quad \mathcal{L}(x, \lambda, \mu) \tag{5.7}$$

subject to

$$a(x) = 0 \tag{5.8}$$
$$b(x) \leq 0. \tag{5.9}$$

The dual function is concave and in general nondifferentiable [23]. This is a fundamental fact that is exploited in the algorithms described in this chapter. The dual problem (DP) is then defined as

$$\underset{\lambda, \mu}{\text{maximize}} \quad \phi(\lambda, \mu) \tag{5.10}$$

subject to

$$\mu \geq 0. \tag{5.11}$$

The LR decomposition procedure is attractive if the dual function is easily evaluated for given values $\bar{\lambda}$ and $\bar{\mu}$ of the multiplier vectors λ and μ, respectively. In other words, if it is easy to solve the so-called relaxed primal problem (RPP) for given $\bar{\lambda}$ and $\bar{\mu}$, i.e., the problem

5.3 Lagrangian Relaxation

$$\underset{x}{\text{minimize}} \quad \mathcal{L}(x, \bar{\lambda}, \bar{\mu}) \quad (5.12)$$

subject to

$$a(x) = 0 \quad (5.13)$$
$$b(x) \leq 0. \quad (5.14)$$

The above problem typically decomposes into subproblems, i.e.,

$$\underset{x_i;\, i=1,\ldots,n}{\text{minimize}} \quad \sum_{i=1}^{n} \mathcal{L}_i(x_i, \bar{\lambda}, \bar{\mu}) \quad (5.15)$$

subject to

$$a_i(x_i) = 0; \quad i = 1, \ldots, n \quad (5.16)$$
$$b_i(x_i) \leq 0; \quad i = 1, \ldots, n. \quad (5.17)$$

This decomposition facilitates its solution, and normally allows physical and economical interpretations. The above problem is called the decomposed primal problem (DPP). Then, the resulting subproblems can be solved in parallel. Under convexity assumptions, the duality theorem says that

$$f(x^*) = \phi(\lambda^*, \mu^*), \quad (5.18)$$

where x^* is the minimizer for the primal problem and (λ^*, μ^*) is the maximizer for the dual problem.

In the nonconvex case, given a feasible solution for the primal problem, x, and a feasible solution for the dual problem, (λ, μ), the weak duality theorem says that

$$f(x^*) \geq \phi(\lambda^*, \mu^*). \quad (5.19)$$

In the convex case, the solution of the dual problem provides the solution of the primal problem. In the nonconvex case the objective function value at the optimal solution of the dual problem provides a lower bound to the objective function value at the optimal solution of the primal problem. The difference between the optimal objective functions of the primal and dual problems is the duality gap. Most engineering and science mathematical programming problems are nonconvex but the duality gap is relatively small in most cases. The solution of the dual problem is called Phase 1 of the LR procedure.

Illustrative Example 5.1 (The LR decomposition). Consider the following optimization problem:

$$\underset{x, y}{\text{minimize}} \quad f(x, y) = x^2 + y^2$$

subject to
$$\begin{aligned} -x - y &\le -4 \\ x &\ge 0 \\ y &\ge 0. \end{aligned}$$

The Lagrangian function is
$$\mathcal{L}(x, y, \mu) = x^2 + y^2 + \mu(-x - y + 4)$$

subject to
$$\begin{aligned} x &\ge 0 \\ y &\ge 0. \end{aligned}$$

Primal and Dual Problems

From the Karush–Kuhn–Tucker conditions we obtain
$$\frac{\partial \mathcal{L}(x, y, \mu)}{\partial x} = 2x - \mu = 0 \Rightarrow x = \frac{\mu}{2}$$
$$\frac{\partial \mathcal{L}(x, y, \mu)}{\partial y} = 2y - \mu = 0 \Rightarrow y = \frac{\mu}{2}.$$

The solution of this problem is
$$x = 2, \quad y = 2, \quad \mu = 4$$
as shown in Figs. 5.1a, b.

To obtain the dual function $\phi(\mu)$ one needs to obtain the infimum of the Lagrangian function,
$$\underset{x, y}{\text{infimum}} \quad \mathcal{L}(x, y, \mu) ,$$
which is attained at the following point:
$$x = \frac{\mu}{2}, \ y = \frac{\mu}{2}.$$

It is a minimum because the Hessian at this point becomes
$$\begin{pmatrix} 2 & 0 \\ 0 & 2 \end{pmatrix},$$
which is positive definite.

Then, the dual function is
$$\phi(\mu) = 4\mu - \mu^2/2 ,$$
and the dual problem becomes

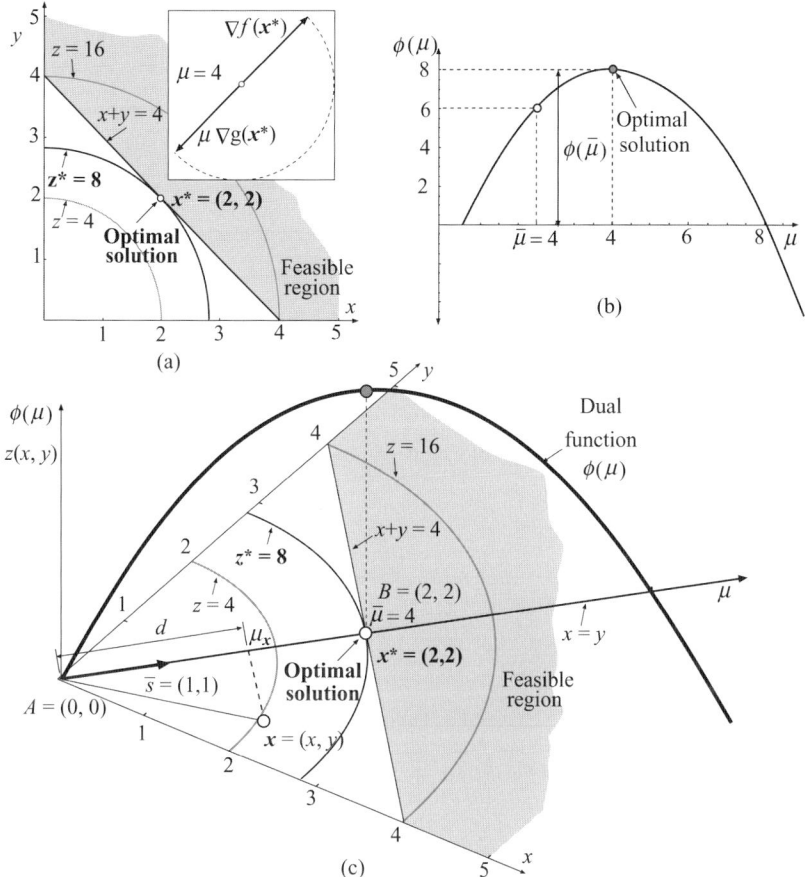

Fig. 5.1. Illustration of the problem in Illustrative Example 5.1; (**a**) minimization on the plane XY; (**b**) dual function; and (**c**) 3-D graphical interpretation of dual and primal problems

$$\text{maximize}_{\mu} \quad 4\mu - \mu^2/2$$

subject to
$$\mu \geq 0 ,$$

whose solution is
$$\mu = 4, \quad z = 8 .$$

Note that the optimal value of the Lagrangian function for a given value of μ is always attained at the point $x = y = \mu/2$. Being both x and y functions of μ, the Lagrangian function $\mathcal{L}(x, y, \mu)$ can be expresed as $\mathcal{L}(x, y)$.

Geometrical Interpretation

An insightful geometrical interpretation is provided below.

For a given point $\boldsymbol{x} = (x, y)$, it might be convenient to compute the corresponding value of $\mu_{\boldsymbol{x}}$ (see Fig. 5.1c). For points $A = (x, y) = (0, 0)$ and $B = (x, y) = (2, 2)$, $\mu_A = 0$ and $\mu_B = 4$, respectively. Thus, for point $\boldsymbol{x}$

$$\mu_{\boldsymbol{x}} = \frac{(\mu_B - \mu_A)\, d}{\overline{AB}}, \tag{5.20}$$

where d is the projection of vector $\boldsymbol{x}$ on the straight line $x = y$ and $\overline{AB}$ is the distance between points A and B. Considering the vector $\boldsymbol{s}^T = (1, 1)$,

$$d = \frac{\boldsymbol{s}^T \boldsymbol{x}}{\sqrt{\boldsymbol{s}^T \boldsymbol{s}}} = \frac{x + y}{\sqrt{2}}. \tag{5.21}$$

Substituting (5.21) into (5.20),

$$\mu_{\boldsymbol{x}} = \frac{(\mu_B - \mu_A)(x + y)}{\overline{AB}\sqrt{2}} = \frac{4(x + y)}{2\sqrt{2}\sqrt{2}} = x + y$$

and the Lagrangian function is

$$\mathcal{L}(x, y) = x^2 + y^2 + (x + y)(-x - y + 4).$$

It is possible to carry out a 3-D graphical interpretation of all functions involved in this problem as shown in Figs. 5.1c and 5.2. In the last one, the function to be optimized $f(x, y) = x^2 + y^2$, the Lagrangian function $\mathcal{L}(x, y, \mu) = \mathcal{L}(x, y)$, the intersection of both functions and the dual function $\phi(\mu)$ are shown. It is clear that the optimal solution of the problem is attained at the saddle point of the Lagrangian function, i.e., the maximum of the dual function and the minimum of the primal one coincides.

Decomposed Solution

The LR decomposition reconstructs and solves the dual problem in a distributed fashion as is illustrated below. Consider the dual problem

$$\underset{\mu}{\text{maximize}} \quad \sum_{i=1}^{2} \phi_i(\mu)$$

subject to

$$\mu \geq 0,$$

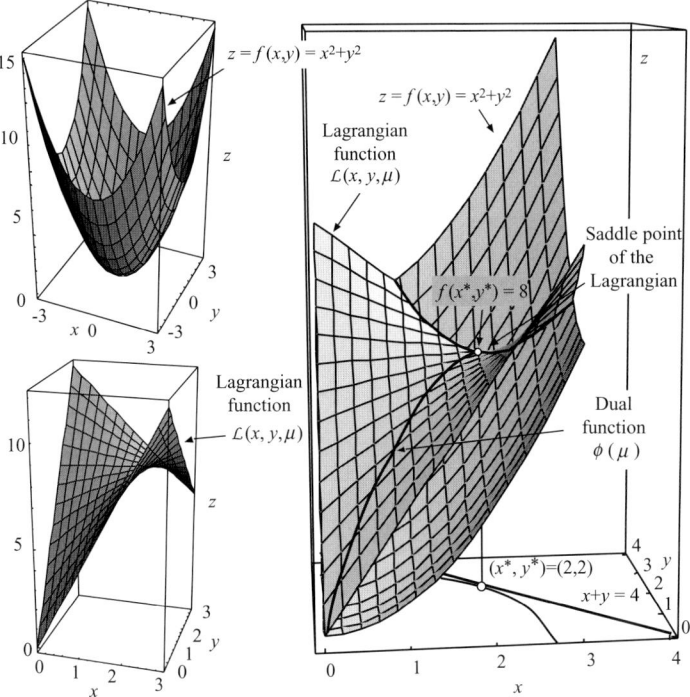

Fig. 5.2. 3-D graphical illustration of all the functions involved in the minimization problem 5.1

where

$$\phi_1(\mu) = \underset{x}{\text{minimize}} \quad \mathcal{L}_1(\mu) = x^2 - \mu x + 2\mu$$

subject to

$$x \geq 0$$

and

$$\phi_2(\mu) = \underset{y}{\text{minimize}} \quad \mathcal{L}_2(\mu) = y^2 - \mu y + 2\mu$$

subject to

$$y \geq 0 .$$

In Fig. 5.3, a 3-D interpretation of the decomposed functions $\mathcal{L}_1(x, \mu)$ and $\mathcal{L}_2(y, \mu)$ for $\mu = 2$ is shown. □

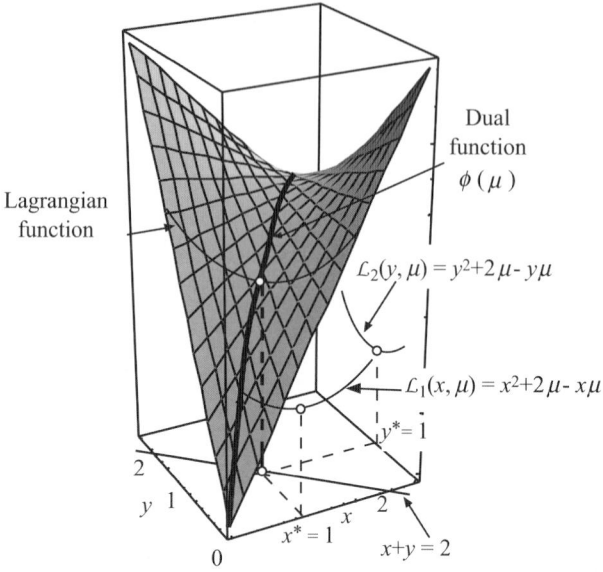

Fig. 5.3. Graphical illustration of the LR decomposition for a given value $\mu = 2$

5.3.2 Algorithm

The Phase 1 Lagrangian relaxation algorithm for solving the dual problem proceeds as follows.

Algorithm 5.1 (The Lagrangian relaxation).

Step 0: Initialization.
   Set $\nu = 1$.
   Initialize dual variables $\boldsymbol{\lambda}^{(\nu)} = \boldsymbol{\lambda}^0$ and $\boldsymbol{\mu}^{(\nu)} = \boldsymbol{\mu}^0$.
   Set $\phi_{\text{down}}^{(\nu-1)} = -\infty$.

Step 1: Solution of the relaxed primal problem.
   Solve the relaxed primal problem (5.12)–(5.14) and get the minimizer $\boldsymbol{x}^{(\nu)}$ and the objective function value at the minimizer $\phi^{(\nu)}$.
   Update the lower bound for the objective function of the primal problem, $\phi_{\text{down}}^{(\nu)} \leftarrow \phi^{(\nu)}$ if $\phi^{(\nu)} > \phi_{\text{down}}^{(\nu-1)}$.

Step 2: Multiplier updating.
   Update multipliers using any of the procedures stated in Subsect. 5.3.4. If possible, update also the objective function upper bound.

Step 3: Convergence checking.

If $||\boldsymbol{\lambda}^{(\nu+1)}-\boldsymbol{\lambda}^{(\nu-1)}||/||\boldsymbol{\lambda}^{(\nu)}|| \leq \varepsilon$ and $||\boldsymbol{\mu}^{(\nu+1)}-\boldsymbol{\mu}^{(\nu-1)}||/||\boldsymbol{\mu}^{(\nu)}|| \leq \varepsilon$, and/or the stopping criterion of Sect. 5.3.4 is met, the ε-optimal solution is $\boldsymbol{x}^* = \boldsymbol{x}^{(\nu)}$, stop. Otherwise set $\nu \leftarrow \nu + 1$, and go to Step 1. □

5.3.3 Dual Infeasibility

The difference between the objective function value of the primal problem for the minimizer and the objective function value of the dual problem at the maximizer is called the duality gap. It is usually the case that the per unit duality gap decreases with the size of the primal problem [30, 31]. Once the solution of the dual problem is achieved, its associated primal problem solution could be nonfeasible and therefore feasibility procedures are required. The procedure used to find a primal feasible near-optimal solution is called Phase 2 of the LR procedure. Phase 2 slightly modifies multiplier values obtained at the end of Phase 1 to achieve primal feasibility. Subgradient procedures (see Sect. 5.3.4) are generally used [32]. In many practical problems they are effective to reach feasibility in a few iterations without altering significantly the objective function value of the dual problem at its maximizer.

5.3.4 Multiplier Updating

Several multiplier updating procedures are explained and compared in this section. In the following, for the sake of clarity, multiplier vectors $\boldsymbol{\lambda}$ and $\boldsymbol{\mu}$ are renamed as $\boldsymbol{\theta} = \text{column}(\boldsymbol{\lambda}, \boldsymbol{\mu})$.

The column vector of constraint mismatches at iteration ν constitutes a subgradient of the dual function [20], i.e.,

$$s^{(\nu)} = \text{column}\left[c(\boldsymbol{x}^{(\nu)}), d(\boldsymbol{x}^{(\nu)})\right] \quad (5.22)$$

is a subgradient vector for the dual function which is used below. Further details can be found in Sect. 4.4 of Chap. 4.

Subgradient (SG)

The multiplier vector is updated as [33]

$$\boldsymbol{\theta}^{(\nu+1)} = \boldsymbol{\theta}^{(\nu)} + k^{(\nu)} \frac{s^{(\nu)}}{||s^{(\nu)}||}, \quad (5.23)$$

where

$$\lim_{\nu \to \infty} k^{(\nu)} \to 0 \quad (5.24)$$

and

$$\sum_{\nu=1}^{\infty} k^{(\nu)} \to \infty \ . \tag{5.25}$$

A typical selection which meets the above requirements is

$$k^{(\nu)} = \frac{1}{a + b\,\nu} \ , \tag{5.26}$$

where a and b are scalar constants.

The subgradient (SG) method is simple to implement and its computational burden is small. However, it progresses slowly to the optimum in an oscillating fashion. This is a consequence of the nondifferentiability of the dual function. Furthermore, the oscillating behavior makes it very difficult to devise an appropriate stopping criterion. It is typically stopped after a prespecified number of iterations.

Illustrative Example 5.2 (SG update). To clarify how the SG method works, Example 5.1 is solved using this multiplier updating procedure. The problem solution is $x^* = y^* = 2$, $f(x^*, y^*) = 8$. The Lagrange multiplier associated with the first constraint has an optimal value $\mu^* = 4$.

The Lagrangian function is

$$\mathcal{L}(x, y, \mu) = x^2 + y^2 + \mu(-x - y + 4) \ .$$

The solution algorithm proceeds as follows.

Step 0: Initialization.
Set $\mu = \mu^{(0)}$.

Step 1: Solution of the relaxed primal problem. The relaxed primal problem decomposes into the two subproblems below (2μ is arbitrarily assigned to each subproblem)

$$\underset{x}{\text{minimize}} \quad x^2 - \mu x + 2\mu$$

subject to

$$x \geq 0$$

and

$$\underset{y}{\text{minimize}} \quad y^2 - \mu y + 2\mu$$

subject to

$$y \geq 0 \ ,$$

whose solutions are denoted, respectively, x^c and y^c.

5.3 Lagrangian Relaxation

Step 2: Multiplier updating. A subgradient procedure with proportionality constant equal to $k^{(\nu)} = \dfrac{1}{a + b\,\nu}$ is used, then

$$\mu \leftarrow \mu + \frac{1}{a + b\nu} \frac{(-x^c - y^c + 4)}{|(-x^c - y^c + 4)|}.$$

Step 3: Convergence checking. If multiplier μ does not change sufficiently, stop; the optimal solution is $x^* = x^c$, $y^* = y^c$. Otherwise the procedure continues in Step 1.

Considering $a = 1$, $b = 0.1$, and an initial multiplier value $\mu^{(0)} = 3$, the algorithm proceeds as shown in Table 5.1. □

Table 5.1. Example: Evolution of the LR algorithm using a subgradient (SG) multiplier updating method

Iteration #	μ	x	y	$f(x,y)$	$\mathcal{L}(x,y,\mu)$
1	3.00	1.50	1.50	4.50	7.50
2	3.91	2.00	2.00	8.00	7.99
3	4.74	2.40	2.40	11.52	7.72
4	3.97	2.00	2.00	8.00	7.99
5	4.69	2.30	2.30	10.58	7.76
6	4.02	2.00	2.00	8.00	7.99
7	3.40	1.70	1.70	5.78	7.82
8	3.98	2.00	2.00	8.00	7.99
9	4.54	2.30	2.30	10.58	7.85
10	4.00	2.00	2.00	8.00	8.00

Cutting Plane Method (CP)

The updated multiplier vector is obtained by solving the linear programming problem below

$$\begin{array}{c} \text{maximize} \\ z, \boldsymbol{\theta} \in C \end{array} \quad z \qquad (5.27)$$

subject to

$$z \leq \phi^{(k)} + \boldsymbol{s}^{(k)T}\left(\boldsymbol{\theta} - \boldsymbol{\theta}^{(k)}\right); \quad k = 1, \ldots, \nu, \qquad (5.28)$$

where C is a convex and compact set. It is made up of the ranges of variations of the multipliers, i.e., $C = \{\boldsymbol{\theta}, \boldsymbol{\theta}_{\text{down}} \leq \boldsymbol{\theta} \leq \boldsymbol{\theta}^{\text{up}}\}$. It should be noted that the above constraints represent half-spaces (hyperplanes) on the multiplier space

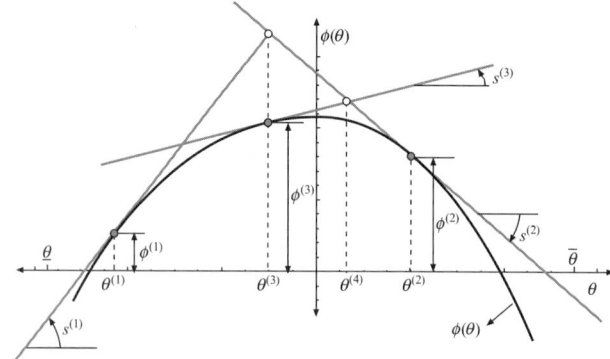

Fig. 5.4. Graphical illustration of the cutting plane method (CP), where reconstruction of the dual function using half-spaces (hyperplanes) is shown

(see Fig. 5.4). Values θ_{down} and θ_{up} are, in general, easily obtained from the physical or economical properties of the system which is modeled. Note that this updating procedure is exactly the same as the master problem of the Benders decomposition technique treated in Chap. 3, but with a concave function and maximizing the objective function instead of minimizing it. It should be noted that the number of constraints of the above problem grows with the number of iterations. The above problem is a relaxed dual problem (RDP), which gets closer to the actual dual problem as the number of iterations grows. The CP method achieves a dual optimum by reconstructing the dual function. It reconstructs the region of interest and regions of no interest. This reconstruction is computationally expensive and therefore the CP method computational burden is high. This algorithm is typically stopped when the multiplier vector difference between two consecutive iterations is below a prespecified threshold.

Illustrative Example 5.3 (The CP update). To clarify how the CP method works, Example 5.1 is solved again using this multiplier updating procedure.

The multiplier updating is as follows.

Step 2: Multiplier updating. The linear programming problem below is solved

$$\underset{z,\mu \in C}{\text{maximize}} \quad z$$

subject to

$$z \leq \phi^{(k)} + s^{(k)}\left(\mu - \mu^{(k)}\right); \qquad k = 1, \ldots, \nu, \tag{5.29}$$

where $\phi^{(k)}$ and $s^{(k)}$ are the dual function and the subgradient at iteration k, respectively,

$$\phi^{(k)} = \left(\left(x^{(k)}\right)^2 + \left(y^{(k)}\right)^2 + \mu^{(k)}\left(-x^{(k)} - y^{(k)} + 4\right)\right)$$

$$s^{(k)} = \left(-x^{(k)} - y^{(k)} + 4\right).$$

Considering $C = \{\mu, 0 \leq \mu \leq 5\}$ and an initial multiplier value $\mu^{(0)} = 3$, the algorithm proceeds as illustrated in Table 5.2. □

Table 5.2. Example: Evolution of the LR algorithm using a cutting plane (CP) multiplier updating method

Iteration #	μ	x	y	$f(x,y)$	$\mathcal{L}(x,y,\mu)$
1	3.00	1.50	1.50	4.50	7.50
2	10.00	5.00	5.00	50.00	−10.00
3	6.50	3.25	3.25	21.12	4.87
4	4.75	2.37	2.37	11.23	7.72
5	3.87	1.94	1.94	7.53	7.99
6	4.31	2.16	2.16	9.33	7.95
7	4.09	2.05	2.05	8.40	7.99
8	4.00	2.00	2.00	8.00	8.00

Bundle Method (BD)

The updated multiplier vector is obtained solving the relaxed dual quadratic programming problem

$$\begin{array}{c} \text{maximize} \\ z, \boldsymbol{\theta} \in C \end{array} \quad z - \alpha^{(\nu)} ||\boldsymbol{\theta} - \boldsymbol{\Theta}^{(\nu)}||^2 \tag{5.30}$$

subject to

$$z \leq \phi^{(k)} + \boldsymbol{s}^{(k)T}\left(\boldsymbol{\theta} - \boldsymbol{\theta}^{(k)}\right); \quad k = 1, \ldots, \nu, \tag{5.31}$$

where α is a penalty parameter, $||\cdot||$ is the 2-norm and $\boldsymbol{\Theta}$, "the center of gravity," is a vector of multipliers centered in the feasibility region so that oscillations are avoided. It should be noted that the number of constraints of the problem above grows with the number of iterations.

The center of gravity is updated in each iteration as stated below.

$$\text{If} \quad \phi\left(\boldsymbol{\theta}^{(\nu)}\right) - \phi\left(\boldsymbol{\Theta}^{(\nu-1)}\right) \geq m\,\delta^{(\nu-1)} \tag{5.32}$$

$$\text{then} \quad \boldsymbol{\Theta}^{(\nu)} = \boldsymbol{\theta}^{(\nu)} \tag{5.33}$$

$$\text{else} \quad \boldsymbol{\Theta}^{(\nu)} = \boldsymbol{\Theta}^{(\nu-1)}, \tag{5.34}$$

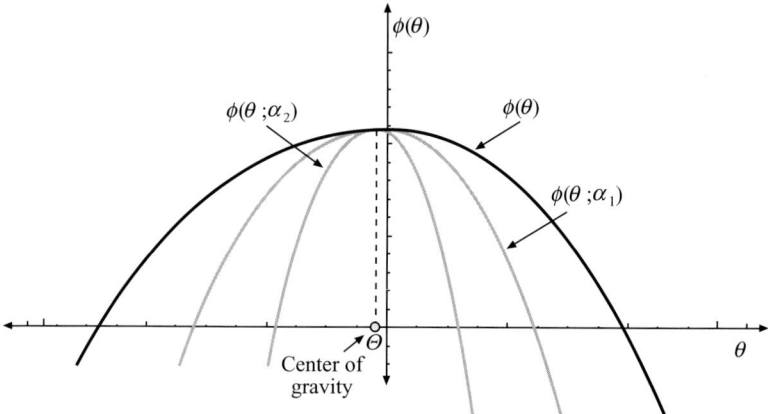

Fig. 5.5. Illustration of the dual and augmented dual functions for the Bundle method (BD)

where m is a parameter whose value is $0 \leq m \leq 1$, and $\delta^{(\nu)}$ is the nominal increase that represents the difference between the objective function value of the relaxed dual problem at iteration ν and the objective function value of the dual problem at iteration ν. This value can be computed as follows:

$$\delta^{(\nu)} = z^{(\nu)} - \alpha^{(\nu)} ||\boldsymbol{\theta}^{(\nu)} - \boldsymbol{\Theta}^{(\nu)}||^2 - \phi(\boldsymbol{\theta}^{(\nu)}) . \tag{5.35}$$

If $\nu = 1$, the center of gravity value is considered to be the initial value of the multiplier vector, $\boldsymbol{\theta}^{(0)}$.

The BD method is a CP method in which the ascent procedure is constrained by an objective function penalty. The target is to center the CP method in the region of interest. In Fig. 5.5 the dual and augmented dual functions for different values of the parameter α are shown. The augmented dual function is the dual function of problem (5.30)–(5.31) However, in order to center the method in the region of interest, it is necessary to carefully tune up the penalty and other parameters. This tune-up is problem dependent and hard to achieve. This algorithm is typically stopped if the multiplier vector difference between two consecutive iterations is below a prespecified threshold. More sophisticated stopping criteria are possible.

Illustrative Example 5.4 (The BD update method). To clarify how the BD method works, Example 5.1 is solved using this multiplier updating procedure. The multiplier updating is as follows.

Step 2: Multiplier updating. The quadratic programming problem below is solved,

$$\underset{z, \mu \in C}{\text{maximize}} \quad z - \alpha^{(\nu)} |\mu - M^{(\nu)}|^2$$

subject to

$$z \leq \phi^{(k)} + s^{(k)}\left(\mu - \mu^{(k)}\right); \qquad k = 1, \ldots, \nu, \qquad (5.36)$$

where $\phi^{(k)}$ and $s^{(k)}$ are the dual function and the subgradient at iteration k, respectively,

$$\phi^{(k)} = \left(\left(x^{(k)}\right)^2 + \left(y^{(k)}\right)^2 + \mu^{(k)}\left(-x^{(k)} - y^{(k)} + 4\right)\right)$$

$$s^{(k)} = \left(-x^{(k)} - y^{(k)} + 4\right).$$

The penalty parameter is calculated as $\alpha^{(\nu)} = d\,\nu$, where d is a constant scalar. If $\nu = 1$, the center of gravity is $M = \mu^0$. In other case, the center of gravity is computed as

$$\begin{aligned}\text{If } \phi\left(\mu^{(\nu)}\right) - \phi\left(M^{(\nu-1)}\right) &\geq m\delta^{(\nu-1)} \\ \text{then} \quad M^{(\nu)} &= \mu^{(\nu)} \\ \text{else} \quad M^{(\nu)} &= M^{(\nu-1)},\end{aligned}$$

where

$$\phi\left(\mu^{(\nu)}\right) = \left((x^{(\nu)})^2 + (y^{(\nu)})^2 + \mu^{(\nu)}(-x^{(\nu)} - y^{(\nu)} + 4)\right)$$

$$\phi\left(M^{(\nu-1)}\right) = \left((x^{(\nu-1)})^2 + (y^{(\nu-1)})^2 + M^{(\nu-1)}(-x^{(\nu-1)} - y^{(\nu-1)} + 4)\right).$$

The normal increase is

$$\delta^{(\nu-1)} = z^{(\nu-1)} - d(\nu - 1)|\mu^{(\nu-1)} - M^{(\nu-1)}|^2 - \phi\left(\mu^{(\nu-1)}\right).$$

Considering $C = \{\mu, 0 \leq \mu \leq 5\}$, $m = 0.5$, $d = 0.02$, and an initial multiplier value $\mu^0 = 3$, the algorithm proceeds as shown in Table 5.3. □

Table 5.3. Example: Evolution of the LR algorithm using a bundle (BD) method multiplier updating procedure

Iteration #	M	μ	x	y	$f(x,y)$	$\mathcal{L}(x,y,\mu)$
1	3.00	3.00	1.50	1.50	4.50	7.50
2	3.00	10.00	5.00	5.00	50.00	−10.00
3	3.00	6.50	3.25	3.25	21.12	4.87
4	3.00	4.75	2.37	2.37	11.23	7.72
5	3.00	3.87	1.94	1.94	7.53	7.99
6	4.31	4.31	2.16	2.16	9.33	7.95
7	4.00	4.09	2.00	2.00	8.00	8.00
8	4.00	4.00	2.00	2.00	8.00	8.00

Trust Region Method (TR)

The updated multiplier vector is obtained solving the relaxed dual linear programming problem below,

$$\begin{array}{c} \text{maximize} \\ z, \boldsymbol{\theta} \in C^{(\nu)} \end{array} \quad z \qquad (5.37)$$

subject to

$$z \leq \phi^{(k)} + \boldsymbol{s}^{(k)^T}\left(\boldsymbol{\theta} - \boldsymbol{\theta}^{(k)}\right); \quad k = 1, \ldots, n; \quad n \leq \overline{n}, \qquad (5.38)$$

where $\overline{n}$ is the maximum number of constraints considered when solving the problem above, and $C^{(\nu)}$ is the dynamically updated set defining the feasibility region for the multipliers [34]. If the number of iterations is larger than the specified maximum number of constraints, the excess constraints are eliminated as stated below. At iteration ν the difference between every hyperplane evaluated at the current multiplier vector and the actual value of the objective function for the current multiplier vector (residual) is computed as (see Fig. 5.6)

$$\varepsilon_i = \phi^{(i)} + \boldsymbol{s}^{(i)^T}\left(\boldsymbol{\theta}^{(\nu)} - \boldsymbol{\theta}^{(i)}\right) - \phi^{(\nu)}; \quad \forall i = 1, \ldots, n . \qquad (5.39)$$

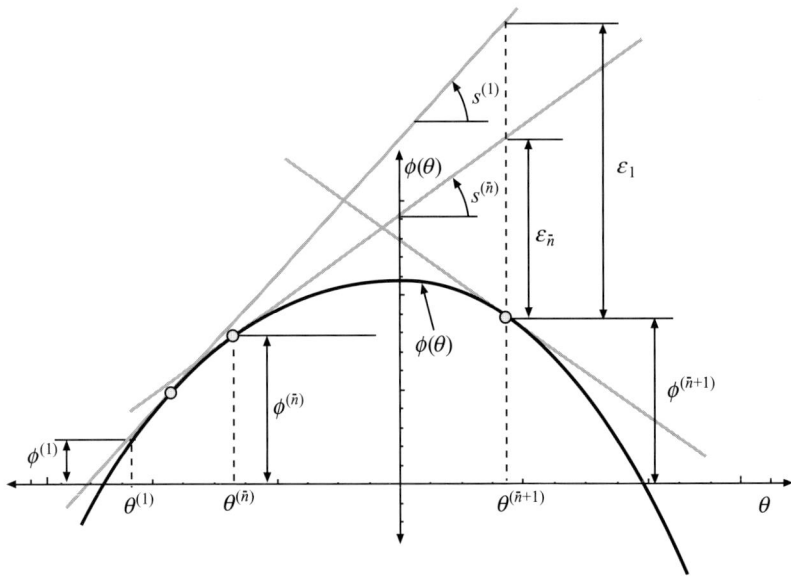

Fig. 5.6. Graphical illustration of the residuals ε_1 and $\varepsilon_{\tilde{n}}$ at iteration $\nu = \overline{n} + 1$ for the trust region (TR) method. As $\varepsilon_1 > \varepsilon_{\tilde{n}}$, hyperplane 1 is not considered for the next iteration

5.3 Lagrangian Relaxation

As soon as n is larger than $\bar{n}$, the "most distant" hyperplanes are not considered, so that the number of hyperplanes is kept constant and equal to $\bar{n}$. It should be noted that the residual ε_i is always positive because the cutting plane reconstruction of the dual function overestimates the actual dual function. This technique to limit the number of hyperplanes considered has proved to be computationally effective. The dynamic updating of the set $C^{(\nu)}$, the feasibility region of the multipliers, is performed as stated below. Let $\phi_i^{(\nu)}$ be the i component of the multiplier vector at iteration ν, then

$$\text{if } \theta_i^{(\nu)} = \overline{\theta}_i^{(\nu)} \tag{5.40}$$
$$\text{then } \overline{\theta}_i^{(\nu+1)} = \overline{\theta}_i^{(\nu)}(1+a) \tag{5.41}$$
$$\text{and } \underline{\theta}_i^{(\nu+1)} = \overline{\theta}_i^{(\nu)}(1-b) \tag{5.42}$$

and

$$\text{if } \theta_i^{(\nu)} = \underline{\theta}_i^{(\nu)} \tag{5.43}$$
$$\text{then } \overline{\theta}_i^{(\nu+1)} = \underline{\theta}_i^{(\nu)}(1+c) \tag{5.44}$$
$$\text{and } \underline{\theta}_i^{(\nu+1)} = \underline{\theta}_i^{(\nu)}(1-d). \tag{5.45}$$

Overlining indicates upper bound and underlining stands for lower bound. The scalar parameters a, b, c, and d allow us to enlarge and shrink the feasibility region of the multiplier vector, i.e., the convex compact set C. This is efficiently accomplished because the above updating procedure is simple. Typical values are $a = c = 2$ and $b = d = 0.8$.

The TR method is a CP method in which the ascent procedure is dynamically constrained by enlarging and shrinking the feasibility region on a coordinate basis. This is possible because the feasibility region is simple: bounds on every multiplier. Through this enlarging/shrinking procedure it is possible to center the algorithm in the area of interest which results in high efficiency. The enlarging/shrinking procedure is not problem dependent and involves straightforward heuristics (see the simple updating procedure above). This method is actually a trust region procedure with dynamic updating of the trust region. This algorithm is typically stopped when the multiplier vector difference between two consecutive iterations is below a prespecified threshold. A small enough difference between an upper bound and a lower bound of the dual optimum is also an appropriate stopping criterion.

Illustrative Example 5.5 (The TR method). To clarify how the TR method works, Example 5.1 is solved using this multiplier updating procedure. The multiplier updating is as follows.

Step 2: Multiplier updating. The linear programming problem below is solved

$$\begin{array}{c} \text{maximize} \\ z, \mu \in C^{(\nu)} \end{array} \quad z$$

subject to

$$z \leq \phi^{(k)} + s^{(k)}\left(\mu - \mu^{(k)}\right); \quad k = 1,\ldots,n; \quad n \leq \overline{n}, \quad (5.46)$$

where $\phi^{(k)}$ and $s^{(k)}$ are the dual function and the subgradient at iteration k, respectively,

$$\phi^{(k)} = \left(\left(x^{(k)}\right)^2 + \left(y^{(k)}\right)^2 + \mu^{(k)}\left(-x^{(k)} - y^{(k)} + 4\right)\right)$$

$$s^{(k)} = \left(-x^{(k)} - y^{(k)} + 4\right).$$

The dynamic set $C^{(\nu)}$ that defines the feasibility region for the multiplier μ is updated as explained above (5.40)–(5.45). If n is larger than $\overline{n}$, the constraints with the highest residual value are not considered so that the number of constraints is kept equal to $\overline{n}$. The residual value, ε_k, is computed using (5.39).

Considering $C^0 = \{\mu, 9 \leq \mu \leq 10\}$, $a = 0.5$, $b = 0.5$, $c = 0.5$, $d = 0.5$, $\overline{n} = 2$, and an initial multiplier value $\mu^0 = 3$, the algorithm proceeds as in Table 5.4 below. □

Table 5.4. Example: Evolution of the LR algorithm using a trust region TR method to update the multipliers

Iteration #	$\overline{\mu}$	$\underline{\mu}$	μ	x	y	$f(x,y)$	$\mathcal{L}(x,y,l)$
1	10.00	9.00	3.00	1.50	1.50	4.50	7.50
2	15.00	4.50	10.00	5.00	5.00	50.00	−10.00
3	15.00	4.50	6.50	3.25	3.25	21.12	4.87
4	15.00	4.50	4.75	2.37	2.37	11.23	7.72
5	22.50	2.25	4.50	2.25	2.25	10.12	7.88
6	22.50	2.25	3.75	1.88	1.88	7.07	7.97
7	22.50	2.25	4.13	2.06	2.06	8.49	7.99
8	22.50	2.25	4.03	2.00	2.00	8.00	8.00
9	22.50	2.25	3.98	2.00	2.00	8.00	8.00
10	22.50	2.25	4.00	2.00	2.00	8.00	8.00

Stopping Criteria

If using the CP method, the objective function value of the relaxed dual problem at every iteration constitutes an upper bound of the optimal dual objective function value. This is so because the piecewise linear reconstruction of the dual function overestimates the actual dual function. On the other hand, the objective function value of the dual problem (evaluated through

the relaxed primal problem) provides at every iteration a lower bound of the optimal dual objective function value. This can be mathematically stated as follows:
$$z^{(\nu)} \geq \phi^* \geq \phi^{(\nu)}, \tag{5.47}$$
where $z^{(\nu)}$ is the objective function value of the relaxed dual problem at iteration ν, ϕ^* is the optimal dual objective function value, and $\phi^{(\nu)}$ is the objective function value of the dual problem at iteration ν. The size of the per unit gap
$$g^{(\nu)} = \frac{z^{(\nu)} - \phi^{(\nu)}}{\phi^{(\nu)}} \tag{5.48}$$
is an appropriate objective function value criterion to stop the search for the dual optimum.

5.4 Augmented Lagrangian Decomposition

The augmented Lagrangian (AL) decomposition procedure is explained below.

5.4.1 Decomposition

The augmented Lagrangian (AL) function of problem (5.1)–(5.5) has the following form:

$$\begin{aligned}\mathcal{A}(\boldsymbol{x}, \boldsymbol{z}, \boldsymbol{\lambda}, \boldsymbol{\mu}, \alpha, \beta) &= f(\boldsymbol{x}) + \boldsymbol{\lambda}^T \boldsymbol{c}(\boldsymbol{x}) + \frac{1}{2}\alpha \|\boldsymbol{c}(\boldsymbol{x})\|^2 \\ &+ \sum_{i=1}^{n_d} \left[\mu_i\left(d_i(\boldsymbol{x}) + z_i^2\right) + \frac{1}{2}\beta\left(d_i(\boldsymbol{x}) + z_i^2\right)^2\right].\end{aligned} \tag{5.49}$$

Penalty parameters α and β are large enough scalars to ensure local convexity and z_i $(i = 1, \ldots, n_d)$ are additional variables to transform inequality constraints into equality constraints. It should be noted that the quadratic terms confer the AL function good convexity properties (see [23, 35]). In Fig. 5.7, a graphical interpretation of the AL function is shown.

For convenience we define $v_i = z_i^2$ $(i = 1, \ldots, n_d)$, therefore, the AL function above is equivalent to

$$\begin{aligned}\mathcal{A}(\boldsymbol{x}, \boldsymbol{v}, \boldsymbol{\lambda}, \boldsymbol{\mu}, \alpha, \beta) &= f(\boldsymbol{x}) + \boldsymbol{\lambda}^T \boldsymbol{c}(\boldsymbol{x}) + \frac{1}{2}\alpha \|\boldsymbol{c}(\boldsymbol{x})\|^2 \\ &+ \sum_{i=1}^{n_d} \left[\mu_i\left(d_i(\boldsymbol{x}) + v_i\right) + \frac{1}{2}\beta\left(d_i(\boldsymbol{x}) + v_i\right)^2\right].\end{aligned} \tag{5.50}$$

The AL decomposition procedure involves minimization of the AL function (5.50). Note that the minimization of (5.50) with respect to $\boldsymbol{v}$ can be carried

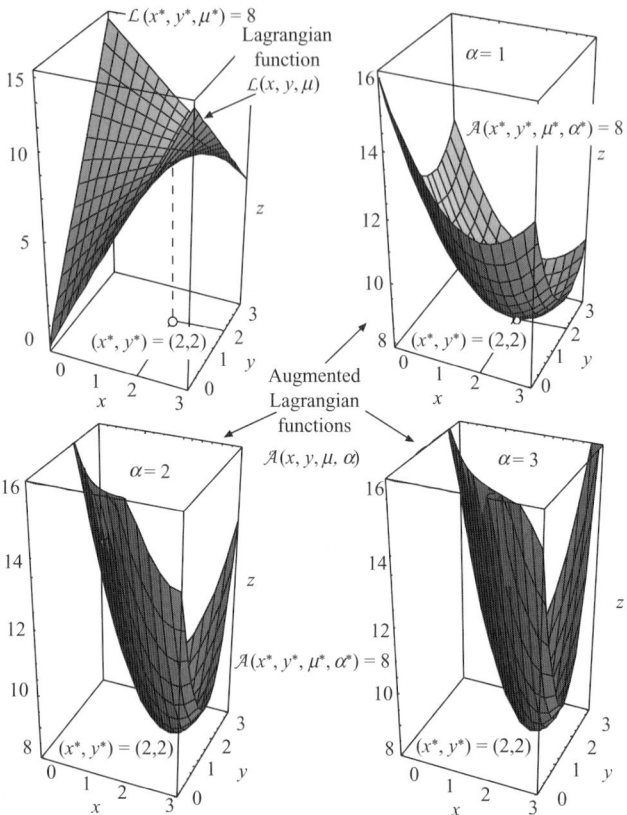

Fig. 5.7. Graphical illustration of the Lagrangian function and the augmented Lagrangian functions for different values of the penalty parameter β for the Illustrative Example 5.6

out analytically in a decomposed manner, leading to a problem that only involves minimization with respect to x. The minimization of (5.50) with respect to v_i is

$$\underset{v_i \geq 0}{\text{minimize}} \quad p_i = \mu_i\big(d_i(x) + v_i\big) + \frac{1}{2}\beta\big(d_i(x) + v_i\big)^2, \tag{5.51}$$

where the variables v_i must be nonnegative.

This problem can be easily solved. The derivative of the objective function above with respect to v_i is $\mu_i + \beta\big(d_i(x) + v_i\big)$. If $v_i > 0$, the derivative must be zero, which implies $v_i = -d_i(x) - \mu_i/\beta$. If $v_i = 0$, this means that $d_i(x)$ is satisfied as an equality, which implies that $\mu_i > 0$, and the derivative must be nonnegative. Therefore,

$$v_i = \max\{0, -d_i(x) - \mu_i/\beta\}; \quad i = 1, \ldots, n_d. \tag{5.52}$$

5.4 Augmented Lagrangian Decomposition

Then the objective function of problem (5.51) is evaluated as follows:

$$p_i = \begin{cases} \dfrac{1}{2\beta}\left[(\mu_i + \beta d_i(\boldsymbol{x}))^2 - \mu_i^2\right] & \text{if} \quad v_i = 0 \\ -\dfrac{1}{2\beta}\mu_i^2 & \text{if} \quad v_i = -d_i(\boldsymbol{x}) - \mu_i/\beta \,. \end{cases}$$

These results can be combined into the formula

$$p_i = \dfrac{1}{2\beta}\left[\left(\max\{0, \mu_i + \beta d_i(\boldsymbol{x})\}\right)^2 - \mu_i^2\right]. \tag{5.53}$$

Finally, we substitute the expression of p_i ($i = 1, \ldots, n_d$) in (5.50) to obtain an explicit expression of the AL function,

$$\begin{aligned} \mathcal{A}(\boldsymbol{x}, \boldsymbol{\lambda}, \boldsymbol{\mu}, \alpha, \beta) &= f(\boldsymbol{x}) + \boldsymbol{\lambda}^T \boldsymbol{c}(\boldsymbol{x}) + \dfrac{1}{2}\alpha \|\boldsymbol{c}(\boldsymbol{x})\|^2 \\ &+ \dfrac{1}{2\beta}\sum_{i=1}^{n_d}\left[\left(\max\{0, \mu_i + \beta d_i(\boldsymbol{x})\}\right)^2 - \mu_i^2\right]. \end{aligned} \tag{5.54}$$

Alternatively, a nonlinear interior point treatment of inequality constraints [36] can be used. This treatment requires adding slack variables to the inequality constraints in (5.49) to convert them into equality constraints. These slack variables are then incorporated to the objective function through logarithmic barrier terms to ensure their positivity.

5.4.2 Algorithm

The decomposition based on the AL is basically similar to the LR decomposition procedure stated in the previous section. The basic difference is that quadratic terms in the AL make the relaxed primal problem nondecomposable. To make it decomposable two procedures are mainly used. They are stated in the Subsect. 5.4.3.

The algorithm proceeds as follows:

Algorithm 5.2 (The AL decomposition).

Step 0: Initialization. Initialize multipliers $\boldsymbol{\lambda}$ and $\boldsymbol{\mu}$ and penalty parameters α and β.

Step 1: Solution of the relaxed primal problem. Solve the relaxed primal problem (5.54). This problem has to be made separable (see Sect. 5.4.3 below).

Step 2: Multiplier updating. Calculate the gradient of the dual function and update multipliers with the target of maximizing the dual function (see Sect. 5.4.4).

If required, update also penalty parameters (see Sect. 5.4.5).

Step 3: Convergence checking. If multipliers do not change significantly in two consecutive iterations, stop, the solution has been reached; otherwise, continue with Step 1. □

5.4.3 Separability

The first procedure to obtain separability linearizes the quadratic terms of the augmented Lagrangian and fixes the minimum number of variables to the values of the previous iteration to achieve separability [37]. The second procedure directly fixes in the augmented Lagrangian the minimum number of variables to the values of the previous iteration to achieve separability.

5.4.4 Multiplier Updating

An appropriate rule to update multipliers $\boldsymbol{\lambda}$ is (see [23, 35])

$$\boldsymbol{\lambda}^{(\nu+1)} = \boldsymbol{\lambda}^{(\nu)} + \alpha\, \boldsymbol{c}(\boldsymbol{x}^{(\nu)})\ . \tag{5.55}$$

Since multipliers $\boldsymbol{\mu}$ are nonnegative, they can be updated as

$$\mu_i^{(\nu+1)} = \max\{0, \mu_i^{(\nu)} + \beta d_i(\boldsymbol{x}^{(\nu)})\}; \qquad i = 1, \ldots, n_d\ . \tag{5.56}$$

Further details on multiplier updating procedures can be found in references [23, 35, 38].

5.4.5 Penalty Parameter Updating

Penalty parameters α and β can be increased with the number of iteration so that convexity is maintained, but in such a way that no numerical ill conditioning appears.

Illustrative Example 5.6 (The AL decomposition). The problem to be solved is

$$\underset{x,y}{\text{minimize}} \quad f(x,y) = x^2 + y^2$$

subject to

$$\begin{aligned} -x - y &\leq -4 \\ x &\geq 0 \\ y &\geq 0, \end{aligned}$$

whose solution is $x^* = y^* = 2$, $f(x^*, y^*) = 8$. The Lagrange multiplier associated with the first constraint has an optimal value $\mu^* = 4$.

5.4 Augmented Lagrangian Decomposition

The AL function is

$$\mathcal{A}(x,y,\mu) = x^2 + y^2 + \frac{1}{2\beta}\left[\left(\max\{0, \mu + \beta(-x-y+4)\}\right)^2 - \mu^2\right].$$

The subproblems to be solved in Step 1 of the decomposition algorithm are

$$\underset{x}{\text{minimize}} \quad x^2 + \frac{1}{2\beta}\left[\left(\max\{0, \mu + \beta(-x-\bar{y}+4)\}\right)^2 - \mu^2\right]$$

subject to

$$x \geq 0$$

and

$$\underset{y}{\text{minimize}} \quad y^2 + \frac{1}{2\beta}\left[\left(\max\{0, \mu + \beta(-\bar{x}-y+4)\}\right)^2 - \mu^2\right]$$

subject to

$$y \geq 0.$$

The algorithm is applied below:

Step 0: Initialization. Variables and multipliers are initialized, i.e.,

$$\bar{x} = 5, \quad \bar{y} = 5, \quad \mu = 3, \quad \beta = 0.3.$$

Step 1: Solution of the relaxed primal problem. This problem decomposes into the two subproblems below. The first subproblem is

$$\underset{x}{\text{minimize}} \quad x^2 + \frac{1}{2 \times 0.3}\left[\left(\max\{0, 3 + 0.3(-x-5+4)\}\right)^2 - 3^2\right]$$

subject to

$$x \geq 0,$$

whose solution is $x = 1.17$.

The second subproblem is

$$\underset{y}{\text{minimize}} \quad y^2 + \frac{1}{2 \times 0.3}\left[\left(\max\{0, 3 + 0.3(-5-y+4)\}\right)^2 - 3^2\right]$$

subject to

$$y \geq 0,$$

whose solution is $y = 1.17$.

Step 2: Multiplier updating. The multiplier is updated as follow:

$$\mu = \max\{0, \mu + \beta(-x - y + 4)\} = \max\{0, 3 + 0.3(-1.17 - 1.17 + 4)\} = 3.50 \ .$$

Step 3: Convergence checking. The multiplier μ does change sufficiently, so the variables are updated

$$\bar{x} = 1.17, \quad \bar{y} = 1.17 \ .$$

The parameter β is increased as

$$\beta \leftarrow 1.2 \, \beta \ .$$

The algorithm continues in Step 1 until convergence is achieved. The algorithm stops for $\nu = 10$, being the solution

$$x = 2.00, \quad y = 2.00, \quad \mu = 4.00, \quad \beta = 1.86, \quad f(x, y) = 8.00 \ .$$

The evolution of the algorithm is shown in Table 5.5 below. □

Table 5.5. Example: Evolution of the AL algorithm for the Illustrative Example 5.6

Iteration #	μ	x	y	$f(x,y)$
1	3.00	1.17	1.17	2.74
2	3.50	1.91	1.91	7.30
3	3.56	1.83	1.83	6.70
4	3.70	1.91	1.91	7.30
5	3.79	1.94	1.94	7.53
6	3.86	1.97	1.97	7.76
7	3.91	1.98	1.98	7.84
8	3.95	1.99	1.99	7.92
9	3.97	1.99	1.99	7.92
10	4.00	2.00	2.00	8.00

5.5 Optimality Condition Decomposition (OCD)

The decomposition technique analyzed in this section can be interpreted as a particular implementation of the LR procedure. It is motivated by a natural decomposition of the optimality conditions of the original problem. An in-depth analysis of this technique can be found in [39].

5.5.1 Motivation: Modified Lagrangian Relaxation

To simplify our motivating analysis, we consider the case in which we have only two groups ($n = 2$) of variables and additionally all constraints are equality ones. The simplified problem has the form

$$\underset{x_1, x_2}{\text{minimize}} \quad f(x_1, x_2) \tag{5.57}$$

subject to

$$h_1(x_1, x_2) = 0 \tag{5.58}$$
$$h_2(x_1, x_2) = 0 \tag{5.59}$$
$$c_j(x_j) = 0; \quad j = 1, 2, \tag{5.60}$$

where the constraints $h_1(x_1, x_2)$ and $h_2(x_1, x_2)$ are complicating constraints. The basic LR procedure (as stated in Sect. 5.3) applied to this problem considers the problem

$$\underset{x_1, x_2}{\text{minimize}} \quad f(x_1, x_2) + \bar{\lambda}_1^T h_1(x_1, x_2) + \bar{\lambda}_2^T h_2(x_1, x_2) \tag{5.61}$$

subject to

$$c_j(x_j) = 0; \quad j = 1, 2, \tag{5.62}$$

defined in terms of multiplier estimates $\bar{\lambda}_1$ and $\bar{\lambda}_2$. Problem (5.61)–(5.62) can be solved by fixing the values of some of the variables ($\bar{x}_2$ and $\bar{x}_1$) to obtain the subproblems

$$\underset{x_1}{\text{minimize}} \quad f(x_1, \bar{x}_2) + \bar{\lambda}_1^T h_1(x_1, \bar{x}_2) \tag{5.63}$$

subject to

$$c_1(x_1) = 0 \tag{5.64}$$

and

$$\underset{x_2}{\text{minimize}} \quad f(\bar{x}_1, x_2) + \bar{\lambda}_2^T h_2(\bar{x}_1, x_2) \tag{5.65}$$

subject to

$$c_2(x_2) = 0 \,. \tag{5.66}$$

Once the solutions for these subproblems have been computed, the multipliers of the complicating constraints can be updated, using for example a subgradient technique,

$$\bar{\boldsymbol{\lambda}}_1^{(\nu+1)} = \bar{\boldsymbol{\lambda}}_1^{(\nu)} + \alpha\, \boldsymbol{h}_1(\boldsymbol{x}_1, \boldsymbol{x}_2) \tag{5.67}$$

$$\bar{\boldsymbol{\lambda}}_2^{(\nu+1)} = \bar{\boldsymbol{\lambda}}_2^{(\nu)} + \alpha\, \boldsymbol{h}_2(\boldsymbol{x}_1, \boldsymbol{x}_2)\,, \tag{5.68}$$

where α is a suitable constant.

Note that the convergence of the procedure requires that the solutions for the subproblems should be computed up to a certain degree of accuracy.

The procedure considered in this section follows a similar approach when applied to problem (5.57)–(5.60). As in the preceding case, to decompose problem (5.61)–(5.62) we require some separable approximation for $f(\boldsymbol{x}_1, \boldsymbol{x}_2)$, $\boldsymbol{h}_1(\boldsymbol{x}_1, \boldsymbol{x}_2)$, and $\boldsymbol{h}_2(\boldsymbol{x}_1, \boldsymbol{x}_2)$ as shown in Fig. 5.8. We also fix some of the variables in these functions to their last computed values, to obtain

$$\underset{\boldsymbol{x}_1}{\text{minimize}} \quad f(\boldsymbol{x}_1, \bar{\boldsymbol{x}}_2) + \bar{\boldsymbol{\lambda}}_2^T \boldsymbol{h}_2(\boldsymbol{x}_1, \bar{\boldsymbol{x}}_2) \tag{5.69}$$

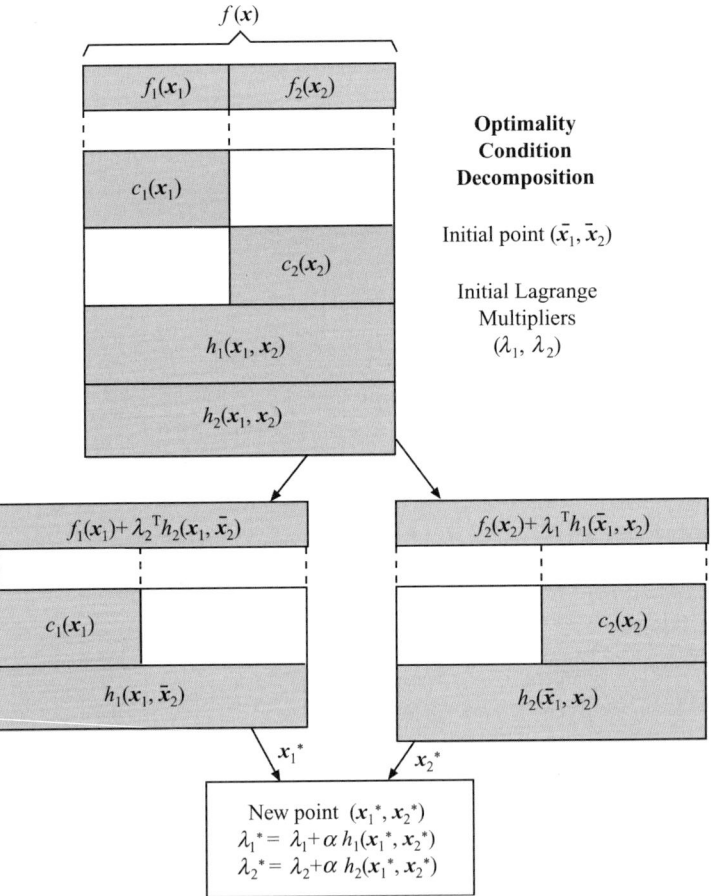

Fig. 5.8. Illustration of the optimality condition decomposition (OCD) procedure

subject to

$$h_1(x_1, \bar{x}_2) = 0 \quad (5.70)$$
$$c_1(x_1) = 0 \quad (5.71)$$

and

$$\underset{x_2}{\text{minimize}} \quad f(\bar{x}_1, x_2) + \bar{\lambda}_1^T h_1(\bar{x}_1, x_2) \quad (5.72)$$

subject to

$$h_2(\bar{x}_1, x_2) = 0 \quad (5.73)$$
$$c_2(x_2) = 0, \quad (5.74)$$

where $\bar{x}_1$ and $\bar{x}_2$ denote the values of the corresponding variables at the last iteration.

To reduce the computational cost, we perform a single iteration for each subproblem before updating the parameters $\bar{x}_1$ and $\bar{x}_2$. This procedure is not very different from a standard Lagrangian approach, except for performing a single iteration for each subproblem, but it presents one significant advantage: it provides efficient information to update the multiplier estimates $\bar{\lambda}_1$ and $\bar{\lambda}_2$. The multipliers corresponding to the subproblem constraints (5.70) and (5.73), $\Delta\lambda_1$ and $\Delta\lambda_2$ have the property that, if the values of $\bar{x}_1$ and $\bar{x}_2$ would be the optimal ones, the best values for λ_1 and λ_2 would be given by $\bar{\lambda}_1 + \Delta\lambda_1$ and $\bar{\lambda}_2 + \Delta\lambda_2$. These updated values can be used for the next iteration. The resulting procedure is very simple to implement, uses few easily updated parameters, and works well in practice for certain classes of problems. It is further analyzed in the following subsections.

5.5.2 Decomposition Structure

For convenience, problem (5.1)–(5.5) can be written as

$$\underset{x_a;\ a=1,\ldots,A}{\text{minimize}} \quad \sum_{a=1}^{A} f_a(x_a) \quad (5.75)$$

subject to

$$h(x_1, \ldots, x_A) \leq 0 \quad (5.76)$$
$$g_a(x_a) \leq 0; \quad a = 1, \ldots, A, \quad (5.77)$$

where x_a are the variables for each block a in which the original problem decomposes. It should be noted that the sets of (5.76) and (5.77) represent both equality and inequality constraints. Equations (5.76) are complicating constraints. These equations contain variables from different blocks and prevent each subproblem from being solved independently. If these equations are removed from problem (5.75)–(5.77), the resulting problem can be trivially decomposed into one subproblem for each block. Constraints (5.77) contain only variables belonging to block a for $a = 1, \ldots, A$. Considering that the optimal values of the Lagrange multipliers in problem (5.75)–(5.77) are known, the problem can be stated in an equivalent form as follows:

$$\underset{\boldsymbol{x}_a;\, a=1,\ldots,A}{\text{minimize}} \quad \sum_{a=1}^{A} f_a(\boldsymbol{x}_a) + \sum_{a=1}^{A} \boldsymbol{\lambda}_a^T \boldsymbol{h}_a(\boldsymbol{x}_1, \ldots, \boldsymbol{x}_A) \tag{5.78}$$

subject to

$$\boldsymbol{h}_a(\boldsymbol{x}_1, \ldots, \boldsymbol{x}_A) \leq \boldsymbol{0}; \quad a = 1, \ldots, A \tag{5.79}$$
$$\boldsymbol{g}_a(\boldsymbol{x}_a) \leq \boldsymbol{0}; \quad a = 1, \ldots, A, \tag{5.80}$$

where constraints (5.76) have been separated in different blocks. Note that the way in which these constraints are distributed does not affect the solution of the problem, i.e., they can be distributed based on engineering insight.

Given trial values to all variables and multipliers (indicated by overlining) different than those in block a, problem (5.78)–(5.80) reduces to

$$\underset{\boldsymbol{x}_a}{\text{minimize}} \quad k + f_a(\boldsymbol{x}_a) + \sum_{b=1,b\neq a}^{A} \overline{\boldsymbol{\lambda}}_b^T \boldsymbol{h}_b(\overline{\boldsymbol{x}}_1, \ldots, \overline{\boldsymbol{x}}_{a-1}, \boldsymbol{x}_a, \overline{\boldsymbol{x}}_{a+1}, \ldots, \overline{\boldsymbol{x}}_A) \tag{5.81}$$

subject to

$$\boldsymbol{h}_a(\overline{\boldsymbol{x}}_1, \ldots, \overline{\boldsymbol{x}}_{a-1}, \boldsymbol{x}_a, \overline{\boldsymbol{x}}_{a+1}, \ldots, \overline{\boldsymbol{x}}_A) \leq \boldsymbol{0} \tag{5.82}$$
$$\boldsymbol{g}_a(\boldsymbol{x}_a) \leq \boldsymbol{0}, \tag{5.83}$$

where $k = \sum_{b=1,b\neq a}^{A} f_b(\overline{\boldsymbol{x}}_b)$ is a constant. The dual variable vector corresponding to constraint (5.82) is denoted by $\boldsymbol{\lambda}_a$. The reduced problem (5.81)–(5.83) can be obtained for every block of the original problem. The proposed decomposition technique is actually based on the solutions of these reduced block-related problems.

5.5.3 Decomposition

The proposed method is based on the decomposition of the optimality conditions for the global problem (5.75)–(5.77), see [20]. Note that from standard optimization theory, the first-order KKT optimality conditions for problem (5.75)–(5.77) are

5.5 Optimality Condition Decomposition (OCD)

$$\nabla_{\boldsymbol{x}_a} f_a(\boldsymbol{x}_a^*) + \sum_{a=1}^{A} \nabla_{\boldsymbol{x}_a}^T \boldsymbol{h}_a(\boldsymbol{x}_1^*, \ldots, \boldsymbol{x}_A^*) \, \boldsymbol{\lambda}_a^* + \nabla_{\boldsymbol{x}_a}^T \boldsymbol{g}_a(\boldsymbol{x}_a^*) \, \boldsymbol{\mu}_a^* = 0 \, ;$$
$$a = 1, \ldots, A \tag{5.84}$$

$$\boldsymbol{h}_a(\boldsymbol{x}_1^*, \ldots, \boldsymbol{x}_A^*) \leq 0; \quad a = 1, \ldots, A \tag{5.85}$$
$$\boldsymbol{h}_a(\boldsymbol{x}_1^*, \boldsymbol{x}_2^*, \ldots, \boldsymbol{x}_n^*)^T \boldsymbol{\lambda}_a^* = 0; \quad a = 1, \ldots, A \tag{5.86}$$
$$\boldsymbol{\lambda}_a^* \geq 0; \quad a = 1, \ldots, A \tag{5.87}$$
$$\boldsymbol{g}_a(\boldsymbol{x}_a^*) \leq 0; \quad a = 1, \ldots, A \tag{5.88}$$
$$\boldsymbol{g}_a(\boldsymbol{x}_a^*)^T \boldsymbol{\mu}_a^* = 0; \quad a = 1, \ldots, A \tag{5.89}$$
$$\boldsymbol{\mu}_a^* \geq 0; \quad a = 1, \ldots, A \, . \tag{5.90}$$

These conditions have been constructed using the optimal values $\boldsymbol{x}_a^*$, $\boldsymbol{\lambda}_a^*$, and $\boldsymbol{\mu}_a^*$ that are assumed known. The values $\boldsymbol{\lambda}_a^*$ are the optimal Lagrange multipliers associated with (5.76) and the values $\boldsymbol{\mu}_a^*$ are the optimal Lagrange multipliers associated with (5.77). For convenience, the block reduced subproblem (5.81)–(5.83) is restated below for optimal values $\boldsymbol{x}_a^*$, $\boldsymbol{\lambda}_a^*$, and $\boldsymbol{\mu}_a^*$:

$$\text{minimize} \quad f_a(\boldsymbol{x}_a) + \sum_{b=1, b \neq a}^{A} \boldsymbol{\lambda}_b^{*T} \boldsymbol{h}_b(\boldsymbol{x}^a) \tag{5.91}$$

subject to

$$\boldsymbol{h}_a(\boldsymbol{x}^a) \leq 0 \tag{5.92}$$
$$\boldsymbol{g}_a(\boldsymbol{x}_a) \leq 0 \, , \tag{5.93}$$

where

$$\boldsymbol{x}^a = (\boldsymbol{x}_1^*, \ldots, \boldsymbol{x}_{a-1}^*, \boldsymbol{x}_a, \boldsymbol{x}_{a+1}^*, \ldots, \boldsymbol{x}_A^*) \, .$$

If the first-order KKT optimality conditions of every block reduced subproblem (5.91)–(5.93) ($a = 1, \ldots, A$) are stuck together, it can be observed that they are identical to the first-order optimality conditions (5.84)–(5.90) of the global problem (5.75)–(5.77). It should be emphasized that this is a relevant result that is exploited in the algorithm below. As previously stated, block subproblem (5.91)–(5.93) is obtained relaxing all the complicating constraints of other blocks, i.e., adding them to the objective function of problem (5.75)–(5.77) and maintaining its own complicating constraints. The reduction is possible once the optimization variables are given trial values. The main difference between the LR algorithm and the proposed decomposition one is that LR adds all the complicating constraints into the objective function. Therefore, it needs auxiliary procedures to update the Lagrange multipliers. On the contrary, the analyzed technique does not need any procedure to update the multipliers because this updating is automatic and results from keeping the complicating constraints (5.92) in every block subproblem. The proposed approach has the advantage that convergence properties do not require to attain an optimal solution of the subproblems at each iteration of the algorithm. It is enough to perform a single iteration for each subproblem, and then to update variable values. As a consequence, computation times can be significantly reduced with respect to other methods that require the computation of the optimum for the subproblems in order to attain convergence. The coordination of the global problem to ensure the satisfaction of the complicating constraints is achieved through the Lagrange multipliers associated with (5.76).

5.5.4 Algorithm

An outline of the proposed algorithm is as follows.

Algorithm 5.3 (The Optimality condition decomposition).

Step 0: Initialization.
Each block ($a = 1, \ldots, A$) initializes its variables and parameters, $\overline{x}_a, \overline{\lambda}_a$.

Step 1: Single iteration.
Each block ($a = 1, \ldots, A$) carries out one single iteration for its corresponding subproblem

$$\underset{x_a}{\text{minimize}} \quad f_a(x_a) + \sum_{b=1, b \neq a}^{A} \overline{\lambda}_b^T h_b(x^a) \tag{5.94}$$

subject to

$$h_a(x^a) \leq 0 \tag{5.95}$$
$$g_a(x_a) \leq 0, \tag{5.96}$$

where $x^a = (\overline{x}_1, \ldots, \overline{x}_{a-1}, x_a, \overline{x}_{a+1}, \ldots, \overline{x}_A)$, and obtains search directions $\Delta x_a, \Delta \lambda_a$.

Step 2: Updating.
Each block ($a = 1, \ldots, A$) updates its variables and parameters

$$\overline{x}_a \leftarrow \overline{x}_a + \Delta x_a, \quad \overline{\lambda}_a \leftarrow \overline{\lambda}_a + \Delta \lambda_a, \quad \text{for } a = 1, \ldots, A \ .$$

Information related to complicating constraints is distributed.

Step 3: Stopping criterion.
The algorithm stops if variables do not change significantly in two consecutive iterations. Otherwise, it continues in Step 1. □

To speed convergence, the search directions obtained in Step 1 can be refined using a Conjugate Gradient procedure [23]. The search directions, $(\Delta x_a, \Delta \lambda_a)$, for subproblems (5.94)–(5.96) can be computed independently of each other, allowing a parallel implementation in a distributed computation environment. A modified Newton procedure can be used, in conjunction with a nonlinear interior point treatment of the inequality constraints [36]. This treatment requires adding slack variables to the inequality constraints (5.95), to convert them into equality constraints. These slack variables are then incorporated to the objective function through logarithmic barrier terms, to ensure their positivity. Step 2 requires a central agent to coordinate the process. This agent receives certain information from the subproblems and returns it to the appropriate subproblems. This information consists of some of the values $\overline{x}_a, \overline{\lambda}_a$, for $a = 1, \ldots, A$. The values $\overline{x}_a$ that have to be distributed are the updated values of the variables associated with the complicating constraints, after one iteration of Newton's method. The values $\overline{\lambda}_a$ that need to be distributed are the updated multipliers corresponding to (5.95) of each area, again after one iteration of Newton's method. It can be noted that the information exchanged between the subproblems and the central agent is minimal. Moreover, in this decomposition algorithm the central agent only distributes information and checks the convergence condition. In other decomposition techniques (such as most common the LR or AL decomposition procedures) the central agent needs to update the exchanged information before distributing it to the different subproblems. In the proposed decomposition algorithm the central agent does not need to update any information, as this information is updated by the subproblems, implying a simpler process.

5.5.5 Convergence Properties

The convergence properties of the decomposition algorithm explained in this section are analyzed below. For the sake of simplicity in this discussion, and without loss of generality, separable constraints (5.77) are omitted. Also, the problem will be represented using only two blocks, a and b. It should be immediate to generalize the following results to more than two blocks. For the centralized approach, the search directions for subproblems a and b, (Δ_a^N, Δ_b^N), are computed by solving in each iteration a system of linear equations of the form

$$\text{KKT} \equiv \begin{pmatrix} \text{KKT}_a & \text{KKT}_{ba} \\ \text{KKT}_{ab} & \text{KKT}_b \end{pmatrix} \begin{pmatrix} \Delta_a^N \\ \Delta_b^N \end{pmatrix} = - \begin{pmatrix} \nabla_{x_a, \lambda_a} \mathcal{L} \\ \nabla_{x_b, \lambda_b} \mathcal{L} \end{pmatrix}, \quad (5.97)$$

where $\Delta_a^N = (\Delta x_a, \Delta \lambda_a)^T$, $\Delta_b^N = (\Delta x_b, \Delta \lambda_b)^T$, and KKT_a, KKT_b, KKT_{ab}, and KKT_{ba} are the Newton matrices [38] for areas a and b, defined as

$$\text{KKT}_a = \begin{pmatrix} \nabla^2_{x_a x_a} \mathcal{L} & \nabla_{x_a} h_a^T \\ \nabla_{x_a} h_a & 0 \end{pmatrix}, \quad \text{KKT}_b = \begin{pmatrix} \nabla^2_{x_b x_b} \mathcal{L} & \nabla_{x_b} h_b^T \\ \nabla_{x_b} h_b & 0 \end{pmatrix}$$

$$\text{KKT}_{ab} = \begin{pmatrix} \nabla^2_{x_a x_b} \mathcal{L} & \nabla_{x_b} h_a^T \\ \nabla_{x_a} h_b & 0 \end{pmatrix}, \quad \text{KKT}_{ba} = \text{KKT}_{ab}^T .$$

The superscript N indicates Newton directions, and $\mathcal{L}$ is the Lagrangian function for problem (5.75)–(5.76), defined as

$$\mathcal{L}(x_a, x_b, \lambda_a, \lambda_b) = f_a(x_a) + f_b(x_b) + \lambda_a^T h_a(x_a, x_b) + \lambda_b^T h_b(x_a, x_b) \quad (5.98)$$

Correspondingly, movement directions for areas a and b, (Δ_a, Δ_b), in Step 1 of the decomposition algorithm can be obtained by solving the decomposable and approximate linear system of equations

$$\overline{\text{KKT}} \equiv \begin{pmatrix} \text{KKT}_a & 0 \\ 0 & \text{KKT}_b \end{pmatrix} \begin{pmatrix} \Delta_a \\ \Delta_b \end{pmatrix} = - \begin{pmatrix} \nabla_{x_a, \lambda_a} \mathcal{L} \\ \nabla_{x_b, \lambda_b} \mathcal{L} \end{pmatrix} . \quad (5.99)$$

From these definitions and from performing Step 1 of the proposed algorithm in parallel, the sufficient condition for convergence of the decomposition algorithm is given below. If at the optimal solution of problem (5.75)–(5.76) it holds that

$$\rho(I - \overline{\text{KKT}}^{-1} \text{KKT}) < 1 , \quad (5.100)$$

then the proposed decomposition algorithm converges locally to the solution at a linear rate. Here $\rho(A)$ denotes the spectral radius of matrix A, matrix I is the identity matrix and it is assumed that functions in (5.75)–(5.77) are twice continuously differentiable. Condition (5.100) is related to the many results reported in the technical literature for the distributed solution of linear systems of equations, see for example [28, 40]. Finally, note that by using Newton method, the local rate of convergence for a centralized approach can be quadratic. Condition (5.100) can be interpreted as a measurement of the coupling between the blocks in the global problem. This measure tends to be smaller for problems with a small number of complicating constraints. These convergence properties seem to be satisfied for most practical cases of interest. If condition (5.100) does not hold, it is possible to modify the proposed decomposition algorithm to attain convergence. For example, a preconditioned Conjugate Gradient method [28] can be applied. This approach would still preserve the property that the operation could be performed allowing each subproblem to maintain its autonomy, i.e., in a decentralized manner.

Note that the OCD algorithm can also be implemented solving the subproblems until optimality (not just one iteration). This results in a clear loss of efficiency, but the implementation becomes much easier, particularly if optimization environments such as GAMS [41] are used.

A simple example that clarifies how the proposed decomposition algorithm works is presented below.

Illustrative Example 5.7 (The Decomposition algorithm). The problem to be solved is

$$\underset{x_1, x_2, y_1, y_2}{\text{minimize}} \quad x_1^2 + x_2^2 + y_1^2 + y_2^2$$

5.5 Optimality Condition Decomposition (OCD)

subject to

$$4x_1 + y_2 - 1 = 0$$
$$x_1 + 4y_2 - 1 = 0.$$

Observe that both constraints are complicating constraints. However, there are only two variables implied in the complicating equations, x_1 and y_2. The solution of this problem is

$$x^* = \begin{pmatrix} 0.2 \\ 0.0 \end{pmatrix}, \quad y^* = \begin{pmatrix} 0.0 \\ 0.2 \end{pmatrix}, \quad \lambda^* = \begin{pmatrix} -0.08 \\ -0.08 \end{pmatrix}.$$

The constraint vector is denoted by $h(x, y)$,

$$h(x, y) = \begin{pmatrix} 4x_1 + y_2 - 1 \\ x_1 + 4y_2 - 1 \end{pmatrix}.$$

Using the proposed methodology, the subproblems to be solved in Step 1 of the decomposition algorithm are, respectively,

$$\underset{x_1, x_2}{\text{minimize}} \quad x_1^2 + x_2^2 + \bar{\lambda}_2 (x_1 + 4\bar{y}_2 - 1)$$

subject to

$$4x_1 + \bar{y}_2 - 1 = 0$$

and

$$\underset{y_1, y_2}{\text{minimize}} \quad y_1^2 + y_2^2 + \bar{\lambda}_1 (4\bar{x}_1 + y_2 - 1)$$

subject to

$$\bar{x}_1 + 4y_2 - 1 = 0.$$

The algorithm is applied below.

Step 0: Initialization. Variables and multipliers are initialized, i.e.,

$$\bar{x} = \begin{pmatrix} 0.4 \\ 0.4 \end{pmatrix}, \quad \bar{y} = \begin{pmatrix} 0.4 \\ 0.4 \end{pmatrix}, \quad \bar{\lambda} = \begin{pmatrix} -0.01 \\ -0.01 \end{pmatrix}.$$

Step 1: Single iteration, system X. System X computes a movement direction for the first decomposed subproblem, using Newton's method, for $x = \bar{x}$. The Lagrangian function for this problem is

$$\mathcal{L}_x(x_1, x_2, \lambda_1) = x_1^2 + x_2^2 - 0.01\ x_1 + \lambda_1(4x_1 + 0.4 - 1)$$

then
$$\nabla_{x_1,x_2,\lambda_1}\mathcal{L}_x(0.4,0.4,-0.01) = \begin{pmatrix} 0.75 \\ 0.80 \\ 1.00 \end{pmatrix}$$

$$\nabla^2_{x_1,x_2,\lambda_1}\mathcal{L}_x(0.4,0.4,-0.01) = \begin{pmatrix} 2 & 0 & 4 \\ 0 & 2 & 0 \\ 4 & 0 & 0 \end{pmatrix}.$$

If the Newton's method is applied

$$\begin{pmatrix} 2 & 0 & 4 \\ 0 & 2 & 0 \\ 4 & 0 & 0 \end{pmatrix} \begin{pmatrix} \Delta x_1 \\ \Delta x_2 \\ \Delta \lambda_1 \end{pmatrix} = - \begin{pmatrix} 0.75 \\ 0.80 \\ 1.00 \end{pmatrix}$$

then

$$\begin{pmatrix} \Delta x_1 \\ \Delta x_2 \\ \Delta \lambda_1 \end{pmatrix} = \begin{pmatrix} -0.25 \\ -0.40 \\ -0.0625 \end{pmatrix}$$

one obtains

$$x = x + \Delta x = \begin{pmatrix} 0.4 \\ 0.4 \end{pmatrix} + \begin{pmatrix} -0.25 \\ -0.40 \end{pmatrix} = \begin{pmatrix} 0.15 \\ 0.00 \end{pmatrix}$$

and
$$\lambda_1 = \lambda_1 + \Delta\lambda_1 = -0.01 + (-0.0625) = -0.0725\ . \tag{5.101}$$

Step 2: Single iteration, system Y. System Y computes a movement direction for the second decomposed subproblem, using Newton's method, for $y = \bar{y}$. The Lagrangian function for this problem is

$$\mathcal{L}_y(y_1,y_2,\lambda_2) = y_1^2 + y_2^2 - 0.01\ y_2 + \lambda_2(0.4 + 4y_2 - 1)$$

then

$$\nabla_{y_1,y_2,\lambda_2}\mathcal{L}_y(0.4,0.4,-0.01) = \begin{pmatrix} 0.80 \\ 0.75 \\ 1.00 \end{pmatrix}$$

$$\nabla^2_{y_1,y_2,\lambda_2}\mathcal{L}_y(0.4,0.4,-0.01) = \begin{pmatrix} 2 & 0 & 0 \\ 0 & 2 & 4 \\ 0 & 4 & 0 \end{pmatrix}.$$

If the Newton method is applied

$$\begin{pmatrix} 2 & 0 & 0 \\ 0 & 2 & 4 \\ 0 & 4 & 0 \end{pmatrix} \begin{pmatrix} \Delta y_1 \\ \Delta y_2 \\ \Delta \lambda_2 \end{pmatrix} = - \begin{pmatrix} 0.80 \\ 0.75 \\ 1.00 \end{pmatrix}$$

then

5.5 Optimality Condition Decomposition (OCD)

one obtains

$$\begin{pmatrix} \Delta y_1 \\ \Delta y_2 \\ \Delta \lambda_2 \end{pmatrix} = \begin{pmatrix} -0.40 \\ -0.25 \\ -0.0625 \end{pmatrix}$$

$$\overline{y} = y + \Delta y = \begin{pmatrix} 0.4 \\ 0.4 \end{pmatrix} + \begin{pmatrix} -0.40 \\ -0.25 \end{pmatrix} = \begin{pmatrix} 0.00 \\ 0.15 \end{pmatrix}$$

and

$$\overline{\lambda}_2 = \lambda_2 + \Delta\lambda_2 = -0.01 + (-0.0625) = -0.0725 \,. \tag{5.102}$$

Step 3: Convergence. Checks if the selected convergence condition $\|h(x,y)\| < 10^{-4}$ is satisfied,

$$h = \begin{pmatrix} -0.25 \\ -0.25 \end{pmatrix}, \quad \|h\| = 0.3536 > 10^{-4} \,.$$

As the convergence condition is not satisfied, variables

$$\overline{x} = x = \begin{pmatrix} 0.15 \\ 0.0 \end{pmatrix}, \quad \overline{y} = y = \begin{pmatrix} 0.0 \\ 0.15 \end{pmatrix}$$

and multipliers

$$\overline{\lambda} = \lambda = \begin{pmatrix} -0.0725 \\ -0.0725 \end{pmatrix}$$

are fixed, the iteration counter is updated, $k = k + 1 = 2$, and Steps 1, 2, and 3 of the algorithm are repeated until convergence is achieved.

In Table 5.6 the evolution of the objective function, variables, and multipliers values as a function of the iteration number are shown. The algorithm stops for $k = 7$, with a tolerance $\|h\| = 8.6317 \times 10^{-5}$. The solution is

$$x = \begin{pmatrix} 0.2 \\ 0.0 \end{pmatrix}, \quad y = \begin{pmatrix} 0.0 \\ 0.2 \end{pmatrix}, \quad \lambda = \begin{pmatrix} -0.08 \\ -0.08 \end{pmatrix}.$$

Table 5.6. Evolution of the optimality condition decomposition OCD algorithm for Illustrative Example 5.7

Iteration #	λ_1	λ_2	x_1	x_2	y_1	y_2	$f(x,y)$
1	−0.010	−0.010	0.150	0.000	0.000	0.150	0.045
2	−0.072	−0.072	0.212	0.000	0.000	0.212	0.090
3	−0.088	−0.088	0.197	0.000	0.000	0.197	0.077
4	−0.076	−0.076	0.201	0.000	0.000	0.201	0.081
5	−0.081	−0.081	0.200	0.000	0.000	0.200	0.079
6	−0.079	−0.079	0.200	0.000	0.000	0.200	0.080
7	−0.080	−0.080	0.200	0.000	0.000	0.200	0.080

222   5 Decomposition in Nonlinear Programming

The initial problem has also been solved using an LR procedure [33] and an augmented LR one [20, 37]. The LR procedure uses a simple subgradient updating of multipliers, and the AL decomposition procedure uses a progressively increasing penalty term and a simple gradient multiplier updating technique. The LR procedure stopped after $k = 53$ iterations, with $||\boldsymbol{h}|| = 4.1772 \times 10^{-5}$. The augmented LR procedure stopped after $k = 15$ iterations, with $||\boldsymbol{h}|| = 9.9172 \times 10^{-5}$.

Figure 5.9 shows the evolution of the objective function (5.101) as a function of the iteration number, for each of the three procedures. The dark gray line represents the evolution of the objective function evaluated at the iterates for the LR procedure. The gray line represents the evolution of the objective function evaluated at the iterates for the AL decomposition procedure. Lastly, the black line represents the evolution of the objective function evaluated at points computed by the proposed decomposition algorithm.

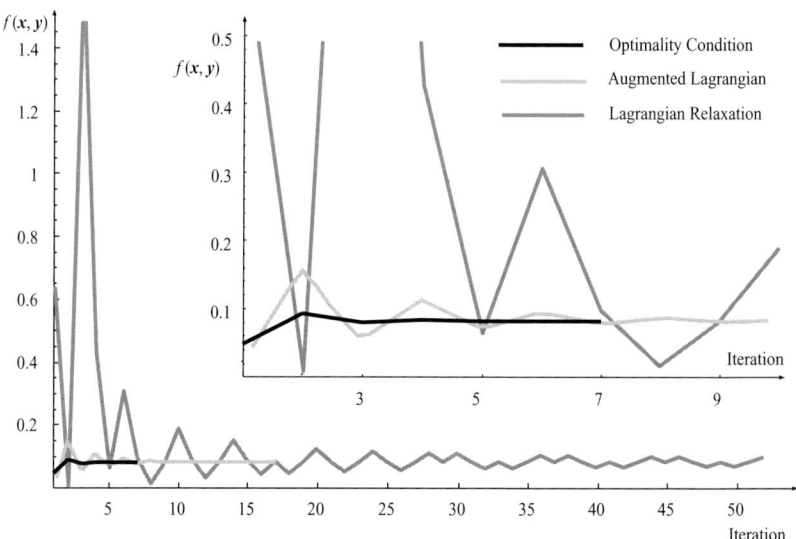

Fig. 5.9. Example: Evolution of the objective function in Illustrative Example 5.7

Note the slow and oscillating behavior of the LR procedure. The quadratic penalty term in the AL procedure corrects this anomaly, although the convergence is still slower than that of the decomposition algorithm analyzed in this section. Figure 5.10 shows the value of multiplier λ_1 at each iteration, for each of the three procedures. The value of multiplier λ_2 is the same for all procedures. As in Fig. 5.9, the dark gray line represents the values of the multiplier computed by the LR procedure; the gray line represents the evolution of the multiplier from the AL procedure; lastly, the black line represents the evolution of the multiplier as obtained by the OCD algorithm.   □

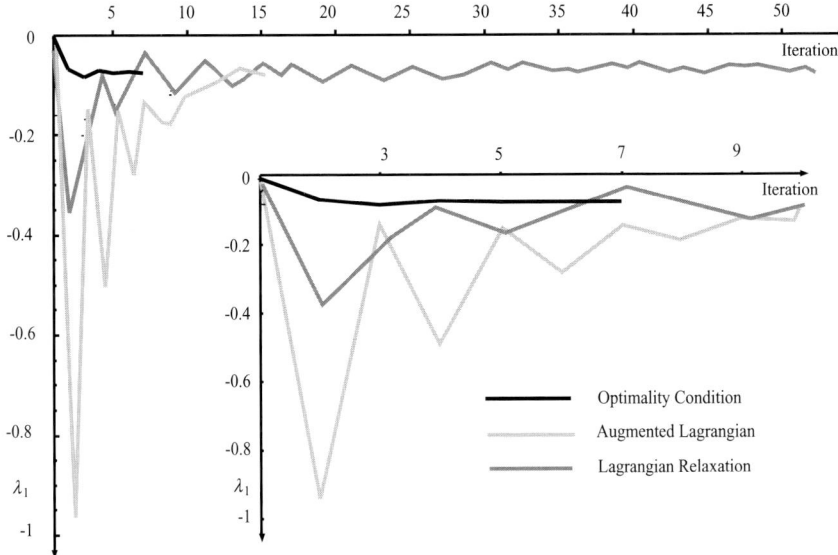

Fig. 5.10. Example: Evolution of the first multiplier

5.6 Complicating Variables

5.6.1 Introduction

The Benders decomposition is analyzed below to address nonlinear problems with decomposable structure and complicating variables.

5.6.2 Benders Decomposition

The problem structure required to apply advantageously the Benders decomposition is

$$\underset{x,y}{\text{minimize}} \quad f(x, y) \tag{5.103}$$

subject to

$$c(x) \leq 0 \tag{5.104}$$
$$d(x, y) \leq 0, \tag{5.105}$$

where $x \in \mathbb{R}^n$, $y \in \mathbb{R}^m$, $f(x,y) : \mathbb{R}^n \times \mathbb{R}^m \to \mathbb{R}$, $c(x) : \mathbb{R}^n \to \mathbb{R}^p$ and $d(x,y) : \mathbb{R}^n \times \mathbb{R}^m \to \mathbb{R}^q$. The problem includes both equality and inequality constraints. Variables x are complicating variables, i.e., variables that if fixed to given values render a simple or decomposable problem.

The auxiliary function $\alpha(\mathbf{x})$ is defined below. Function $\alpha(\mathbf{x})$ expresses the objective function of the original problem (total costs) as a function solely of the complicating variables.

$$\alpha(\mathbf{x}) = \underset{\mathbf{y}}{\text{minimum}} \quad f(\mathbf{x}, \mathbf{y}) \tag{5.106}$$

subject to

$$\mathbf{d}(\mathbf{x}, \mathbf{y}) \leq \mathbf{0} . \tag{5.107}$$

Using function $\alpha(\mathbf{x})$, the original problem can be expressed as

$$\underset{\mathbf{x}}{\text{minimize}} \quad \alpha(\mathbf{x}) \tag{5.108}$$

subject to

$$\mathbf{c}(\mathbf{x}) \leq \mathbf{0} . \tag{5.109}$$

The procedure explained below produces iteratively better and better approximations to function $\alpha(\mathbf{x})$. If complicating variables are fixed to specific values using constraints of the form, $\mathbf{x} = \mathbf{x}^{(\nu)}$, and so that $\mathbf{c}(\mathbf{x}^{(\nu)}) \leq \mathbf{0}$, the resulting problem is easy to solve. This problem has the following form:

$$\underset{\mathbf{y}}{\text{minimize}} \quad f(\mathbf{x}, \mathbf{y}) \tag{5.110}$$

subject to

$$\mathbf{d}(\mathbf{x}, \mathbf{y}) \leq \mathbf{0} \tag{5.111}$$

$$\mathbf{x} = \mathbf{x}^{(\nu)} : \boldsymbol{\lambda}^{(\nu)} . \tag{5.112}$$

The problem above is denominated subproblem. Typically, it decomposes in many subproblems. The solution of the problem above provides values for the noncomplicating variables, $\mathbf{y}^{(\nu)}$, and the dual variable vector associated with those constraints that fix the complicating variables to given values. This sensitivity vector is denoted by $\boldsymbol{\lambda}^{(\nu)}$. An upper bound of the optimal objective function value is readily available because problem (5.110)–(5.112) is more constrained than the original one. This upper bound is $z_{\text{up}}^{(\nu)} = f(\mathbf{x}^{(\nu)}, \mathbf{y}^{(\nu)})$. The information obtained solving the subproblem allows reproducing more and more accurately the original problem. Moreover, if function $\alpha(\mathbf{x})$ is convex, the following problem approximates from below the original one:

$$\underset{\alpha, \mathbf{x}}{\text{minimize}} \quad \alpha \tag{5.113}$$

subject to

$$c(x) \leq 0 \tag{5.114}$$

$$\alpha \geq f(x^{(\nu)}, y^{(\nu)}) + \sum_{k=1}^{n} \lambda_k^{(\nu)}(x_k - x_k^{(\nu)}). \tag{5.115}$$

The last constraint of the problem above is called the Benders cut. The problem itself is denominated master problem. Note that the optimal objective function value of this problem is a lower bound of the optimal objective function value of the original problem. This is so because problem (5.113)–(5.115) is a relaxation of the original problem. The solution of this master problem provides new values for the complicating variables that are used for solving a new subproblem. In turn, this subproblem provides information to formulate a more accurate master problem that provides new values of complicating variables. The procedure continues until upper and lower bounds of the objective function optimal value are close enough.

5.6.3 Algorithm

The Benders decomposition works as follows.

Algorithm 5.4 (The Benders decomposition algorithm).

Input. An NLPP with complicating variables, and a small tolerance value ε to control convergence.
Output. The solution of the NLPP problem obtained after using the Benders decomposition algorithm.

Step 0: Initialization. Find feasible values for the complicating variables x_0, so that $c(x_0) \leq 0$.
Set $\nu = 1$, $x^{(\nu)} = x_0$, $z_{\text{down}}^{(\nu)} = -\infty$.

Step 1: Subproblem solution. Solve subproblem or subproblems

$$\underset{y}{\text{minimize}} \quad f(x, y)$$

subject to

$$d(x, y) \leq 0$$
$$x = x^{(\nu)} : \lambda^{(\nu)}.$$

The solution of this subproblem provides $y^{(\nu)}$, $f(x^{(\nu)}, y^{(\nu)})$, and $\lambda^{(\nu)}$.
Update the objective function upper bound, $z_{\text{up}}^{(\nu)} = f(x^{(\nu)}, y^{(\nu)})$.

Step 2: Convergence check. If $|z_{\text{up}}^{(\nu)} - z_{\text{down}}^{(\nu)}|/|z_{\text{down}}^{(\nu)}| \leq \varepsilon$, the solution with a level of accuracy ε of the objective function is

$$x^* = x^{(\nu)}$$
$$y^* = y^{(\nu)} \ .$$

Otherwise, the algorithm continues with the next step.

Step 3: Master problem solution. Update the iteration counter, $\nu \leftarrow \nu+1$. Solve the master problem

$$\underset{\alpha, x}{\text{minimize}} \quad \alpha$$

subject to

$$\alpha \geq f(x^{(i)}, y^{(i)}) + \sum_{k=1}^{n} \lambda_k^{(i)}(x_k - x_k^{(i)}); \quad \forall i = 1, \ldots, \nu - 1$$

$$c(x) \leq 0 \ .$$

Note that at every iteration a new constraint is added. The solution of the master problem provides $x^{(\nu)}$ and $\alpha^{(\nu)}$.

Update objective function lower bound, $z_{\text{down}}^{(\nu)} = \alpha^{(\nu)}$. The algorithm continues in Step 1.

It should be noted that the behavior of the master problem above depends on the iterative evolution of $\alpha^{(\nu)}$. If $\alpha^{(\nu)}$ ($\nu = 1, 2, \ldots$) presents a convex envelope, the above master problem properly reproduces the original problem and the procedure converges to the solution of the original problem. □

Illustrative Example 5.8 (The Benders decomposition for an NLPP).

Consider the problem

$$\underset{x, y_1, y_2}{\text{minimize}} \quad z = 3x \, y_2 + y_1 \, y_2$$

subject to

$$4y_1 - x^2 \leq 5$$
$$2y_2^2 - \frac{x \, y_2}{3} \leq 2$$
$$y_1 \geq 0$$
$$x \geq 2$$
$$x \leq 12 \ .$$

The optimal values of this example are $x^* = 12$, $y_1^* = 37.25$, and $y_2^* = -0.41$ with an objective function value equal to $z^* = -30.34$. If variable x is considered to be a complicating variable, the above problem is solved using the Benders decomposition algorithm.

Step 0: Initialization. The iteration counter is initialized, $\nu = 1$. The initial value for the complicating variable x is $x^{(1)} = 2$. The lower bound of the objective function is $z_{\text{down}}^{(1)} = -\infty$.

5.6 Complicating Variables

Step 1: Subproblem solution. The subproblem below is solved.

$$\underset{x, y_1, y_2}{\text{minimize}} \quad z = 3x\, y_2 + y_1\, y_2$$

subject to

$$\begin{aligned}
4y_1 - x^2 &\leq 5 \\
2y_2^2 - \frac{x\, y_2}{3} &\leq 2 \\
y_1 &\geq 0 \\
x &= 2 : \lambda,
\end{aligned}$$

whose solution is $y_1^{(1)} = 2.25$, $y_2^{(1)} = -0.85$, and $\lambda^{(1)} = -2.81$ with an objective function value $z = -6.99$. The upper bound of the objective function optimal value is $z_{\text{up}}^{(1)} = -6.99$.

Step 2: Convergence checking. The expression $|z_{\text{up}}^{(1)} - z_{\text{down}}^{(1)}|/|z_{\text{down}}^{(1)}| = 1$ is not small enough, therefore, the procedure continues in Step 3.

Step 3: Master problem solution. The iteration counter is updated, $\nu = 1 + 1 = 2$. The master problem below is solved.

$$\underset{\alpha}{\text{minimize}} \quad \alpha$$

subject to

$$\begin{aligned}
-6.99 - 2.81(x - 2) &\leq \alpha \\
x &\geq 2 \\
x &\leq 12.
\end{aligned}$$

The solution of this problem is $x^{(2)} = 12.00$ and $\alpha^{(2)} = -35.13$. The lower bound of the objective function optimal value is $z_{\text{down}}^{(2)} = \alpha^{(2)} = -35.13$. The procedure continues in Step 1.

Step 1: Subproblem solution. The subproblem below is solved.

$$\underset{x, y_1, y_2}{\text{minimize}} \quad z = 3x\, y_2 + y_1\, y_2$$

subject to

$$\begin{aligned}
4y_1 - x^2 &\leq 5 \\
2y_2^2 - \frac{x\, y_2}{3} &\leq 2 \\
y_1 &\geq 0 \\
x &= 12 : \lambda,
\end{aligned}$$

whose solution is $y_1^{(2)} = 37.25$, $y_2^{(2)} = -0.41$, and $\lambda^{(2)} = -1.94$ with an objective function value $z = -30.34$. The upper bound of the objective function optimal value is $z_{\text{up}}^{(2)} = -30.34$.

Step 2: Convergence checking. The expression $|z_{\text{up}}^{(2)} - z_{\text{down}}^{(2)}|/|z_{\text{down}}^{(2)}| = 0.1363$ is not small enough, therefore, the procedure continues in Step 3.

Step 3: Master problem solution. The iteration counter is updated, $\nu = 2 + 1 = 3$. The master problem below is solved.

$$\text{minimize} \quad \alpha$$
$$\alpha$$

subject to
$$\begin{aligned}
-6.99 - 2.81 \times (x - 2) &\leq \alpha \\
-30.34 - 1.94 \times (x - 12) &\leq \alpha \\
x &\geq 2 \\
x &\leq 12 .
\end{aligned}$$

The solution of this problem is $x^{(3)} = 12$ and $\alpha^{(3)} = -30.34$. The lower bound of the objective function optimal value is $z_{\text{down}}^{(3)} = -30.34$.

Step 1: Subproblem solution. The subproblem below is solved.

$$\underset{x, y_1, y_2}{\text{minimize}} \quad z = 3x\, y_2 + y_1\, y_2$$

subject to
$$\begin{aligned}
4y_1 - x^2 &\leq 5 \\
2y_2^2 - \frac{x\, y_2}{3} &\leq 2 \\
y_1 &\geq 0 \\
x &= 12 : \lambda ,
\end{aligned}$$

whose solution is $y_1^{(3)} = 37.25$, $y_2^{(3)} = -0.41$, and $\lambda^{(3)} = -1.94$ with an objective function value $z = -30.34$. The upper bound of the objective function optimal value is $z_{\text{up}}^{(3)} = -30.34$.

Step 2: Convergence checking. The expression $|z_{\text{up}}^{(3)} - z_{\text{down}}^{(3)}|/|z_{\text{down}}^{(3)}| = 0$ is small enough, therefore, the optimal solution has been found.

The solution is $x^* = 12$, $y_1^* = -37.25$, and $y_2^* = -0.41$ with an optimal objective function value -30.34. The convergence behavior of this example is illustrated in Fig. 5.11. □

An engineering-based computational example is detailed below.

Computational Example 5.1 (The Reliability-based optimization of a rubblemound breakwater).

In Sect. 1.5.4, p. 48, the reliability-based optimization of a rubblemound breakwater was presented. In this subsection we give a possible way of solving the corresponding optimization problem using the Benders decomposition.

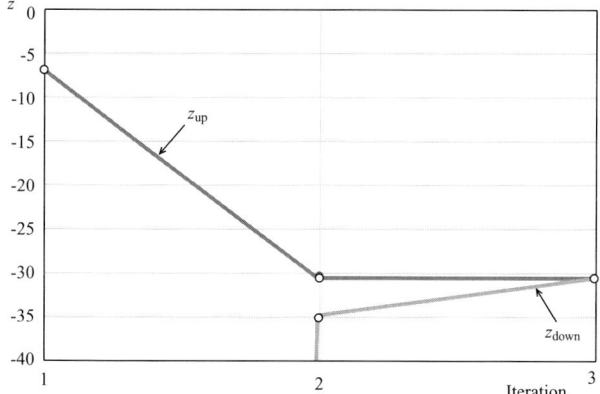

Fig. 5.11. The Benders algorithm evolution in Illustrative Example 5.8

The method proceeds as follows:

Step 0: Initialization. Select initial values for the design variables (complicating variables) $\boldsymbol{d}^{(0)} = \{F_c^{(0)}, \tan \alpha_s^{(0)}\}$.

Set the iteration counter to $\nu = 1$, $\boldsymbol{d}^{(\nu)} = \boldsymbol{d}^{(0)}$, and the total cost lower bound to $C_{\text{down}}^{(\nu)} = -\infty$.

Step 1: Evaluating the reliability index β. Based on the actual design values (complicating variables), the reliability index is calculated solving the problem

$$\underset{H,T}{\text{minimize}} \quad \beta^{(\nu)} = \sqrt{z_1^2 + z_2^2}$$

subject to

$$\frac{R_u}{H} = A_u \left(1 - e^{B_u I_r}\right)$$

$$I_r = \frac{\tan \alpha_s}{\sqrt{H/L}}$$

$$\left(\frac{2\pi}{T}\right)^2 = g \frac{2\pi}{L} \tanh \frac{2\pi D_{wl}}{L}$$

$$\Phi(z_1) = 1 - \exp(-2(H/H_s)^2)$$

$$\Phi(z_2) = 1 - \exp(-0.675(T/\bar{T})^4)$$

$$R_u = F_c$$

$$\boldsymbol{d} = \boldsymbol{d}^{(\nu)} : \boldsymbol{\lambda}^{(\nu)}.$$

The solution of this subproblem provides $\beta^{(\nu)}$, and the partial derivatives of the reliability index with respect to the design variables $\boldsymbol{\lambda}^{(\nu)}$.

Then, it is possible to evaluate the probability of failure of one wave (P_f), the probability of failure in the design sea state (P_f^D), the volumes of material (v_c, v_a), the construction cost C_co, the insurance cost C_in, and the total cost $C_\text{to}^{(\nu)}$ (see Fig. 5.12) for the actual values of the complicating (design) variables $\boldsymbol{d}^{(\nu)}$ and the reliability index $\beta^{(\nu)}$ as

$$\begin{aligned}
v_\text{c} &= 10h \\
v_\text{a} &= \frac{1}{2}(D_\text{wl}+2)(46+D_\text{wl}+\frac{(D_\text{wl}+2)}{\tan\alpha_\text{s}^{(\nu)}}) \\
F_\text{c}^{(\nu)} &= 2+h \\
P_\text{f} &= \Phi(-\beta^{(\nu)}) \\
P_\text{f}^D &= 1-(1-P_\text{f})^{d_\text{st}/\bar{T}} \\
C_\text{co} &= c_\text{c}v_\text{c}+c_\text{a}v_\text{a} \\
C_\text{in} &= 5000+1.25\times 10^6 P_\text{f}^D \\
C_\text{to}^{(\nu)} &= C_\text{co}+C_\text{in} \ .
\end{aligned}$$

Thus, we have just calculated one point of the total cost function, for reconstructing the function using hyperplanes, we need the partial derivatives of function C_to with respect to the design variables $\boldsymbol{d}^{(\nu)}$, $\boldsymbol{\Omega}^{(\nu)}$ shown in Fig. 5.12. These partial derivatives can be calculated using the following expression:

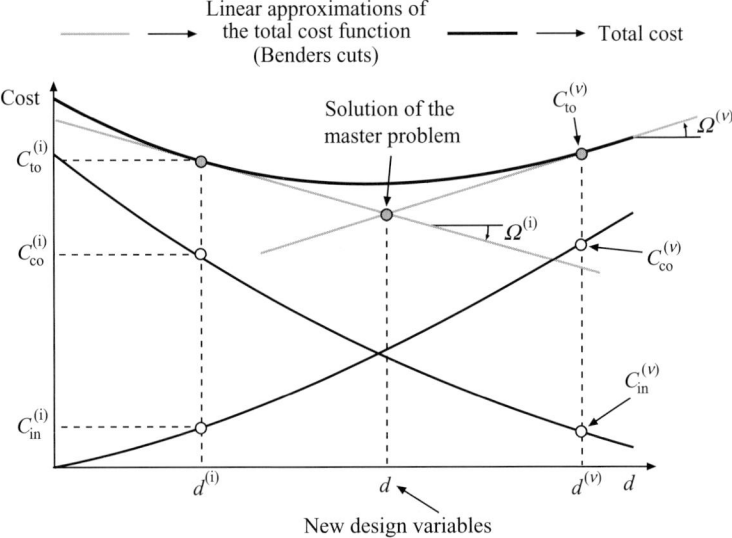

Fig. 5.12. Graphical illustration of the reconstruction of the total cost function using the Benders cuts

$$\Omega^{(\nu)} = \frac{\partial C_{\text{to}}^{(\nu)}}{\partial \boldsymbol{d}^{(\nu)}} = \frac{\partial C_{\text{co}}}{\partial \boldsymbol{d}^{(\nu)}} + \frac{\partial C_{\text{in}}}{\partial \boldsymbol{d}^{(\nu)}},$$

where $\dfrac{\partial C_{\text{co}}}{\partial \boldsymbol{d}^{(\nu)}}$ is the partial derivative of the construction cost, given in analytical form or calculated using the auxiliary problem,

$$\underset{F_{\text{c}},\, \tan \alpha}{\text{minimize}} \quad C_{\text{co}} = c_{\text{c}} v_{\text{c}} + c_{\text{a}} v_{\text{a}}$$

subject to

$$\begin{aligned}
v_{\text{c}} &= 10h \\
v_{\text{a}} &= \frac{1}{2}(D_{\text{wl}} + 2)(46 + D_{\text{wl}} + \frac{D_{\text{wl}} + 2}{\tan \alpha_{\text{s}}}) \\
F_{\text{c}} &= 2 + h \\
\boldsymbol{d} &= \boldsymbol{d}^{(\nu)} : \boldsymbol{\theta}^{(\nu)},
\end{aligned}$$

where $\boldsymbol{\theta}^{(\nu)}$ are the required derivatives.

The partial derivative of the insurance cost is obtained as follows:

$$\begin{aligned}
\frac{\partial C_{\text{in}}}{\partial \boldsymbol{d}^{(\nu)}} &= \frac{\partial C_{\text{in}}}{\partial P_{\text{f}}^D} \frac{\partial P_{\text{f}}^D}{\partial \boldsymbol{d}^{(\nu)}} \\
&= \frac{\partial C_{\text{in}}}{\partial P_{\text{f}}^D} \frac{d_{\text{st}}}{\bar{T}} (1 - P_{\text{f}})^{(d_{\text{st}}/\bar{T} - 1)} \frac{\partial P_{\text{f}}}{\partial \boldsymbol{d}^{(\nu)}} \\
&= \frac{\partial C_{\text{in}}}{\partial P_{\text{f}}^D} \frac{d_{\text{st}}}{\bar{T}} (1 - P_{\text{f}})^{(d_{\text{st}}/\bar{T} - 1)} \frac{\exp(-\beta^2/2)}{\sqrt{2\pi}} \frac{\partial \beta}{\partial \boldsymbol{d}^{(\nu)}},
\end{aligned}$$

where for this example

$$\frac{\partial C_{\text{in}}}{\partial P_{\text{f}}^D} = 2.5 \times 10^6 P_{\text{f}}^D \quad \text{and} \quad \frac{\partial \beta}{\partial \boldsymbol{d}^{(\nu)}} = \boldsymbol{\lambda}^{(\nu)}.$$

Set the total cost upper bound to $C_{\text{up}}^{(\nu)} = C_{\text{to}}^{(\nu)}$.

Step 2: Convergence check. If $|C_{\text{up}}^{(\nu)} - C_{\text{down}}^{(\nu)}|/|C_{\text{up}}^{(\nu)}| \leq \varepsilon$, the solution with a level of accuracy ε of the objective function and design variables is

$$C_{\text{to}}^* = C_{\text{to}}^{(\nu)}, \quad \boldsymbol{d}^* = \boldsymbol{d}^{(\nu)}.$$

Otherwise, set $\nu \leftarrow \nu + 1$ and go to Step 3.

Step 3: Master problem. The hyperplane reconstruction of the total cost function is used for calculating the new values of the complicating variables $\boldsymbol{d}$.

$$\underset{\alpha_{\text{cost}},\, \boldsymbol{d}}{\text{minimize}} \quad \alpha_{\text{cost}}$$

5 Decomposition in Nonlinear Programming

subject to

$$\alpha_{\text{cost}} \geq C_{\text{to}}^{(\nu)} + \sum_{i=1}^{n} \Omega_i^{(\nu)} \left(d_i - d_i^{(\nu)}\right); \quad \forall i = 1, \ldots, \nu - 1$$

$$2 \leq \frac{1}{\tan \alpha} \leq 5$$

$$\alpha_{\text{cost}} \geq 5000 .$$

The solution of this master problem provides the new values of the design variables $d^{(\nu)}$ for iteration ν, and the corresponding lower bound of the total cost function $C_{\text{down}}^{(\nu)} = \alpha_{\text{cost}}$.

The algorithm continues in Step 1.

As it has been shown, the reliability-based optimization problems characterized as bi-level optimization problems, can be solved easily using the Benders decomposition procedure.

A Numerical Example

To perform a reliability-based design of an individual rubblemound breakwater, assume the following values for the variables involved:

$$D_{\text{wl}} = 20 \text{ m}, \ A_u = 1.05 , \ B_u = -0.67 , \ c_c = 60 \text{ \$/m}^3,$$

$$g = 9.81 \text{ m/s}^2, \ c_a = 2.4 \text{ \$/m}^3, \ H_s = 5 \text{ m}, \ \bar{T} = 10 \text{ s}, \ d_{\text{st}} = 1 \text{ h} .$$

The solution of this problem, using the method above, is shown in Table 5.7. It turns out that convergence of the process requires only 11 iterations. The evolution of the total cost function bounds during the process is illustrated in Fig. 5.13. The optimal reliability index and probability of run-up for a single wave and during the design sea state are, respectively,

Table 5.7. Illustration of the iterative procedure

ν	F_c	$\tan \alpha_s$	C_{co}	C_{in}	C_{to}	C_{down}	C_{up}	Error
1	7.00	0.33	6484.8	8297.8	14782.6	5000.0	14782.6	1.9565
2	5.65	0.20	6836.4	5000.0	11836.4	5000.0	11836.4	1.3673
3	9.32	0.50	7296.0	5000.0	12296.0	9682.5	12296.0	0.2699
4	6.52	0.29	6489.7	5542.8	12032.5	11116.5	12032.5	0.0824
5	6.66	0.29	6571.1	5077.2	11648.3	11197.9	11648.3	0.0402
6	7.02	0.29	6786.8	5000.0	11786.8	11413.5	11786.8	0.0327
7	5.98	0.24	6598.6	5007.5	11606.1	11521.9	11606.1	0.0073
8	6.40	0.26	6583.4	5021.2	11604.5	11570.4	11604.5	0.0030
9	6.00	0.24	6553.6	5033.9	11587.5	11571.8	11587.5	0.0014
10	5.67	0.22	6578.7	5020.4	11599.1	11584.6	11599.1	0.0013
11	5.88	0.23	6571.3	5019.8	11591.1	11585.8	11591.1	0.0005

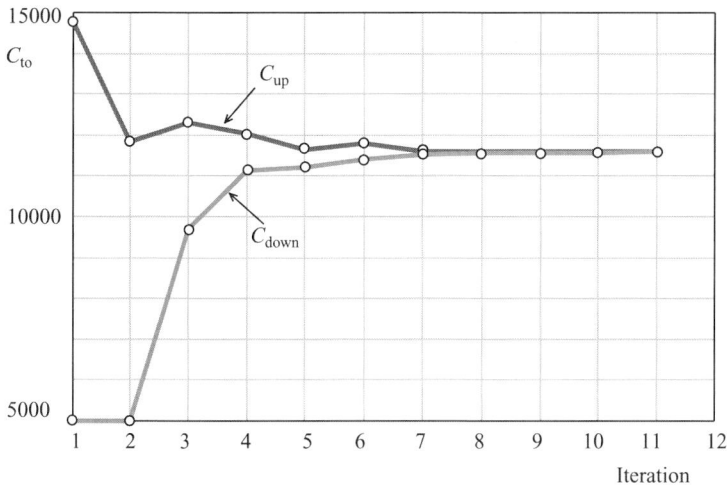

Fig. 5.13. Evolution of the objective function for the rubblemound breakwater example (Computational Example 5.1)

$$\beta^* = 4.738,\ P_{\text{f}}^* = 0.00000111,\ \text{and}\ P_{\text{f}}^{\text{D}^*} = 0.00039845\ .$$

The GAMS code for solving this rubblemound breakwater example is given in the Appendix A, p. 407. □

5.7 From Lagrangian Relaxation to Dantzig-Wolfe Decomposition

In Chap. 2 we have stated that linear programming problems with the following structure:

$$\underset{x_j;\, j=1,\ldots,n}{\text{minimize}} \quad z = \sum_{j=1}^{n} c_j x_j \tag{5.116}$$

subject to

$$\sum_{j=1}^{n} e_{ij} x_j = f_i; \quad i = 1,\ldots,q \tag{5.117}$$

$$\sum_{j=1}^{n} a_{ij} x_j = b_i; \quad i = 1,\ldots,m \tag{5.118}$$

$$0 \le x_j \le x_j^{\text{up}}; \quad j = 1,\ldots,n, \tag{5.119}$$

where constraints (5.117) have a decomposable structure in r blocks and constraints (5.118) are complicating constraints, can be solved using the Dantzig-Wolfe decomposition technique.

A different alternative considering Theorem 3.1 is to obtain the dual of (5.116)–(5.119), which has a decomposable structure with complicating variables, and solve it using the Benders decomposition.

Additionally, in this section we show how problem (5.116)–(5.119) can be solved using the Lagrangian relaxation, and how the Dantzig-Wolfe decomposition can be derived from the LR approach. These questions are addressed below.

5.7.1 Lagrangian Relaxation in LP

Problems (5.116)–(5.119) have the adequate structure to apply the LR advantageously because if the complicating constraints (5.118) are relaxed, it becomes drastically simplified.

Considering the Lagrangian function as

$$\mathcal{L}(\boldsymbol{x}, \boldsymbol{\lambda}) = \sum_{j=1}^{n} c_j x_j + \sum_{i=1}^{m} \lambda_i h_i(\boldsymbol{x})$$

$$= \sum_{j=1}^{n} c_j x_j + \sum_{i=1}^{m} \lambda_i \left(b_i - \sum_{j=1}^{n} a_{ij} x_j \right)$$

$$= \sum_{j=1}^{n} \left(c_j - \sum_{i=1}^{m} \lambda_i a_{ij} \right) x_j + \sum_{i=1}^{m} \lambda_i b_i = y + \sum_{i=1}^{m} \lambda_i b_i$$

where $\boldsymbol{h}(\boldsymbol{x})$ are the equality complicating constraints mismatches, and y is an auxiliary variable. Then, for convenience, the dual function $\phi(\boldsymbol{\lambda})$ is defined as

$$\phi(\boldsymbol{\lambda}) = \underset{\boldsymbol{x}}{\text{minimum}} \quad \mathcal{L}(\boldsymbol{x}, \boldsymbol{\lambda}) = \sum_{j=1}^{n} \left(c_j - \sum_{i=1}^{m} \lambda_i a_{ij} \right) x_j + \sum_{i=1}^{m} \lambda_i b_i \tag{5.120}$$

subject to

$$\sum_{j=1}^{n} e_{ij} x_j = f_i; \quad i = 1, \ldots, q \tag{5.121}$$

$$0 \leq x_j \leq x_j^{\text{up}}; \quad j = 1, \ldots, n, \tag{5.122}$$

which is a concave function [22].

If primal problem (5.116)–(5.119) has an optimal solution (has not unbounded optimum), then the optimal solution of the primal and dual problems coincide.

5.7 From Lagrangian Relaxation to Dantzig-Wolfe Decomposition

The dual problem is then defined as

$$\begin{array}{c} \text{maximize} \\ \lambda_i; i=1,\ldots,m \end{array} \phi(\boldsymbol{\lambda}) \ . \qquad (5.123)$$

In what follows the LR algorithm is applied.

Consider given values for the multipliers $\lambda_1^{(k)}, \lambda_2^{(k)}, \ldots, \lambda_m^{(k)}$, and consider the following subproblem t:

Subproblem t:

$$\begin{array}{c} \text{minimize} \\ x_j; j = n_{t-1}+1, \ldots, n_t \end{array} y_t^{(k)} = \sum_{j=n_{t-1}+1}^{n_t} \left(c_j - \sum_{i=1}^{m} \lambda_i^{(k)} a_{ij} \right) x_j \qquad (5.124)$$

subject to

$$\sum_{j=n_{t-1}+1}^{n_t} e_{ij} x_j = f_i; \quad i = q_{t-1}+1, \ldots, q_t \qquad (5.125)$$

$$0 \leq x_j \leq x_j^{\text{up}}; \quad j = n_{t-1}+1, \ldots, n_t \ . \qquad (5.126)$$

This problem is denominated subproblem for the block t, and is the same as the relaxed subproblem (2.45)–(2.47) of the Dantzig-Wolfe decomposition algorithm.

Once the subproblems for all blocks ($t = 1, \ldots, r$) are solved, we evaluate the dual function using the objective function values of the block subproblems $\left(y_t^{(k)} \right)$,

$$\phi\left(\boldsymbol{\lambda}^{(k)}\right) - \sum_{t=1}^{r} y_t^{(k)} + \sum_{i=1}^{m} \lambda_i^{(k)} b_i - y^{(k)} + \sum_{i=1}^{m} \lambda_i^{(k)} b_i \ . \qquad (5.127)$$

Value (5.127) is the solution of a particular instance of the original problem, i.e., a problem more restricted than the original one. Therefore, value (5.127) is a lower bound of the optimal value of the objective function of the original problem, i.e.,

$$\phi_{\text{down}}^{(k)} = y^{(k)} + \sum_{i=1}^{m} \lambda_i^{(k)} b_i \ . \qquad (5.128)$$

This lower bound is identical to the one obtained while deriving the Dantzig-Wolfe decomposition algorithm (2.76).

Note that the gradient of the dual function with respect to the multiplier vector $(\boldsymbol{\lambda})$ is the vector of mismatches of the corresponding constraints $\boldsymbol{h}(\boldsymbol{x})$,

$$\theta_i^{(k)} = b_i - \sum_{j=1}^{n} a_{ij} x_j^{(k)} = b_i - r_i^{(k)}; \quad i = 1, \ldots, m \ , \qquad (5.129)$$

where $\boldsymbol{\theta}$ is the gradient of the dual function (objective function) and $\boldsymbol{x}^{(k)}$ is the solution of the subproblems.

The following step consists of updating the multiplier vector by solving the linear programming problem:

Master problem:

$$\underset{\lambda_1,\ldots,\lambda_m,\alpha}{\text{maximize}} \quad \alpha \tag{5.130}$$

subject to

$$\alpha \leq \phi\left(\boldsymbol{\lambda}^{(k)}\right) + \boldsymbol{\theta}^{(k)^T}\left(\boldsymbol{\lambda} - \boldsymbol{\lambda}^{(k)}\right); \quad k = 1,\ldots,\nu-1, \tag{5.131}$$

where α is a scalar. Constraints (5.131) represent half-spaces (hyperplanes) on the multiplier space.

This is a relaxed version of problem (5.123) because it approximates this problem from above. Therefore, the solution is an upper bound of the optimal solution of the original problem

$$\phi_{\text{up}}^{(k)} = \alpha . \tag{5.132}$$

Up to this point, we have illustrated the solution of (5.116)–(5.119) by Lagragian relaxation.

5.7.2 Dantzig-Wolfe from Lagrangian Relaxation

In the preceding subsection we have illustrated how to use the Lagrangian relaxation to solve LP problems with complicating constraints. But, what about deriving the Dantzig-Wolfe method from the Lagrangian relaxation? This question is addressed in this subsection.

Using Corollary 4.1 we obtain the dual problem of the master problem (5.130).

Dual master problem:

$$\underset{u_k;\, k=1,\ldots,\nu-1}{\text{minimize}} \quad \sum_{k=1}^{\nu-1}\left(\phi\left(\boldsymbol{\lambda}^{(k)}\right) - \sum_{i=1}^{m}\theta_i^{(k)}\lambda_i^{(k)}\right)u_k \tag{5.133}$$

subject to

$$\sum_{k=1}^{\nu-1} u_k = 1 \tag{5.134}$$

$$-\sum_{k=1}^{\nu-1}\theta_i^{(k)}u_k = 0; \quad i = l+1,\ldots,m \tag{5.135}$$

$$u_k \geq 0; \quad k = 1,\ldots,\nu-1 . \tag{5.136}$$

5.7 From Lagrangian Relaxation to Dantzig-Wolfe Decomposition   237

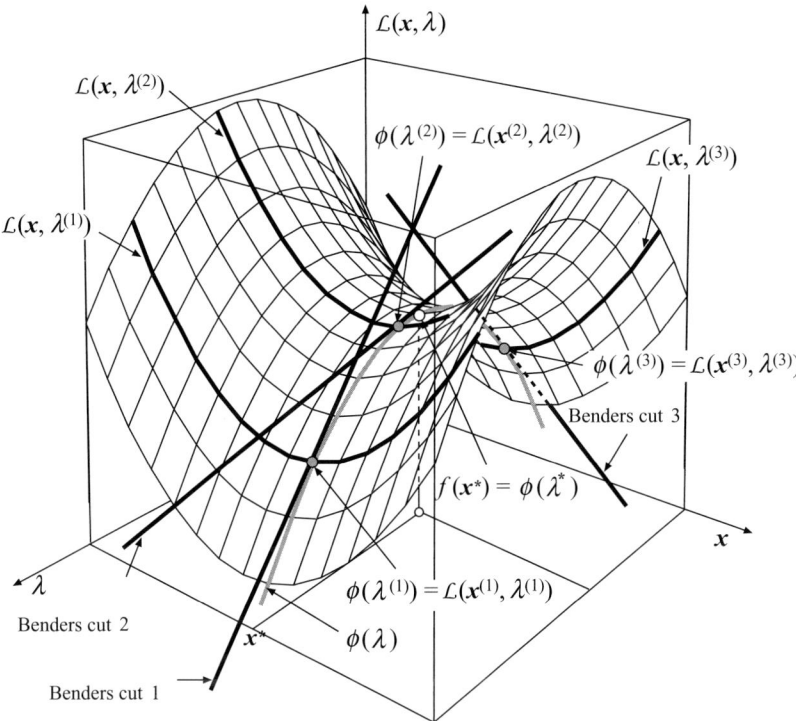

Fig. 5.14. Graphical illustration of the reconstruction by hyperplanes of the dual function $\phi(\lambda)$

This dual master problem does not seem to be similar to the Dantzig-Wolfe master problem (2.29)–(2.32). However, taking into account (5.127) and (5.129) the following simplifications are possible:

$$
\begin{aligned}
-\sum_{k=1}^{\nu-1}\theta_i^{(k)}u_k &= -\sum_{k=1}^{\nu-1}\left(b_i - \sum_{j=1}^{n}a_{ij}x_j^{(k)}\right)u_k \\
&= -\sum_{k=1}^{\nu-1}\left(b_i - r_i^{(k)}\right)u_k = -b_i\sum_{k=1}^{\nu-1}u_k + \sum_{k=1}^{\nu-1}r_i^{(k)}u_k \\
&= -b_i + \sum_{k=1}^{\nu-1}r_i^{(k)}u_k
\end{aligned}
\tag{5.137}
$$

$$
\phi\left(\boldsymbol{\lambda}^{(k)}\right) - \sum_{i=1}^{m}\theta_i^{(k)}\lambda_i^{(k)} = y^{(k)} + \sum_{i=1}^{m}\lambda_i^{(k)}b_i - \sum_{i=1}^{m}\theta_i^{(k)}\lambda_i^{(k)}
$$

$$= \sum_{j=1}^{n}\left(c_j - \sum_{i=1}^{m}\lambda_i^{(k)} a_{ij}\right) x_j^{(k)} + \sum_{i=1}^{m}\lambda_i^{(k)} b_i - \sum_{i=1}^{m}\theta_i^{(k)} \lambda_i^{(k)}$$

$$= \sum_{j=1}^{n}\left(c_j - \sum_{i=1}^{m}\lambda_i^{(k)} a_{ij}\right) x_j^{(k)} + \sum_{i=1}^{m}\lambda_i^{(k)} b_i - \sum_{i=1}^{m}\left(b_i - \sum_{j=1}^{n} a_{ij} x_j^{(k)}\right) \lambda_i^{(k)}$$

$$= \sum_{j=1}^{n} c_j x_j^{(k)} = z^{(k)}. \tag{5.138}$$

Substituting (5.137) into (5.135), and (5.138) into (5.133), respectively, we obtain

$$\underset{u_k;\, k=1,\ldots,\nu-1}{\text{minimize}} \quad \sum_{k=1}^{\nu-1} z^{(k)} u_k \tag{5.139}$$

subject to

$$\sum_{k=1}^{\nu-1} u_k = 1 \tag{5.140}$$

$$\sum_{k=1}^{\nu-1} r_i^{(k)} u_k = b_i: \quad \mu_i; \quad i = \ell+1,\ldots,m \tag{5.141}$$

$$u_k \geq 0; \quad k = 1,\ldots,\nu-1. \tag{5.142}$$

This dual master problem is equal to the Dantzig-Wolfe master problem (2.29)–(2.32).

Note that dealing with linear programming, the solution of this dual master problem is the same as the master problem (5.130)–(5.131). Therefore an upper bound of the original problem (5.132) is

$$\phi_{\text{up}}^{(k)} = \alpha = \sum_{k=1}^{\nu-1} z^{(k)} u_k,$$

which is the same as the expression (2.68) obtained in the Dantzig-Wolfe decomposition.

So it can be concluded that the LR and the Dantzig-Wolfe decompositions are equivalent procedures, being the only difference the formulation of the master problem, where the LR uses dual variables, whereas the Dantzig-Wolfe decomposition uses primal variables.

5.8 Concluding Remarks

This chapter considers different techniques to decompose nonlinear programming problems that do have decomposable structure. For the complicating constraint case, three procedures are analyzed, being the OCD technique the one that presents better convergence and efficiency properties. For the complicating variable case, the Benders decomposition is reviewed. It is shown that the application of the LR to a linear problem renders the Dantzig-Wolfe decomposition algorithm. Several examples illustrate the decomposition principles and procedures studied in this chapter.

The problem considered may have both complicating variables and constraints. In such situation a nested decomposition can be used. For instance, an outer Benders decomposition algorithm may deal with complicating variables whereas an inner Lagrangian relaxation procedure may deal with complicating constraints.

5.9 Exercises

Exercise 5.1. Consider 3 h in which demands are 150, 300, and 500 units, respectively. Consider three electricity plants whose minimum outputs are zero. Ramp-up limits are 200, 100, and 100 units per hour, respectively, and ramp-down limits 300, 150, and 100 units per hour, respectively, and variable costs are, respectively, 0.100, 0.125, and 0.150 $ per unit.

Solve this multiperiod production planning problem using LR and AL decomposition. Solve the subproblems using linear programming.

Exercise 5.2. Consider the following problem

$$\underset{x,y}{\text{minimize}} \quad f(x,y) = x^2 + y^2$$

subject to

$$x + y - 10 = 0$$
$$x \geq 0$$
$$y \geq 0.$$

1. Write the Lagrangian function associated with this problem.
2. Solve the problem using an LR procedure and a subgradient updating of multipliers.
3. Solve the problem using an LR procedure and a bundle method to update multipliers.
4. Compare the solutions of 2 and 3.

Exercise 5.3. Consider the problem stated in Exercise 5.2,

$$\underset{x,y}{\text{minimize}} \quad f(x,y) = x^2 + y^2$$

240   5 Decomposition in Nonlinear Programming

subject to
$$x + y - 10 = 0$$
$$x \geq 0$$
$$y \geq 0 \ .$$

1. Write the AL function associated with this problem.
2. Solve the problem using an AL decomposition.

Exercise 5.4. The production-scheduling problem formulated in Sect. 1.5.1, p. 39, has such a structure that it can be solved using either LR or AL decomposition.

Solve the numerical example stated in Sect. 1.5.1 using both LR and AL decomposition, and verify that the results obtained coincide with those provided in that section. Compare the numerical behavior the LR algorithm with that of the AL decomposition algorithm.

Exercise 5.5. Consider the water supply network in Fig. 5.15. It consists of two cities communicated by a single channel, a set of nodes and a set of connections. The nodes have been numbered in an optimal order, so that if the flow balance equations are written, the associated matrix exhibits a nice block and banded pattern.

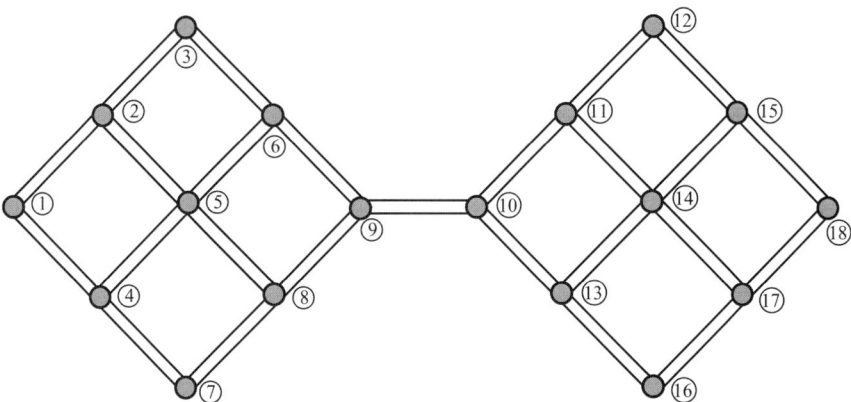

Fig. 5.15. A water supply network consisting of two cities communicated by a single channel

1. Assign to each connection i an arbitrary direction (arrow) corresponding to the flow direction (from higher to lower height) and a variable x_i.
2. Assign to each node j an arbitrary supply or consumption amount q_j, but be sure to assign values such that the total consumption coincides with the

total supply, i.e., $\sum_{i \in I} q_i = \sum_{j \in J} q_j$, where I is the set of supply nodes, and J is the set of consumption nodes.
3. Write the flow balance equations for all nodes (input amount of water equal to output amount of water including supply and consumption).
4. Write the resulting system of equations in matrix form and identify its banded character and the blocks associated with the two cities.
5. Propose a nonlinear objective function to calculate the cost of the water supply system.
6. If the node numbering were done at random, what would it be the aspect of the associated matrix?
7. Propose some decomposition procedures to solve the problem.

Exercise 5.6. The optimal operation of a multiarea electricity network is addressed in Sect. 1.5.2, p. 42. This problem presents such a structure that it can be solved through the use of the OCD algorithm.

Solve the numerical example discussed in Sect. 1.5.2 using the OCD method, and compare the results obtained with those provided in that section. Analyze the numerical behavior of this decomposition algorithm for this particular problem.

Exercise 5.7. A cardboard box is used to store lamps. Knowing that the upper, bottom, and front sides should be built with double quantity of material, determine the dimension of the box that contains maximum volume for a surface of available cardboard equal to 10.

Consider the equality constraint of this problem a complicating constraint and solve it using the LR.

Consider the height of the cardboard box a complicating variable and solve this problem using the Benders decomposition.

Exercise 5.8. Using appropriate control systems the energy flows through three parallel lines between nodes A y B can be fixed. Losses incurred in transmission are given by

$$p(x_1, x_2, x_3) = \frac{1}{2}\left(x_1^2 + x_2^2 + \frac{x_3^2}{10}\right) + x_3 ,$$

where x_i ($i = 1, 2, 3$) represents the volume of energy transmitted through line i. Considering that the total energy transmitted is 10, compute the flow to be send through every line so that losses are minimized.

Consider the equality constraint of this problem a complicating constraint and solve it using the LR.

Consider x_3 as a complicating variable and solve this problem using the Benders decomposition.

Exercise 5.9. Consider two production devices serving a 3 period demand of 100, 40, and 60 units. The operating range of devices 1 and 2 are within

0 and 150, and 0 and 180 units, respectively. The production of each device cannot change above 60 units from one period to the next one. The quadratic production cost of device 1 is characterized by fixed, linear, and quadratic coefficients of values 5, 10, and 0.1, respectively. Coefficients for device 2 are 6, 12, and 0.08, respectively. The devices are not working before the study horizon.

1. Formulate the optimal scheduling problem that allows determining the start-up and shut-down sequence of the production devices that minimize production cost while serving the demand.
2. Considering the conditions of demand supply complicating constraints, solve the problem using the LR.
3. Repeat 2 using the AL decomposition.
4. Repeat 2 using the OCD.

Exercise 5.10. The investment planning problem considered in Sect. 1.6.1, p. 53, presents such a structure that it can be conveniently solved using the Benders decomposition.

Solve the numerical example related to this investment-planning problem that is stated in Sect. 1.6.1 using the Benders decomposition, and verify that the solution obtained is the one provided in that section.

6
Decomposition in Mixed-Integer Programming

6.1 Introduction

This chapter considers mixed-integer mathematical programming problems, both linear and nonlinear.

Mixed-integer linear programming (MILP) problems can be solved in a centralized fashion using the powerful solvers nowadays available. Branch and cut techniques that have been developed during the last decade of the twentieth century allow us, using personal computers, to solve problems at least two orders of magnitude larger than those problems solvable before the development of such branch and cut techniques [42]. Alternatively, MILP problems can be decomposed to separate integer and continuous variables, which is equivalent to considering the integer variables as complicating variables. The resulting continuous subproblem may be decomposed by blocks. In this case, such decomposable structure can be usually exploited computationally to develop efficient algorithms. This situation often arises in practice, particularly, in long-term multiperiod investment planning problems. Investment decisions are integer while operation decisions are continuous and often separable by a time period.

The case of complicating constraints in MILP problems is not so common in practice. A decomposition technique similar to the Dantzig-Wolfe decomposition for such type of problems is denominated "Branch and Price." This rather specific decomposition technique is computationally involved and is not addressed in this book. It is described in detail in [43].

Mixed-integer nonlinear programming (MINLP) problems can be analyzed from two different perspectives: considering the integer variables as complicating variables, and considering the nonlinear constraints as complicating constraints. If integer variables are considered as complicating variables and, additionally, the problem has a decomposable structure, the decomposition allows a distributed solution of the original problem. This situation is often encountered in practice. Similar considerations to those made for the case of linear decomposable problems are applicable for these problems. Actually, long-term multiperiod investment problems usually fall under this category.

In general, it is assumed that the continuous nonlinear programming (NLP) problem resulting from fixing integer variables to given values in a MINLP problem is convex. This assumption is fundamental to guarantee the convergence of the decomposition procedures proposed in this chapter. In practice, local convexity in the neighborhood of the optimal solution normally suffices to tackle realistic problems, for which good guesses of the neighborhood of the optimal solution are available.

On the other hand, considering nonlinear constraints as complicating constraints and treating them through linearization procedures do not lead generally to a decomposed problem. Nevertheless, these techniques are also analyzed in this chapter.

If Lagrangian relaxation techniques as described in Chap. 5 are applied to MINLP problems, two unfortunate circumstances arise. First, the optimal solution of the dual problem is different than the optimal solution of the primal problem, the difference being the duality gap. Second, the solution of the primal problem associated with the optimal solution of the dual problem is generally infeasible for the primal problem. Nevertheless, from a practical point of view, the above two unfortunate facts have commonly low impact as duality gaps are relatively small and a simple mechanism can be used to make primal infeasible solutions feasible. A practical case study that exhibits the duality gap and primal infeasibility is described in Sect. 9.5.

In the following, MILP problems are considered first. Then, MINLP problems are analyzed.

6.2 Mixed-Integer Linear Programming

A general MILP problem has the form

$$\underset{x_1,\ldots,x_n; y_1,\ldots,y_m}{\text{minimize}} \quad \sum_{i=1}^{n} c_i \, x_i + \sum_{j=1}^{m} d_j \, y_j \qquad (6.1)$$

subject to

$$\sum_{i=1}^{n} a_{\ell i} \, x_i + \sum_{j=1}^{m} e_{\ell j} \, y_j = b_\ell; \quad \ell = 1, \ldots, q \qquad (6.2)$$

$$x_i^{\text{down}} \leq x_i \leq x_i^{\text{up}}, \qquad x_i \in \mathbb{N}; \qquad i = 1, \ldots, n \qquad (6.3)$$

$$y_j^{\text{down}} \leq y_j \leq y_j^{\text{up}}, \qquad y_j \in \mathbb{R}; \qquad j = 1, \ldots, m \,. \qquad (6.4)$$

Note that upper and lower bounds have been imposed on optimization variables. This reflects what happens in most engineering and science problems and simplifies the mathematical treatment required.

6.2 Mixed-Integer Linear Programming

The most common integer variables in real world applications are binary variables. Note that any integer variable can be substituted by a set of binary variables, as shown below.

The integer variable

$$x = \{a_1, a_2, \ldots, a_n\} \tag{6.5}$$

can be substituted by n binary variables as follows:

$$x = \sum_{i=1}^{n} a_i \, u_i \tag{6.6}$$

$$\sum_{i=1}^{n} u_i = 1 \tag{6.7}$$

$$u_i \in \{0, 1\}; \quad i = 1, \ldots, n \,. \tag{6.8}$$

If a centralized solution of problem (6.1)–(6.4) is not advisable, the integer variables can be considered as complicating variables and the Benders decomposition scheme is used. This is detailed below.

6.2.1 The Benders Decomposition for MILP Problems

To solve problem (6.1)–(6.4), the Benders decomposition scheme, as explained in Chap. 3, works as follows:

1. Fix integer variables to given feasible integer values.
2. For fixed-integer variable values, solve the resulting continuous LP problem (or subproblems), and obtain its optimal objective function value and the sensitivities associated with constraints fixing integer variables to specific values. Obtain also an upper bound of the objective function optimal value.
3. Solve the MILP master problem to determine improved values of the integer variables. Obtain also a lower bound of the objective function optimal value.
4. If bounds of the objective function optimal value are close enough, stop, the optimal solution has been reached; otherwise, the algorithm continues in Step 2.

A formal description of the Benders decomposition algorithm for MILP problems is as follows.

Algorithm 6.1 (The Benders decomposition algorithm to solve MILP problems).

Input. Data for the MILP problem (6.1)–(6.4).
Output. The solution of problem (6.1)–(6.4) obtained after using the Benders decomposition algorithm.

Step 0: Initialization. Initialize the iteration counter, $\nu = 1$, and let

$$x_i^{(\nu)} = \begin{cases} x_i^{\text{down}} & \text{if } c_i \geq 0 \\ x_i^{\text{up}} & \text{if } c_i < 0 \end{cases}$$

$$\alpha^{(\nu)} = \alpha^{\text{down}}$$

because it is the trivial solution of the master problem

$$\underset{x_1,\ldots,x_n,\alpha}{\text{minimize}} \quad \sum_{i=1}^{n} c_i x_i + \alpha$$

subject to

$$x_i^{\text{down}} \leq x_i \leq x_i^{\text{up}}, \quad x_i \in \mathbb{N}; \quad i = 1, \ldots, n$$

$$\alpha \geq \alpha^{\text{down}}.$$

Step 1: Subproblem solution. Solve the LP subproblem

$$\underset{y_1,\ldots,y_m}{\text{minimize}} \quad \sum_{j=1}^{m} d_j y_j$$

subject to

$$\sum_{j=1}^{m} e_{\ell j} y_j = b_\ell - \sum_{i=1}^{n} a_{\ell i} x_i; \quad \ell = 1, \ldots, q$$

$$y_j^{\text{down}} \leq y_j \leq y_j^{\text{up}}, \quad y_j \in \mathbb{R}; \quad j = 1, \ldots, m$$

$$x_i = x_i^{(\nu)}: \quad \lambda_i; \quad i = 1, \ldots, n .$$

The solution of this problem is $y_1^{(\nu)}, \ldots, y_m^{(\nu)}$ with dual variable values $\lambda_1^{(\nu)}, \ldots, \lambda_n^{(\nu)}$.

The problem above may be decomposed by blocks. If this is the case, it is solved by blocks. If it is infeasible, additional variables and objective function penalties can be used to avoid infeasibility. This is done in a similar manner as in the continuous case, as explained in Subsect. 3.3.4 of Chap. 3.

Step 2: Convergence checking. Compute upper and lower bounds of the optimal value of the objective function of the original problem

$$z_{\text{up}}^{(\nu)} = \sum_{i=1}^{n} c_i x_i^{(\nu)} + \sum_{j=1}^{m} d_j y_j^{(\nu)}$$

$$z_{\text{down}}^{(\nu)} = \sum_{i=1}^{n} c_i x_i^{(\nu)} + \alpha^{(\nu)} .$$

If $z_{up}^{(\nu)} - z_{down}^{(\nu)}$ is smaller than a pre-specified tolerance, stop, the optimal solution is $x_1^{(\nu)}, \ldots, x_n^{(\nu)}$ and $y_1^{(\nu)}, \ldots, y_m^{(\nu)}$. Otherwise, the algorithm continues with the next step.

Step 3: Master problem solution. Update the iteration counter, $\nu \leftarrow \nu+1$. Solve the MILP master problem

$$\underset{x_1, \ldots, x_n, \alpha}{\text{minimize}} \quad \sum_{i=1}^{n} c_i x_i + \alpha$$

subject to

$$\alpha \geq \sum_{j=1}^{m} d_j y_j^{(k)} + \sum_{i=1}^{n} \lambda_i^{(k)}(x_i - x_i^{(k)}); \quad k = 1, \ldots, \nu - 1$$

$$x_i^{\text{down}} \leq x_i \leq x_i^{\text{up}}, \quad x_i \in \mathbb{N}; \quad i = 1, \ldots, n$$

$$\alpha \geq \alpha^{\text{down}}.$$

The solution of this problem is $x_1^{(\nu)}, \ldots, x_n^{(\nu)}$ and $\alpha^{(\nu)}$. The algorithm continues with Step 1. □

A computational example is solved below. The problem considered in this example decomposes into two continuous subproblems once its single integer variable is fixed to a given integer value.

Computational Example 6.1 (The Benders decomposition for MILP problems). Consider the problem

$$\underset{y_1, y_2, y_3, x}{\text{minimize}} \quad z = -\frac{3}{2} y_1 - 2y_2 - 2y_3 - 2x$$

subject to

$$\begin{aligned}
-y_1 - 3y_2 \phantom{{}+y_3} + 2x &\leq 2 \\
y_1 + 3y_2 \phantom{{}+y_3} - x &\leq 3 \\
y_3 - 3x &\leq \tfrac{7}{2} \\
x &\leq 100 \\
y_1, y_2, y_3, x &\geq 0 \\
y_1, y_2, y_3 &\in \mathbb{R} \\
x &\in \mathbb{N},
\end{aligned}$$

whose optimal solution is $y_1^* = 8$, $y_2^* = 0$, $y_3^* = \frac{37}{2}$, and $x^* = 5$ with an optimal objective function value $z^* = -59$.

If variable x is considered to be a complicating variable, the above problem can be solved using the Benders decomposition algorithm. This is done below.

6 Decomposition in Mixed-Integer Programming

Step 0: Initialization. The iteration counter is initialized, $\nu = 1$.
The initial master problem is

$$\text{minimize} \quad -2x + \alpha$$
$$x, \alpha$$

subject to

$$x \leq 100$$
$$\alpha \geq -50$$
$$x \in \mathbb{N} .$$

Its optimal solution is $x^{(1)} = 100$ and $\alpha^{(1)} = -50$.

Step 1: Subproblem solution. If complicating variable x is fixed to a given value, the original problem decomposes into two subproblems. These subproblems are solved below.

The first subproblem is

$$\text{minimize} \quad z_{s1} = -\frac{3}{2}y_1 - 2y_2 + 40w$$
$$y_1, y_2, w$$

subject to

$$\begin{array}{rrrrrl} -y_1 & -3y_2 & +2x & -w & \leq & 2 \\ y_1 & +3y_2 & -x & -w & \leq & 3 \\ & & x & & = & 100 : \lambda_1 \\ & y_1, y_2 & & & \geq & 0 \\ & y_1, y_2 & & & \in & \mathbb{R} . \end{array}$$

Its optimal solution is $y_1^{(1)} = \frac{301}{2}$, $y_2^{(1)} = 0$, $w^{(1)} = \frac{95}{2}$, and $\lambda_1^{(1)} = \frac{71}{4}$ with an optimal objective function value $z_{s1}^{(1)} = \frac{6{,}697}{4}$. Note that the artificial variable w has been included to attain feasibility.

The second subproblem is

$$\text{minimize} \quad z_{s2} = -2y_3$$
$$y_3$$

subject to

$$\begin{array}{rrrl} y_3 & -3x & \leq & \frac{7}{2} \\ & x & = & 100 : \lambda_2 \\ & y_3 & \geq & 0 \\ & y_3 & \in & \mathbb{R} . \end{array}$$

Its optimal solution is $y_3^{(1)} = \frac{607}{2}$ and $\lambda_2^{(1)} = -6$ with an objective function value $z_{s2} = -607$.

Step 2: Convergence checking. An upper bound of the objective function optimal value is

6.2 Mixed-Integer Linear Programming

$$z_{\text{up}}^{(1)} = -\frac{3}{2}y_1^{(1)} - 2y_2^{(1)} + 40w^{(1)} - 2y_3^{(1)} - 2x^{(1)} = \frac{3{,}469}{4}$$

and a lower bound is

$$z_{\text{down}}^{(1)} = -2x^{(1)} + \alpha^{(1)} = -250.$$

The difference $z_{\text{up}}^{(1)} - z_{\text{down}}^{(1)} = \frac{4{,}469}{4}$ is not small enough; therefore, the procedure continues with Step 3.

Step 3: Master problem solution. The iteration counter is updated, $\nu = 1 + 1 = 2$.

The current master problem is

$$\underset{x,\,\alpha}{\text{minimize}} \quad z_m = -2x + \alpha$$

subject to

$$\begin{aligned}
\frac{6{,}697}{4} - 607 + \frac{71}{4} \times (x - 100) - 6 \times (x - 100) &\leq \alpha \\
x &\leq 100 \\
\alpha &\geq -50 \\
x &\in \mathbb{N}.
\end{aligned}$$

Its optimal solution is $x^{(2)} = 5$ and $\alpha^{(2)} = -49$ with an optimal objective function value $z_m^{(2)} = -59$.

The algorithm continues with Step 1.

Step 1: Subproblem solution. The subproblems are solved below.

The first subproblem is

$$\underset{y_1,\,y_2}{\text{minimize}} \quad z_{s1} = -\frac{3}{2}y_1 - 2y_2$$

subject to

$$\begin{aligned}
-y_1 - 3y_2 + 2x &\leq 2 \\
y_1 + 3y_2 - x &\leq 3 \\
x &= 5 : \lambda_1 \\
y_1, y_2 &\geq 0 \\
y_1, y_2 &\in \mathbb{R}.
\end{aligned}$$

Its optimal solution is $y_1^{(2)} = 8$, $y_2^{(2)} = 0$, and $\lambda_1^{(2)} = \frac{-3}{2}$ with an optimal objective function value $z_{s1} = -12$.

The second subproblem is

$$\underset{y_3}{\text{minimize}} \quad z_{s2} = -2y_3$$

subject to

$$y_3 - 3x \le \frac{7}{2}$$
$$x = 5 : \lambda_2$$
$$y_3 \ge 0$$
$$y_3 \in \mathbb{R} .$$

Its optimal solution is $y_3^{(2)} = \frac{37}{2}$ and $\lambda_2^{(2)} = -6$ with an optimal objective function value $z_{s2} = -37$.

Step 2: Convergence checking. An upper bound of the objective function optimal value is

$$z_{\text{up}}^{(2)} = -\frac{3}{2} y_1^{(2)} - 2 y_2^{(2)} - 2 y_3^{(2)} - 2 x^{(2)} = -59$$

and a lower bound is

$$z_{\text{down}}^{(2)} = -2 x^{(2)} + \alpha^{(2)} = -59.$$

The difference $z_{\text{up}}^{(2)} - z_{\text{down}}^{(2)} = 0$ is small enough; therefore, the optimal solution has been found. It is $y_1^* = 8$, $y_2^* = 0$, $y_3^* = \frac{37}{2}$, and $x^* = 5$ with an optimal objective function value equal to -59. □

6.2.2 Convergence

The convergence of the Benders decomposition algorithm for MILP problems is guaranteed as long as the envelope of function $\alpha(x_1, \ldots, x_n)$ is convex. This function is defined as

$$\alpha(x_1, \ldots, x_n) = \underset{y_1, \ldots, y_m}{\text{minimum}} \sum_{j=1}^{m} d_j \, y_j \qquad (6.9)$$

subject to

$$\sum_{j=1}^{m} e_{\ell j} \, y_j = b_\ell - \sum_{i=1}^{n} a_{\ell i} \, x_i; \quad \ell = 1, \ldots, q \qquad (6.10)$$

$$y_j^{\text{down}} \le y_j \le y_j^{\text{up}}, y_j \in \mathbb{R}; \quad j = 1, \ldots, m. \qquad (6.11)$$

This convergence property follows directly from the convexity proof of function $\alpha(x_1, \ldots, x_n)$ corresponding to the continuous case, as stated in Sect. 3.3.1.

MINLP problems are analyzed in the following sections.

6.3 Mixed-Integer Nonlinear Programming

The considered MINLP problem is

$$\underset{x_1,\ldots,x_n;y_1,\ldots,y_m}{\text{minimize}} \quad f(x_1,\ldots,x_n;y_1,\ldots,y_m) \tag{6.12}$$

subject to

$$h_k(x_1,\ldots,x_n;y_1,\ldots,y_m) = 0; \quad k = 1,\ldots,q \tag{6.13}$$

$$g_l(x_1,\ldots,x_n;y_1,\ldots,y_m) \leq 0; \quad l = 1,\ldots,r \tag{6.14}$$

$$x_i^{\text{down}} \leq x_i \leq x_i^{\text{up}}, \quad x_i \in \mathbb{N}; \quad i = 1,\ldots,n \tag{6.15}$$

$$y_j^{\text{down}} \leq y_j \leq y_j^{\text{up}}, \quad y_j \in \mathbb{R}; \quad j = 1,\ldots,m. \tag{6.16}$$

Note that upper and lower bounds are imposed on optimization variables to reflect physical limits, which results in a simpler mathematical treatment.

As previously stated, MINLP problems can be addressed either considering the integer variables as complicating variables or the nonlinear constrains as complicating constraints, the former approach being more common in practice. These two approaches are described in detail in the following subsections.

It is assumed that the continuous NLP problem resulting from fixing in the original MINLP problem the integer variables to any given feasible values is convex; otherwise, the convergence of the procedures analyzed in this section cannot be guaranteed. However, local convexity in a neighborhood of the optimal solution is sufficient to guarantee convergence in most practical applications.

6.4 Complicating Variables: Nonlinear Case

If the integer variables of an MINLP problem are considered as complicating variables, the Benders decomposition algorithm can be used straightforwardly. This is what is done below.

6.4.1 The Benders Decomposition

The Benders decomposition algorithm to solve MINLP problems is as follows.

Algorithm 6.2 (The Benders decomposition algorithm to solve MINLP problems).

Input. Data for the MILP problem (6.12)–(6.16).
Output. The solution of problem (6.12)–(6.16) obtained after using the Benders decomposition algorithm.

Step 0: Initialization. Initialize the iteration counter, $\nu = 1$. Solve the initial MILP master problem below.

$$\text{minimize} \quad \alpha$$
$$\alpha$$

subject to
$$x_i^{\text{down}} \leq x_i \leq x_i^{\text{up}}, \quad x_i \in \mathbb{N}; \quad i = 1, \ldots, n$$
$$\alpha \geq \alpha^{\text{down}}.$$

Its trivial solution is $x_1^{(\nu)}, \ldots, x_n^{(\nu)}$; $\alpha^{(\nu)} = \alpha^{\text{down}}$.

Step 1: Subproblem solution. Solve the continuous NLP subproblem:

$$\underset{y_1, \ldots, y_m}{\text{minimize}} \quad f(x_1, \ldots, x_n; y_1, \ldots, y_m)$$

subject to
$$h_k(x_1, \ldots, x_n; y_1, \ldots, y_m) = 0; \quad k = 1, \ldots, q$$
$$g_l(x_1, \ldots, x_n; y_1, \ldots, y_m) \leq 0; \quad l = 1, \ldots, r$$
$$y_j^{\text{down}} \leq y_j \leq y_j^{\text{up}}, \quad y_j \in \mathbb{R}; \quad j = 1, \ldots, m$$
$$x_i = x_i^{(\nu)} : \quad \lambda_i; \quad i = 1, \ldots, n.$$

The solution of this problem is $y_1^{(\nu)}, \ldots, y_n^{(\nu)}$ with dual variable values $\lambda_1^{(\nu)}, \ldots, \lambda_n^{(\nu)}$.

The above problem may be decomposed by blocks. If this is the case, it is solved by blocks. This is a situation often encountered in practice.

Step 2: Convergence checking. Compute upper and lower bounds of the optimal value of the objective function of the original problem:

$$z_{\text{up}}^{(\nu)} = f(x_1^{(\nu)}, \ldots, x_n^{(\nu)}; y_1^{(\nu)}, \ldots, y_m^{(\nu)})$$

$$z_{\text{down}}^{(\nu)} = \alpha^{(\nu)}.$$

If $z_{\text{up}}^{(\nu)} - z_{\text{down}}^{(\nu)}$ is smaller, than a pre-specified tolerance, stop, the optimal solution is $x_1^{(\nu)}, \ldots, x_n^{(\nu)}$ and $y_1^{(\nu)}, \ldots, y_m^{(\nu)}$. Otherwise, the algorithm continues with the next step.

Step 3: Master problem solution. Update the iteration counter, $\nu \leftarrow \nu + 1$. Solve the MILP master problem

$$\text{minimize} \quad \alpha$$
$$\alpha$$

subject to

$$\alpha \geq f(x_1^{(k)}, \ldots, x_n^{(k)}; y_1^{(k)}, \ldots, y_m^{(k)}) + \sum_{i=1}^{n} \lambda_i^{(k)}(x_i - x_i^{(k)}); \quad k = 1, \ldots, \nu - 1$$

$$x_i^{\text{down}} \leq x_i \leq x_i^{\text{up}}, \qquad x_i \in \mathbb{N}; \quad i = 1, \ldots, n$$

$$\alpha \geq \alpha^{\text{down}}.$$

The solution of this problem is $x_1^{(\nu)}, \ldots, x_n^{(\nu)}$ and $\alpha^{(\nu)}$. The algorithm continues with Step 1.

□

6.4.2 Subproblem Infeasibility

The infeasibility of the subproblem or subproblems can be treated using artificial variables and objective function penalties as stated for the linear continuous case analyzed in Subsect. 3.3.4 p. 128. However, constraints specific to the problem that force feasibility not altering the optimal solution are advisable. This is so because penalty terms in nonlinear problems may result in slow or no convergence. Constraints based on engineering or science facts that avoid infeasibility while not altering the optimal solution are denominated "feasibility cuts."

A computational example on the application of the Benders decomposition algorithm to MINLP problems is described next.

Computational Example 6.2 (The Benders decomposition for MINLP problems). Consider the problem

$$\begin{array}{c} \text{minimize} \\ x, y \end{array} \quad z = -x - y$$

subject to

$$\begin{aligned} \tfrac{1}{2} \exp(2y) - x &\leq \tfrac{1}{4} \\ 0 \leq y &\leq \tfrac{1}{2} \\ x &\in \{0, 1\}, \end{aligned}$$

whose optimal solution is $x^* = 1$ and $y^* = 0.4581$ with an optimal objective function value equal to $z^* = -1.4581$ as shown in Fig. 6.1.

If variable x is considered to be a complicating variable, the above problem can be solved using the Benders decomposition algorithm. This is illustrated further.

Step 0: Initialization. The iteration counter is initialized, $\nu = 1$.
The initial master problem is

$$\begin{array}{c} \text{minimize} \\ \alpha \end{array} \quad \alpha$$

subject to

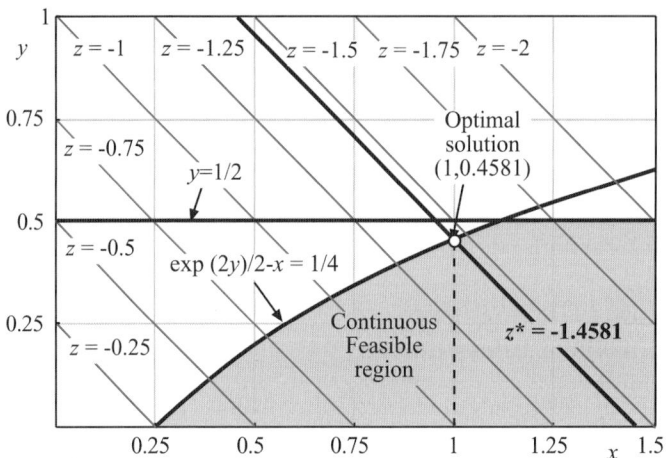

Fig. 6.1. Graphical illustration of the Computational Example 6.2

$$\alpha \geq -50$$
$$x \in \{0, 1\}.$$

The solution of this problem is $x^{(1)} = 0$ and $\alpha^{(1)} = -50$.

Step 1: Subproblem solution. The subproblem is

$$\underset{x,y}{\text{minimize}} \quad z = -x - y + 40w$$

subject to

$$\tfrac{1}{2}\exp(2y) - x - w \leq \tfrac{1}{4}$$
$$0 \leq y \leq \tfrac{1}{2}$$
$$x = 0 : \lambda.$$

Its optimal solution is $y^{(1)} = 0$, $w^{(1)} = \tfrac{1}{4}$ and $\lambda^{(1)} = -41$ with an optimal objective function value $z = 10$. Note that an artificial variable w has been included to attain feasibility (see Fig. 6.2).

Step 2: Convergence checking. Upper and lower bounds of the objective function optimal value are

$$z_{\text{up}}^{(1)} = -x^{(1)} - y^{(1)} + 40w^{(1)} = 10$$

and

$$z_{\text{down}}^{(1)} = \alpha^{(1)} = -50.$$

The difference $z_{\text{up}}^{(1)} - z_{\text{down}}^{(1)} = 60$ is not small enough; therefore, the procedure continues with the next step.

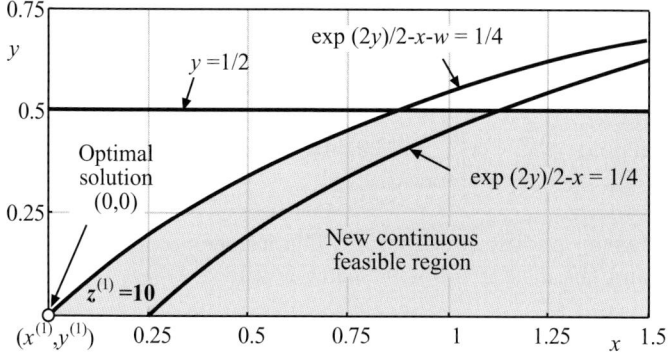

Fig. 6.2. Graphical illustration of how the feasible region is modified to make the subproblem feasible

Step 3: Master problem solution. The iteration counter is updated, $\nu = 1 + 1 = 2$.

The master problem is

$$\underset{\alpha}{\text{minimize}} \quad \alpha$$

subject to

$$\begin{aligned} 10 - 41 \times (x - 0) &\leq \alpha \\ \alpha &\geq -50 \\ x &\in \{0, 1\} . \end{aligned}$$

Its optimal solution is $x^{(2)} = 1$ and $\alpha^{(2)} = -31$. The objective function optimal value is $z^{(2)} = -31$.

The algorithm continues with Step 1.

Step 1: Subproblem solution. The subproblem is

$$\underset{x, y}{\text{minimize}} \quad z = -x - y$$

subject to

$$\begin{aligned} \tfrac{1}{2} \exp(2y) - x &\leq \tfrac{1}{4} \\ 0 \leq y &\leq \tfrac{1}{2} \\ x &= 1 : \lambda , \end{aligned}$$

whose solution is $y^{(2)} = 0.458$ and $\lambda^{(2)} = -1.400$ with an optimal objective function value $z = -1.458$.

Step 2: Convergence checking. Upper and lower bounds of the objective function optimal value are

and
$$z_{\text{up}}^{(2)} = -x^{(2)} - y^{(2)} = -1.458$$

$$z_{\text{down}}^{(2)} = \alpha^{(2)} = -31 \ .$$

The difference $z_{\text{up}}^{(2)} - z_{\text{down}}^{(2)} = 29.542$ is not small enough; therefore, the procedure continues with the next step.

Step 3: Master problem solution. The iteration counter is updated, $\nu = 1 + 1 = 3$, and the master problem below is solved:

$$\underset{\alpha}{\text{minimize}} \quad \alpha$$

subject to
$$\begin{aligned} 10 - 41 \times (x - 0) &\leq \alpha \\ -1.458 - 1.400 \times (x - 1) &\leq \alpha \\ \alpha &\geq -50 \\ x &\in \{0, 1\} \ . \end{aligned}$$

Its optimal solution is $x^{(3)} = 1$ and $\alpha^{(3)} = -1.458$ with an optimal objective function value $z^{(3)} = -1.458$

The algorithm continues with Step 1.

Step 1: Subproblem solution. The subproblem is

$$\underset{x,y}{\text{minimize}} \quad z = -x - y$$

subject to
$$\begin{aligned} \tfrac{1}{2} \exp(2y) - x &\leq \tfrac{1}{4} \\ 0 \leq y &\leq \tfrac{1}{2} \\ x &= 1 : \lambda \ , \end{aligned}$$

whose solution is $y^{(3)} = 0.458$ and $\lambda^{(3)} = -1.400$ with an objective function value $z = -1.458$.

Step 2: Convergence checking. Upper and lower bounds of the objective function optimal value are

$$z_{\text{up}}^{(3)} = -x^{(3)} - y^{(3)} = -1.458$$

and
$$z_{\text{down}}^{(3)} = \alpha^{(3)} = -1.458 \ .$$

The difference $z_{\text{up}}^{(1)} - z_{\text{down}}^{(1)} = 0$ is small enough; therefore, the optimal solution has been found. It is $x^* = 1$ and $y^* = 0.458$ with an optimal objective function value equal to -1.458. In Fig. 6.3 the evolution of the bounds $z_{\text{up}}^{(\nu)}$ and $z_{\text{down}}^{(\nu)}$ is shown. □

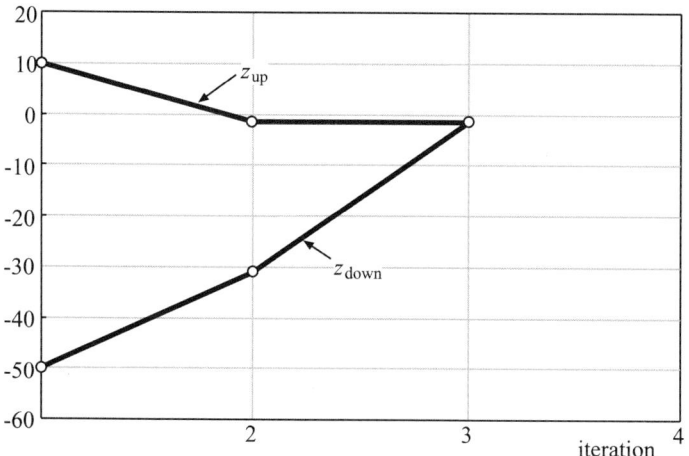

Fig. 6.3. Evolution of the upper and lower bounds of the Computational Example 6.2

6.4.3 Convergence

The convergence of the Benders decomposition algorithm for MINLP problems is guaranteed as long as the envelope of function $\alpha(x_1,\ldots,x_n)$ is convex.

This function is defined as

$$\alpha(x_1,\ldots,x_n) = \min_{y_1,\ldots,y_m} \; f(x_1,\ldots,x_n;y_1,\ldots,y_m) \quad (6.17)$$

subject to

$$h_k(x_1,\ldots,x_n;y_1,\ldots,y_m) = 0; \quad k=1,\ldots,q \quad (6.18)$$

$$g_l(x_1,\ldots,x_n;y_1,\ldots,y_m) \leq 0; \quad l=1,\ldots,r \quad (6.19)$$

$$y_j^{\text{down}} \leq y_j \leq y_j^{\text{up}}, \quad y_j \in \mathbb{R}; \quad j=1,\ldots,m. \quad (6.20)$$

Convexity, using cutting hyperplanes, allows a reconstruction from below of the function $\alpha(x_1,\ldots,x_n)$, as accurate as needed, and this is a key factor that guarantees the convergence of the algorithm. In practice, local convexity (vs. global convexity) normally suffices because in engineering or science applications variable limits usually tightly bound the feasibility region.

6.5 Complicating Constraints: Nonlinear Case

The nonlinear constraints of an MINLP problem can be considered as complicating constraints. This leads to the Outer Linearization algorithm described in this section.

6.5.1 Outer Linearization Algorithm

The outer linearization procedure is only applicable to problems whose nonlinear constraints are inequalities.

This algorithm also requires the considered problem to have a linear objective function. However, this is achieved in a simple manner using the following transformation.

The problem

$$\underset{x_1,\ldots,x_n;y_1,\ldots,y_m}{\text{minimize}} \quad f(x_1,\ldots,x_n;y_1,\ldots,y_m)$$

subject to

$$g_l(x_1,\ldots,x_n;y_1,\ldots,y_m) \leq 0; \quad l=1,\ldots,r$$
$$x_i^{\text{down}} \leq x_i \leq x_i^{\text{up}}, \quad x_i \in \mathbb{N}; \quad i=1,\ldots,n$$
$$y_j^{\text{down}} \leq y_j \leq y_j^{\text{up}}, \quad y_j \in \mathbb{R}; \quad j=1,\ldots,m$$

is equivalent to the problem

$$\underset{x_1,\ldots,x_n;y_1,\ldots,y_m,z}{\text{minimize}} \quad z$$

subject to

$$f(x_1,\ldots,x_n;y_1,\ldots,y_m) - z \leq 0$$
$$g_l(x_1,\ldots,x_n;y_1,\ldots,y_m) \leq 0; \quad l=1,\ldots,r$$
$$x_i^{\text{down}} \leq x_i \leq x_i^{\text{up}}, \quad x_i \in \mathbb{N}; \quad i=1,\ldots,n$$
$$y_j^{\text{down}} \leq y_j \leq y_j^{\text{up}}, \quad y_j \in \mathbb{R}; \quad j=1,\ldots,m \ .$$

The basic functioning of the Outer Linearization algorithm is as follows:

1. Ignore the nonlinear constraints, solve the resulting initial MILP problem, and obtain the initial solution.
2. Determine the most violated nonlinear constraint in the current solution, and linearize it in the current solution.
3. Add the linear constraint obtained in the preceding step to the current MILP problem, solve it, and update the current solution.
4. If all constraints in the current solution are sufficiently satisfied, stop, the optimal solution has been reached; otherwise, continue with Step 2.

A formal statement of this algorithm is presented in the following.

Consider the problem

$$\underset{x_1,\ldots,x_n;y_1,\ldots,y_m}{\text{minimize}} \quad \sum_{i=1}^{n} c_i\, x_i + \sum_{j=1}^{m} d_j\, y_j$$

subject to

$$\sum_{i=1}^{n} a_{\ell i}\, x_i + \sum_{j=1}^{m} e_{\ell j}\, y_j = b_\ell; \quad \ell = 1, \ldots, q$$

$$g_l(x_1, \ldots, x_n; y_1, \ldots, y_m) \leq 0; \quad l = 1, \ldots, r$$

$$x_i^{\text{down}} \leq x_i \leq x_i^{\text{up}}, \quad x_i \in \mathbb{N}; \quad i = 1, \ldots, n$$

$$y_j^{\text{down}} \leq y_j \leq y_j^{\text{up}}, \quad y_j \in \mathbb{R}; \quad j = 1, \ldots, m \,.$$

Note that optimization variables are bounded above and below reflecting physical limits which are always present in engineering and science problems.

The algorithm works as follows.

Algorithm 6.3 (The outer linearization algorithm).

Input. Data for the MILP problem.

Output. The solution of problem obtained after using the Outer Linearization algorithm.

Step 0: Initialization. Initialize the iteration counter, $\nu = 1$. Solve the initial MILP problem below.

$$\underset{x_1, \ldots, x_n;\, y_1, \ldots, y_m}{\text{minimize}} \quad \sum_{i=1}^{n} c_i\, x_i + \sum_{j=1}^{m} d_j\, y_j$$

subject to

$$\sum_{i=1}^{n} a_{\ell i}\, x_i + \sum_{j=1}^{m} e_{\ell j}\, y_j = b_\ell; \quad \ell = 1, \ldots, q$$

$$x_i^{\text{down}} \leq x_i \leq x_i^{\text{up}}, \quad x_i \in \mathbb{N}; \quad i = 1, \ldots, n$$

$$y_j^{\text{down}} \leq y_j \leq y_j^{\text{up}}, \quad y_j \in \mathbb{R}; \quad j = 1, \ldots, m \,.$$

The solution obtained is $x_1^{(\nu)}, \ldots, x_n^{(\nu)}; y_1^{(\nu)}, \ldots, y_m^{(\nu)}$.

Step 1: Determining constraint violations. Identify the most violated nonlinear constraint in the current solution $x_1^{(\nu)}, \ldots, x_n^{(\nu)}, y_1^{(\nu)}, \ldots, y_m^{(\nu)}$, i.e.,

$$g_{l_\nu}(x_1^{(\nu)}, \ldots, x_n^{(\nu)}; y_1^{(\nu)}, \ldots, y_m^{(\nu)}) = \underset{l}{\text{maximum}} \quad g_l(x_1^{(\nu)}, \ldots, x_n^{(\nu)}; y_1^{(\nu)}, \ldots, y_m^{(\nu)}).$$

Step 2: Convergence check. If $g_{l_\nu}(x_1^{(\nu)}, \ldots, x_n^{(\nu)}; y_1^{(\nu)}, \ldots, y_m^{(\nu)}) \leq 0$, stop, the current solution $x_1^{(\nu)}, \ldots, x_n^{(\nu)}, y_1^{(\nu)}, \ldots, y_m^{(\nu)}$ is the optimal solution; otherwise the algorithm continues with the next step.

Step 3: Linearization. Linearize the most violated constraint in the current solution.

$$l_\nu(x_1,\ldots,x_n;y_1,\ldots,y_m) = g_{l_\nu}(x_1^{(\nu)},\ldots,x_n^{(\nu)};y_1^{(\nu)},\ldots,y_m^{(\nu)})$$

$$+ \left(\nabla g_{l_\nu}(x_1^{(\nu)},\ldots,x_n^{(\nu)};y_1^{(\nu)},\ldots,y_m^{(\nu)})\right)^T \begin{pmatrix} x_1 - x_1^{(\nu)} \\ \vdots \\ x_n - x_n^{(\nu)} \\ y_1 - y_1^{(\nu)} \\ \vdots \\ y_m - y_m^{(\nu)} \end{pmatrix}.$$

Step 4: Solution of the linearized problem. Solve the MILP problem

$$\underset{x_1,\ldots,x_n;y_1,\ldots,y_m}{\text{minimize}} \quad \sum_{i=1}^n c_i\, x_i + \sum_{j=1}^m d_j\, y_j$$

subject to

$$\sum_{i=1}^n a_{\ell i}\, x_i + \sum_{j=1}^m e_{\ell j}\, y_j = b_\ell; \qquad \ell = 1,\ldots,q$$

$$l_k(x_1,\ldots,x_n;y_1,\ldots,y_m) \le 0; \qquad k = 1,\ldots,\nu$$

$$x_i^{\text{down}} \le x_i \le x_i^{\text{up}}, \qquad x_i \in \mathbb{N}; \qquad i = 1,\ldots,n$$

$$y_j^{\text{down}} \le y_j \le y_j^{\text{up}}, \qquad y_j \in \mathbb{R}; \qquad j = 1,\ldots,m\,.$$

The solution obtained is $x_1^{(\nu+1)},\ldots,x_n^{(\nu+1)}; y_1^{(\nu+1)},\ldots,y_m^{(\nu+1)}$. Update iteration counter, $\nu \leftarrow \nu + 1$, and continue with Step 1. □

A computational example is solved in detail below. The problem of this example is the same one used in Computational Example 6.2.

Computational Example 6.3 (The outer linearization for MINLP problems). Consider the problem previously analyzed in Computational Example 6.2

$$\underset{x,y}{\text{minimize}} \quad z = -x - y$$

subject to

$$\begin{array}{rcl} g(x,y) = \tfrac{1}{2}\exp(2y) - x - \tfrac{1}{4} & \le & 0 \\ 0 \le y & \le & \tfrac{1}{2} \\ x & \in & \{0,1\}\,, \end{array}$$

whose optimal solution is $x^* = 1$ and $y^* = 0.458$ with an optimal objective function value equal to $z^* = -1.458$ as shown in Fig. 6.1.

This problem can be solved using the previously stated outer linearization algorithm. A tolerance of 1×10^{-4} is considered. The solution procedure is illustrated below.

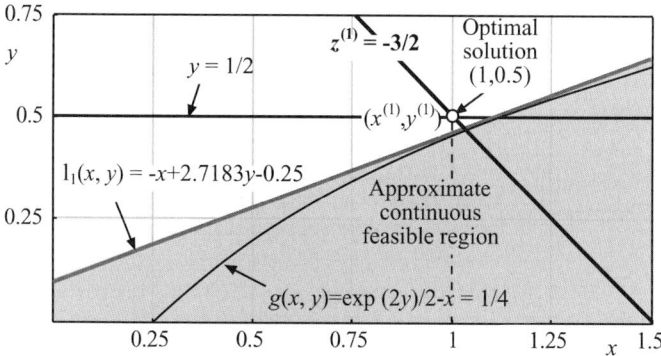

Fig. 6.4. Graphical illustration of the decomposition algorithm and the linear approximation $l_1(x,y)$ of the nonlinear constraint $g(x,y)$ for iteration $\nu = 1$

Step 0: Initialization. The iteration counter is initialized, $\nu = 1$.
The initial MILP problem is

$$\underset{x,y}{\text{minimize}} \quad z = -x - y$$

subject to

$$0 \leq y \leq \frac{1}{4}$$
$$x \in \{0, 1\}.$$

Its optimal solution is $x^{(1)} = 1$, $y^{(1)} = \frac{1}{2}$ with an optimal objective function value $z^{(1)} = \frac{-3}{2}$ (see Fig. 6.4).

Step 1: Determining the most violated constraint. The considered problem has a single nonlinear constraint; therefore, that constraint is the most violated one

$$g(x^{(1)}, y^{(1)}) = \frac{1}{2}\exp\left(2 \times \frac{1}{2}\right) - 1 - \frac{1}{4} = 0.109 \;.$$

Step 2: Convergence check. $g(x^{(1)}, y^{(1)}) = 0.109$ is not small enough, the algorithm continues with the next step.

Step 3: Linearization. The nonlinear constraint is linearized. Its gradient in the current solution is

$$\nabla g(x^{(1)}, y^{(1)}) = \begin{pmatrix} -1 & \exp(2 \times \frac{1}{2}) \end{pmatrix}^T = \begin{pmatrix} -1 & 2.718 \end{pmatrix}^T \;.$$

The corresponding linear constraint is

$$l_1(x,y) = g(x^{(1)}, y^{(1)}) + \left(\nabla g(x^{(1)}, y^{(1)})\right)^T \begin{pmatrix} x - x^{(1)} \\ y - y^{(1)} \end{pmatrix}$$

or

$$l_1(x,y) = 0.109 + \begin{pmatrix} -1 & 2.718 \end{pmatrix} \begin{pmatrix} x - 1 \\ y - \frac{1}{2} \end{pmatrix},$$

and finally

$$l_1(x,y) = -x + 2.718\,y - 0.250 .$$

Step 4: Solution of the linearized problem. The current MILP problem is

$$\begin{array}{cl} \text{minimize} & z = -x - y \\ x, y & \end{array}$$

subject to

$$\begin{array}{rcl} -x + 2.718\,y - 0.250 & \leq & 0 \\ 0 \leq y & \leq & \frac{1}{2} \\ x & \in & \{0, 1\} . \end{array}$$

Its optimal solution is $x^{(2)} = 1$, $y^{(2)} = 0.46$ with an optimal objective function value $z^{(2)} = -1.46$ as shown in Fig. 6.5.
Update iteration counter, $\nu = 1 + 1 = 2$, and continue with Step 1.

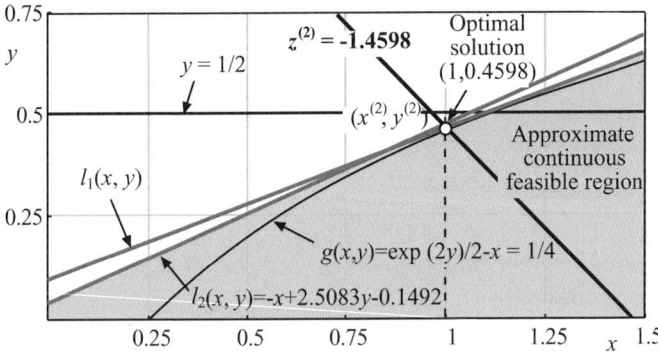

Fig. 6.5. Graphical illustration of the decomposition algorithm and the linear approximations $l_1(x,y)$ and $l_2(x,y)$ of the nonlinear constraint $g(x,y)$ for iteration $\nu = 2$

6.5 Complicating Constraints: Nonlinear Case

Step 1: Determining the most violated constraint. The considered problem has a single nonlinear constraint; therefore, that constraint is the most violated one

$$g(x^{(2)}, y^{(2)}) = \frac{1}{2}\exp(2 \times 0.46) - 1 - \frac{1}{4} = 0.004.$$

Step 2: Convergence check. Since $g(x^{(2)}, y^{(2)}) = 0.004$ is not small enough, the algorithm continues with the next step.

Step 3: Linearization. The nonlinear constraint is linearized. Its gradient in the current solution is

$$\nabla g(x^{(2)}, y^{(2)}) = \begin{pmatrix} -1 & \exp(2 \times 0.46) \end{pmatrix}^T = \begin{pmatrix} -1 & 2.508 \end{pmatrix}^T.$$

The corresponding linear constraint is

$$l_2(x, y) = g(x^{(2)}, y^{(2)}) + \left(\nabla g(x^{(2)}, y^{(2)})\right)^T \begin{pmatrix} x - x^{(2)} \\ y - y^{(2)} \end{pmatrix}$$

or

$$l_2(x, y) = 0.004 + \begin{pmatrix} -1 & 2.508 \end{pmatrix} \begin{pmatrix} x - 1 \\ y - 0.4 \end{pmatrix},$$

and finally

$$l_2(x, y) = -x + 2.508\, y - 0.149.$$

Step 4: Solution of the linearized problem. The current MILP problem is

$$\underset{x, y}{\text{minimize}} \quad z = -x - y$$

subject to

$$\begin{aligned}
-x + 2.508y - 0.149 &\leq 0 \\
-x + 2.718y - 0.250 &\leq 0 \\
0 \leq y &\leq \tfrac{1}{2} \\
x &\in \{0, 1\}.
\end{aligned}$$

Its optimal solution is $x^{(3)} = 1$, $y^{(3)} = 0.458$ with an optimal objective function value $z^{(3)} = -1.458$.

Update the iteration counter, $\nu = 2 + 1 = 3$, and continue with Step 1.

Step 1: Determining the most violated constraint. The considered problem has a single nonlinear constraint; therefore, that constraint is the most violated one.

$$g(x^{(3)}, y^{(3)}) = \frac{1}{2}\exp(2 \times 0.458) - 1 - \frac{1}{4} = 0.0001 \ .$$

Step 2: Convergence check.
$g(x^{(3)}, y^{(3)}) = 0.0001$ is small enough ($\leq 1 \times 10^{-4}$); therefore, the algorithm terminates, and the optimal solution is found. It is $x^* = 1$, $y^* = 0.458$ with an optimal objective function value $z^* = -1.458$. □

6.5.2 Convergence

The converge proof of the outer linearization algorithm can be found in Floudas [27]. The basic fact to show the convergence of the outer linearization algorithm relies on the analysis of the series of solutions generated by the algorithm. This series either converges to a locally optimal solution or never finds a feasible solution if the problem is infeasible.

6.6 Concluding Remarks

This chapter presents a set of solution techniques for mixed-integer linear and nonlinear problems that present in general a decomposable structure. The complicating variable case is the one often encountered in practical engineering and science problems.

The Benders decomposition algorithm, due to Benders [24], is used to address problems with complicating variables. The Benders procedure was generalized and publicized in the technical literature by Geoffrion [25, 44].

The outer linearization algorithm used for MILP problems is based on the pioneering work of Kelly on cutting plane algorithms [45]. It has been analyzed in detail by Floudas [27]. A state-of-the-art MILP solvers, using branch and cut techniques, are reported in [42]. These solvers constitute the base of the previous solution procedures.

Among other applications, the solution of long-term multiperiod investment problems using the Benders decomposition are reported in [46, 47, 48, 49, 50], for example.

6.7 Exercises

Exercise 6.1. Given the problem

$$\begin{array}{ll} \text{minimize} & 4y_1 + 3y_2 + y_3 + 0.5y_4 + 4x_1 + 5x_2 \\ y_1, y_2, y_3, y_4, x_1, x_2 & \end{array}$$

subject to

$$\begin{array}{rrrrr}
y_1 & -y_2 & +x_1 & -8x_2 \le & 1.5 \\
-2y_1 & +3y_2 & -x_1 & +2x_2 \le & -5 \\
& 7y_3 +2y_4 & +x_1 & -5x_2 \le & 2 \\
& 4y_3 -y_4 & +3x_1 & -x_2 \le & -2.5 \\
& & x_1 & \le & 2 \\
& & & x_2 \le & 3.5 \\
& & & x_1, x_2 \in & \mathbb{N} .
\end{array}$$

Find the optimal solution considering x_1 and x_2 as complicating variables.

Exercise 6.2. Consider the following problem

$$\underset{x, y}{\text{minimize}} \quad 2x + 3y$$

subject to

$$\begin{array}{rl}
xy \ge & 6 \\
\exp(y) +x \ge & 2 \\
x \le & 10 \\
y \le & 3 \\
x \in & \mathbb{N} .
\end{array}$$

Check that the following vector (x, y) is a solution:

$$x = 3, \ y = 2, \ z = 12 .$$

Using the Benders decomposition algorithm, obtain the final solution.

Exercise 6.3. In Sect. 1.7.1, p. 55, the unit commitment problem of thermal power plants is analyzed, whose target is to supply at minimum cost the demand for electricity throughout a multiperiod planning horizon, such as a week divided in hours. This problem can be properly solved using the Benders decomposition.

Solve the numerical example stated in Sect. 1.7.1 applying the Benders decomposition and check that the results obtained coincide with those provided in that section.

Exercise 6.4. Consider the following problem

$$\underset{x, y}{\text{maximize}} \quad -7x + 4y$$

subject to

$$x^2 + y^2 \le 1$$
$$y \le 2$$
$$x \in \mathbb{N}.$$

1. Check that the following vector (x, y) is a solution:

$$x = 0, \ y = 1, \ z = 4 \ .$$

2. Using the outer linearization algorithm, obtain the optimal solution.

Exercise 6.5. In Sect. 1.8.2, p. 60, the problem of designing and operating a water supply system is considered. This problem has such a structure that it can be solved using the Benders decomposition.

Solve the numerical problem stated in Sect. 1.8.2 using the Benders decomposition and check the obtained results against those provided in that section.

Exercise 6.6. Consider an electric energy system that includes three nodes and three lines. Data for lines are given in Table 6.1.

Table 6.1. Line data for Exercise 6.6

Line	Susceptance	Capacity limit (unit)
1–2	2.5	0.3
1–3	3.5	0.7
2–3	3.0	0.7

Nodes 1 and 2 are production nodes and the data for the production facilities are provided in Table 6.2. Node 3 is a demand node and its demand value is 0.85 MW.

Table 6.2. Production plan data for Exercise 6.6

Plant	Maximum capacity (unit)	Minimum output (unit)
1	0.9	0
2	0.9	0

The production cost c_i (in dollars) of each production plant has the following form:

$$c_i = \begin{cases} 0 & \text{if } P_i = 0 \\ f_i + v_i \ P_i & \text{if } P_i > 0, \end{cases}$$

where f_i is the fixed cost, v_i the variable production cost, and P_i the production output. Numerical data are given in Table 6.3.

Formulate this single-period minimum production cost problem as an MILP problem.

Table 6.3. Production cost data for Exercise 6.6

Plant	f_i ($)	v_i ($/unit)
1	10	6
2	5	7

Considering the binary variables complicating variables, solve this problem using the Benders decomposition.

Exercise 6.7. The capacity expansion planning problem consists in determining the investment additions in every period of a planning horizon in such a way as to minimize both investment and operating costs. This problem is addressed in Sect. 1.8.1, p. 57, and has such a structure that it can be solved using the Benders decomposition.

Solve the numerical example analyzed in Sect. 1.8.1 using the Benders decomposition and verify that the results obtained coincide with those provided in that section.

Exercise 6.8. Consider the same water supply network as in Exercise 5.5, p. 240. It consists of two cities communicated by a single channel (see Fig. 6.6), a set of nodes and a set of connections. The nodes have been numbered in an optimal order, so that if the flow balance equations are written, the associated matrix exhibits a nice block and banded pattern. Nodes 1 and 18 are assumed to be the water supply nodes and the rest are assumed to be consumption points.

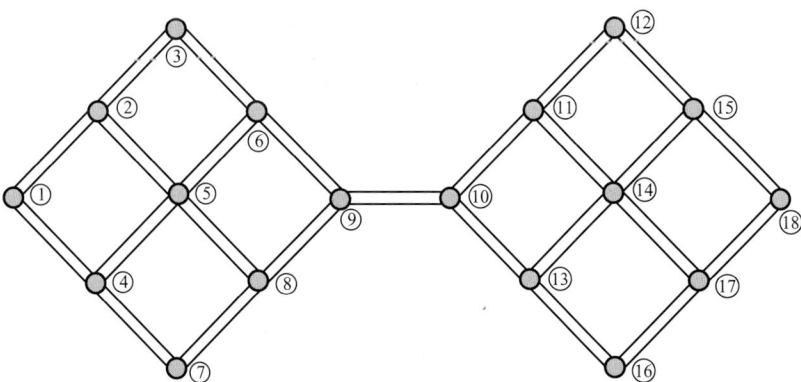

Fig. 6.6. A water supply network consisting of two cities communicated by a single channel

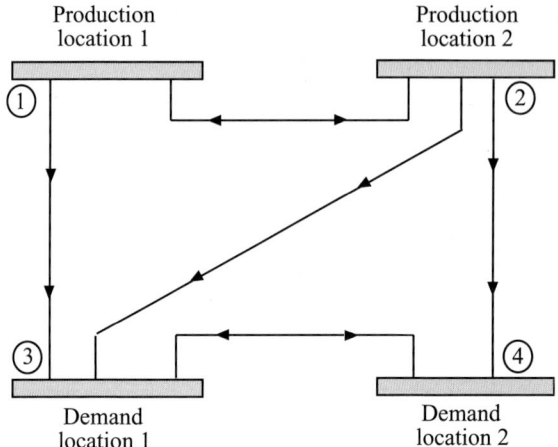

Fig. 6.7. Transportation network for Exercise 6.9

1. Add some constraints to consider the failure of some connections while optimizing the nonlinear objective function.
2. Propose a decomposition procedure to solve the problem using mixed-integer programming and decomposing the problem by city.
3. Is it foreseeable to have problems solving the subproblems?

Exercise 6.9. Consider the capacity expansion planning of two production facilities to supply the demand of two cities during a 2-year time horizon. The interconnection of production and demand locations are specified in Fig. 6.7. Demands at location 1 for years 1 and 2 are 8 and 6 units, respectively; and at location 2, 11 and 9 units, respectively. The maximum capacity to be built at locations 1 and 2 are 10 and 12 units, respectively. Expansion alternatives are discrete and include values 4, 6, and 10 units at location 1, and 3, 8, and 12 units at location 2. Building costs are provided in Table 6.4. Transportation costs and capacities are provided in Table 6.5.

Formulate a problem to determine the production capacity to be built at each location each year. Show that this problem has a complicating variable decomposable structure. Solve it using the Benders decomposition.

Table 6.4. Building cost for the problem of Exercise 6.9

	Building cost ($/unit)	
Period t	Location 1	Location 2
1	2.0	3.5
2	2.5	3.0

Table 6.5. Transportation cost and capacities for Exercise 6.9

Road (i–j)	Capacity (unit)	Cost ($/unit)
1–2	11	0.5
1–3	9	0.6
2–3	5	0.7
2–4	5	0.8
3–4	4	0.4

Exercise 6.10. Consider an electric energy system including 4 nodes and 4 lines. The generating nodes are 1 and 2 while the consumption nodes are 3 and 4. Production plants located at nodes 1 and 2 are denominated C1 and C2, respectively. A 3-h planning horizon is considered.

Solve the corresponding multiperiod network-constrained production planning problem using the Benders decomposition and determine

1. Status (on-line or off-line) of every plant in every time period.
2. Production of every plant in every time period.
3. Height (angle) of every node.
4. Sensitivity of every power balance equation in every time period.

System data are provided in Tables 6.6, 6.7, and 6.8.

Table 6.6. Production plant data for Exercise 6.10

Plant	Maximum capacity (MW)	Minimum output (MW)	Variable cost ($/MWh)	Start-up cost ($)	Fixed cost ($)
C1	1.30	0.02	0.100	10	20
C2	2.50	0.02	0.125	17	18

Table 6.7. Line data for Exercise 6.10

From/to	Susceptance	Maximum capacity (MW)
1–2	1.2	1.50
1–3	1.5	1.50
2–4	1.7	1.80
3-4	1.1	1.75

Table 6.8. Demand data for Exercise 6.10

Period	Demand (MW)		
	1	2	3
Node 3	0.20	2.50	0.10
Node 4	0.60	0.10	0.10

Note that equations describing how electricity is transmitted through a transmission line are explained in Sect. 1.5.2, p. 42.

Exercise 6.11. Consider 3 h when demands are 150, 300, and 500 MW, respectively. Consider three electricity plants with the maximum output powers 350, 200, and 140 MW, and the corresponding minimum output powers 50, 80, and 40 MW, respectively. Ramp-up limits are 200, 100, and 100 MW/h, respectively and the ramp-down limits 300, 150, and 100 MW/h, respectively. Fixed costs are 5, 7, and 6 $/h, respectively; start-up cost $20, $18, and $5, respectively; shut-down costs $0.5, $0.3, and $1.0, respectively; and variable costs 0.100, 0.125, and 0.150 $/MWh, respectively. All plants are off-line at the beginning of the planning horizon.

Solve this multiperiod production planning problem using the Lagrangian relaxation and augmented Lagrangian decomposition. Solve subproblems using mixed-integer linear programming.

7
Other Decomposition Techniques

In previous chapters several standard decomposition techniques have been described for linear, nonlinear, and mixed-integer problems, for the cases of complicating constraints and variables. In this chapter, other techniques are described and illustrative examples are given, including the important engineering design method (see Castillo et al. [15, 16, 17, 18, 19] and Mínguez [13]), which has clear practical interest.

7.1 Bilevel Decomposition

Consider the problem
$$\underset{\boldsymbol{x}}{\text{minimize}} \quad c(\boldsymbol{x}) \tag{7.1}$$

subject to
$$\boldsymbol{h}(\boldsymbol{x}) \leq \boldsymbol{h}_0 \tag{7.2}$$
$$\boldsymbol{g}(\boldsymbol{x}) \leq \boldsymbol{0}, \tag{7.3}$$

where the function $\boldsymbol{h}(\boldsymbol{x})$ cannot be easily evaluated. For example,
$$h_i(\boldsymbol{x}) = \underset{\boldsymbol{u}}{\text{minimum}} \quad \ell_i(\boldsymbol{x};\boldsymbol{u}); \quad \forall i \tag{7.4}$$

subject to
$$r_j(\boldsymbol{u}) = k_j; \quad \forall j, \tag{7.5}$$

or perhaps $\boldsymbol{h}(\boldsymbol{x})$ is the output of a complicated finite element program. Note that constraints (7.3) can be both equality and inequality constraints.

The main difficulty of this problem is that the constraint (7.2) cannot be incorporated into standard optimization frameworks, so that decomposition techniques are required. Since two related optimization problems are solved, it is called a bilevel decomposition. Nevertheless, the structure of the problem suggests a Benders type decomposition mechanism.

In this section two different alternative methods for solving this type of problems are discussed.

7.1.1 A Relaxation Method

Under some circumstances, the solution of the problem (7.1)–(7.5) can be obtained as the limit of the solutions of a sequence $\{P^{(\nu)} : \nu = 1, 2, \ldots\}$ of problems, where $P^{(\nu)}$ consists of

$$\text{minimize} \quad c(\boldsymbol{x}) \qquad (7.6)$$
$$\boldsymbol{x}$$

subject to

$$\boldsymbol{r}(\boldsymbol{x}) \leq \boldsymbol{r}_0^{(\nu)} \qquad (7.7)$$
$$\boldsymbol{g}(\boldsymbol{x}) \leq \boldsymbol{0}, \qquad (7.8)$$

where $\boldsymbol{h}(\boldsymbol{x})$ has been replaced by $\boldsymbol{r}(\boldsymbol{x})$, which is easily implementable in standard optimization algorithms, and a rule for obtaining $\boldsymbol{r}_0^{(\nu+1)}$ is given

$$\boldsymbol{r}_0^{(\nu+1)} = \boldsymbol{r}_0^{(\nu)} + \rho(\boldsymbol{h}_0 - \boldsymbol{h}^{(\nu)}),$$

where $\boldsymbol{h}^{(\nu)}$ is the vector of solutions of the problems (7.4)–(7.5) for the optimal solution $\boldsymbol{x}^{(\nu)}$ of the problem $P^{(\nu)}$.

This requires the existence of a regular one-to-one (increasing) unknown correspondence, $\boldsymbol{h}(\boldsymbol{x}) = \boldsymbol{q}(\boldsymbol{r}(\boldsymbol{x}))$, between $\boldsymbol{h}(\boldsymbol{x})$ and $\boldsymbol{r}(\boldsymbol{x})$, which is denoted by $\boldsymbol{h} = \boldsymbol{q}(\boldsymbol{r})$.

More precisely, to solve the initial problem (7.1)–(7.5) we propose the algorithm below.

Algorithm 7.1 (The relaxation algorithm).

Input. The problem (7.1)–(7.5), a relaxation factor ρ, initial values $\boldsymbol{r}_0^{(1)}$ for $\boldsymbol{r}_0$, and the tolerance ε to check convergence.
Output. The solution of the problem (7.1)–(7.5) within the given tolerance.

Step 0: Initialization. Initialize the iteration counter, $\nu = 1$, and fix the values of $\boldsymbol{r}_0$ for the first iteration to the initial values $\boldsymbol{r}_0^{(1)}$.

Step 1: Master problem solution. Solve the master problem

$$\text{minimize} \quad c(\boldsymbol{x}) \qquad (7.9)$$
$$\boldsymbol{x}$$

subject to
$$r(x) \leq r_0^{(\nu)} \qquad (7.10)$$
$$g(x) \leq 0 \qquad (7.11)$$

and obtain its optimal solution $x^{(\nu)}$.

Step 2: Subproblems solution. Obtain, for all i, the values
$$h_i^{(\nu)} = h_i(x^{(\nu)}) = \underset{u}{\text{minimum}} \quad \ell_i(x^{(\nu)}; u) \qquad (7.12)$$

subject to
$$r_i(u) = k_i. \qquad (7.13)$$

Step 3: Check convergence. If $\|h^{(\nu)} - h^{(\nu-1)}\| < \varepsilon$, stop the procedure and output the solution. Otherwise, continue with Step 4.

Step 4: Update the r_0 bounds. Use the formula
$$r_0^{(\nu+1)} = r_0^{(\nu)} + \rho(h_0 - h^{(\nu)}),$$
increase by 1 the iteration counter $\nu = \nu + 1$, and continue with Step 1. □

Theorem 7.1 (Convergence of Algorithm 7.1). *Under some regularity conditions, which guarantee the Taylor series expansion of $q(r)$ and $\|I_m - \rho \nabla q(r_+^{(\nu)})\| < 1$, the above algorithm leads to the solution of the problem (7.1)–(7.5).* □

Proof. Assuming some regularity conditions, the convergence of the algorithm can be justified as follows:
$$\begin{aligned} h^{(\nu+1)} &= q(r_0^{(\nu+1)}) = q(r_0^{(\nu)} + \rho(h_0 - h^{(\nu)})) \\ &= q(r_0^{(\nu)}) + \rho \nabla q(r_+^{(\nu)})(h_0 - h^{(\nu)}) \\ &= h^{(\nu)} + \rho \nabla q(r_+^{(\nu)})(h_0 - h^{(\nu)}) \\ &= \left(I_m - \rho \nabla q(r_+^{(\nu)})\right) h^{(\nu)} + \rho \nabla q(r_+^{(\nu)}) h_0 \\ &= (I_m - \rho q_\nu) h^{(\nu)} + \rho q_\nu h_0, \end{aligned}$$

where $r_+^{(\nu)}$ is the usual intermediate point of the Taylor series, I_m is the identity matrix whose dimension m is that of h, and $q_\nu = \nabla q(r_+^{(\nu)})$. Then, we have

$$h^{(\nu)} = \left(\prod_{i=1}^{\nu-1}(I_m - \rho q_i)\right) h^{(1)} + \rho \left(\sum_{j=1}^{\nu-1} q_j \prod_{s=j+1}^{\nu-1}(I_m - \rho q_s)\right) h_0 ,$$

which for adequate values of ρ and regularity conditions, including $||I_m - \rho q_i|| < 1$, converges to h_0. □

Illustrative Example 7.1 (The relaxation method). Consider the problem

$$\underset{x_1, x_2, x_3, x_4}{\text{minimize}} \quad z = (x_1 + x_2 - 2)^2 + (x_3 + x_4 - 2)^2 \qquad (7.14)$$

subject to

$$h_1(x) = 1.2 \qquad (7.15)$$
$$h_2(x) = 6 \qquad (7.16)$$
$$x_1 - x_2 = 3 , \qquad (7.17)$$

where

$$h_1(x) = \underset{u_1, u_2, u_3, u_4}{\text{minimum}} \quad \sum_{i=1}^{4}(u_i - x_i)^2 \qquad (7.18)$$

subject to

$$3u_1 + u_2 + 2u_3 + u_4 = 6 \qquad (7.19)$$

and

$$h_2(x) = \underset{u_1, u_2, u_3, u_4}{\text{minimum}} \quad \sum_{i=1}^{4}(u_i - x_i)^2 \qquad (7.20)$$

subject to

$$u_1 + u_2 + u_3 + 2u_4 = 7. \qquad (7.21)$$

The solution of the problem (7.14)–(7.21) can be obtained as the limit of the solutions of the sequence $\{P^{(\nu)} : \nu = 1, 2, \ldots\}$ of problems (7.6)–(7.8), i.e., using the iterative process.

Step 0: Initialization. Initialize the iteration counter, $\nu = 1$, and let $r_1^{(1)} = r_1^{(0)}$ and $r_2^{(1)} = r_2^{(0)}$.

Step 1: Solve the Master problem.

$$\underset{x_1, x_2, x_3, x_4}{\text{minimize}} \quad z = (x_1 + x_2 - 2)^2 + (x_3 + x_4 - 2)^2 \qquad (7.22)$$

subject to

$$3x_1 + x_2 + 2x_3 + x_4 = r_1^{(\nu)} \tag{7.23}$$
$$x_1 + x_2 + x_3 + 2x_4 = r_2^{(\nu)} \tag{7.24}$$
$$x_1 - x_2 = 3. \tag{7.25}$$

In this step the values of the x variables, $x^{(\nu)}$, are obtained.

Step 2: Solve the subproblems. The problems (7.18)–(7.19) and (7.20)–(7.21) are solved for fixed values of the x variables, i.e., for $x = x^{(\nu)}$.

Step 3: Check convergence. If the error is lower than or equal to the tolerance

$$\left(1.2 - h_1^{(\nu)}\right)^2 + \left(6 - h_2^{(\nu)}\right)^2 \leq \varepsilon ,$$

stop the process. Otherwise go to Step 4.

Step 4: Update values. Use the following rule to obtain $r^{(\nu+1)}$:

$$\begin{pmatrix} r_1^{(\nu+1)} \\ r_2^{(\nu+1)} \end{pmatrix} = \begin{pmatrix} r_1^{(\nu)} \\ r_2^{(\nu)} \end{pmatrix} + 0.9 \begin{pmatrix} 1.2 - h_1^{(\nu)} \\ 6.0 - h_2^{(\nu)} \end{pmatrix} ,$$

increase by 1 the iteration counter $\nu = \nu + 1$, and go to Step 1. Table 7.1 illustrates the evolution of the iterative process showing the values of the variables and the error, for the following data:

$$\rho = 0.9, \quad r_1^{(0)} = 7, \quad r_2^{(0)} = 4, \quad \varepsilon = 0.00001 .$$

□

Table 7.1. Iterative process until the solution is obtained

ν	$z^{(\nu)}$	$x_1^{(\nu)}$	$x_2^{(\nu)}$	$x_3^{(\nu)}$	$x_4^{(\nu)}$	$r_1^{(\nu)}$	$r_2^{(\nu)}$	$h_1^{(\nu)}$	$h_2^{(\nu)}$	Error$^{(\nu)}$
1	0.89	2.17	−0.83	0.00	1.33	7.00	4.00	0.07	1.29	1.000000
2	0.09	2.61	−0.39	−1.61	3.82	8.02	8.24	0.27	0.22	23.508934
3	2.96	3.11	0.11	−3.79	7.01	8.86	13.44	0.54	5.93	34.261783
4	3.51	3.16	0.16	−3.53	6.85	9.45	13.50	0.79	6.04	0.435572
5	3.81	3.19	0.19	−3.33	6.71	9.81	13.46	0.97	5.97	0.168727
6	4.03	3.21	0.21	−3.24	6.65	10.02	13.49	1.08	6.02	0.054007
7	4.11	3.22	0.22	−3.17	6.61	10.13	13.47	1.14	5.99	0.015312
8	4.18	3.22	0.22	−3.15	6.59	10.19	13.49	1.17	6.01	0.004069
9	4.20	3.22	0.22	−3.13	6.58	10.22	13.48	1.18	5.99	0.001055
10	4.22	3.23	0.23	−3.13	6.58	10.23	13.48	1.19	6.00	0.000275
11	4.22	3.23	0.23	−3.12	6.57	10.24	13.48	1.20	6.00	0.000074
12	4.23	3.23	0.23	−3.12	6.57	10.24	13.48	1.20	6.00	0.000021

Illustrative Example 7.2 (The relaxation method: The wall design). In Chap. 1, Sect. 1.5.3, the wall design problem was stated and three different design methods were presented: the classical, the modern, and mixed approaches. However, no algorithms were given to solve them. In this example we illustrate the use of the relaxation method by its application to the wall problem.

In fact, the relaxation method consists of repeating a sequence of three steps: (1) an optimal (in the sense of optimizing an objective function) classic design, based on given safety factors, is done; (2) reliability indices or bounds for all failures modes are determined; and (3) all mode safety factor bounds are adjusted. The three steps are repeated until convergence, i.e., until the safety factors lower bounds and the failure mode probability upper bounds are satisfied. More precisely, using the numerical values stated in Sect. 1.5.3, p. 45, the method proceeds as follows:

Step 0: Initialization. Initialize the iteration counter, $\nu = 1$, and let $F_o^{(1)} = 1.5$.

Step 1: Solve the classical problem.

$$\underset{a,b}{\text{Minimize}} \quad \text{cost} = ab \tag{7.26}$$

subject to

$$\frac{23a^2 b}{300} \geq F_o^{(\nu)} \tag{7.27}$$

$$b \geq 4 . \tag{7.28}$$

Step 2: Solve the subproblem.

$$\beta^{(\nu)} = \underset{h,t}{\text{minimize}} \quad \sqrt{z_1^2 + z_2^2} \tag{7.29}$$

subject to

$$z_1 = \frac{t - 50}{15}$$

$$z_2 = \frac{h - 3}{0.2}$$

$$\frac{23a^2 b}{300} = 1 .$$

Step 3: Check convergence. If $|\beta^{(\nu)} - \beta^{(\nu-1)}| < \epsilon$, stop. Otherwise, continue with Step 4.

Step 4: Update safety factors. Using

$$F_o^{(\nu+1)} = \max\left(F_o^{(\nu)} + \rho(\beta^0 - \beta^{(\nu)}), F_o^0\right),$$

the safety factors are updated. Next, increase by 1 the iteration counter $\nu = \nu + 1$, and go to Step 1.

Note that values of the actual reliability index $\beta^{(\nu)}$ below the desired bound levels β^0, lead to an increase of the associated safety factor bound. Note also that if the safety factor required by the reliability indices becomes smaller than the associated lower bound, it is kept equal to F_o^0.

Using the above algorithm, the results shown in Table 7.2 are obtained. This table shows the progress and convergence of the algorithm, which requires only six iterations. Note that only the β constraint is active and the constraint associated with F_o is inactive. This is illustrated in the last row of Table 7.2, where the active value has been boldfaced. This means that the reliability index for overturning is more restrictive than the corresponding safety factor.

Table 7.2. Illustration of the iterative process of the original algorithm

ν	Cost$^{(\nu)}$	$a^{(\nu)}$	$b^{(\nu)}$	$F_{ro}^{(\nu)}$	$\beta^{(\nu)}$	Error$^{(\nu)}$
1	9.148	2.139	4.277	1.500	1.586	0.8915578
2	10.800	2.324	4.648	1.924	2.860	0.4455132
3	10.957	2.341	4.681	1.966	2.983	0.0409869
4	10.976	2.343	4.685	1.971	2.998	0.0050819
5	10.979	2.343	4.686	1.972	3.000	0.0006479
6	10.979	2.343	4.686	1.972	**3.000**	0.0000829

Note that the optimal design is more expensive than the initial one because at iteration $\nu = 1$ the wall does not hold the reliability constraint $\beta \geq \beta^0$. F_{ro} is the actual safety factor related to overturning failure. □

7.1.2 The Cutting Hyperplane Method

The iterative method presented in Sect. 7.1.1 for solving the problem (7.1)–(7.5), requires a relaxation factor ρ that needs to be fixed by trial and error. An adequate selection leads to a fast convergence of the process, but an inadequate selection can lead to lack of convergence. In this section an alternative method (see Castillo et al. [51]) that solves this shortcoming is given, and in addition it exhibits a better convergence. The method is explained in the following algorithm.

Algorithm 7.2 (The cutting plane algorithm).

Input. The problem (7.1)–(7.5), initial values $x^{(0)}$ for x, and the tolerance ε to check convergence.

Output. The solution of the problem (7.1)–(7.5) within the given tolerance.

Step 0: Initialization. Initialize the iteration counter, $\nu = 1$.

Step 1: Master problem solution. Solve the master problem

$$\underset{x}{\text{minimize}} \quad c(x) \qquad (7.30)$$

subject to

$$h^{(s)} + \lambda^{(s)T}(x - x^{(s)}) \leq h_0; \quad s = 1, 2, \ldots, \nu - 1 \qquad (7.31)$$
$$g(x) \leq 0 \qquad (7.32)$$

and obtain its optimal solution $x^{(\nu)}$.

Step 2: Subproblems solution. Obtain, for all i, the values

$$h_i^{(\nu)} = h_i(x^{(\nu)}) = \underset{u}{\text{minimum}} \quad \ell_i(x^{(\nu)}; u) \qquad (7.33)$$

subject to

$$r_j(u) = k_j \qquad (7.34)$$
$$x = x^{(\nu)} : \lambda^{(\nu)} . \qquad (7.35)$$

Step 3: Check convergence. If $||h^{(\nu)} - h^{(\nu-1)}|| < \varepsilon$, stop the procedure and output the solution. Otherwise, update the iteration counter $\nu = \nu + 1$, and continue with Step 1. □

It should be noted that (7.31) constitute a hyperplane reconstruction of the original constraints $h(x) \geq h_0$.

Illustrative Example 7.3 (Using the cutting hyperplanes method). The cutting hyperplanes method applied to the wall example is as follows:

Step 0: Initialization. Let

$$\nu = 1, \quad F_o^{(1)} = 1.5 .$$

Step 1: Solve the classical problem.

$$\underset{a,b}{\text{minimize}} \quad ab \qquad (7.36)$$

subject to

$$\frac{23a^2b}{300} \geq F_o^{(\nu)} \tag{7.37}$$

$$\beta^{(s)} + \begin{pmatrix} \lambda_1^{(s)} \\ \lambda_2^{(s)} \end{pmatrix}^T \left(\begin{pmatrix} a \\ b \end{pmatrix} - \begin{pmatrix} a^{(s)} \\ b^{(s)} \end{pmatrix} \right) \geq \beta^0; \quad \forall s = 1, 2, \ldots, \nu - 1 \tag{7.38}$$

$$b \leq 4. \tag{7.39}$$

Step 2: Solve the subproblem.

$$\beta^{(\nu)} = \underset{h,t,a,b}{\text{minimum}} \quad z_1^2 + z_2^2 \tag{7.40}$$

subject to

$$a = a^{(\nu)} : \lambda_1^{(\nu)} \tag{7.41}$$

$$b = b^{(\nu)} : \lambda_2^{(\nu)} \tag{7.42}$$

$$z_1 = \frac{t - 50}{15} \tag{7.43}$$

$$z_2 = \frac{h - 3}{0.2} \tag{7.44}$$

$$\frac{23a^2b}{300} = 1. \tag{7.45}$$

Step 3: Check convergence. If $|\beta^{(\nu)} - \beta^{(\nu-1)}| < \epsilon$, stop. Otherwise, update the iteration counter $\nu = \nu + 1$, and continue with Step 1.

The iterative procedure leads to the results shown in Table 7.3 that provides the same information as Table 7.2 using the alternative procedure. In this case the process converges after five iterations.

It should be noted that problem (7.36)–(7.39) is a relaxation of the initial problem in the sense that functions $\beta(\cdot)$ are approximated using cutting hyperplanes. Function $\beta(\cdot)$ becomes more precisely approximated as the iterative procedure progresses, which implies that problem (7.36)–(7.39) reproduces

Table 7.3. Illustration of the iterative process for the alternative algorithm

ν	Cost$^{(\nu)}$	$a^{(\nu)}$	$b^{(\nu)}$	$F_{ro}^{(\nu)}$	$\beta^{(\nu)}$	Error$^{(\nu)}$
1	9.148	2.139	4.277	1.500	1.586	0.8915578
2	11.108	2.357	4.713	2.007	3.101	0.4885284
3	10.979	2.343	4.686	1.972	3.000	0.0334907
4	10.979	2.343	4.686	1.972	3.000	0.0001204
5	10.979	2.343	4.686	1.972	3.000	0.0000000

more exactly the initial problem (see Kelly [45]). Observe, additionally, that cutting hyperplanes are constructed using the dual variable vector associated with constraints (7.41) and (7.42) in problems (7.40)–(7.45) (the subproblems). The constraints (7.41) and (7.42) in problem (7.40)–(7.45) fix to given values the optimization variables of problem (7.36)–(7.39) (the master problem). □

7.2 Bilevel Programming

Consider the following bilevel programming problem:

$$\underset{x}{\text{minimize}} \quad f^U(x, y^*) \tag{7.46}$$

subject to

$$h^U(x, y^*) = 0 \tag{7.47}$$
$$g^U(x, y^*) \leq 0 \tag{7.48}$$
$$y^* = \arg\underset{y}{\text{minimize}} \quad f^L(x, y) \tag{7.49}$$

subject to
$$h^L(x, y) = 0 \tag{7.50}$$
$$g^L(x, y) \leq 0, \tag{7.51}$$

where superscripts 'U' and 'L' denote upper-level and lower-level, respectively.

The above problem consists of an upper-level optimization problem, (7.46)–(7.48), associated with a lower-level optimization problem (7.49)–(7.51). The lower-level problem considers x as a parameter and obtains the optimal value of y that depends on parameter x. The upper-level problem obtains the optimal value of x using the optimal value of y computed in the lower-level problem.

It is not possible to solve the bilevel problem (7.46)–(7.51) in this implicit form. The most common algorithmic approach to attack bilevel problems is based on solving the nonlinear problem obtained by replacing the lower-level problem with its Karush–Kuhn–Tucker conditions, see [52, 53].

The KKT conditions of the lower-level problem are

$$\nabla_y f^L(x, y) + \lambda^T \nabla_y h^L(x, y) + \mu^T \nabla_y g^L(x, y) = 0 \tag{7.52}$$
$$\mu^T g^L(x, y) = 0 \tag{7.53}$$
$$\mu \geq 0 \tag{7.54}$$
$$h^L(x, y) = 0 \tag{7.55}$$
$$g^L(x, y) \leq 0. \tag{7.56}$$

Therefore, the bilevel programming problem can now be expressed as the following nonlinear programming problem:

7.2 Bilevel Programming

$$\begin{array}{c} \text{minimize} \\ x, y \end{array} \quad f^U(x, y) \qquad (7.57)$$

subject to

$$\begin{align}
h^U(x, y) &= 0 & (7.58) \\
g^U(x, y) &\leq 0 & (7.59) \\
\nabla_y f^L(x, y) + \lambda^T \nabla_y h^L(x, y) + \mu^T \nabla_y g^L(x, y) &= 0 & (7.60) \\
\mu^T g^L(x, y) &= 0 & (7.61) \\
\mu &\geq 0 & (7.62) \\
h^L(x, y) &= 0 & (7.63) \\
g^L(x, y) &\leq 0. & (7.64)
\end{align}$$

The bilevel problems may present a decomposable structure that can be exploited through decomposition techniques. This bilevel formulation is of interest for real-word problems. For example [54] formulates and solves the terrorist threat problem in a electric energy system as a general bi-level programming problem.

Illustrative Example 7.4 (The bilevel programming problem). Consider the following bilevel problem to be solved:

$$\begin{array}{c} \text{minimize} \\ x \end{array} \quad 4y^* - x$$

subject to

$$\begin{align}
y^* + 2x &\leq 8 \\
y^* &= \arg \begin{array}{c} \text{minimize} \\ y \end{array} \quad -y - x
\end{align}$$

subject to
$$\begin{align}
-y &\leq 0 \\
y + x &\leq 7 \\
-x &\leq 0 \\
x &\leq 4.
\end{align}$$

To solve this problem, the first-order optimality conditions of the lower-level problem are included as constraints of the upper-level problem, as it is done below,

$$\begin{array}{c} \text{minimize} \\ x, y \end{array} \quad 4y - x$$

subject to

$$\begin{aligned}
y + 2x &\le 8 \\
-1 - \mu_1 + \mu_2 &= 0 \\
-y\mu_1 &= 0 \\
(y + x - 7)\mu_2 &= 0 \\
-y &\le 0 \\
y + x &\le 7 \\
-x &\le 0 \\
x &\le 4 \\
\mu_1, \mu_2 &\ge 0 \,.
\end{aligned}$$

The solution of this nonlinear programming problem is

$$x = 1,\ y = 6\,.$$

□

7.3 Equilibrium Problems

In an economic equilibrium, the demands of consumers and the supplies of producers are balanced at a price level. Consider a particular market equilibrium in which the cost for the supply activities is represented by the function $c(\boldsymbol{x})$. The demand function $\boldsymbol{d}(\boldsymbol{p})$ is a function of prices. The equilibrium problem can be formulated as

$$\underset{\boldsymbol{x}}{\text{minimize}} \quad c(\boldsymbol{x}) - \boldsymbol{p}^T \boldsymbol{x} \tag{7.65}$$

subject to

$$\begin{aligned}
\boldsymbol{b} - \boldsymbol{A}\boldsymbol{x} &\le \boldsymbol{0} : \boldsymbol{\alpha} & (7.66) \\
\boldsymbol{d}(\boldsymbol{p}) - \boldsymbol{E}\boldsymbol{x} &\le \boldsymbol{0} : \boldsymbol{\mu} & (7.67) \\
-\boldsymbol{x} &\le \boldsymbol{0} : \boldsymbol{\beta}\,, & (7.68)
\end{aligned}$$

where $\boldsymbol{x}$ represents the production levels and $\boldsymbol{p}$ prices. Equation (7.65) is the minus profit of selling the production $\boldsymbol{x}$. Equation (7.66) represent the operating constraints for production devices whose dual vector is $\boldsymbol{\alpha}$; (7.67) represent demand requirements being $\boldsymbol{\mu}$ the corresponding dual vector. Finally, (7.68) states that production levels are positive.

Equilibrium conditions relate the vector of prices, $\boldsymbol{p}$, with the dual vector corresponding to constraints (7.67), $\boldsymbol{\mu}$, that is $\boldsymbol{p} = \boldsymbol{\mu}$.

Therefore, the solution of the equilibrium problem can be obtained by solving its first-order KKT optimality conditions, i.e., by solving the following nonlinear system of equalities and inequalities:

7.3 Equilibrium Problems

$$\nabla_x c(x) - p - A^T \alpha - E^T \mu - \beta = 0 \quad (7.69)$$
$$(b - Ax)\alpha = 0 \quad (7.70)$$
$$(d(p) - Ex)\mu = 0 \quad (7.71)$$
$$(-x)\beta = 0 \quad (7.72)$$
$$\alpha \geq 0 \quad (7.73)$$
$$\mu \geq 0 \quad (7.74)$$
$$\beta \geq 0 \quad (7.75)$$
$$b - Ax \leq 0 \quad (7.76)$$
$$d(p) - Ex \leq 0 \quad (7.77)$$
$$-x \leq 0. \quad (7.78)$$

This system of equations can be reduced eliminating variable β and replacing variable μ by variable p. Equations (7.69), (7.72), and (7.75) can be expressed through the (7.79) and (7.80). The resulting system is

$$\nabla_x c(x) - p - A^T \alpha - E^T p \geq 0 \quad (7.79)$$
$$\left(\nabla_x c(x) - p - A^T \alpha - E^T p\right) x = 0 \quad (7.80)$$
$$(Ax - b)\alpha = 0 \quad (7.81)$$
$$(Ex - d(p))p = 0 \quad (7.82)$$
$$\alpha \geq 0 \quad (7.83)$$
$$p \geq 0 \quad (7.84)$$
$$b - Ax \leq 0 \quad (7.85)$$
$$d(p) - Ex \leq 0 \quad (7.86)$$
$$-x \leq 0. \quad (7.87)$$

The above system of inequalities and equalities can be solved through the solution of the following quadratic programming problem.

$$\underset{x,p,\alpha}{\text{minimize}} \quad \left(\nabla_x c(x) - p - A^T \alpha - E^T p\right) x + (Ax - b)\alpha + (Ex - d(p))p$$
$$(7.88)$$

subject to

$$\nabla_x c(x) - p - A^T \alpha - E^T p \geq 0 \quad (7.89)$$
$$Ax - b \geq 0 \quad (7.90)$$
$$Ex - d(p) \geq 0 \quad (7.91)$$
$$x \geq 0 \quad (7.92)$$
$$\alpha \geq 0 \quad (7.93)$$
$$p \geq 0. \quad (7.94)$$

The objective function is the product of the nonnegative constraints and the nonnegative variables, being always bounded below. Therefore, the solution of this quadratic programming problem is the solution of the system (7.79)–(7.87) if and only if it is a global minimum whose objective function value is equal to zero.

It should be noted that equilibrium problems may present a decomposable structure that can be exploited through decomposition techniques.

The following example illustrates the structure of equilibrium problems.

Illustrative Example 7.5 (Equilibrium problem). Consider a market equilibrium problem with two producers and two demands. The equilibrium problem is

$$\underset{x_1, x_2}{\text{minimize}} \quad 3x_1 + 5x_2 - p_1 x_1 - p_2 x_2 \tag{7.95}$$

subject to

$$1 \leq x_1 \leq 10 \tag{7.96}$$
$$3 \leq x_2 \leq 12 \tag{7.97}$$
$$2 + 3p_1 \leq x_1 \tag{7.98}$$
$$4 + 2p_2 \leq x_2, \tag{7.99}$$

which can be written as

$$\underset{x_1, x_2}{\text{minimize}} \quad 3x_1 + 5x_2 - p_1 x_1 - p_2 x_2 \tag{7.100}$$

subject to

$$1 - x_1 \leq 0 : \alpha_1 \tag{7.101}$$
$$-10 + x_1 \leq 0 : \alpha_2 \tag{7.102}$$
$$3 - x_2 \leq 0 : \alpha_3 \tag{7.103}$$
$$-12 + x_2 \leq 0 : \alpha_4 \tag{7.104}$$
$$3p_1 + 2 - x_1 \leq 0 : p_1 \tag{7.105}$$
$$2p_2 + 4 - x_2 \leq 0 : p_2. \tag{7.106}$$

The quadratic programming problem equivalent to the equilibrium problem is formulated as follows:

$$\begin{aligned}&\underset{x, p, \alpha}{\text{minimize}}\\&(3 - p_1 - \alpha_1 + \alpha_2 - p_1)x_1 + (5 - p_2 - \alpha_3 + \alpha_4 - p_2)x_2 \\&+ (x_1 - 1)\alpha_1 + (10 - x_1)\alpha_2 + (x_2 - 3)\alpha_3 + (12 - x_2)\alpha_4 \\&+ (x_1 - 3p_1 + 2)p_1 + (x_2 - 2p_2 + 4)p_2\end{aligned} \tag{7.107}$$

subject to

$$3 - p_1 - \alpha_1 + \alpha_2 - p_1 \geq 0 \quad (7.108)$$
$$5 - p_2 - \alpha_3 + \alpha_4 - p_2 \geq 0 \quad (7.109)$$
$$1 - x_1 \geq 0 \quad (7.110)$$
$$-10 + x_1 \geq 0 \quad (7.111)$$
$$3 - x_2 \geq 0 \quad (7.112)$$
$$-12 + x_2 \geq 0 \quad (7.113)$$
$$3p_1 + 2 - x_1 \geq 0 \quad (7.114)$$
$$2p_2 + 4 - x_2 \geq 0 \quad (7.115)$$
$$\alpha_1, \alpha_2, \alpha_3, \alpha_4 \geq 0 \quad (7.116)$$
$$p_1, p_2 \geq 0. \quad (7.117)$$

The solution of the problem is

$$x_1 = 6.5, \ x_2 = 9.0, \ p_1 = 1.5, \ p_2 = 2.5.$$

□

7.4 Coordinate Descent Decomposition

In this section the coordinate descent decomposition method is described. It should be emphasized that the convergence of this algorithm is not guaranteed. Nevertheless, it usually behaves properly in practical applications.

Consider the problem

$$\underset{x_1, x_2, \ldots, x_n}{\text{minimize}} \quad f(x_1, x_2, \ldots, x_n), \quad (7.118)$$

where $x_i \in \mathbb{R}^{k_i}$ ($i = 1, 2, \ldots, n$) subject to

$$h(x_1, x_2, \ldots, x_n) = 0 \quad (7.119)$$
$$g(x_1, x_2, \ldots, x_n) \leq 0, \quad (7.120)$$

then, the coordinate descent decomposition method is described using the following algorithm.

Algorithm 7.3 (The coordinate descent decomposition algorithm).

Input. The problem (7.118)–(7.120), initial values $x_i^{(0)}$ for $i = 1, 2, \ldots, n$, and the error tolerance ε to check convergence.

Output. The solution of the problem (7.118)–(7.120) within the given error.

Step 0: Initialization. Let $\nu = 1$.

Step 1: Solve the subproblems. For $i = 1 - n$, solve the problem

$$\underset{x_i}{\text{minimize}} \quad f\left(x_1^{(0)}, x_2^{(0)}, \ldots, x_{i-1}^{(0)}, x_i, x_{i+1}^{(0)}, \ldots, x_n^{(0)}\right) \tag{7.121}$$

subject to

$$h\left(x_1^{(0)}, x_2^{(0)}, \ldots, x_{i-1}^{(0)}, x_i, x_{i+1}^{(0)}, \ldots, x_n^{(0)}\right) = 0 \tag{7.122}$$

$$g\left(x_1^{(0)}, x_2^{(0)}, \ldots, x_{i-1}^{(0)}, x_i, x_{i+1}^{(0)}, \ldots, x_n^{(0)}\right) \leq 0, \tag{7.123}$$

obtain its optimal solution $x_i^{(\nu)}$, and let $x_i^{(0)} = x_i^{(\nu)}$.

Step 2: Check convergence. If $\|x^{(\nu)} - x^{(\nu-1)}\| < \varepsilon$, stop the procedure and output the solution. Otherwise, continue with Step 1. □

If dealing with over-constrained subproblems, it might be convenient replacing Step 1 by the following alternative Step 1:

Step 1: Solve the subproblems. For $i = 1 - n$, solve the problem

$$\underset{x_i, \varepsilon, \delta}{\text{minimize}} \quad f\left(x_1^{(0)}, x_2^{(0)}, \ldots, x_{i-1}^{(0)}, x_i, x_{i+1}^{(0)}, \ldots, x_n^{(0)}\right) + \kappa^{(\nu)} \sum_{i=1}^{n} (\|\varepsilon\| + \|\delta\|), \tag{7.124}$$

where $\kappa > 1$, subject to

$$h\left(x_1^{(0)}, x_2^{(0)}, \ldots, x_{i-1}^{(0)}, x_i, x_{i+1}^{(0)}, \ldots, x_n^{(0)}\right) = \varepsilon \tag{7.125}$$

$$g\left(x_1^{(0)}, x_2^{(0)}, \ldots, x_{i-1}^{(0)}, x_i, x_{i+1}^{(0)}, \ldots, x_n^{(0)}\right) \leq \delta, \tag{7.126}$$

and obtain its optimal solution $x_i^{(\nu)}$, and let $x_i^{(0)} = x_i^{(\nu)}$.

The ε and δ variables are used to avoid infeasibility of the subproblems, because the number of unknowns is smaller than the number of constraints.

This method can be interpreted as one method that proceeds by using partial derivatives with respect to the block variables (the remaining variables are kept constant).

Though this method converges under some regularity assumptions, its convergence is slow. However, it can be easily implemented in parallel procedures. This method is the coordinate descent method in which different optimization problems are solved in an iterative manner. In each iteration, the objective function is optimized with respect to a single variable or groups of variables, while the remaining variables are kept constant (thus, the name of coordinate descent). There are as many optimization problems as variables or groups, and all the variables must be considered either individually or in one group. This way, the partial effect of each variable or group is considered, which is equivalent to considering partial derivatives.

7.4.1 Banded Matrix Structure Problems

The optimization problems with banded matrix structure are well suited for the coordinate descent decomposition algorithm:

$$\minimize_{\boldsymbol{x},\boldsymbol{y}} \sum_{i=1}^{\nu} f_i(\boldsymbol{x}_i, \boldsymbol{y}_i)$$

subject to

$$\begin{bmatrix} \boldsymbol{K}_1 & \boldsymbol{C}_1 & 0 & \cdots & 0 & 0 & 0 \\ \boldsymbol{C}_1^T & \boldsymbol{K}_2 & \boldsymbol{C}_2 & \cdots & 0 & 0 & 0 \\ \vdots & \vdots & \vdots & \ddots & \vdots & \vdots & \vdots \\ 0 & 0 & 0 & \cdots & \boldsymbol{C}_{\nu-2}^T & \boldsymbol{K}_{\nu-1} & \boldsymbol{C}_{\nu-1} \\ 0 & 0 & 0 & \cdots & 0 & \boldsymbol{C}_{\nu-1}^T & \boldsymbol{K}_\nu \end{bmatrix} \begin{bmatrix} \boldsymbol{x}_1 \\ \boldsymbol{x}_2 \\ \vdots \\ \boldsymbol{x}_{\nu-1} \\ \boldsymbol{x}_\nu \end{bmatrix} = \begin{bmatrix} \boldsymbol{y}_1 \\ \boldsymbol{y}_2 \\ \vdots \\ \boldsymbol{y}_{\nu-1} \\ \boldsymbol{y}_\nu \end{bmatrix}. \quad (7.127)$$

This problem can be solved iteratively, by decomposing it into ν problems $(i = 1, 2, \ldots, \nu)$, as follows:

$$\minimize_{\boldsymbol{x}_i, \boldsymbol{y}_i} \quad f_i(\boldsymbol{x}_i, \boldsymbol{y}_i) + \mu_\nu(||\boldsymbol{\varepsilon}_{i-1}|| + ||\boldsymbol{\varepsilon}_i|| + ||\boldsymbol{\varepsilon}_{i+1}||) \qquad (7.128)$$

subject to

$$\boldsymbol{C}_{i-2}^T \tilde{\boldsymbol{x}}_{i-2} + \boldsymbol{K}_{i-1} \tilde{\boldsymbol{x}}_{i-1} + \boldsymbol{C}_{i-1} \boldsymbol{x}_i = \tilde{\boldsymbol{y}}_{i-1} + \boldsymbol{\varepsilon}_{i-1} \qquad (7.129)$$

$$\boldsymbol{C}_{i-1}^T \tilde{\boldsymbol{x}}_{i-1} + \boldsymbol{K}_i \boldsymbol{x}_i + \boldsymbol{C}_i \tilde{\boldsymbol{x}}_{i+1} = \boldsymbol{y}_i + \boldsymbol{\varepsilon}_i \qquad (7.130)$$

$$\boldsymbol{C}_i^T \boldsymbol{x}_i + \boldsymbol{K}_{i+1} \tilde{\boldsymbol{x}}_{i+1} + \boldsymbol{C}_{i+1} \tilde{\boldsymbol{x}}_{i+2} = \tilde{\boldsymbol{y}}_{i+1} + \boldsymbol{\varepsilon}_{i+1}, \qquad (7.131)$$

where μ_ν is a large constant that tends to infinity with ν, the tildes refer to fixing the values of the corresponding variables, and the constraints (7.129) for $i = 1$ and (7.131) for $i = \nu$ must be eliminated.

Note that each of the subproblems incorporates all the constraints that depend on the variables of a single block, and then they are strongly conditioned.

The ε variables are used to avoid infeasibility of the subproblems, because the number of unknowns is smaller than the number of constraints.

There are many examples where this banded structure appears, as finite element problems, finite difference problems. Two illustrative examples are given below.

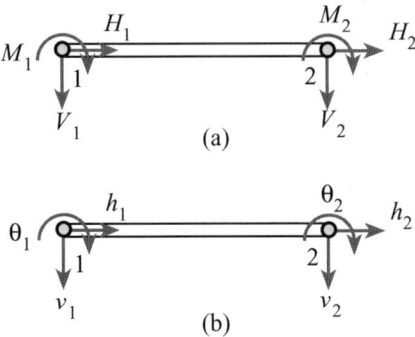

Fig. 7.1. (a) Node forces and moments, and (b) node displacements and rotations for an elementary horizontal piece

Illustrative Example 7.6 (The matrix analysis of structures). Consider the structural piece in Fig. 7.1, where

$$H_i, V_i, M_i; \quad i = 1, 2$$

are the horizontal and vertical forces, and the moments applied to nodes 1 and 2, and

$$h_i, v_i, \theta_i; \quad i = 1, 2$$

are the horizontal and vertical displacements, and the rotations, respectively, of these nodes.

Using the well-known two cases of clamped at both ends and clamped and supported beams in Figs. 7.2a and b, respectively, and the simple axial compression well-known law, the following system of equations establishes the relationships between these variables.

$$\begin{pmatrix} H_1 \\ V_1 \\ M_1 \\ -- \\ H_2 \\ V_2 \\ M_2 \end{pmatrix} = K \begin{pmatrix} h_1 \\ v_1 \\ \theta_1 \\ -- \\ h_2 \\ v_2 \\ \theta_2 \end{pmatrix} = \begin{pmatrix} K_1 & | & K_2 \\ -- & + & -- \\ K_2^T & | & K_3 \end{pmatrix} \begin{pmatrix} h_1 \\ v_1 \\ \theta_1 \\ -- \\ h_2 \\ v_2 \\ \theta_2 \end{pmatrix}$$

$$= \begin{pmatrix} \frac{AE}{L} & 0 & 0 & | & -\frac{AE}{L} & 0 & 0 \\ 0 & \frac{12EI}{L^3} & \frac{6EI}{L^2} & | & 0 & -\frac{12EI}{L^3} & \frac{6EI}{L^2} \\ 0 & \frac{6EI}{L^2} & \frac{4EI}{L} & | & 0 & -\frac{6EI}{L^2} & \frac{2EI}{L} \\ -- & -- & -- & + & -- & -- & -- \\ -\frac{AE}{L} & 0 & 0 & | & \frac{AE}{L} & 0 & 0 \\ 0 & -\frac{12EI}{L^3} & -\frac{6EI}{L^2} & | & 0 & \frac{12EI}{L^3} & -\frac{6EI}{L^2} \\ 0 & \frac{6EI}{L^2} & \frac{2EI}{L} & | & 0 & -\frac{6EI}{L^2} & \frac{4EI}{L} \end{pmatrix} \begin{pmatrix} h_1 \\ v_1 \\ \theta_1 \\ -- \\ h_2 \\ v_2 \\ \theta_2 \end{pmatrix},$$

(7.132)

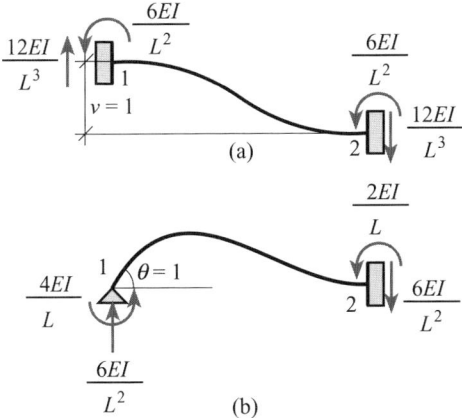

Fig. 7.2. Basic cases of clamped at both ends and clamped and supported beams used for building the stiffness matrix. The relative displacement in (**a**) and the rotation in (**b**) are unity

where the matrix K of the system is known as the *stiffness matrix*, A is the cross section, E is the Young modulus, I is the moment of inertia of the cross section of the piece, L is the length, K_2^T is the transpose of K_2, and the meaning of the block matrices K_1, K_2, and K_3 becomes clear from system (7.132).

The above relationships between forces and moments and displacements and rotations for a horizontal piece can be generalized to include pieces in any position by means of the following considerations.

Consider the piece in Fig. 7.3, where h' and v', and H' and V' are the displacements and forces in the piece and its orthogonal direction, respectively.

Then, we have

$$\begin{pmatrix} h \\ v \\ m \end{pmatrix} = G \begin{pmatrix} h' \\ v' \\ m' \end{pmatrix}, \quad \begin{pmatrix} H \\ V \\ M \end{pmatrix} = G \begin{pmatrix} H' \\ V' \\ M' \end{pmatrix}, \qquad (7.133)$$

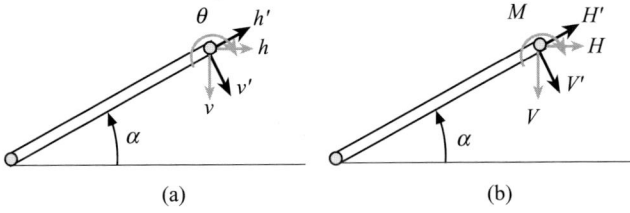

Fig. 7.3. Piece rotated an angle α with respect to the horizontal line

where
$$G = \begin{pmatrix} \cos\alpha & \sin\alpha & 0 \\ -\sin\alpha & \cos\alpha & 0 \\ 0 & 0 & 1 \end{pmatrix} \qquad (7.134)$$

is the rotation matrix.

Thus, for the piece in Fig. 7.3 we have

$$\begin{pmatrix} H_1 \\ V_1 \\ M_1 \\ H_2 \\ V_2 \\ M_2 \end{pmatrix} = \begin{pmatrix} G & 0 \\ 0 & G \end{pmatrix} K \begin{pmatrix} G^{-1} & 0 \\ 0 & G^{-1} \end{pmatrix} \begin{pmatrix} h_1 \\ v_1 \\ \theta_1 \\ h_2 \\ v_2 \\ \theta_2 \end{pmatrix}. \qquad (7.135)$$

If we have a structure composed of several pieces, its stiffness matrix can be obtained from the stiffness matrices of all its pieces.

For example, consider the structure in Fig. 7.4a. In the upper part of Fig. 7.5, the building process of the stiffness matrix is illustrated. First, the stiffness matrices K^I, K^{II}, and K^{III} of the three pieces I, II, and III are calculated, and next, their corresponding blocks (see system of (7.132)) are assembled in the corresponding places (see the node numbers and the block matrices in the upper right matrix in Fig. 7.5).

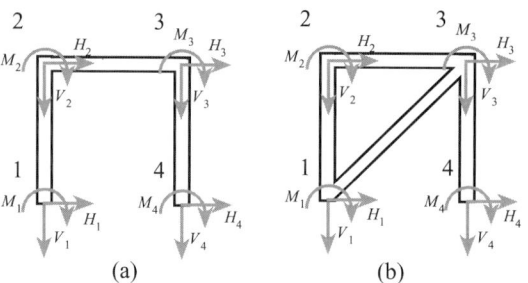

Fig. 7.4. Two one-level structures

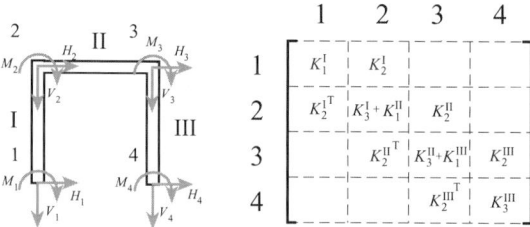

Fig. 7.5. Stiffness matrix for the structure in Fig. 7.4a, and illustration of how to build it

7.4 Coordinate Descent Decomposition

The corresponding system of equations relating forces and moments, $H_i, V_i, M_i (i = 1, 2, 3, 4)$ with displacements and rotations, $h_i, v_i, \theta_i (i = 1, 2, 3, 4)$ for $A_i = E_i = I_i = L_i = 1 (i = 1, 2, 3)$ becomes

$$\begin{pmatrix} H_1 \\ V_1 \\ M_1 \\ - \\ H_2 \\ V_2 \\ M_2 \\ - \\ H_3 \\ V_3 \\ M_3 \\ - \\ H_4 \\ V_4 \\ M_4 \end{pmatrix} = \left(\begin{array}{ccc|ccc|ccc|ccc} 12 & 0 & 6 & -12 & 0 & 6 & 0 & 0 & 0 & 0 & 0 & 0 \\ 0 & 1 & 0 & 0 & -1 & 0 & 0 & 0 & 0 & 0 & 0 & 0 \\ 6 & 0 & 4 & -6 & 0 & 2 & 0 & 0 & 0 & 0 & 0 & 0 \\ \hline -12 & 0 & -6 & 13 & 0 & -6 & -1 & 0 & 0 & 0 & 0 & 0 \\ 0 & -1 & 0 & 0 & 13 & 6 & 0 & -12 & 6 & 0 & 0 & 0 \\ 6 & 0 & 2 & -6 & 6 & 8 & 0 & -6 & 2 & 0 & 0 & 0 \\ \hline 0 & 0 & 0 & -1 & 0 & 0 & 13 & 0 & -6 & -12 & 0 & -6 \\ 0 & 0 & 0 & 0 & -12 & -6 & 0 & 13 & -6 & 0 & -1 & 0 \\ 0 & 0 & 0 & 0 & 6 & 2 & -6 & -6 & 8 & 6 & 0 & 2 \\ \hline 0 & 0 & 0 & 0 & 0 & 0 & -12 & 0 & 6 & 12 & 0 & 6 \\ 0 & 0 & 0 & 0 & 0 & 0 & 0 & -1 & 0 & 0 & 1 & 0 \\ 0 & 0 & 0 & 0 & 0 & 0 & -6 & 0 & 2 & 6 & 0 & 4 \end{array} \right) \begin{pmatrix} h_1 \\ v_1 \\ \theta_1 \\ - \\ h_2 \\ v_2 \\ \theta_2 \\ - \\ h_3 \\ v_3 \\ \theta_3 \\ - \\ h_4 \\ v_4 \\ \theta_4 \end{pmatrix},$$

(7.136)

where the banded matrix structure becomes apparent.

Any optimization problem based on this structure must incorporate these constraints (7.136). For example, one can

$$\underset{\boldsymbol{h}, \boldsymbol{v}, \boldsymbol{\theta}}{\text{minimize}} \quad H_3$$

subject to

$$h_1 = 0$$
$$v_1 = 0$$
$$\theta_1 = 0$$
$$-0.03 \leq h_2 \leq 0.03$$
$$-0.001 \leq v_2 \leq 0.001$$
$$-0.02 \leq \theta_2 \leq 0.02$$
$$H_2 = 0$$
$$V_2 = 0$$
$$M_2 = 0$$
$$H_4 = 0$$
$$V_4 = 0$$
$$M_4 = 0$$

and constraints (7.136). This means obtaining the minimum horizontal force in node 3 subject to some displacements and rotation constraints at nodes 1 and 2.

Table 7.4. Iterative process using the method described in constraints (7.129)–(7.131)

	\multicolumn{9}{c}{iteration}								
	1	2	3	4	5	6	7	8	9
H_1	0.487	0.445	0.456	0.464	0.468	0.472	0.474	0.476	0.477
V_1	0.010	0.001	0.001	0.001	0.001	0.001	0.001	0.001	0.001
M_1	0.212	0.208	0.212	0.215	0.216	0.217	0.218	0.219	0.219
H_2	0.000	0.000	0.000	0.000	0.000	0.000	0.000	0.000	0.000
V_2	0.000	0.000	0.000	0.000	0.000	0.000	0.000	0.000	0.000
M_2	0.000	0.000	0.000	0.000	0.000	0.000	0.000	0.000	0.000
H_3	−0.465	−0.484	−0.483	−0.482	−0.482	−0.482	−0.482	−0.482	−0.481
V_3	−0.008	−0.030	−0.016	−0.009	−0.005	−0.003	−0.002	−0.001	−0.001
M_3	0.250	0.277	0.269	0.265	0.263	0.262	0.262	0.262	0.262
H_4	0.000	0.000	0.000	0.000	0.000	0.000	0.000	0.000	0.000
V_4	0.000	0.000	0.000	0.000	0.000	0.000	0.000	0.000	0.000
M_4	0.000	0.000	0.000	0.000	0.000	0.000	0.000	0.000	0.000
h_1	0.000	0.000	0.000	0.000	0.000	0.000	0.000	0.000	0.000
v_1	0.000	0.000	0.000	0.000	0.000	0.000	0.000	0.000	0.000
θ_1	0.000	0.000	0.000	0.000	0.000	0.000	0.000	0.000	0.000
h_2	−0.023	−0.030	−0.030	−0.030	−0.030	−0.030	−0.030	−0.030	−0.030
v_2	−0.001	−0.001	−0.001	−0.001	−0.001	−0.001	−0.001	−0.001	−0.001
θ_2	0.014	0.014	0.016	0.017	0.018	0.019	0.019	0.019	0.019
h_3	−0.516	−0.513	−0.513	−0.513	−0.512	−0.512	−0.512	−0.512	−0.511
v_3	0.136	0.144	0.145	0.146	0.147	0.147	0.148	0.148	0.148
θ_3	0.266	0.277	0.277	0.277	0.278	0.278	0.279	0.279	0.279
h_4	−0.786	−0.788	−0.789	−0.789	−0.790	−0.790	−0.790	−0.791	−0.791
v_4	0.164	0.154	0.149	0.148	0.147	0.147	0.147	0.148	0.148
θ_4	0.269	0.273	0.275	0.276	0.277	0.278	0.278	0.279	0.279

Table 7.4 shows the evolution of the iterative process using the method described in constraints (7.129)–(7.131), until convergence. □

Illustrative Example 7.7 (A flow application). In this example we consider the water flow under a dam. It is well known that water flows in small amounts under the dam structures due to the gradient caused by the different levels of water up and down stream as shown in Fig. 7.6. The stability analysis of the dam requires the knowledge of the water pressures at the dam foundation. The flow is governed by the differential equation

$$\Delta\phi = \frac{\partial^2\phi}{\partial x^2} + \frac{\partial^2\phi}{\partial y^2} = 0,$$

which can be approximated by finite differences using the net in Fig. 7.6 and the set of equations

$$\phi_r + \phi_l + \phi_u + \phi_b - 4\phi_c = 0, \quad \text{for all node } c \tag{7.137}$$

7.4 Coordinate Descent Decomposition

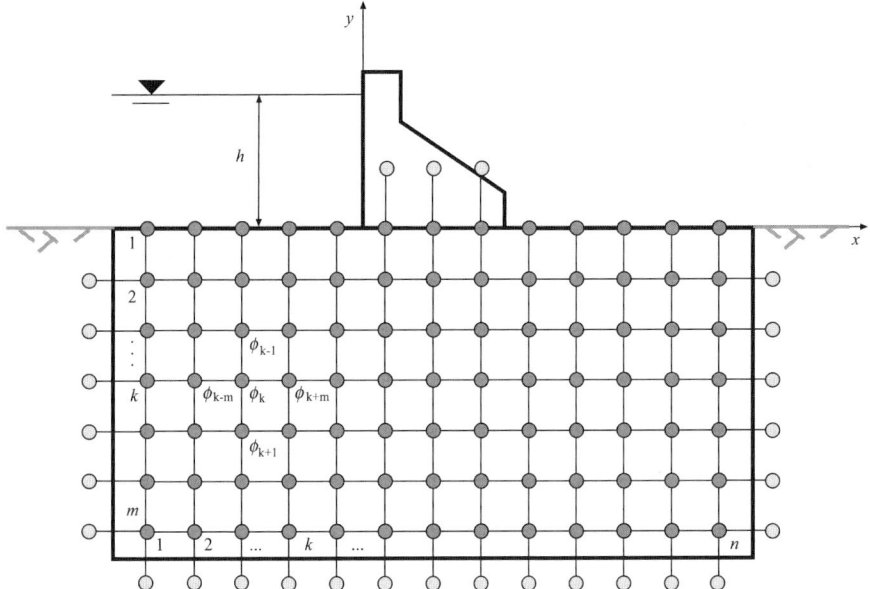

Fig. 7.6. Finite difference net for the approximation of the differential equation for the water flow under a dam

where r, l, u, b refer to the right, left, up, and bottom nodes of the center node c.

Potential ϕ at any point (x, y) is defined as

$$\phi = y + \frac{p}{\gamma}, \tag{7.138}$$

where p is the water pressure and γ is the water unit weight.

We assume that ϕ is a constant on both sides of the dam and takes values h and 0 (note that atmospheric pressure implies $p = 0$) at the up and down stream, respectively, and that the soil is limited by its right, left, and bottom sides by impervious materials. We also assume that the dam foundation is impervious. This implies that the contours of ϕ must be perpendicular to the impervious boundaries. To this end we assume that the values of the potential ϕ at the artificial white nodes in the boundary are coincident with the values of the potential at the nodes they are connected. Thus, for the bottom nodes in the impervious boundaries (7.137) becomes

$$\phi_r + \phi_l + \phi_u - 3\phi_c = 0; \quad c \in D_{in}, \tag{7.139}$$

and similarly, for the right, left, and up (below the dam) boundaries.

For the right and left corner nodes, (7.137) becomes

$$\phi_l + \phi_u - 2\phi_c = 0; \quad c \in D_{rb} \tag{7.140}$$

and

$$\phi_r + \phi_u - 2\phi_c = 0; \quad c \in D_{lb}, \tag{7.141}$$

respectively.

If one is interested in maximizing the height in the dam, the problem can be stated as

$$\underset{h, \boldsymbol{\phi}}{\text{maximize}} \quad h$$

subject to

$$\begin{aligned}
\phi_r + \phi_l + \phi_u + \phi_b - 4\phi_c &= 0; \quad c \in D_{in} \\
\phi_r + \phi_l + \phi_b - 3\phi_c &= 0; \quad c \in D_{df} \\
\phi_r + \phi_l + \phi_b - 3\phi_c &= h; \quad c \in D_{us} \\
\phi_r + \phi_l + \phi_b - 3\phi_c &= 0; \quad c \in D_{ds} \\
\phi_r + \phi_l + \phi_u - 3\phi_c &= 0; \quad c \in D_b \\
\phi_l + \phi_u + \phi_b - 3\phi_c &= 0; \quad c \in D_r \\
\phi_r + \phi_u + \phi_b - 3\phi_c &= 0; \quad c \in D_l \\
\phi_l + \phi_u - 2\phi_c &= 0; \quad c \in D_{rbc} \\
\phi_r + \phi_u - 2\phi_c &= 0; \quad c \in D_{lbc} \\
\sum_{c \in D_{df}} a_c \phi_c &\leq a_0 \\
\sum_{c \in D_{df}} b_c \phi_c &\leq b_0,
\end{aligned}$$

where $D_{in}, D_{df}, D_{us}, D_{ds}, D_b, D_r, D_l, D_{rbc}, D_{lbc}$ are the sets of interior, dam foundation, up-stream, down-stream, bottom, right, left, right-bottom corner, and left-bottom corner, respectively, and the last two constraints refer to the dam overturning and sliding constraints.

For the purpose of illustration, in Fig. 7.7 we show a simple example and the corresponding system of equations that allows calculating the potentials in all nodes.

$$\left(\begin{smallmatrix}
1 & 0 & 0 & 0 & 0 & 0 & 0 & 0 & 0 & 0 & 0 & 0 & 0 & 0 & 0 & 0 & 0 & 0 & 0 & 0 & 0 & 0 & 0 & 0 & 0\\
1 & -3 & 1 & 0 & 0 & 0 & 1 & 0 & 0 & 0 & 0 & 0 & 0 & 0 & 0 & 0 & 0 & 0 & 0 & 0 & 0 & 0 & 0 & 0 & 0\\
0 & 1 & -3 & 1 & 0 & 0 & 0 & 1 & 0 & 0 & 0 & 0 & 0 & 0 & 0 & 0 & 0 & 0 & 0 & 0 & 0 & 0 & 0 & 0 & 0\\
0 & 0 & 1 & -3 & 1 & 0 & 0 & 0 & 1 & 0 & 0 & 0 & 0 & 0 & 0 & 0 & 0 & 0 & 0 & 0 & 0 & 0 & 0 & 0 & 0\\
0 & 0 & 0 & 1 & -2 & 0 & 0 & 0 & 0 & 1 & 0 & 0 & 0 & 0 & 0 & 0 & 0 & 0 & 0 & 0 & 0 & 0 & 0 & 0 & 0\\
0 & 0 & 0 & 0 & 0 & 1 & 0 & 0 & 0 & 0 & 0 & 0 & 0 & 0 & 0 & 0 & 0 & 0 & 0 & 0 & 0 & 0 & 0 & 0 & 0\\
0 & 1 & 0 & 0 & 0 & 1 & -4 & 1 & 0 & 0 & 0 & 1 & 0 & 0 & 0 & 0 & 0 & 0 & 0 & 0 & 0 & 0 & 0 & 0 & 0\\
0 & 0 & 1 & 0 & 0 & 0 & 1 & -4 & 1 & 0 & 0 & 0 & 1 & 0 & 0 & 0 & 0 & 0 & 0 & 0 & 0 & 0 & 0 & 0 & 0\\
0 & 0 & 0 & 1 & 0 & 0 & 0 & 1 & -4 & 1 & 0 & 0 & 0 & 1 & 0 & 0 & 0 & 0 & 0 & 0 & 0 & 0 & 0 & 0 & 0\\
0 & 0 & 0 & 0 & 1 & 0 & 0 & 0 & 1 & -3 & 0 & 0 & 0 & 0 & 1 & 0 & 0 & 0 & 0 & 0 & 0 & 0 & 0 & 0 & 0\\
0 & 0 & 0 & 0 & 0 & 1 & 0 & 0 & 0 & 0 & -3 & 1 & 0 & 0 & 0 & 1 & 0 & 0 & 0 & 0 & 0 & 0 & 0 & 0 & 0\\
0 & 0 & 0 & 0 & 0 & 0 & 1 & 0 & 0 & 0 & 1 & -4 & 1 & 0 & 0 & 0 & 1 & 0 & 0 & 0 & 0 & 0 & 0 & 0 & 0\\
0 & 0 & 0 & 0 & 0 & 0 & 0 & 1 & 0 & 0 & 0 & 1 & -4 & 1 & 0 & 0 & 0 & 1 & 0 & 0 & 0 & 0 & 0 & 0 & 0\\
0 & 0 & 0 & 0 & 0 & 0 & 0 & 0 & 1 & 0 & 0 & 0 & 1 & -4 & 1 & 0 & 0 & 0 & 1 & 0 & 0 & 0 & 0 & 0 & 0\\
0 & 0 & 0 & 0 & 0 & 0 & 0 & 0 & 0 & 1 & 0 & 0 & 0 & 1 & -3 & 0 & 0 & 0 & 0 & 1 & 0 & 0 & 0 & 0 & 0\\
0 & 0 & 0 & 0 & 0 & 0 & 0 & 0 & 0 & 0 & 0 & 0 & 0 & 0 & 0 & 1 & 0 & 0 & 0 & 0 & 0 & 0 & 0 & 0 & 0\\
0 & 0 & 0 & 0 & 0 & 0 & 0 & 0 & 0 & 0 & 1 & 0 & 0 & 0 & 0 & 1 & -4 & 1 & 0 & 0 & 1 & 0 & 0 & 0 & 0\\
0 & 0 & 0 & 0 & 0 & 0 & 0 & 0 & 0 & 0 & 0 & 1 & 0 & 0 & 0 & 0 & 1 & -4 & 1 & 0 & 0 & 1 & 0 & 0 & 0\\
0 & 0 & 0 & 0 & 0 & 0 & 0 & 0 & 0 & 0 & 0 & 0 & 1 & 0 & 0 & 0 & 1 & -4 & 1 & 0 & 0 & 0 & 1 & 0 & 0\\
0 & 0 & 0 & 0 & 0 & 0 & 0 & 0 & 0 & 0 & 0 & 0 & 0 & 1 & 0 & 0 & 0 & 1 & -3 & 0 & 0 & 0 & 0 & 1 & 0\\
0 & 0 & 0 & 0 & 0 & 0 & 0 & 0 & 0 & 0 & 0 & 0 & 0 & 0 & 0 & 0 & 0 & 0 & 0 & 1 & 0 & 0 & 0 & 0 & 0\\
0 & 0 & 0 & 0 & 0 & 0 & 0 & 0 & 0 & 0 & 0 & 0 & 0 & 0 & 0 & 0 & 1 & 0 & 0 & 0 & 1 & -3 & 1 & 0 & 0\\
0 & 0 & 0 & 0 & 0 & 0 & 0 & 0 & 0 & 0 & 0 & 0 & 0 & 0 & 0 & 0 & 0 & 1 & 0 & 0 & 0 & 1 & -3 & 1 & 0\\
0 & 0 & 0 & 0 & 0 & 0 & 0 & 0 & 0 & 0 & 0 & 0 & 0 & 0 & 0 & 0 & 0 & 0 & 1 & 0 & 0 & 0 & 1 & -3 & 1\\
0 & 0 & 0 & 0 & 0 & 0 & 0 & 0 & 0 & 0 & 0 & 0 & 0 & 0 & 0 & 0 & 0 & 0 & 0 & 1 & 0 & 0 & 0 & 1 & -2
\end{smallmatrix}\right)
\times
\begin{pmatrix}\phi_1\\ \phi_2\\ \phi_3\\ \phi_4\\ \phi_5\\ \phi_6\\ \phi_7\\ \phi_8\\ \phi_9\\ \phi_{10}\\ \phi_{11}\\ \phi_{12}\\ \phi_{13}\\ \phi_{14}\\ \phi_{15}\\ \phi_{16}\\ \phi_{17}\\ \phi_{18}\\ \phi_{19}\\ \phi_{20}\\ \phi_{21}\\ \phi_{22}\\ \phi_{23}\\ \phi_{24}\\ \phi_{25}\end{pmatrix}
=
\begin{pmatrix}10\\ 0\\ 0\\ 0\\ 0\\ 10\\ 0\\ 0\\ 0\\ 0\\ 0\\ 0\\ 0\\ 0\\ 0\\ 0\\ 0\\ 0\\ 0\\ 0\\ 0\\ 0\\ 0\\ 0\\ 0\end{pmatrix}
\qquad (7.142)$$

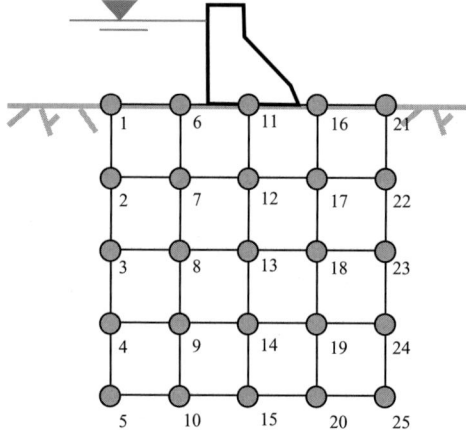

Fig. 7.7. A simple example of finite difference net for analyzing the water flow under a dam

$$\begin{pmatrix} 0 & 0 & 0 & 0 & 0 & 1 & 0 & 0 & 0 & 0 & 1 & 0 & 0 & 0 & 0 & 1 & 0 & 0 & 0 & 0 & 0 & 0 & 0 & 0 & 0 \\ 0 & 0 & 0 & 0 & 0 & b_1 & 0 & 0 & 0 & 0 & b_2 & 0 & 0 & 0 & 0 & b_3 & 0 & 0 & 0 & 0 & 0 & 0 & 0 & 0 & 0 \end{pmatrix} \begin{pmatrix} \phi_1 \\ \phi_2 \\ \phi_3 \\ \phi_4 \\ \phi_5 \\ \phi_6 \\ \phi_7 \\ \phi_8 \\ \phi_9 \\ \phi_{10} \\ \phi_{11} \\ \phi_{12} \\ \phi_{13} \\ \phi_{14} \\ \phi_{15} \\ \phi_{16} \\ \phi_{17} \\ \phi_{18} \\ \phi_{19} \\ \phi_{20} \\ \phi_{21} \\ \phi_{22} \\ \phi_{23} \\ \phi_{24} \\ \phi_{25} \end{pmatrix} \leq \begin{pmatrix} a_0 \\ b_0 \end{pmatrix},$$

(7.143)

where $b_1, b_2, b_3, a_0,$ and b_0 are known constants.

The coordinate descent method applied to this problem consists of maximizing the objective function assuming that one of the ϕ variables is free and the remaining variables are fixed to their actual values, and repeating the process for all variables until convergence. Thus, only the subsets of constraints in systems (7.142) and (7.143) in which the free variable has coefficients different from zero, are considered in the corresponding problem. □

7.5 Exercises

Exercise 7.1. Consider the transnational soda company problem in p. 8 and explain how can it be solved using the banded matrix structure of the constraints.

Exercise 7.2. Consider the following problem:

$$\underset{x_1, x_2}{\text{minimize}} \quad \left(\frac{x_1}{y_1}\right)^2 + \left(\frac{x_2}{y_2}\right)^2$$

subject to

$$g_1(\boldsymbol{x}, \boldsymbol{y}) = \frac{x_1 x_2}{y_1 y_2} \geq g_1^{(0)}$$

$$g_2(\boldsymbol{x}, \boldsymbol{y}) = \frac{x_2}{y_2}\sqrt{\frac{y_1}{x_1}} \geq g_2^{(0)}$$

$$\boldsymbol{h}(\boldsymbol{x}, \boldsymbol{y}) \geq \boldsymbol{h}_0 ,$$

where function $\boldsymbol{h}(\boldsymbol{x}, \boldsymbol{y})$ is the solution of the following problems for $i = 1, 2$

$$h_i(\boldsymbol{x}, \boldsymbol{y}) = \underset{u_1, u_2, v_1, v_2}{\text{minimum}} \quad \sqrt{\sum_{j=1}^{2}\left(\frac{u_j - x_j}{x_j v_{x_j}}\right)^2 + \sum_{j=1}^{2}\left(\frac{v_j - y_j}{y_j v_{y_j}}\right)^2}$$

subject to

$$g_i(\boldsymbol{u}, \boldsymbol{v}) = 1 ,$$

where

$$g_1(\boldsymbol{u}, \boldsymbol{v}) = \frac{u_1 u_2}{v_1 v_2} \quad \text{and} \quad g_2(\boldsymbol{u}, \boldsymbol{v}) = \frac{u_2}{v_2}\sqrt{\frac{v_1}{u_1}} .$$

Supposing the following data values:

$$y_1 = 1.0, \ y_2 = 1.0, \ v_{x_1} = 0.02, \ v_{x_2} = 0.02, \ v_{y_1} = 0.1, \ v_{y_2} = 0.1 ,$$

$$g_1^{(0)} = 1.2, \ g_2^{(0)} = 1.6, \ h_1^{(0)} = 3.0, \ h_2^{(0)} = 4.0 ,$$

and taking as starting point $\boldsymbol{x} = (5.0, 5.0)^T$:

1. Solve the problem using the relaxation method.
2. Solve the problem using the cutting plane method.

Exercise 7.3. Apply the banded matrix structure technique to solve the water supply system example in Sect. 1.4.3, p. 36. Write a GAMS problem to solve it globally and using this technique.

Exercise 7.4. Consider the structure on the right of Fig. 7.4, where a diagonal piece is added to it. Rewrite the constraints (stiffness matrix) after adding this piece. Discuss how this change complicates the problem.

Exercise 7.5. Consider the net in Fig. 7.6.

1. Number the nodes up–down and left–right.
2. Write the finite difference constraints for interior, boundary, and corner nodes.
3. Write the stability (overturning and sliding) constraints.
4. Number the nodes left–right and up–down.
5. Repeat item 2 and 3 for this numbering.
6. Discuss which of the two numberings is the most convenient.
7. Solve the optimization problem discussed in Example 7.7.

Exercise 7.6. In the rubblemound breakwater problem in Sect. 1.5.4, p. 48, the goal is an optimal design of the breakwater based on minimizing the construction and the insurance costs against overtopping damage of the internal structures and ships. Consider the following problem where just the minimization of the construction cost is considered:

$$\underset{F_c, \tan \alpha_s}{\text{minimize}} \quad C_{co} = c_c v_c + c_a v_a$$

subject to

$$H = 1.8 H_s$$
$$T = 1.1 T_z$$
$$1/5 \leq \tan \alpha_s \leq 1/2$$
$$F_c = 2 + d$$
$$v_c = 10d$$
$$v_a = \frac{1}{2}(D_{wl} + 2)\left(46 + D_{wl} + \frac{D_{wl} + 2}{\tan \alpha_s}\right)$$
$$\frac{R_U}{H} = A_U \left(1 - e^{B_U I_r}\right)$$
$$I_r = \frac{\tan \alpha_s}{\sqrt{H/L}}$$
$$\left(\frac{2\pi}{T}\right)^2 = g \frac{2\pi}{L} \tanh \frac{2\pi D_{wl}}{L}$$
$$F_c / R_U \geq F_0$$
$$\beta \geq \beta_0,$$

where the last two constraints are safety constraints associated with the classical approach based on safety factors and the probabilistic approach based on reliability indices, respectively. F_0 and β_0 are the minimum required safety factor and reliability index against overtopping, respectively. Note that the reliability index is calculated the same way as in the problem in Sect. 1.5.4, p. 48.

Solve this problem using the relaxation method described in Sect. 7.2 considering the same data and $F_0 = 1.15$ and $\beta_0 = 4.5$. Note that this problem is similar to the wall problem.

Exercise 7.7. Give a detailed explanation of how to apply the coordinate descent decomposition to the stochastic hydro scheduling problem in p. 12.

Exercise 7.8. Consider the following bilevel programming problem

$$\underset{x}{\text{minimize}} \quad 4x + 8y^*$$

subject to

$$xy^* \leq 17$$
$$5x + 6y^* \geq 43$$
$$x \geq 0$$
$$y^* = \underset{y}{\arg\text{maximize}} \quad -2x + 4y$$

subject to
$$x + 3y \leq 19$$
$$y \leq 18$$
$$y \geq 0.$$

1. Obtain the KKT conditions of the lower-level problem.
2. Compute the solution of the bilevel problem.

Exercise 7.9. Consider an electricity network that includes one generator and one demand. The production cost of the generator is equal to 3 $/MWh and the maximum and minimum output are 8 and 2 MW, respectively. The demand function depends on the energy price and is equal to $d(p) = 11 - 0.5p$. Compute the power produced by the generator and the energy price at equilibrium.

Part III

Local Sensitivity Analysis

8
Local Sensitivity Analysis

8.1 Introduction

In engineering practice, people use mathematical models to describe their problems. However, mathematical models are not exact replicas, but simplifications of reality. Frequently, when we specify a model, we act as if the model were true and the associated assumptions were valid. Similarly, when estimating the parameters of the model we use data, which are not exact but subject to errors, lack of precision, etc. Consequently, conclusions drawn from an analysis are sensitive to models and data. In some applications, even small changes in the data can have a substantial effects on the results. It is therefore essential for data analysts to be able to assess the sensitivity of their conclusions to model and data. This is known as sensitivity analysis. Sensitivity analysis allows the analyst to assess the effects on inferences of changes in the data values, to detect outliers or wrong data values, to define testing strategies, to optimize resources, reduce costs, etc.

Sensitivity analysis is the study of how the variation in the output of a model can be apportioned, qualitatively or quantitatively, to different sources of variation, and aims to determine how the model depends upon the data or information fed into it, upon its structure, and upon the framing assumptions made to build it. As a whole, sensitivity analysis is used to increase the confidence in the model and its predictions, by providing an understanding of how the model response variables respond to changes in the inputs. Adding a sensitivity analysis to an study means adding quality to it.

In this chapter, some methods for sensitivity analysis are discussed. The chapter is structured as follows. In Sect. 8.2 the problem of sensitivity is stated. Section 8.3 derives some formulas from duality theory that are applicable to the sensitivity of the objective function. In Sect. 8.4 a general formula for obtaining all the sensitivities at once, i.e., the sensitivities of the objective function and the primal and dual variables with respect to data, is given, and all the methods are illustrated by their application to particular examples. In Sect. 8.5 interesting particular cases, including the LP case are discussed and some examples are given. In Sect. 8.6 sensitivities of active constraints to data are analyzed.

8.2 Statement of the Problem

Consider the following primal NLPP

$$\underset{x}{\text{minimize}} \quad z_P = f(x, a) \tag{8.1}$$

subject to

$$h(x, a) = b : \lambda \tag{8.2}$$
$$g(x, a) \leq c : \mu, \tag{8.3}$$

where $f : \mathbb{R}^n \times \mathbb{R}^p \to \mathbb{R}$, $h : \mathbb{R}^n \times \mathbb{R}^p \to \mathbb{R}^\ell$, $g : \mathbb{R}^n \times \mathbb{R}^p \to \mathbb{R}^m$ with $h(x,a) = (h_1(x,a), \ldots, h_\ell(x,a))^T$ and $g(x,a) = (g_1(x,a), \ldots, g_m(x,a))^T$ are regular enough for the mathematical developments to be valid over the feasible region $S(a) = \{x | h(x,a) = b, g(x,a) \leq c\}$ and $f, h, g \in C^2$. It is also assumed that the problem (8.1)–(8.3) has an optimum at x^*.

As indicated in Chap. 4, the Primal problem P, as stated in problem (8.1)–(8.3), has an associated dual problem D, which is defined as

$$\underset{\lambda, \mu}{\text{maximize}} \; z_D = \underset{x}{\text{Inf}} \; \{\mathcal{L}(x, \lambda, \mu, a, b, c)\} \tag{8.4}$$

subject to

$$\mu \geq 0, \tag{8.5}$$

where

$$\mathcal{L}(x, \lambda, \mu, a, b, c) = f(x, a) + \lambda^T(h(x,a) - b) + \mu^T(g(x,a) - c) \tag{8.6}$$

is the Lagrangian function associated with the primal problem (8.1)–(8.3), and λ and μ, the dual variables, are vectors of dimensions ℓ and m, respectively. Note that only the dual variables (μ in this case) associated with the inequality constraints [$g(x)$ in this case], must be nonnegative.

Given some regularity conditions on local convexity (see Luenberger [23], Castillo et al. [55]), if the primal problem (8.1)–(8.3) has a locally optimal solution x^*, the dual problem (8.4)–(8.5) also has a locally optimal solution (λ^*, μ^*), and the optimal values of the objective functions of both problems coincide. Note that if J is the set of indices j for which $g_j(x^*, a) = c_j$, a local solution x^* is a regular point of the constraints $h(x, a) = b$ and $g(x, a) \leq c$ if the gradient vectors $\nabla_x h_k(x^*, a), \nabla_x g_j(x^*, a)$, where $k = 1, \ldots, \ell; j \in J$, are linearly independent.

When dealing with the optimization problem (8.1)–(8.3), the following questions regarding sensitivity analysis are of interest:

1. What is the local sensitivity of $z_P^* = f(\boldsymbol{x}^*, \boldsymbol{a})$ to changes in $\boldsymbol{a}$, $\boldsymbol{b}$, and $\boldsymbol{c}$? That is, the sensitivity of the objective function at the optimal point when the parameters or data are modified.
2. What is the local sensitivity of $\boldsymbol{x}^*$ to changes in $\boldsymbol{a}$, $\boldsymbol{b}$, and $\boldsymbol{c}$? That is, the sensitivity of the primal variables at their optimal values if the parameters or data are locally modified.
3. What are the local sensitivity of $\boldsymbol{\lambda}^*$ and $\boldsymbol{\mu}^*$ to changes in $\boldsymbol{a}$, $\boldsymbol{b}$, and $\boldsymbol{c}$? That is, the local sensitivities of the dual variables with respect to data or parameters.

The answers to these questions are given in the following sections.

8.3 Sensitivities Based on Duality Theory

In this section some sensitivities are calculated using duality theory. We first remind the reader the Karush–Kuhn–Tucker (KKT) conditions (see Chap. 4).

8.3.1 Karush–Kuhn–Tucker Conditions

The primal (8.1)–(8.3) and the dual (8.4)–(8.5) problems, respectively, can be solved using the Karush–Kuhn–Tucker first order necessary conditions (KKTCs) (see, for example, Luenberger [23], Bazaraa, Sherali, and Shetty [20], Castillo et al. [21]):

$$\nabla_{\boldsymbol{x}} f(\boldsymbol{x}^*, \boldsymbol{a}) + \boldsymbol{\lambda}^{*T} \nabla_{\boldsymbol{x}} h(\boldsymbol{x}^*, \boldsymbol{a}) + \boldsymbol{\mu}^{*T} \nabla_{\boldsymbol{x}} g(\boldsymbol{x}^*, \boldsymbol{a}) = \boldsymbol{0} \quad (8.7)$$

$$h(\boldsymbol{x}^*, \boldsymbol{a}) = \boldsymbol{b} \quad (8.8)$$

$$g(\boldsymbol{x}^*, \boldsymbol{a}) \leq \boldsymbol{c} \quad (8.9)$$

$$\boldsymbol{\mu}^{*T}(g(\boldsymbol{x}^*, \boldsymbol{a}) - \boldsymbol{c}) = \boldsymbol{0} \quad (8.10)$$

$$\boldsymbol{\mu}^* \geq \boldsymbol{0}, \quad (8.11)$$

where $\boldsymbol{x}^*$ and $(\boldsymbol{\lambda}^*, \boldsymbol{\mu}^*)$ are the primal and dual optimal solutions, $\nabla_{\boldsymbol{x}} f(\boldsymbol{x}^*, \boldsymbol{a})$ is the gradient (vector of partial derivatives) of $f(\boldsymbol{x}^*, \boldsymbol{a})$ with respect to $\boldsymbol{x}$, evaluated at the optimal point $\boldsymbol{x}^*$. The vectors $\boldsymbol{\mu}^*$ and $\boldsymbol{\lambda}^*$ are also called the *Kuhn–Tucker multipliers*. Condition (8.7) says that the gradient of the Lagrangian function in (8.6) evaluated at the optimal solution $\boldsymbol{x}^*$ must be zero. Conditions (8.8)–(8.9) are called *the primal feasibility* conditions. Condition (8.10) is known as the *complementary slackness condition*. Condition (8.11) requires the nonnegativity of the multipliers of the inequality constraints, and is referred to as the *dual feasibility conditions*.

The following example illustrates the use of the KKT conditions.

Illustrative Example 8.1 (Dual problems and KKT conditions). Consider the following optimization problem:

$$\text{minimize} \quad z = (x_1 - 1)^2 + (x_2 - 1)^2 \qquad (8.12)$$
$$x_1, x_2$$

subject to
$$\begin{aligned} x_1 + x_2 &\leq 1 \\ x_1 &\geq 0 \\ x_2 &\geq 0. \end{aligned} \qquad (8.13)$$

The Lagrangian function is

$$\mathcal{L}(\boldsymbol{x}, \boldsymbol{\mu}) = (x_1 - 1)^2 + (x_2 - 1)^2 + \mu_1(x_1 + x_2 - 1) + \mu_2(-x_1) + \mu_3(-x_2)$$

and the KKT conditions become

$$\frac{\partial \mathcal{L}(\boldsymbol{x}, \boldsymbol{\mu})}{\partial x_1} = 2(x_1 - 1) + \mu_1 - \mu_2 = 0 \Rightarrow x_1 = \frac{\mu_2 - \mu_1}{2} + 1 \quad (8.14)$$

$$\frac{\partial \mathcal{L}(\boldsymbol{x}, \boldsymbol{\mu})}{\partial x_2} = 2(x_2 - 1) + \mu_1 - \mu_3 = 0 \Rightarrow x_2 = \frac{\mu_3 - \mu_1}{2} + 1 \quad (8.15)$$

$$x_1 + x_2 \leq 1 \qquad (8.16)$$
$$-x_1 \leq 0 \qquad (8.17)$$
$$-x_2 \leq 0 \qquad (8.18)$$
$$\mu_1(x_1 + x_2 - 1) = 0 \qquad (8.19)$$
$$\mu_2(-x_1) = 0 \qquad (8.20)$$
$$\mu_3(-x_2) = 0 \qquad (8.21)$$
$$\mu_1, \mu_2, \mu_3 \geq 0, \qquad (8.22)$$

which have one solution

$$x_1 = 1/2, \ x_2 = 1/2, \ \mu_1 = 1, \ \mu_2 = 0, \ \mu_3 = 0.$$

This is illustrated in Fig. 8.1, where the optimal point is shown together with the gradients of the objective function and the active constraint. Note that they have opposite directions.

To obtain the dual function $\phi(\mu_1, \mu_2, \mu_3)$ one needs to obtain the Infimum:

$$\underset{x_1, x_2}{\text{Infimum}} \quad \mathcal{L}(x_1, x_2, \mu_1, \mu_2, \mu_3),$$

which is attained at the point [see (8.14) and (8.15)]:

$$x_1 = \frac{\mu_2 - \mu_1}{2} + 1, \ x_2 = \frac{\mu_3 - \mu_1}{2} + 1. \qquad (8.23)$$

It is a minimum because the Hessian of $\mathcal{L}(\boldsymbol{x}, \boldsymbol{\mu})$ at this point becomes

$$\begin{pmatrix} 2 & 0 \\ 0 & 2 \end{pmatrix},$$

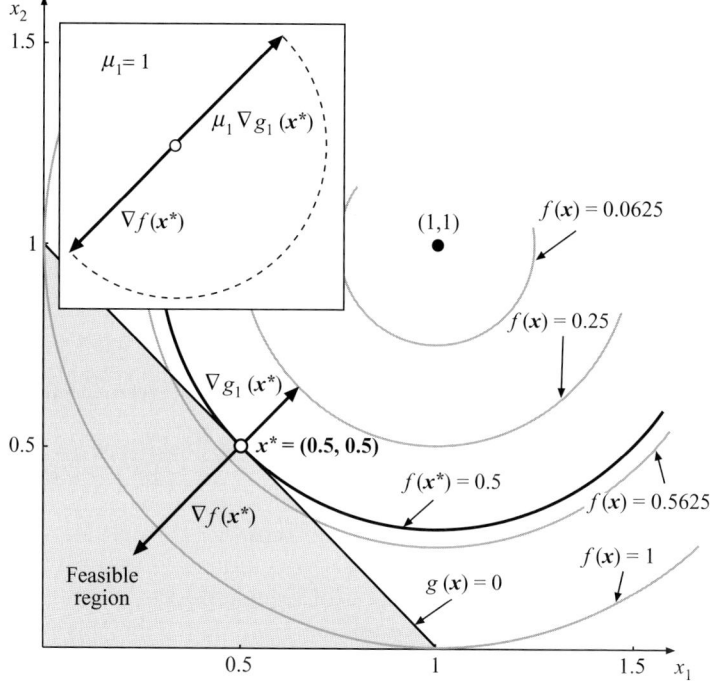

Fig. 8.1. Illustration of the minimization problem in Example 8.1

which is positive definite.
Then, the dual function is

$$\phi(\mu) = \left(-2\mu_1^2 - 4\mu_2 - \mu_2^2 - \mu_3(4+\mu_3) + 2\mu_1(2+\mu_2+\mu_3)\right)/4$$

and the dual problem becomes

maximize $\left(-2\mu_1^2 - 4\mu_2 - \mu_2^2 - \mu_3(4+\mu_3) + 2\mu_1(2+\mu_2+\mu_3)\right)/4$
μ_1, μ_2, μ_3

subject to

$$\mu_1, \mu_2, \mu_3 \geq 0 ,$$

whose solution is

$$\mu_1 = 1, \quad \mu_2 = 0, \quad \mu_3 = 0, \quad z = 1/2 .$$

□

8.3.2 Obtaining the Set of All Dual Variable Values

Once the primal problem has been solved, the values of the optimal dual variables (the solution of the dual problem) can be easily obtained, as shown in the following section. In fact, when asked for the optimal solution of a primal problem, most algorithms embedded in computer packages (GAMS-SNOPT, GAMS-CONOPT, GAMS-MINOS, MATLAB, etc.) also give the optimal solution of the associated dual problem, with practically no extra computational cost. However, if one is interested in deriving analytical expressions for the optimal values of the dual variables or in calculating these values, one can proceed as follows.

The KKTCs in conditions (8.7)–(8.11) allow us to obtain the multipliers (values of the dual variables $\boldsymbol{\lambda}^*$ and $\boldsymbol{\mu}^*$) once the optimal solution $\boldsymbol{x}^*$ of the primal problem (8.1)–(8.3) has been obtained using the subset of linear equations in $\boldsymbol{\lambda}^*$ and $\boldsymbol{\mu}^*$:

$$\nabla_{\boldsymbol{x}} f(\boldsymbol{x}^*, \boldsymbol{a}) + \boldsymbol{\lambda}^{*T} \nabla_{\boldsymbol{x}} h(\boldsymbol{x}^*, \boldsymbol{a}) + \boldsymbol{\mu}^{*T} \nabla_{\boldsymbol{x}} g(\boldsymbol{x}^*, \boldsymbol{a}) = \boldsymbol{0} \quad (8.24)$$

$$\boldsymbol{\mu}^{*T}(g(\boldsymbol{x}^*, \boldsymbol{a}) - \boldsymbol{c}) = \boldsymbol{0} \quad (8.25)$$

$$\boldsymbol{\mu}^* \geq \boldsymbol{0}. \quad (8.26)$$

To this end, we proceed with the following steps:

Step 1. Determine the subset of indices of inequality constraints (8.3) which are active, i.e., those j such that $g_j(\boldsymbol{x}^*, \boldsymbol{a}) = c_j$. Let J be this set, and let $\mu_j = 0$ for all $j \notin J$.

Step 2. Solve in $\boldsymbol{\lambda}^*$ and $\boldsymbol{\mu}^*$ the system of equations and inequalities:

$$\nabla_{\boldsymbol{x}} f(\boldsymbol{x}^*, \boldsymbol{a}) + \boldsymbol{\lambda}^{*T} \nabla_{\boldsymbol{x}} h(\boldsymbol{x}^*, \boldsymbol{a}) + \sum_{j \in J} \mu_j \nabla_{\boldsymbol{x}} g_j(\boldsymbol{x}^*, \boldsymbol{a}) = \boldsymbol{0} \quad (8.27)$$

$$\mu_j^* \geq 0, \ \forall J. \quad (8.28)$$

Since the unknowns in system (8.27) and (8.28) are the sensitivity vectors $\boldsymbol{\lambda}^*$ and $\boldsymbol{\mu}^*$, this system is linear, and therefore easy to solve. This system can be solved using the procedures in Castillo et al. [56, 57].

Furthermore, the solution of the system (8.27) and (8.28) provides analytical expressions to calculate vectors $\boldsymbol{\lambda}^*$ and $\boldsymbol{\mu}^*$ as a function of the primal solution $\boldsymbol{x}^*$ of problem (8.1)–(8.3). This is an important result because this analytical expression is not provided by practical solution algorithms.

8.3.3 Some Sensitivities of the Objective Function

The practical importance of the dual solutions derives from the fact that the values of the dual variables give the sensitivities of the optimal objective function value with respect to the parameters $\boldsymbol{b}$ and $\boldsymbol{c}$ appearing on the right-hand side of the constraints, as stated in the following theorem.

Theorem 8.1 (Sensitivities). *Consider the optimization problem (8.1)–(8.3) whose solution $\boldsymbol{x}^*$ is a regular point (see Definition 4.2) and that no* degenerate (see Definition 4.3) inequality constraints exist. Assume also that sufficient conditions (4.4) for a minimum hold. Then, we have

$$\frac{\partial f(\boldsymbol{x}^*,\boldsymbol{a})}{\partial b_i} = -\lambda_i^*; \quad i=1,2,\ldots,\ell; \quad \frac{\partial f(\boldsymbol{x}^*,\boldsymbol{a})}{\partial c_j} = -\mu_j^*; \quad j=1,2,\ldots,m,$$

i.e., the sensitivities of the optimal objective function value of the problem (8.1)–(8.3) with respect to changes in the terms appearing on the right-hand side of the constraints are the negative of the optimal values of the corresponding dual variables. □

The proof of this theorem can be found, for instance, in Luenberger [23], and can be considered as a corollary of Theorem 8.2.

For this important result to be applicable to practical cases of sensitivity analysis, the parameters for which the sensitivities are sought after must appear on the right-hand side of the primal problem constraints. But what about parameters not satisfying this condition, as $\boldsymbol{a}$, for example? The answer to this question is given in Sect. 8.4.

To illustrate Theorem 8.1, it is applied to a linear programming problem.

Illustrative Example 8.2 (The Linear programming sensitivities). Consider the following linear programming problem:

$$\underset{x_1, x_2, \ldots, x_n}{\text{minimize}} \quad z_P = \sum_{i=1}^{n} c_i x_i \qquad (8.29)$$

subject to

$$\sum_{i=1}^{n} p_{ji} x_i = r_j : \lambda_j; \quad j=1,2,\ldots,\ell \qquad (8.30)$$

$$\sum_{i=1}^{n} q_{ki} x_i \le s_k : \mu_k; \quad k=1,2,\ldots,m, \qquad (8.31)$$

where λ_j and μ_k are the corresponding dual variables.

The sensitivities of the optimal value of the objective function to r_j and s_k, following Theorem 8.1, are

$$\frac{\partial z_P^*}{dr_j} = -\lambda_j^* \qquad (8.32)$$

$$\frac{\partial z_P^*}{ds_k} = -\mu_k^*. \qquad (8.33)$$

□

The following section uses a simple but efficacious trick to obtain all possible sensitivities of the objective function.

8.3.4 A Practical Method for the Sensitivities of the Objective Function

In this section we show how the duality methods can be applied to derive the objective function sensitivities in a straightforward manner. The basic idea is simple. Assume that we desire to know the sensitivity of the objective function to changes in some data values. If we convert the data into artificial variables and set them, by means of constraints, to their actual values, we obtain a problem that is equivalent to the initial optimization problem but has a constraint such that the values of the dual variables associated with them give the desired sensitivities.

To be more precise, the primal optimization problem (8.1)–(8.3) is equivalent to the following one:

$$\underset{x, \tilde{a}}{\text{minimize}} \quad z_P = f(x, \tilde{a}) \tag{8.34}$$

subject to

$$h(x, \tilde{a}) = b \tag{8.35}$$
$$g(x, \tilde{a}) \leq c \tag{8.36}$$
$$\tilde{a} = a : \eta. \tag{8.37}$$

It is clear that problems (8.1)–(8.3) and (8.34)–(8.37) are equivalent, but for the second the sensitivities with respect to a are readily available. Note that to be able to use the important result of Theorem 8.1, we convert the data a into artificial variables $\tilde{a}$ and set them to their actual values a as in constraint (8.37). Then, by Theorem 8.1, the negative values of the dual variables η associated with constraint (8.37) are the sensitivities sought after, i.e., the partial derivatives $\partial z_P / \partial a_i$ $(i = 1, 2, \ldots, p)$.

Remark 8.1. Note that this equivalence gives a powerful method to obtain the objective function sensitivities.

8.3.5 A General Formula for the Sensitivities of the Objective Function

The following theorem provides a closed form formula for local sensitivities of the objective function (see Castillo et al. [58]). It gives an alternative to the method developed in the preceding section.

Theorem 8.2 (The objective function sensitivities with respect to the parameter a). *Consider the primal problem (8.1)–(8.3) whose solution x^* is a regular point (see Definition 4.2) and that no degenerate (see Definition 4.3) inequality constraints exist. Assume also that sufficient conditions (4.4) for a minimum hold. Then, the sensitivity of the objective function with respect to the components of the parameter a is given by*

8.3 Sensitivities Based on Duality Theory   311

$$\frac{\partial z_P^*}{\partial \boldsymbol{a}} = \nabla_{\boldsymbol{a}} \mathcal{L}(\boldsymbol{x}^*, \boldsymbol{\lambda}^*, \boldsymbol{\mu}^*, \boldsymbol{a}, \boldsymbol{b}, \boldsymbol{c}) , \quad (8.38)$$

which is the gradient vector of the Lagrangian function

$$\mathcal{L}(\boldsymbol{x}, \boldsymbol{\lambda}, \boldsymbol{\mu}, \boldsymbol{a}, \boldsymbol{b}, \boldsymbol{c}) = f(\boldsymbol{x}, \boldsymbol{a}) + \boldsymbol{\lambda}^T (\boldsymbol{h}(\boldsymbol{x}, \boldsymbol{a}) - \boldsymbol{b}) + \boldsymbol{\mu}^T (\boldsymbol{g}(\boldsymbol{x}, \boldsymbol{a}) - \boldsymbol{c}) \quad (8.39)$$

with respect to $\boldsymbol{a}$ evaluated at the optimal solution $\boldsymbol{x}^*, \boldsymbol{\lambda}^*$, and $\boldsymbol{\mu}^*$. □

Proof. Since the problem (8.1)–(8.3) is equivalent to the problem (8.34)–(8.37), from the first KKT condition for this problem one gets

$$\begin{aligned}
\nabla_{\boldsymbol{a}} \mathcal{L}(\boldsymbol{x}^*, \boldsymbol{\lambda}^*, \boldsymbol{\mu}^*, \boldsymbol{a}, \boldsymbol{b}, \boldsymbol{c}) &= \nabla_{\boldsymbol{a}} \mathcal{L}(\boldsymbol{x}^*, \boldsymbol{\lambda}^*, \boldsymbol{\mu}^*, \boldsymbol{\eta}^*, \tilde{\boldsymbol{a}}, \boldsymbol{a}, \boldsymbol{b}, \boldsymbol{c}) \quad (8.40) \\
&= \nabla_{\boldsymbol{a}} f(\boldsymbol{x}^*, \tilde{\boldsymbol{a}}) + \boldsymbol{\lambda}^{*T} \nabla_{\boldsymbol{a}} (\boldsymbol{h}(\boldsymbol{x}^*, \tilde{\boldsymbol{a}}) - \boldsymbol{b}) \\
&\quad + \boldsymbol{\mu}^{*T} \nabla_{\boldsymbol{a}} (\boldsymbol{g}(\boldsymbol{x}^*, \tilde{\boldsymbol{a}}) - \boldsymbol{c}) + \boldsymbol{\eta}^{*T} \nabla_{\boldsymbol{a}} (\tilde{\boldsymbol{a}} - \boldsymbol{a}) \\
&= -\boldsymbol{\eta}^* , \quad (8.41)
\end{aligned}$$

i.e., the sensitivities of the objective function of the problem (8.1)–(8.3) with respect to the parameters $\boldsymbol{a}$ are the partial derivative of its Lagrangian function with respect to the $\boldsymbol{a}$ components at the optimal point. □

Remark 8.2. The practical importance of Theorem 8.2 is that it supplies an analytical expression for the objective function sensitivities.

Note that Theorem 8.1 is just a corollary of Theorem 8.2.
To illustrate Theorem 8.2, it is applied to two different LP problems.

Illustrative Example 8.3 (The linear programming case). The simplest and very important case where Theorem 8.2 can be applied is the case of linear programming.

Consider the same linear programming problem problem as in constraints (8.29)–(8.31). The Lagrangian function becomes

$$\begin{aligned}
\mathcal{L}(\boldsymbol{x}, \boldsymbol{\lambda}, \boldsymbol{\mu}, \boldsymbol{c}, \boldsymbol{r}, \boldsymbol{s}) &= \sum_{i=1}^{n} c_i x_i + \sum_{j=1}^{\ell} \lambda_j \left(\sum_{i=1}^{n} p_{ji} x_i - r_j \right) \\
&\quad + \sum_{k=1}^{m} \mu_k \left(\sum_{i=1}^{n} q_{ki} x_i - s_k \right) . \quad (8.42)
\end{aligned}$$

To obtain the sensitivities of the optimal value of the objective function to r_j, s_k, c_i, p_{ji}, or q_{ki}, following Theorem 8.2, we simply obtain the partial derivatives of the Lagrangian function with respect to the corresponding parameter, i.e.,

$$\frac{\partial z_P^*}{\partial r_j} = \frac{\partial \mathcal{L}(\boldsymbol{x}, \boldsymbol{\lambda}, \boldsymbol{\mu})}{\partial r_j} = -\lambda_j^* \qquad (8.43)$$

$$\frac{\partial z_P^*}{\partial s_k} = \frac{\partial \mathcal{L}(\boldsymbol{x}, \boldsymbol{\lambda}, \boldsymbol{\mu})}{\partial s_k} = -\mu_k^* \qquad (8.44)$$

$$\frac{\partial z_P^*}{\partial c_i} = \frac{\partial \mathcal{L}(\boldsymbol{x}, \boldsymbol{\lambda}, \boldsymbol{\mu})}{\partial c_i} = x_i^* \qquad (8.45)$$

$$\frac{\partial z_P^*}{\partial p_{ji}} = \frac{\partial \mathcal{L}(\boldsymbol{x}, \boldsymbol{\lambda}, \boldsymbol{\mu})}{\partial p_{ji}} = \lambda_j^* x_i^* \qquad (8.46)$$

$$\frac{\partial z_P^*}{\partial q_{ki}} = \frac{\partial \mathcal{L}(\boldsymbol{x}, \boldsymbol{\lambda}, \boldsymbol{\mu})}{\partial q_{ki}} = \mu_k^* x_i^*. \qquad (8.47)$$

This is a simple case that leads to very neat results. □

Illustrative Example 8.4 (The dependence on a common parameter). Consider also the case of all parameters depending on a common parameter u, i.e., the problem

$$\underset{x_1, x_2, \ldots, x_n}{\text{minimize}} \quad \sum_{i=1}^{n} c_i(u) x_i \qquad (8.48)$$

subject to

$$\sum_{i=1}^{n} p_{ji}(u) x_i = r_j(u) : \lambda_j; \quad j = 1, 2, \ldots, \ell \qquad (8.49)$$

$$\sum_{i=1}^{n} q_{ki}(u) x_i \leq s_k(u) : \mu_k; \quad k = 1, 2, \ldots, m. \qquad (8.50)$$

Then, the sensitivity of the optimal value of the objective function to u is given by [see (8.42)]:

$$\frac{\partial \mathcal{L}(\boldsymbol{x}, \boldsymbol{\lambda}, \boldsymbol{\mu}, u)}{\partial u} = \sum_{i=1}^{n} \frac{\partial c_i(u)}{\partial u} x_i + \sum_{j=1}^{\ell} \lambda_j \left(\sum_{i=1}^{n} \frac{\partial p_{ji}(u)}{\partial u} x_i - \frac{\partial r_j(u)}{\partial u} \right)$$

$$+ \sum_{k=1}^{m} \mu_k \left(\sum_{i=1}^{n} \frac{\partial q_{ki}(u)}{\partial u} x_i - \frac{\partial s_k(u)}{\partial u} \right). \qquad (8.51)$$

Note that the cases in constraints (8.43)–(8.47) are particular cases of (8.51). □

Note also that Theorem 8.2 is applicable to any NLPP as straightforwardly as to the two LP problems above, as illustrated in the following example.

Illustrative Example 8.5 (The sensitivities in NLPP). Consider the following optimization problem:

8.3 Sensitivities Based on Duality Theory

$$\underset{x,y}{\text{minimize}} \quad z = a_1 x^2 + a_2 y^2 \tag{8.52}$$

subject to

$$\begin{aligned} a_4 x + y &= a_5 \\ xy &\geq a_3 \\ (x-4)^2 + a_6(y-2)^2 &\leq a_7, \end{aligned} \tag{8.53}$$

where $a_1 = 1$, $a_2 = 1$, $a_3 = 4$, $a_4 = 1$, $a_5 = 5$, $a_6 = 1$, and $a_7 = 1$.
The Lagrangian function is

$$\mathcal{L}(x,y,\lambda,\mu_1,\mu_2) = x^2 + y^2 + \lambda(x+y-5) + \mu_1(4-xy) + \mu_2((x-4)^2 + (y-2)^2 - 1),$$

and the KKTC are

$$\begin{aligned} \frac{\partial \mathcal{L}}{\partial x} &= 2x + \lambda - \mu_1 y + 2\mu_2(x-4) = 0 & (8.54) \\ \frac{\partial \mathcal{L}}{\partial y} &= 2y + \lambda - \mu_1 x + 2\mu_2(y-2) = 0 & (8.55) \\ x + y &= 5 & (8.56) \\ xy &\geq 4 & (8.57) \\ (x-4)^2 + (y-2)^2 &\leq 1 & (8.58) \\ \mu_1(4 - xy) &= 0 & (8.59) \\ \mu_2((x-4)^2 + (y-2)^2 - 1) &= 0 & (8.60) \\ \mu_1 &\geq 0 & (8.61) \\ \mu_2 &\geq 0. & (8.62) \end{aligned}$$

The only KKT point that satisfies the KKT conditions is the point $(x^*, y^*) = (3, 2)$, which is the optimal solution, as illustrated in Fig. 8.2. The corresponding values of the multipliers are $\lambda = -4$, $\mu_1 = 0$, and $\mu_2 = 1$.

To study the sensitivity of the objective function optimal value to the $\boldsymbol{a}$-parameters, the problem (8.52)–(8.53) can be written as

$$\underset{x,y,\boldsymbol{a}}{\text{minimize}} \quad z = a_1 x^2 + a_2 y^2$$

subject to

$$\begin{aligned} xy &\geq a_3 \\ a_4 x + y &= a_5 \\ (x-4)^2 + a_6(y-2)^2 &\leq a_7 \\ \boldsymbol{a} &= \boldsymbol{a}_0, \end{aligned} \tag{8.63}$$

where, since the parameters a_3, a_5, and a_7 already appear on the right-hand sides of some constraints, we let $\boldsymbol{a} = (a_1, a_2, a_4, a_6)^T$ and $\boldsymbol{a}_0 = (1, 1, 1, 1)^T$.

The resulting values of the sensitivities, i.e., the negative of the dual variable values are

314   8 Local Sensitivity Analysis

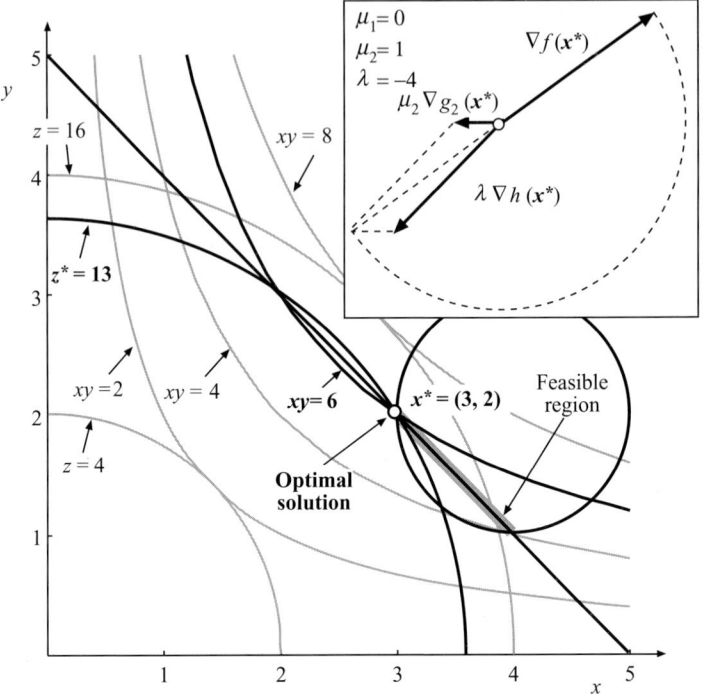

Fig. 8.2. Illustration of the minimization problem in Example 8.5

$$\frac{\partial z^*}{\partial a_1} = 9, \ \frac{\partial z^*}{\partial a_2} = 4, \ \frac{\partial z^*}{\partial a_3} = 0, \ \frac{\partial z^*}{\partial a_4} = -12, \ \frac{\partial z^*}{\partial a_5} = 4, \ \frac{\partial z^*}{\partial a_6} = 0, \ \frac{\partial z^*}{\partial a_7} = -1,$$

where the asterisks refer to the optimal values.

Alternatively, one can proceed as follows. The Lagrangian function of the problem (8.52)–(8.53) in terms of the parameters is

$$\begin{aligned}\mathcal{L}(x, y, \lambda, \mu_1, \mu_2) &= a_1 x^2 + a_2 y^2 + \lambda(a_4 x + y - a_5) + \mu_1(a_3 - xy) \\ &\quad + \mu_2((x-4)^2 + a_6(y-2)^2 - a_7),\end{aligned}$$

then, using Theorem 8.2, one gets

$$\frac{\partial z^*}{\partial a_1} = \frac{\partial \mathcal{L}(x, y, \lambda, \mu_1, \mu_2)}{\partial a_1} = (x^*)^2 = 9$$

$$\frac{\partial z^*}{\partial a_2} = \frac{\partial \mathcal{L}(x, y, \lambda, \mu_1, \mu_2)}{\partial a_2} = (y^*)^2 = 4$$

$$\frac{\partial z^*}{\partial a_3} = \frac{\partial \mathcal{L}(x, y, \lambda, \mu_1, \mu_2)}{\partial a_3} = \mu_1^* = 0$$

$$\frac{\partial z^*}{\partial a_4} = \frac{\partial \mathcal{L}(x, y, \lambda, \mu_1, \mu_2)}{\partial a_4} = \lambda^* x^* = -12$$

$$\frac{\partial z^*}{\partial a_5} = \frac{\partial \mathcal{L}(x,y,\lambda,\mu_1,\mu_2)}{\partial a_5} = -\lambda^* = 4$$

$$\frac{\partial z^*}{\partial a_6} = \frac{\partial \mathcal{L}(x,y,\lambda,\mu_1,\mu_2)}{\partial a_6} = \mu_2^*(y^*-2)^2 = 0$$

$$\frac{\partial z^*}{\partial a_7} = \frac{\partial \mathcal{L}(x,y,\lambda,\mu_1,\mu_2)}{\partial a_7} = -\mu_2^* = -1 \;,$$

i.e., the same results as before. □

8.4 A General Method for Obtaining All Sensitivities

The method developed in the preceding section was limited to determining the sensitivities of the objective function. In what follows, and in order to simplify the mathematical derivations, the parameter vectors b and c, used in constraints (8.1)–(8.3) and (8.7)–(8.11), are assumed to be subsumed by a. In this section we present a powerful method that allows us to determine all sensitivities at once, i.e., the sensitivities of the optimal solutions $(x^*, \lambda^*, \mu^*, z^*)$ of the problems (8.1)–(8.3) and (8.4)–(8.5) to local changes in the parameters a. Pioneering work leading to this method is due to Fiacco [83]. To this end, we perturb or modify a, x, λ, μ, and z in such a way that the KKTC still hold. Thus, to obtain the sensitivity equations we differentiate problems (8.1) and (8.7)–(8.11), as follows:

$$(\nabla_{\boldsymbol{x}} f(\boldsymbol{x}^*, \boldsymbol{a}))^T d\boldsymbol{x} + (\nabla_{\boldsymbol{a}} f(\boldsymbol{x}^*, \boldsymbol{a}))^T d\boldsymbol{a} - dz = 0 \quad (8.64)$$

$$\left(\nabla_{\boldsymbol{xx}} f(\boldsymbol{x}^*, \boldsymbol{a}) + \sum_{k=1}^{\ell} \lambda_k^* \nabla_{\boldsymbol{xx}} h_k(\boldsymbol{x}^*, \boldsymbol{a}) + \sum_{j=1}^{m} \mu_j^* \nabla_{\boldsymbol{xx}} g_j(\boldsymbol{x}^*, \boldsymbol{a}) \right) d\boldsymbol{x}$$
$$+ \left(\nabla_{\boldsymbol{xa}} f(\boldsymbol{x}^*, \boldsymbol{a}) + \sum_{k=1}^{\ell} \lambda_k^* \nabla_{\boldsymbol{xa}} h_k(\boldsymbol{x}^*, \boldsymbol{a}) + \sum_{j=1}^{m} \mu_j^* \nabla_{\boldsymbol{xa}} g_j(\boldsymbol{x}^*, \boldsymbol{a}) \right) d\boldsymbol{a}$$
$$+ \nabla_{\boldsymbol{x}} \boldsymbol{h}(\boldsymbol{x}^*, \boldsymbol{a}) d\boldsymbol{\lambda} + \nabla_{\boldsymbol{x}} \boldsymbol{g}(\boldsymbol{x}^*, \boldsymbol{a}) d\boldsymbol{\mu} = \boldsymbol{0}_n \quad (8.65)$$

$$(\nabla_{\boldsymbol{x}} \boldsymbol{h}(\boldsymbol{x}^*, \boldsymbol{a}))^T d\boldsymbol{x} + (\nabla_{\boldsymbol{a}} \boldsymbol{h}(\boldsymbol{x}^*, \boldsymbol{a}))^T d\boldsymbol{a} = \boldsymbol{0}_\ell \quad (8.66)$$

$$(\nabla_{\boldsymbol{x}} g_j(\boldsymbol{x}^*, \boldsymbol{a}))^T d\boldsymbol{x} + (\nabla_{\boldsymbol{a}} g_j(\boldsymbol{x}^*, \boldsymbol{a}))^T d\boldsymbol{a} = 0; \quad \text{if } \mu_j^* \neq 0; \; j \in J \quad (8.67)$$

$$(\nabla_{\boldsymbol{x}} g_j(\boldsymbol{x}^*, \boldsymbol{a}))^T d\boldsymbol{x} + (\nabla_{\boldsymbol{a}} g_j(\boldsymbol{x}^*, \boldsymbol{a}))^T d\boldsymbol{a} \leq 0; \quad \text{if } \mu_j^* = 0; \; j \in J \quad (8.68)$$

$$-d\mu_j \leq 0; \quad \text{if } \mu_j^* = 0; \; j \in J \quad (8.69)$$

$$d\mu_j \left[(\nabla_{\boldsymbol{x}} g_j(\boldsymbol{x}^*, \boldsymbol{a}))^T d\boldsymbol{x} + (\nabla_{\boldsymbol{a}} g_j(\boldsymbol{x}^*, \boldsymbol{a}))^T d\boldsymbol{a} \right] = 0; \quad \text{if } \mu_j^* = 0; \; j \in J, \quad (8.70)$$

where all the matrices are evaluated at the optimal solution, and redundant constraints have been removed. More precisely, the constraints (8.67)–(8.70) are simplifications of the constraints that result directly from differentiating constraints (8.9)–(8.11), i.e., from

$$(\nabla_{\boldsymbol{x}} g_j(\boldsymbol{x}^*, \boldsymbol{a}))^T d\boldsymbol{x} + (\nabla_{\boldsymbol{a}} g_j(\boldsymbol{x}^*, \boldsymbol{a}))^T d\boldsymbol{a} \leq 0; \ j \in J, \tag{8.71}$$

and

$$(\mu_j^* + d\mu_j)(g_j(\boldsymbol{x}^*, \boldsymbol{a}) + dg_j(\boldsymbol{x}^*, \boldsymbol{a})) = \mu_j^* dg_j(\boldsymbol{x}^*, \boldsymbol{a}) + d\mu_j(g_j(\boldsymbol{x}^*, \boldsymbol{a}) + dg_j(\boldsymbol{x}^*, \boldsymbol{a})); \ j \in J. \tag{8.72}$$

Since all these inequality constraints are active, we have $g_j(\boldsymbol{x}^*, \boldsymbol{a}) = 0; \ \forall j \in J$ and then (8.72) results in (8.67) for $\mu_j^* \neq 0$, and in (8.70) for $\mu_j^* = 0$.

Finally, since (8.67) implies (8.71), for $\mu_j^* \neq 0$, (8.71) must be written only for $\mu_j^* = 0$, i.e., (8.68).

Note that constraint (8.67) forces the constraints $g_j(\boldsymbol{x}^*, \boldsymbol{a}) = 0$ whose multipliers are different from zero ($\mu_j^* \neq 0$) to remain active, constraint (8.68) allows the optimal point to move inside the feasible region, constraint (8.69) forces the Lagrange multipliers to be greater than or equal to zero, and constraint (8.70) forces a new point to hold the *complementary slackness condition* for $\mu_j^* = 0$. This last constraint is a second-order constraint that implies that one of the constraints (8.68) or (8.69) has to be an equality constraint.

In matrix form, the system (8.64)–(8.69) can be written as

$$\boldsymbol{M}\delta\boldsymbol{p} = \begin{bmatrix} \boldsymbol{F_x} & \boldsymbol{F_a} & 0 & 0 & -1 \\ \hline \boldsymbol{F_{xx}} & \boldsymbol{F_{xa}} & \boldsymbol{H_x}^T & \boldsymbol{G_x}^T & 0 \\ \hline \boldsymbol{H_x} & \boldsymbol{H_a} & 0 & 0 & 0 \\ \hline \boldsymbol{G_x^1} & \boldsymbol{G_a^1} & 0 & 0 & 0 \end{bmatrix} \begin{bmatrix} d\boldsymbol{x} \\ d\boldsymbol{a} \\ d\boldsymbol{\lambda} \\ d\boldsymbol{\mu} \\ dz \end{bmatrix} = 0 \tag{8.73}$$

$$\boldsymbol{N}\delta\boldsymbol{p} = \begin{bmatrix} \boldsymbol{G_x^0} & \boldsymbol{G_a^0} & 0 & 0 & 0 \\ \hline 0 & 0 & 0 & -\boldsymbol{I}_{m_J}^0 & 0 \end{bmatrix} \begin{bmatrix} d\boldsymbol{x} \\ d\boldsymbol{a} \\ d\boldsymbol{\lambda} \\ d\boldsymbol{\mu} \\ dz \end{bmatrix} \leq 0, \tag{8.74}$$

where $m_J = \text{card}(J)$ is the number of active inequality constraints and the meaning of matrices $\boldsymbol{M}$ and $\boldsymbol{N}$ becomes clear from the system (8.64)–(8.69), and the submatrices are defined below (corresponding dimensions in parenthesis)

$$\boldsymbol{F_x}_{(1 \times n)} = (\nabla_{\boldsymbol{x}} f(\boldsymbol{x}^*, \boldsymbol{a}))^T \tag{8.75}$$

$$\boldsymbol{F_a}_{(1 \times p)} = (\nabla_{\boldsymbol{a}} f(\boldsymbol{x}^*, \boldsymbol{a}))^T \tag{8.76}$$

$$\boldsymbol{F_{xx}}_{(n \times n)} = \nabla_{\boldsymbol{xx}} f(\boldsymbol{x}^*, \boldsymbol{a}) + \sum_{k=1}^{\ell} \lambda_k^* \nabla_{\boldsymbol{xx}} h_k(\boldsymbol{x}^*, \boldsymbol{a})$$
$$+ \sum_{j=1}^{m_J} \mu_j^* \nabla_{\boldsymbol{xx}} g_j(\boldsymbol{x}^*, \boldsymbol{a}) \tag{8.77}$$

8.4 A General Method for Obtaining All Sensitivities

$$\boldsymbol{F}\boldsymbol{xa}_{(n\times p)} = \nabla_{\boldsymbol{xa}}f(\boldsymbol{x}^*,\boldsymbol{a}) + \sum_{k=1}^{\ell}\lambda_k^*\nabla_{\boldsymbol{xa}}h_k(\boldsymbol{x}^*,\boldsymbol{a})$$

$$+ \sum_{j=1}^{m_J}\mu_j^*\nabla_{\boldsymbol{xa}}g_j(\boldsymbol{x}^*,\boldsymbol{a}) \quad (8.78)$$

$$\boldsymbol{Hx}_{(\ell\times n)} = (\nabla_{\boldsymbol{x}}\boldsymbol{h}(\boldsymbol{x}^*,\boldsymbol{a}))^T \quad (8.79)$$

$$\boldsymbol{Ha}_{(\ell\times p)} = (\nabla_{\boldsymbol{a}}\boldsymbol{h}(\boldsymbol{x}^*,\boldsymbol{a}))^T \quad (8.80)$$

$$\boldsymbol{Gx}_{(m_J\times n)} = (\nabla_{\boldsymbol{x}}\boldsymbol{g}(\boldsymbol{x}^*,\boldsymbol{a}))^T \quad (8.81)$$

$$\boldsymbol{Ga}_{(m_J\times p)} = (\nabla_{\boldsymbol{a}}\boldsymbol{g}(\boldsymbol{x}^*,\boldsymbol{a}))^T, \quad (8.82)$$

where $\boldsymbol{G_x^0}$ and $\boldsymbol{G_a^0}$ refer to the submatrices of $\boldsymbol{G_x}$ and $\boldsymbol{G_a}$, respectively, associated with the null μ-multipliers of active constraints, $\boldsymbol{G_x^1}$ and $\boldsymbol{G_a^1}$ refer to the submatrices of $\boldsymbol{G_x}$ and $\boldsymbol{G_a}$, respectively, associated with the non-null μ-multipliers of active constraints, and $-\boldsymbol{I}_{m_J}^0$ is the negative of a unit matrix after removing all rows $j \in J$ such that $\mu_j^* \neq 0$.

The dimensions of all the above matrices are given in Table 8.1.

Table 8.1. Main matrices and their respective dimensions

$\nabla_{\boldsymbol{x}}f(\boldsymbol{x}^*,\boldsymbol{a})$	$\nabla_{\boldsymbol{a}}f(\boldsymbol{x}^*,\boldsymbol{a})$	$\nabla_{\boldsymbol{xx}}f(\boldsymbol{x}^*,\boldsymbol{a})$	$\nabla_{\boldsymbol{xx}}h_k(\boldsymbol{x}^*,\boldsymbol{a})$	$\nabla_{\boldsymbol{xx}}g_j(\boldsymbol{x}^*,\boldsymbol{a})$
$n \times 1$	$p \times 1$	$n \times n$	$n \times n$	$n \times n$
$\nabla_{\boldsymbol{xa}}f(\boldsymbol{x}^*,\boldsymbol{a})$	$\nabla_{\boldsymbol{xa}}h_k(\boldsymbol{x}^*,\boldsymbol{a})$	$\nabla_{\boldsymbol{xa}}g_j(\boldsymbol{x}^*,\boldsymbol{a}))$	$\nabla_{\boldsymbol{x}}\boldsymbol{h}(\boldsymbol{x}^*,\boldsymbol{a})$	$\nabla_{\boldsymbol{a}}\boldsymbol{h}(\boldsymbol{x}^*,\boldsymbol{a})$
$n \times p$	$n \times p$	$n \times p$	$n \times \ell$	$p \times l$
$\nabla_{\boldsymbol{x}}\boldsymbol{g}(\boldsymbol{x}^*,\boldsymbol{a})$	$\nabla_{\boldsymbol{a}}\boldsymbol{g}(\boldsymbol{x}^*,\boldsymbol{a})$	$\nabla_{\boldsymbol{x}}g_j(\boldsymbol{x}^*,\boldsymbol{a})$	$\nabla_{\boldsymbol{a}}g_j(\boldsymbol{x}^*,\boldsymbol{a})$	$g(\boldsymbol{x}^*,\boldsymbol{a})$
$n \times m_J$	$p \times m_J$	$n \times 1$	$m_J \times 1$	$m_J \times 1$
$d\boldsymbol{x}$	$d\boldsymbol{a}$	$d\boldsymbol{\lambda}$	$d\boldsymbol{\mu}$	dz
$n \times 1$	$p \times 1$	$\ell \times 1$	$m_J \times 1$	1×1

In order to consider the second-order condition (8.70) the system (8.73)–(8.74) has to be modified extracting from (8.74) and adding to (8.73) the row associated with either the term $\boldsymbol{G}^0$ or $-\boldsymbol{I}_{m_J}^0$ for each constraint such that $\mu_j^* = 0$; $j \in J$. The interpretation is simple, we add into (8.73) the term related to $\boldsymbol{G}^0$ for the constraints we want to remain active after the perturbation, or the term associated with $-\boldsymbol{I}_{m_J}^0$ for the constraints we want to allow to become inactive. Note that 2^{m_0} combinations (systems) are possible, where m_0 in the number of constraints whose $\mu_j^* = 0$. In what follows we initially consider the system (8.73)–(8.74) and later we take into account constraint (8.70).

8.4.1 Determining the Set of All Feasible Perturbations

Conditions (8.73)–(8.74) define the set of feasible perturbations $\delta\boldsymbol{p} = (d\boldsymbol{x}, d\boldsymbol{a}, d\boldsymbol{\lambda}, d\boldsymbol{\mu}, dz)^T$, i.e., for moving from one KKT solution to another KKT solution.

Since constraints (8.73)–(8.74) constitute an homogeneous linear system of equalities and inequalities in $d\boldsymbol{x}, d\boldsymbol{a}, d\boldsymbol{\lambda}, d\boldsymbol{\mu}$, and $d\boldsymbol{z}$, its general solution is a polyhedral cone (see Padberg [59] and Castillo et al. [21, 60]):

$$\delta \boldsymbol{p} = \sum_{i=1}^{t} \rho_i \boldsymbol{v}_i + \sum_{j=1}^{q} \pi_j \boldsymbol{w}_j , \qquad (8.83)$$

where $\rho_i \in \mathbb{R} \, (i = 1, 2, \cdots, t)$ and $\pi_j \in \mathbb{R}^+ (j = 1, 2, \cdots, q)$, and $\boldsymbol{v}_i$ and $\boldsymbol{w}_j$ are vectors that generate the linear space and the proper cone parts of the polyhedral cone, respectively.

It should be noted that since a linear space is a particular case of a cone, one can obtain a linear space as the solution of a homogeneous system of linear inequalities.

The vertex cone representation (8.83) of the feasible perturbations can be obtained using the Γ-algorithm (see Padberg [59] and Castillo et al. [21, 56]), which is known to be computationally intensive for large problems. However, one can obtain first the solution of constraint (8.73) (the corresponding null space), and then use the Γ-algorithm to incorporate the constraints in constraint (8.74), which are only a reduced number (active inequality constraints with null μ-multipliers) or none. Note that the null space computation is a standard procedure whose associated computational burden is similar to that of solving a linear homogeneous system of N equations ($O(N^3)$) [61].

Nevertheless, as we shall see, the obtention of the vertex cone representation (8.83), though convenient, could be unnecessary.

Once constraint (8.83) is known, all feasible perturbations become available. Note that if we want to take into account constraint (8.70) all possible combinations of the system (8.73)–(8.74) must be solved so that several solutions (8.83) exist. Any selection of $\rho_i \in \; i = 1, 2, \ldots, t$ and $\pi_j \in \mathbb{R}^+ (j = 1, 2, \ldots, q)$ in any solution leads to a feasible perturbation and all of them can be obtained in this form.

8.4.2 Discussion of Directional and Partial Derivatives

Conditions (8.73)–(8.74) can be written as

$$\boldsymbol{U} \begin{bmatrix} d\boldsymbol{x} \\ d\boldsymbol{\lambda} \\ d\boldsymbol{\mu} \\ d\boldsymbol{z} \end{bmatrix} = \boldsymbol{S} d\boldsymbol{a} \qquad (8.84)$$

$$\boldsymbol{V} \begin{bmatrix} d\boldsymbol{x} \\ d\boldsymbol{\lambda} \\ d\boldsymbol{\mu} \\ d\boldsymbol{z} \end{bmatrix} \leq \boldsymbol{T} d\boldsymbol{a} , \qquad (8.85)$$

where the matrices $\boldsymbol{U}, \boldsymbol{V}, \boldsymbol{S}$, and $\boldsymbol{T}$ are

8.4 A General Method for Obtaining All Sensitivities

$$U = \left[\begin{array}{c|c|c|c} F_x & 0 & 0 & -1 \\ \hline F_{xx} & H_x^T & G_x^T & 0 \\ \hline H_x & 0 & 0 & 0 \\ \hline G_x^1 & 0 & 0 & 0 \end{array}\right], \quad S = -\left[\begin{array}{c} F_a \\ \hline F_{xa} \\ \hline H_a \\ \hline G_a^1 \end{array}\right], \quad (8.86)$$

$$V = \left[\begin{array}{c|c|c|c} G_x^0 & 0 & 0 & 0 \\ \hline 0 & 0 & -I_{m_J}^0 & 0 \end{array}\right], \quad T = -\left[\begin{array}{c} G_a^0 \\ \hline 0 \end{array}\right]. \quad (8.87)$$

Note that as system (8.84)–(8.85) comes from (8.73)–(8.74) and due to condition (8.70), several systems (8.84)–(8.85) corresponding to the different combinations may exist.

An optimal point $(x^*, \lambda^*, \mu^*, z^*)$ can be classified as follows:

Regular point: The solution x^*, λ^*, μ^*, and z^* is a regular point if the gradient vectors of the active constraints are linearly independent. Under this circumstance, the optimal point can be nondegenerate or degenerate:

1. **Nondegenerate:** The Lagrange multipliers μ^* of active inequality constraints are different from zero, there is no matrix V and U^{-1} exists.
2. **Degenerate:** The Lagrange multipliers μ^* of active inequality constraints are different from zero, there is no matrix V and U^{-1} does not exist. Alternatively, some of the Lagrange multipliers of active inequality constraints in μ^* are equal to zero and matrix U^{-1} does not exist because U is not a square matrix.

Nonregular point: The gradient vectors of the active constraints are linearly dependent. Note that the KKT conditions do not characterize adequately this case because there exist infinite Lagrange multiplier value combinations that hold. However, the method also provides the sensitivities for given values of the Lagrange multipliers. In this case no difference is made between nondegenerate and degenerate cases because matrix U is never invertible.

Note that the most common situation occurs if we have a regular nondegenerate point. The cases of regular degenerate and nonregular points are exceptional. However, since we deal with a set of parametric optimization problems (we use parameters a), normally there exist particular values for the parameters such that these two cases occur as important transition situations.

Finally, it should be noted that expressions (8.84) and (8.85) allow us to determine

(1) directional derivatives if they exist.
(2) partial derivatives if they exist.
(3) all partial derivatives at once if they exist.

Note that existence means that there is a feasible perturbation where the KKT conditions still hold. We deal with all these problems in the following subsections.

8.4.3 Determining Directional Derivatives if They Exist

To check if a directional derivative exists, we replace $d\boldsymbol{a}$ by the corresponding unit vector and solve all possible combinations of the system (8.84)–(8.85). If at least for one of the combinations it exists (existence) and the solution is unique (uniqueness), then the directional derivative exists.

One can obtain first the solution of constraint (8.84) (the corresponding null space), and then use the Γ-algorithm [56] to incorporate the constraints in (8.85), which are only a reduced number (active inequality constraints with null μ-multipliers).

8.4.4 Partial Derivatives

A partial derivative is a special case of directional derivative. The partial derivative of u with respect to a_k means the increment in u due to a unit increment in a_k and null increments in $a_r, r \neq k$. Then, in a feasible perturbation $\delta \boldsymbol{p}$ that contains a unit component da_k together with null values for components $da_i, \forall i \neq k$, the remaining perturbation components contain the corresponding right-derivatives (sensitivities) with respect to a_k, i.e.,

$$\left(\frac{\partial x_1}{\partial a_k^+}, \ldots, \frac{\partial x_n}{\partial a_k^+}, 0, \ldots, 0, 1, 0, \ldots, 0, \frac{\partial \lambda_1}{\partial a_k^+}, \ldots, \frac{\partial \lambda_p}{\partial a_k^+}, \frac{\partial \mu_1}{\partial a_k^+}, \ldots, \frac{\partial \mu_{m_J}}{\partial a_k^+}, \frac{\partial z}{\partial a_k^+} \right)^T. \tag{8.88}$$

Similarly, a feasible perturbation of the form

$$\left(\frac{\partial x_1}{\partial a_k^-}, \ldots, \frac{\partial x_n}{\partial a_k^-}, 0, \ldots, 0, -1, 0, \ldots, 0, \frac{\partial \lambda_1}{\partial a_k^-}, \ldots, \frac{\lambda_p}{\partial a_k^-}, \frac{\partial \mu_1}{\partial a_k^-}, \ldots, \frac{\partial \mu_{m_J}}{\partial a_k^-}, \frac{\partial z}{\partial a_k^-} \right)^T \tag{8.89}$$

contains as the remaining components all the left-derivatives with respect to a_k. If both exist and coincide in absolute value but not in sign, the corresponding partial derivative exists.

The partial derivative is obtained solving the directional derivatives for $d\boldsymbol{a}_k$ and $-d\boldsymbol{a}_k$, respectively, and checking if both exist, and coincide in absolute value but not in sign. If the answer is positive, the corresponding partial derivative exists.

Note that this procedure also allows us to know if there are directional derivatives for any arbitrary vector $d\boldsymbol{a}$ in both directions $d\boldsymbol{a}$ and $-d\boldsymbol{a}$.

8.4.5 Obtaining All Sensitivities at Once

If the solution x^*, λ^*, μ^*, and z^* is a nondegenerate regular point, then the matrix U is invertible and the solution of the system (8.84)–(8.85) is unique and it becomes

$$\begin{bmatrix} dx \\ d\lambda \\ d\mu \\ dz \end{bmatrix} = U^{-1} S \, da, \qquad (8.90)$$

where (8.85) is satisfied trivially since V does not exist.

Several partial derivatives can be simultaneously obtained if the vector da in (8.90) is replaced by a matrix including several vectors (columns) with the corresponding unit directions. In particular, replacing da by the unit matrix I_p in (8.90) all the partial derivatives are obtained. The matrix with all partial derivatives becomes

$$\begin{bmatrix} \dfrac{\partial x}{\partial a} \\ \dfrac{\partial \lambda}{\partial a} \\ \dfrac{\partial \mu}{\partial a} \\ \dfrac{\partial z}{\partial a} \end{bmatrix} = U^{-1} S. \qquad (8.91)$$

For any vector da the derivatives in both directions da and $-da$ are obtained simultaneously.

8.5 Particular Cases

There are some important particular cases of the feasible perturbation equations (8.73)–(8.74).

8.5.1 No Constraints

In the particular case of an optimization problem with no constraints the system (8.65)–(8.70) transforms to

$$\nabla_{xx} f(x^*, a) dx + \nabla_{xa} f(x^*, a) da = 0 \qquad (8.92)$$

and if $\nabla_{xx} f(\bar{x}, a)$ is invertible, one gets

$$\frac{\partial x}{\partial a}_{(n \times p)} = -\left(\nabla_{xx} f(x^*, a)\right)^{-1} \nabla_{xa} f(x^*, a), \qquad (8.93)$$

where $\dfrac{\partial x}{\partial a}$ is the matrix containing all the sensitivities of x with respect to a.

Illustrative Example 8.6 (The maximum likelihood method). Consider the following sample:

$$a = (1.341, 3.171, 3.629, 0.964, 5.904, -3.07, 2.573, -0.432, 2.1, 0.886),$$

drawn from a normal distribution $N(\mu, \sigma^2)$, which depends on the set of parameters $x = (\mu, \sigma)$.

The maximum likelihood estimates $\hat{\mu}$ and $\hat{\sigma}$ of the parameters μ and σ are the solutions of the optimization problem

$$\underset{\mu, \sigma}{\text{minimize}} \quad z = \frac{1}{2} \sum_{i=1}^{p} \left(\frac{a_i - \mu}{\sigma} \right)^2 + p \log \sigma + p \log(2\pi)/2,$$

whose solution is

$$\mu^* = 1.7067, \quad \sigma^* = 2.3047, \quad z^* = 22.5387.$$

Since

$$\frac{\partial^2 z}{\partial \mu^2} = \frac{p}{\sigma^2}$$

$$\frac{\partial^2 z}{\partial \mu \partial \sigma} = \frac{2 \sum_{i=1}^{p} (a_i - \mu)}{\sigma^3}$$

$$\frac{\partial^2 z}{\partial \sigma^2} = \frac{3}{\sigma^4} \sum_{i=1}^{p} (a_i - \mu)^2 - \frac{p}{\sigma^2}$$

$$\frac{\partial^2 z}{\partial \mu \partial a_i} = -\frac{1}{\sigma^2}; \quad i = 1, \ldots, p$$

$$\frac{\partial^2 z}{\partial \sigma \partial a_i} = \frac{2(\mu - a_i)}{\sigma^3}; \quad i = 1, \ldots, p,$$

we get

$$\nabla_{xx} f(x, a) = \begin{pmatrix} \dfrac{p}{\sigma^2} & \dfrac{2 \sum_{i=1}^{p} (a_i - \mu)}{\sigma^3} \\ \dfrac{2 \sum_{i=1}^{p} (a_i - \mu)}{\sigma^3} & \dfrac{3 \sum_{i=1}^{p} (a_i - \mu)^2 - p\sigma^2}{\sigma^4} \end{pmatrix} = \begin{pmatrix} 1.88282 & 0 \\ 0 & 3.76565 \end{pmatrix}$$

$$\nabla_{xa} f(x, a) = \begin{pmatrix} -\dfrac{1}{\sigma^2} & \ldots & -\dfrac{1}{\sigma^2} \\ \dfrac{2(\mu - a_1)}{\sigma^3} & \ldots & \dfrac{2(\mu - a_p)}{\sigma^3} \end{pmatrix}$$

and then from (8.93) one obtains the sensitivities

$$\begin{pmatrix} \dfrac{\partial \mu}{\partial a_1} & \cdots & \dfrac{\partial \mu}{\partial a_p} \\ \dfrac{\partial \sigma}{\partial a_1} & \cdots & \dfrac{\partial \sigma}{\partial a_p} \end{pmatrix} = \dfrac{1}{p} \begin{pmatrix} -1 & \cdots & -1 \\ \dfrac{a_1 - \mu}{\sigma} & \cdots & \dfrac{a_p - \mu}{\sigma} \end{pmatrix}. \quad (8.94)$$

In addition, from Theorem 8.2 one gets

$$\dfrac{\partial z}{\partial a_i} = \dfrac{a_i - \mu}{\sigma^2}; \quad i = 1, \ldots, p.$$

Table 8.2 gives all sensitivities for this example. □

Table 8.2. Sensitivities

i	1	2	3	4	5	6	7	8	9	10
$\partial \mu / \partial a_i$	-0.1	-0.1	-0.1	-0.1	-0.1	-0.1	-0.1	-0.1	-0.1	-0.1
$\partial \sigma / \partial a_i$	0.016	-0.064	-0.083	0.032	-0.182	0.207	-0.038	0.093	-0.017	0.036
$\partial z / \partial a_i$	-0.069	0.276	0.362	-0.140	0.790	-0.900	0.163	-0.403	0.074	-0.154

8.5.2 Same Active Constraints

In this section we assume that the active inequality constraints remain active, i.e., the perturbed case has the same active constraints as the initial problem. Assume that the nonlinear problem has been solved and that one knows its optimal solution x^* and its dual solution λ^*, and that one has removed the inactive constraints and all active inequality constraints have been converted to equality constraints. Then, U and S can be written as

$$U = \begin{bmatrix} F_x & | & 0 & | & -1 \\ F_{xx} & | & H_x^T & | & 0 \\ H_x & | & 0 & | & 0 \end{bmatrix}, \quad S = - \begin{bmatrix} F_a \\ F_{xa} \\ H_a \end{bmatrix}, \quad (8.95)$$

and

$$\begin{bmatrix} \dfrac{\partial x}{\partial a} & \dfrac{\partial \lambda}{\partial a} & \dfrac{\partial z}{\partial a} \end{bmatrix}^T = U^{-1} S. \quad (8.96)$$

Note that (8.96) allows us to compute all sensitivities at once. However, several particular cases are discussed below.

If the number of constraints equals the number of variables and matrix H_x is invertible, we have

$$\begin{bmatrix}\frac{\partial x}{\partial a}\\\frac{\partial \lambda}{\partial a}\\\frac{\partial z}{\partial a}\end{bmatrix} = U^{-1}S = -\begin{bmatrix}0_{n\times 1} & 0_{n\times n} & H_x^{-1}{}_{n\times \ell}\\ 0_{\ell\times 1} & (H_x^T)_{\ell\times n}^{-1} & -(H_x^T)^{-1}F_{xx}H_x^{-1}{}_{\ell\times \ell}\\ -1_{1\times 1} & 0_{1\times n} & F_xH_x^{-1}{}_{1\times \ell}\end{bmatrix}\begin{bmatrix}F_{a\,1\times p}\\ F_{xa\,n\times p}\\ H_{a\,\ell\times p}\end{bmatrix},$$
(8.97)

from which we get the closed formulas

$$\frac{\partial x}{\partial a} = -H_x^{-1}H_a \qquad (8.98)$$

$$\frac{\partial \lambda}{\partial a} = (H_x^{-1})^T\left[F_{xx}H_x^{-1}H_a - F_{xa}\right] \qquad (8.99)$$

$$\frac{\partial z}{\partial a} = F_a - F_xH_x^{-1}H_a = F_a + \lambda^T H_a. \qquad (8.100)$$

Alternatively, if the matrix F_{xx} is invertible, then $B = -H_xF_{xx}^{-1}H_x^T$ under the given assumptions is also invertible and we have

$$U^{-1} = \begin{bmatrix} 0_{n\times 1} & \left(I + F_{xx}^{-1}H_x^TB^{-1}H_x\right)F_{xx}^{-1} & -F_{xx}^{-1}H_x^TB^{-1}\\ 0_{\ell\times 1} & -B^{-1}H_xF_{xx}^{-1} & B^{-1}\\ -1_{1\times 1} & F_x\left(I + F_{xx}^{-1}H_x^TB^{-1}H_x\right)F_{xx}^{-1} & -F_xF_{xx}^{-1}H_x^TB^{-1}\end{bmatrix},$$
(8.101)

from which we get the alternative closed formulas

$$\frac{\partial x}{\partial a} = -\left(I + F_{xx}^{-1}H_x^TB^{-1}H_x\right)F_{xx}^{-1}F_{xa}$$
$$+F_{xx}^{-1}H_x^TB^{-1}H_a \qquad (8.102)$$

$$\frac{\partial \lambda}{\partial a} = B^{-1}H_xF_{xx}^{-1}F_{xa} - B^{-1}H_a \qquad (8.103)$$

$$\frac{\partial z}{\partial a} = F_a - F_x\left(I + F_{xx}^{-1}H_x^TB^{-1}H_x\right)F_{xx}^{-1}F_{xa}$$
$$+F_xF_{xx}^{-1}H_x^TB^{-1}H_a. \qquad (8.104)$$

Note that (8.104) is exactly Theorem 8.2.

Illustrative Example 8.7 (The LP problem). Consider the following LP problem:

$$\underset{x}{\text{minimize}} \quad z = c^T x \qquad (8.105)$$

subject to

$$Ax = b \;:\; \lambda, \qquad (8.106)$$

where $c = (c_1, c_2, \ldots, c_n)$, $x = (x_1, x_2, \ldots, x_n)$, $b = (b_1, b_2, \ldots, b_m) \geq 0$, A is a matrix of dimensions $m \times n$ with elements a_{ij} ($i = 1, 2, \ldots m; j = 1, 2, \ldots, n$) and λ are the dual variables.

8.5 Particular Cases

We are interested in determining the sensitivities of z^*, $\boldsymbol{x}^*$, and $\boldsymbol{\lambda}^*$ with respect to all the data

$$\boldsymbol{a} = (c_1, \ldots, c_n; a_{11}, \ldots, a_{m1}, a_{12}, \ldots, a_{m2}, \ldots, a_{1n}, \ldots, a_{mn}, b_1, \ldots, b_m).$$

Assume that all the constraints in (8.106) are active [this is easy to achieve, especially if one has already solved the LP problem (8.105)–(8.106)] and non-basic (zero) variables are eliminated, i.e., $n = m$, and $\boldsymbol{A}$ reduces to an invertible matrix. Then, the problem is regular and the matrix $\boldsymbol{U}$ becomes

$$\boldsymbol{U} = \begin{pmatrix} \boldsymbol{c}^T_{1 \times n} & | & \boldsymbol{0}_{1 \times n} & | & -\boldsymbol{1}_{1 \times 1} \\ \boldsymbol{0}_{n \times n} & | & \boldsymbol{A}^T_{n \times n} & | & \boldsymbol{0}_{n \times 1} \\ \boldsymbol{A}_{n \times n} & | & \boldsymbol{0}_{n \times n} & | & \boldsymbol{0}_{n \times 1} \end{pmatrix}, \tag{8.107}$$

whose inverse is

$$\boldsymbol{U}^{-1} = \begin{pmatrix} \boldsymbol{0}_{n \times 1} & | & \boldsymbol{0}_{n \times n} & | & \boldsymbol{A}^{-1}_{n \times n} \\ \boldsymbol{0}_{n \times 1} & | & \left(\boldsymbol{A}^T\right)^{-1}_{n \times n} & | & \boldsymbol{0}_{n \times n} \\ -\boldsymbol{1}_{1 \times 1} & | & \boldsymbol{0}_{1 \times n} & | & \left(\boldsymbol{c}^T \boldsymbol{A}^{-1}\right)_{1 \times n} \end{pmatrix}. \tag{8.108}$$

The matrix $\boldsymbol{S}$ is

$$\boldsymbol{S} = - \begin{pmatrix} \boldsymbol{x}^T_{1 \times n} & | & \boldsymbol{0}_{1 \times n^2} & | & \boldsymbol{0}_{1 \times n} \\ \boldsymbol{I}_n & | & \left(\boldsymbol{I}_n \otimes \boldsymbol{\lambda}^T\right)_{n \times n^2} & | & \boldsymbol{0}_{n \times n} \\ \boldsymbol{0}_{n \times n} & | & \left(\boldsymbol{x}^T \otimes \boldsymbol{I}_n\right)_{n \times n^2} & | & -\boldsymbol{I}_n \end{pmatrix}, \tag{8.109}$$

where $\otimes$ refers to the tensor or Kronecker's product of matrices.

Then, the local sensitivities become

$$\frac{\partial z}{\partial c_j} = x_j; \quad \frac{\partial z}{\partial a_{ij}} = \lambda_i x_j; \quad \frac{\partial z}{\partial b_i} = -\lambda_i,$$

$$\frac{\partial x_j}{\partial c_k} = 0; \quad \frac{\partial x_j}{\partial a_{ik}} = -a^{ji} x_k; \quad \frac{\partial x_j}{\partial b_i} = a^{ji}, \tag{8.110}$$

$$\frac{\partial \lambda_i}{\partial c_j} = -a^{ji}; \quad \frac{\partial \lambda_i}{\partial a_{\ell j}} = -a^{ji} \lambda_\ell; \quad \frac{\partial \lambda_i}{\partial b_\ell} = 0,$$

where a^{ji} are the elements of $\boldsymbol{A}^{-1}$.

Observe that (8.110) provide the sensitivities of the objective function, the primal variables and the dual variables with respect to all parameters in (8.105)–(8.106). □

Illustrative Example 8.8 (The sensitivity with respect to parameters). Consider that all the data are functions of a set of parameters $\boldsymbol{\theta}$, i.e.,

$$\text{minimize} \quad z = c^T(\boldsymbol{\theta})x \qquad (8.111)$$

subject to
$$A(\boldsymbol{\theta})x = b(\boldsymbol{\theta}) : \boldsymbol{\lambda} \qquad (8.112)$$

then, using the chain rule and (8.110), the sensitivities become

$$\frac{\partial z}{\partial \theta_r} = \sum_j x_j \frac{\partial c_j(\boldsymbol{\theta})}{\partial \theta_r} + \sum_{i,j} \lambda_i x_j \frac{\partial a_{ij}(\boldsymbol{\theta})}{\partial \theta_r} - \sum_i \lambda_i \frac{\partial b_i(\boldsymbol{\theta})}{\partial \theta_r}$$

$$\frac{\partial x_j}{\partial \theta_r} = -\sum_{i,k} a^{ji} x_k \frac{\partial a_{ik}(\boldsymbol{\theta})}{\partial \theta_r} + \sum_i a^{ji} \frac{\partial b_i(\boldsymbol{\theta})}{\partial \theta_r} \qquad (8.113)$$

$$\frac{\partial \lambda_i}{\partial \theta_r} = -\sum_j a^{ji} \frac{\partial c_j(\boldsymbol{\theta})}{\partial \theta_r} - \sum_{\ell,j} a^{ji} \lambda_\ell \frac{\partial a_{\ell j}(\boldsymbol{\theta})}{\partial \theta_r}$$

or in matrix form

$$\frac{\partial z}{\partial \boldsymbol{\theta}} = x^T \frac{\partial c}{\partial \boldsymbol{\theta}} - \boldsymbol{\lambda}^T \left(\frac{\partial b}{\partial \boldsymbol{\theta}} - \frac{\partial A}{\partial \boldsymbol{\theta}} x \right)$$

$$\frac{\partial x}{\partial \boldsymbol{\theta}} = A^{-1} \left(\frac{\partial b}{\partial \boldsymbol{\theta}} - \frac{\partial A}{\partial \boldsymbol{\theta}} x \right) \qquad (8.114)$$

$$\frac{\partial \boldsymbol{\lambda}}{\partial \boldsymbol{\theta}} = (A^{-1})^T \left(\frac{\partial c}{\partial \boldsymbol{\theta}} - \left(\frac{\partial A}{\partial \boldsymbol{\theta}} \right)^T \boldsymbol{\lambda} \right).$$

Note that this is the general case and that the sensitivities in (8.110) are particular cases. □

8.5.3 The General Case

The general case is illustrated in the following examples.

Illustrative Example 8.9 (The regular nondegenerate case). Consider the following parametric optimization problem:

$$\underset{x_1, x_2}{\text{minimize}} \quad z = (x_1 - 1)^2 + (x_2 - 1)^2 \qquad (8.115)$$

subject to
$$\begin{array}{rcl} x_1 + x_2 & \leq & a_1 \\ x_1 & \geq & a_2 \\ x_2 & \geq & a_3, \end{array} \qquad (8.116)$$

which for $a_1 = 1, a_2 = 0, a_3 = 0$ leads to the following optimal solution (see Example 8.1):

8.5 Particular Cases   327

$$x_1 = 1/2, \quad x_2 = 1/2, \quad \mu_1 = 1, \quad \mu_2 = 0, \quad \mu_3 = 0, \quad z = 1/2 \,.$$

Note that the only active constraint is the first one, and then, the remaining constraints are inactive and can be eliminated from the analysis.

To analyze the existence of partial derivatives we solve the system of inequalities (8.73)–(8.74), which in this case becomes

$$\left(\begin{array}{cc|ccc|c|c} -1 & -1 & 0 & 0 & 0 & 0 & -1 \\ \hline 2 & 0 & 0 & 0 & 0 & 1 & 0 \\ 0 & 2 & 0 & 0 & 0 & 1 & 0 \\ \hline 1 & 1 & -1 & 0 & 0 & 0 & 0 \end{array}\right) \left(\begin{array}{c} dx_1 \\ dx_2 \\ \hline da_1 \\ da_2 \\ da_3 \\ \hline d\mu_1 \\ \hline dz \end{array}\right) = \mathbf{0} \,, \qquad (8.117)$$

whose solution is the linear space

$$\left(\begin{array}{c} dx_1 \\ dx_2 \\ \hline da_1 \\ da_2 \\ da_3 \\ \hline d\mu_1 \\ \hline dz \end{array}\right) = \left(\begin{array}{ccc} 0 & 0 & -1 \\ 0 & 0 & -1 \\ \hline 0 & 0 & -2 \\ 0 & 1 & 0 \\ 1 & 0 & 0 \\ \hline 0 & 0 & 2 \\ \hline 0 & 0 & 2 \end{array}\right) \left(\begin{array}{c} \rho_1 \\ \rho_2 \\ \rho_3 \end{array}\right) \,. \qquad (8.118)$$

Expression (8.117) can be transformed [see (8.84)–(8.85)] as follows:

$$\left(\begin{array}{cc|c|c} -1 & -1 & 0 & -1 \\ \hline 2 & 0 & 1 & 0 \\ 0 & 2 & 1 & 0 \\ \hline 1 & 1 & 0 & 0 \end{array}\right) \left(\begin{array}{c} dx_1 \\ dx_2 \\ \hline d\mu_1 \\ \hline dz \end{array}\right) = \left(\begin{array}{ccc} 0 & 0 & 0 \\ \hline 0 & 0 & 0 \\ 0 & 0 & 0 \\ \hline -1 & 0 & 0 \end{array}\right) \left(\begin{array}{c} da_1 \\ da_2 \\ da_3 \end{array}\right) \,. \qquad (8.119)$$

Since the matrix $\mathbf{U}$ on the left-hand side of expression (8.119) is invertible, all the sensitivities with respect to a_1, a_2, and a_3 can be obtained at once using (8.91):

$$\left(\begin{array}{ccc} \dfrac{\partial x_1^*}{\partial a_1} & \dfrac{\partial x_1^*}{\partial a_2} & \dfrac{\partial x_1^*}{\partial a_3} \\ \dfrac{\partial x_2^*}{\partial a_1} & \dfrac{\partial x_2^*}{\partial a_2} & \dfrac{\partial x_2^*}{\partial a_3} \\ \hline \dfrac{\partial \mu_1^*}{\partial a_1} & \dfrac{\partial \mu_1^*}{\partial a_2} & \dfrac{\partial \mu_1^*}{\partial a_3} \\ \hline \dfrac{\partial z^*}{\partial a_1} & \dfrac{\partial z^*}{\partial a_2} & \dfrac{\partial z^*}{\partial a_3} \end{array}\right) = \left(\begin{array}{cc|c|c} -1 & -1 & 0 & -1 \\ \hline 2 & 0 & 1 & 0 \\ 0 & 2 & 1 & 0 \\ \hline 1 & 1 & 0 & 0 \end{array}\right)^{-1} \left(\begin{array}{ccc} 0 & 0 & 0 \\ \hline 0 & 0 & 0 \\ 0 & 0 & 0 \\ \hline -1 & 0 & 0 \end{array}\right) = \left(\begin{array}{ccc} \frac{1}{2} & 0 & 0 \\ \frac{1}{2} & 0 & 0 \\ \hline -1 & 0 & 0 \\ \hline -1 & 0 & 0 \end{array}\right) \,.$$

Note that the sensitivities with respect to the parameters a_2 and a_3, appearing only in the inactive constraints, are null. □

Illustrative Example 8.10 (The regular degenerate case). Consider the following parametric optimization problem (see Fig. 8.3):

$$\underset{x,y}{\text{minimize}} \quad z = x^2 + y^2 \tag{8.120}$$

subject to

$$\begin{aligned}(x-2)^2 + (y-2)^2 &\le 2 \\ -x+a &\le 0. \end{aligned} \tag{8.121}$$

The KKT conditions for this problem are

$$\begin{aligned}\begin{pmatrix} 2x \\ 2y \end{pmatrix} + \begin{pmatrix} 2(x-2) \\ 2(y-2) \end{pmatrix}\mu_1 + \begin{pmatrix} -1 \\ 0 \end{pmatrix}\mu_2 &= \begin{pmatrix} 0 \\ 0 \end{pmatrix} \\ (x-2)^2 + (y-2)^2 - 2 &\le 0 \\ -x + a &\le 0 \\ \mu_1\left((x-2)^2 + (y-2)^2 - 2\right) &= 0 \\ \mu_2(-x+a) &= 0 \\ \mu_1, \mu_2 &\ge 0, \end{aligned} \tag{8.122}$$

and the corresponding solution, for the particular case $a = 1$ is

$$x_1^* = 1, \quad x_2^* = 1, \quad \mu_1^* = 1, \quad \mu_2^* = 0, \quad z^* = 2. \tag{8.123}$$

In Fig. 8.3 the optimal solution, the feasible region of the problem (8.120)–(8.121), as well as the graphical interpretation of the first equation in (8.122) are shown. Note that the second constraint, $-x + a \le 0$, it is not necessary for getting the optimal solution (8.123), this means that it could be removed and the same optimal solution would still remain.

A vector of changes

$$\delta \boldsymbol{p} = (dx, dy, da, d\mu_1, d\mu_2, dz)^T$$

must satisfy the system (8.64)–(8.70), which for this example becomes

$$(2 \ \ 2)\begin{pmatrix} dx \\ dy \end{pmatrix} + 0 da - dz = 0 \tag{8.124}$$

$$\left(\begin{pmatrix} 2 & 0 \\ 0 & 2 \end{pmatrix} + \mu_1 \begin{pmatrix} 2 & 0 \\ 0 & 2 \end{pmatrix} + \mu_2 \begin{pmatrix} 0 & 0 \\ 0 & 0 \end{pmatrix}\right)\begin{pmatrix} dx \\ dy \end{pmatrix}$$

$$+ \left(\begin{pmatrix} 0 \\ 0 \end{pmatrix} + \mu_1 \begin{pmatrix} 0 \\ 0 \end{pmatrix} + \mu_2 \begin{pmatrix} 0 \\ 0 \end{pmatrix}\right) da$$

$$+ \begin{pmatrix} -2 & -1 \\ -2 & 0 \end{pmatrix}\begin{pmatrix} d\mu_1 \\ d\mu_2 \end{pmatrix} = \begin{pmatrix} 0 \\ 0 \end{pmatrix} \tag{8.125}$$

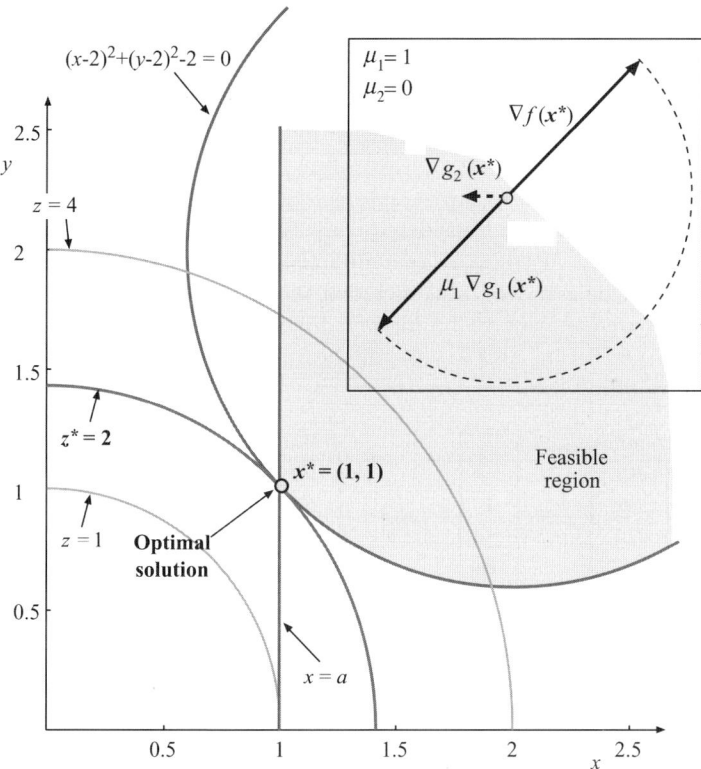

Fig. 8.3. Illustration of the feasible region, the optimal values, and the KKT multipliers for the regular degenerate example

$$(-2 \quad -2) \begin{pmatrix} dx \\ dy \end{pmatrix} = 0 \quad (8.126)$$

$$(-1 \quad 0) \begin{pmatrix} dx \\ dy \end{pmatrix} + da \leq 0 \quad (8.127)$$

$$d\mu_2 \geq 0. \quad (8.128)$$

In this case, the gradients of the constraints are linearly independent and one of the Lagrange multipliers is null; so, we have a regular degenerate case. The system (8.73)–(8.74), using expressions (8.124) and (8.128), becomes

$$\boldsymbol{M\delta p} = \begin{pmatrix} 2 & 2 & 0 & 0 & 0 & -1 \\ 4 & 0 & 0 & -2 & -1 & 0 \\ 0 & 4 & 0 & -2 & 0 & 0 \\ -2 & -2 & 0 & 0 & 0 & 0 \end{pmatrix} \begin{pmatrix} dx \\ dy \\ da \\ d\mu_1 \\ d\mu_2 \\ dz \end{pmatrix} = \boldsymbol{0} \quad (8.129)$$

and

$$N\delta p = \left(\begin{array}{cc|c|cc|c} -1 & 0 & 1 & 0 & 0 & 0 \\ 0 & 0 & 0 & 0 & -1 & 0 \end{array}\right) \begin{pmatrix} dx \\ dy \\ da \\ d\mu_1 \\ d\mu_2 \\ dz \end{pmatrix} \leq \mathbf{0}. \quad (8.130)$$

Note that we have not considered constraint (8.70) yet. If we want (1) the second inequality constraint to remain active the first equation in constraint (8.130) should be removed and included in (8.129), whereas if we want (2) the inequality constraint to be allowed to become inactive then the second equation in constraint (8.130) should be removed and included in (8.129). The corresponding solutions are the cones

$$\begin{pmatrix} dx \\ dy \\ da \\ d\mu_1 \\ d\mu_2 \\ dz \end{pmatrix} = \pi \begin{bmatrix} 1 \\ -1 \\ 1 \\ -2 \\ 8 \\ 0 \end{bmatrix} \quad \text{and} \quad \begin{pmatrix} dx \\ dy \\ da \\ d\mu_1 \\ d\mu_2 \\ dz \end{pmatrix} = \pi \begin{bmatrix} 0 \\ 0 \\ -1 \\ 0 \\ 0 \\ 0 \end{bmatrix}, \quad (8.131)$$

respectively, where $\pi \in \mathbb{R}^+$, which give all feasible perturbations. Note, for example, that the optimal objective function does not depend on the parameter a because dz is null for all possible values of π. Note also that in the solution that allows the second inequality constraint to become inactive, the only element different from zero is da, which can only be negative, this means that decreasing the parameter a makes the second constraint inactive as it is shown in Fig. 8.4b.

In order to study the existence of directional derivatives with respect to a we use the directions $d\mathbf{a} = 1$ and $d\mathbf{a} = -1$, and solve the two possible combinations of (8.84)–(8.85) that lead to

$$\begin{pmatrix} \frac{\partial x}{\partial a^+} \\ \frac{\partial y}{\partial a^+} \\ \frac{\partial \mu_1}{\partial a^+} \\ \frac{\partial \mu_2}{\partial a^+} \\ \frac{\partial z}{\partial a^+} \end{pmatrix} = \begin{bmatrix} 1 \\ -1 \\ -2 \\ 8 \\ 0 \end{bmatrix}, \quad \begin{pmatrix} \frac{\partial x}{\partial a^-} \\ \frac{\partial y}{\partial a^-} \\ \frac{\partial \mu_1}{\partial a^-} \\ \frac{\partial \mu_2}{\partial a^-} \\ \frac{\partial z}{\partial a^-} \end{pmatrix} = [\emptyset], \text{ and } \begin{pmatrix} \frac{\partial x}{\partial a^+} \\ \frac{\partial y}{\partial a^+} \\ \frac{\partial \mu_1}{\partial a^+} \\ \frac{\partial \mu_2}{\partial a^+} \\ \frac{\partial z}{\partial a^+} \end{pmatrix} = [\emptyset], \quad \begin{pmatrix} \frac{\partial x}{\partial a^-} \\ \frac{\partial y}{\partial a^-} \\ \frac{\partial \mu_1}{\partial a^-} \\ \frac{\partial \mu_2}{\partial a^-} \\ \frac{\partial z}{\partial a^-} \end{pmatrix} = \begin{bmatrix} 0 \\ 0 \\ 0 \\ 0 \\ 0 \end{bmatrix},$$
(8.132)

respectively, which implies that only right-or-left hand side derivatives exist. Increasing the parameter a makes the second constraint to remain active with an associated Lagrange multiplier different from zero (note that the multiplier μ_2 only can increase holding the last constraint in (8.122)) as shown in Fig. 8.4a. The partial derivative of z with respect to a is zero. Decreasing the parameter a makes the second constraint inactive whereas the solution of the problem remains the same evidencing that the second constraint is redundant, as shown in Fig. 8.4b. □

Illustrative Example 8.11 (The nonregular case). Consider the following parametric optimization problem (see Fig. 8.5):

$$\underset{x,y}{\text{minimize}} \quad z = -2a_1 x - y \qquad (8.133)$$

subject to

$$\begin{aligned} x + y &= a_2 \\ x &\leq a_3 \\ y &\leq a_4 \\ x + 4y/3 &\leq 4. \end{aligned} \qquad (8.134)$$

The KKT conditions for this problem are

$$\begin{pmatrix} -2a_1 \\ -1 \end{pmatrix} + \begin{pmatrix} 1 \\ 1 \end{pmatrix} \lambda + \begin{pmatrix} 1 \\ 0 \end{pmatrix} \mu_1 + \begin{pmatrix} 0 \\ 1 \end{pmatrix} \mu_2 + \begin{pmatrix} 1 \\ 4/3 \end{pmatrix} \mu_3 = \begin{pmatrix} 0 \\ 0 \end{pmatrix}$$

$$\begin{aligned} x + y - a_2 &= 0 \\ x - a_3 &\leq 0 \\ y - a_4 &\leq 0 \\ x + 4y/3 - 4 &\leq 0 \\ \mu_1(x - a_3) &= 0 \\ \mu_2(y - a_4) &= 0 \\ \mu_3(x + 4y/3 - 4) &= 0 \\ \mu_1, \mu_2, \mu_3 &\geq 0, \end{aligned}$$
(8.135)

and one of the possible solutions, for the particular case $a_1 = 1, a_2 = 2, a_3 = 1$, and $a_4 = 1$ is

$$x = 1, \ y = 1, \ \lambda = 1, \ \mu_1 = 1, \ \mu_2 = 0, \ \mu_3 = 0, \ z = -3.$$

In Fig. 8.5 the optimal solution, the feasible region of the problem (8.133)–(8.134), and the graphical interpretation of the first equation in constraint (8.135) are shown. Note that the constraints associated with null multipliers are not necessary for getting the optimal solution, this means that they could be removed and the same optimal solution would still remain.

Note that in this example the dual problem has infinite solutions, because the gradients of the active constraints are linearly dependent. Since the two first inequality constraints are active, they will remain either active or inactive in a neighborhood of the optimum depending on the values of the Lagrange multipliers.

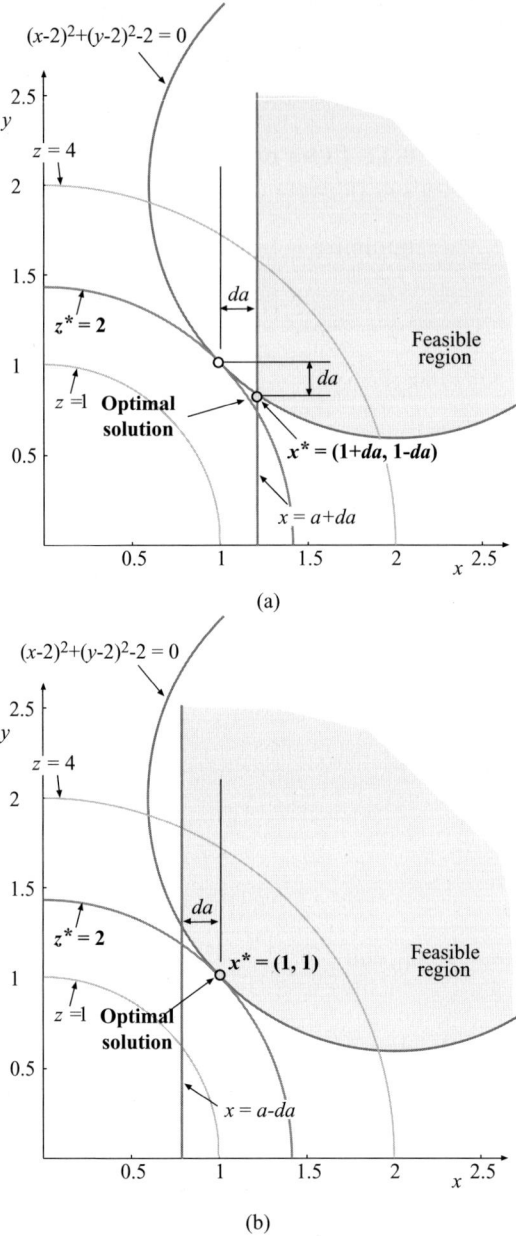

Fig. 8.4. Illustration of the feasible regions, and optimal values of the modified problems due to changes in the parameter a, for the regular degenerate case: (**a**) positive increment, (**b**) negative increment

8.5 Particular Cases   333

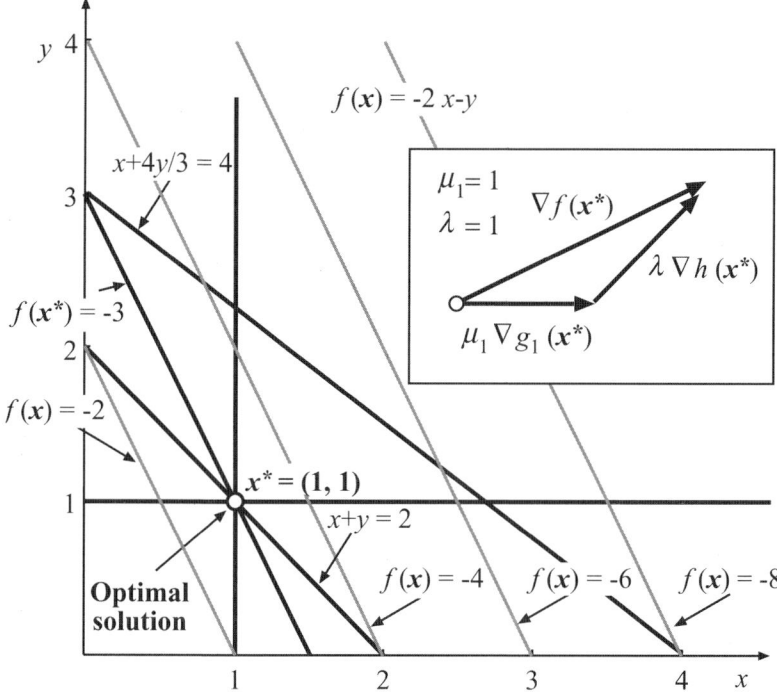

Fig. 8.5. Illustration of the minimization problem in Example 8.11

We analyze only the case $\lambda = 1, \mu_1 = 1$, and $\mu_2 = 0$. A vector of changes

$$\delta p = (dx, dy, da_1, da_2, da_3, da_4, d\lambda, d\mu_1, d\mu_2, dz)^T$$

must satisfy the system (8.64)–(8.70). The system (8.73)–(8.74) becomes

$$M\delta p = \begin{pmatrix} -2 & -1 & -2 & 0 & 0 & 0 & 0 & 0 & 0 & -1 \\ 0 & 0 & -2 & 0 & 0 & 0 & 1 & 1 & 0 & 0 \\ 0 & 0 & 0 & 0 & 0 & 0 & 1 & 0 & 1 & 0 \\ 1 & 1 & 0 & -1 & 0 & 0 & 0 & 0 & 0 & 0 \\ 1 & 0 & 0 & 0 & -1 & 0 & 0 & 0 & 0 & 0 \end{pmatrix} \begin{pmatrix} dx \\ dy \\ da_1 \\ da_2 \\ da_3 \\ da_4 \\ d\lambda \\ d\mu_1 \\ d\mu_2 \\ dz \end{pmatrix} = \mathbf{0} \qquad (8.136)$$

$$\boldsymbol{N}\delta\boldsymbol{p} = \begin{pmatrix} 0 & 1 & 0 & 0 & 0 & -1 & 0 & 0 & 0 & 0 \\ 0 & 0 & 0 & 0 & 0 & 0 & 0 & 0 & -1 & 0 \end{pmatrix} \begin{pmatrix} dx \\ dy \\ da_1 \\ da_2 \\ da_3 \\ da_4 \\ d\lambda \\ d\mu_1 \\ d\mu_2 \\ dz \end{pmatrix} \leq \boldsymbol{0}. \tag{8.137}$$

Note that we have not considered constraint (8.70) yet. If we desire (1) the inequality constraint related to μ_2 to remain active, the first equation in constraint (8.137) should be removed and included in (8.136), whereas if we desire (2) the inequality constraint to be allowed to become inactive, then the second equation in constraint (8.137) should be removed and included in (8.136). The corresponding solutions are the cones

$$\begin{pmatrix} dx \\ dy \\ da_1 \\ da_2 \\ da_3 \\ da_4 \\ d\lambda \\ d\mu_1 \\ d\mu_2 \\ dz \end{pmatrix} = \rho_1 \begin{bmatrix} -1 \\ 0 \\ 1 \\ -1 \\ -1 \\ 0 \\ 0 \\ 2 \\ 0 \\ 0 \end{bmatrix} + \rho_2 \begin{bmatrix} -1 \\ 2 \\ 0 \\ 1 \\ -1 \\ 2 \\ 0 \\ 0 \\ 0 \\ 0 \end{bmatrix} + \rho_3 \begin{bmatrix} -1 \\ 0 \\ 0 \\ -1 \\ -1 \\ 0 \\ 0 \\ 0 \\ 0 \\ 2 \end{bmatrix} + \pi \begin{bmatrix} 1 \\ 0 \\ -1 \\ 1 \\ 1 \\ 0 \\ -2 \\ 0 \\ 2 \\ 0 \end{bmatrix} \tag{8.138}$$

and

$$\begin{pmatrix} dx \\ dy \\ da_1 \\ da_2 \\ da_3 \\ da_4 \\ d\lambda \\ d\mu_1 \\ d\mu_2 \\ dz \end{pmatrix} = \rho_1 \begin{bmatrix} -1 \\ 0 \\ 1 \\ -1 \\ -1 \\ 0 \\ 0 \\ 2 \\ 0 \\ 0 \end{bmatrix} + \rho_2 \begin{bmatrix} -1 \\ 2 \\ 0 \\ 1 \\ -1 \\ 2 \\ 0 \\ 0 \\ 0 \\ 0 \end{bmatrix} + \rho_3 \begin{bmatrix} -1 \\ 0 \\ 0 \\ -1 \\ -1 \\ 0 \\ 0 \\ 0 \\ 0 \\ 2 \end{bmatrix} + \pi \begin{bmatrix} 1 \\ -2 \\ 0 \\ -1 \\ 1 \\ 0 \\ 0 \\ 0 \\ 0 \\ 0 \end{bmatrix}, \tag{8.139}$$

respectively, where $\rho_1, \rho_2, \rho_3 \in \mathbb{R}$ and $\pi \in \mathbb{R}^+$. The vector associated with π for the first hypothesis corresponds to the feasible changes in the Lagrange multipliers owing to the linearly dependence on the constraint gradients but only positive increments are allowed because as $\mu_2 = 0$, a negative increment would imply a negative multiplier which does not satisfy (8.135).

8.5 Particular Cases   335

Figure 8.5 shows a graphical interpretation of the first equation in constraint (8.135) particularized for this case. Note that constraints related to μ_2 and μ_3 are not necessary for obtaining the optimal solution. This means that they could be removed and the same optimal solution would still remain.

In order to study the existence of partial derivatives with respect to a_1, it is possible to consider the directions in which the desired directional derivatives are looked for, $d\boldsymbol{a} = (1\ 0\ 0\ 0)^T$ and $d\boldsymbol{a} = (-1\ 0\ 0\ 0)^T$, respectively, and solve the two possible combinations of (8.84)–(8.85) leading to

$$\begin{pmatrix} \frac{\partial x}{\partial a_1^+} \\ \frac{\partial y}{\partial a_1^+} \\ \frac{\partial \lambda}{\partial a_1^+} \\ \frac{\partial \mu_1}{\partial a_1^+} \\ \frac{\partial \mu_2}{\partial a_1^+} \\ \frac{\partial z}{\partial a_1^+} \\ \frac{\partial}{\partial a_1^+} \end{pmatrix} = \begin{bmatrix} 0 \\ 0 \\ 0 \\ 2 \\ 0 \\ -2 \end{bmatrix} + \pi \begin{bmatrix} 0 \\ 0 \\ -1 \\ 1 \\ 1 \\ 0 \end{bmatrix}, \quad \begin{pmatrix} \frac{\partial x}{\partial a_1^-} \\ \frac{\partial y}{\partial a_1^-} \\ \frac{\partial \lambda}{\partial a_1^-} \\ \frac{\partial \mu_1}{\partial a_1^-} \\ \frac{\partial \mu_2}{\partial a_1^-} \\ \frac{\partial z}{\partial a_1^-} \\ \frac{\partial}{\partial a_1^-} \end{pmatrix} = \begin{bmatrix} 0 \\ 0 \\ 0 \\ -2 \\ 0 \\ 2 \end{bmatrix} + \pi \begin{bmatrix} 0 \\ 0 \\ -1 \\ 1 \\ 1 \\ 0 \end{bmatrix}, \quad \pi \in \mathbb{R}^+, \tag{8.140}$$

and

$$\begin{pmatrix} \frac{\partial x}{\partial a_1^+} \\ \frac{\partial y}{\partial a_1^+} \\ \frac{\partial \lambda}{\partial a_1^+} \\ \frac{\partial \mu_1}{\partial a_1^+} \\ \frac{\partial \mu_2}{\partial a_1^+} \\ \frac{\partial z}{\partial a_1^+} \\ \frac{\partial}{\partial a_1^+} \end{pmatrix} = \begin{bmatrix} 0 \\ 0 \\ 0 \\ 2 \\ 0 \\ -2 \end{bmatrix}, \quad \begin{pmatrix} \frac{\partial x}{\partial a_1^-} \\ \frac{\partial y}{\partial a_1^-} \\ \frac{\partial \lambda}{\partial a_1^-} \\ \frac{\partial \mu_1}{\partial a_1^-} \\ \frac{\partial \mu_2}{\partial a_1^-} \\ \frac{\partial z}{\partial a_1^-} \\ \frac{\partial}{\partial a_1^-} \end{pmatrix} = \begin{bmatrix} 0 \\ 0 \\ 0 \\ -2 \\ 0 \\ 2 \end{bmatrix}, \tag{8.141}$$

respectively, which imply that the following partial derivatives exist: $\frac{\partial x}{\partial a_1} = \frac{\partial y}{\partial a_1} = 0$ and $\frac{\partial z}{\partial a_1} = -2$, because they are unique and have the same absolute value and different sign. However, the partial derivatives $\frac{\partial \lambda}{\partial a_1}$, $\frac{\partial \mu_1}{\partial a_1}$, and $\frac{\partial \mu_2}{\partial a_1}$ do not exist, because the corresponding $d\lambda$, $d\mu_1$, $d\mu_2$ are not unique [they depend on the arbitrary real number π in contraint (8.140)].

Note that the feasible region (Figs. 8.6 and 8.7) degenerates to a single point $(x, y) = (1, 1)$ [see Fig. 8.7a], so that the optimal solution of problem (8.133)–(8.134) corresponds to this point independently of the objective

336   8 Local Sensitivity Analysis

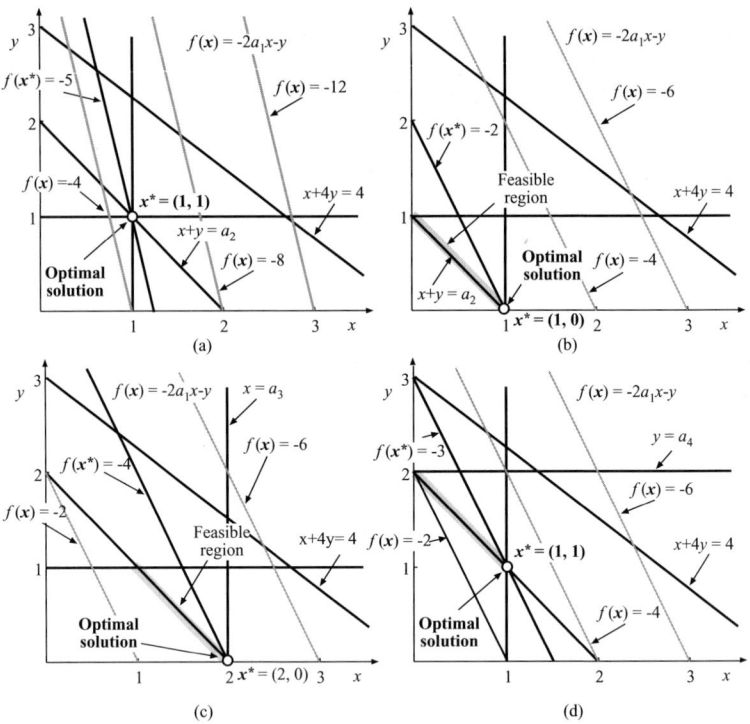

Fig. 8.6. First illustration of Example 8.11

function used. Thus, as parameter a_1 affects only the objective function, the derivatives of x_1 and x_2 with respect to this parameter are null (see Fig. 8.6a).

In order to study the existence of partial derivatives with respect to a_2, we solve (8.84)–(8.85) using the directions $d\boldsymbol{a} = \begin{pmatrix} 0 & 1 & 0 & 0 \end{pmatrix}^T$ and $d\boldsymbol{a} = \begin{pmatrix} 0 & -1 & 0 & 0 \end{pmatrix}^T$:

$$\begin{pmatrix} \dfrac{\partial x}{\partial a_2^+} \\ \dfrac{\partial y}{\partial a_2^+} \\ \dfrac{\partial \lambda}{\partial a_2^+} \\ \dfrac{\partial \mu_1}{\partial a_2^+} \\ \dfrac{\partial \mu_2}{\partial a_2^+} \\ \dfrac{dz}{\partial a_2^+} \end{pmatrix} = \pi \begin{bmatrix} 0 \\ 0 \\ -1 \\ 1 \\ 1 \\ 0 \end{bmatrix}, \begin{pmatrix} \dfrac{\partial x}{\partial a_2^-} \\ \dfrac{\partial y}{\partial a_2^-} \\ \dfrac{\partial \lambda}{\partial a_2^-} \\ \dfrac{\partial \mu_1}{\partial a_2^-} \\ \dfrac{\partial \mu_2}{\partial a_2^-} \\ \dfrac{dz}{\partial a_2^-} \end{pmatrix} = \pi \begin{bmatrix} 0 \\ 0 \\ -1 \\ 1 \\ 1 \\ 0 \end{bmatrix}, \begin{pmatrix} \dfrac{\partial x}{\partial a_2^+} \\ \dfrac{\partial y}{\partial a_2^+} \\ \dfrac{\partial \lambda}{\partial a_2^+} \\ \dfrac{\partial \mu_1}{\partial a_2^+} \\ \dfrac{\partial \mu_2}{\partial a_2^+} \\ \dfrac{dz}{\partial a_2^+} \end{pmatrix} = [\emptyset], \begin{pmatrix} \dfrac{\partial x}{\partial a_2^-} \\ \dfrac{\partial y}{\partial a_2^-} \\ \dfrac{\partial \lambda}{\partial a_2^-} \\ \dfrac{\partial \mu_1}{\partial a_2^-} \\ \dfrac{\partial \mu_2}{\partial a_2^-} \\ \dfrac{dz}{\partial a_2^-} \end{pmatrix} = \begin{bmatrix} 0 \\ -1 \\ 0 \\ 0 \\ 0 \\ 1 \end{bmatrix},$$

(8.142)

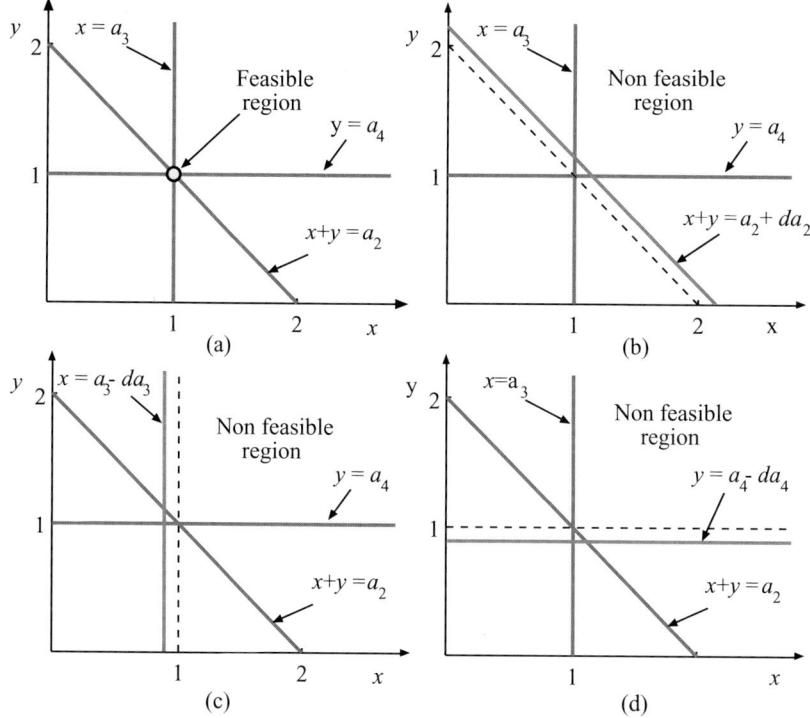

Fig. 8.7. Second illustration of Example 8.11

respectively, which imply that no partial derivatives exist. Note that the vectors associated with the π-values correspond to the possible changes in the multipliers that make the solution to remain valid without changing the optimal values of x, y, and z. This is possible because the dual problem has an infinite number of solutions. Note, as well, that no right-hand side derivatives exist because increasing a_2 makes the problem infeasible but the left-hand side derivatives exist as shown in Fig. 8.7b: $\frac{\partial x}{\partial a_2^-} = 0$, $\frac{\partial y}{\partial a_2^-} = -1$ and $\frac{\partial z}{\partial a_2^-} = 1$. In Fig. 8.6b the new optimal point is shown if a_2 is decreased in one unit, where $x = 1$ remains the same, $y = 0$ decreases in one unit, and $z = -2$ increases in one unit (note that the unit increase is valid because we are dealing with an LP problem, but in NLPP only small perturbations are allowed). Note that the second inequality constraint $(y \leq a_4)$ becomes inactive.

If we consider directions $d\mathbf{a} = (0 \ 0 \ 1 \ 0)^T$ and $d\mathbf{a} = (0 \ 0 \ -1 \ 0)^T$, respectively, and solve the two possible combinations of (8.84)–(8.85), the right and left-hand side derivatives with respect to a_3 are obtained:

$$\begin{pmatrix} \frac{\partial x}{\partial a_3^+} \\ \frac{\partial y}{\partial a_3^+} \\ \frac{\partial \lambda}{\partial a_3^+} \\ \frac{\partial \mu_1}{\partial a_3^+} \\ \frac{\partial \mu_2}{\partial a_3^+} \\ \frac{\partial z}{\partial a_3^+} \\ \frac{\partial}{\partial a_3^+} \end{pmatrix} = \pi \begin{bmatrix} 0 \\ 0 \\ -1 \\ 1 \\ 1 \\ 0 \end{bmatrix}, \begin{pmatrix} \frac{\partial x}{\partial a_3^-} \\ \frac{\partial y}{\partial a_3^-} \\ \frac{\partial \lambda}{\partial a_3^-} \\ \frac{\partial \mu_1}{\partial a_3^-} \\ \frac{\partial \mu_2}{\partial a_3^-} \\ \frac{\partial z}{\partial a_3^-} \\ \frac{\partial}{\partial a_3^-} \end{pmatrix} = \pi \begin{bmatrix} 0 \\ 0 \\ -1 \\ 1 \\ 1 \\ 0 \end{bmatrix}, \begin{pmatrix} \frac{\partial x}{\partial a_3^+} \\ \frac{\partial y}{\partial a_3^+} \\ \frac{\partial \lambda}{\partial a_3^+} \\ \frac{\partial \mu_1}{\partial a_3^+} \\ \frac{\partial \mu_2}{\partial a_3^+} \\ \frac{\partial z}{\partial a_3^+} \\ \frac{\partial}{\partial a_3^+} \end{pmatrix} = \begin{bmatrix} 1 \\ -1 \\ 0 \\ 0 \\ 0 \\ -1 \end{bmatrix}, \begin{pmatrix} \frac{\partial x}{\partial a_3^-} \\ \frac{\partial y}{\partial a_3^-} \\ \frac{\partial \lambda}{\partial a_3^-} \\ \frac{\partial \mu_1}{\partial a_3^-} \\ \frac{\partial \mu_2}{\partial a_3^-} \\ \frac{\partial z}{\partial a_3^-} \\ \frac{\partial}{\partial a_3^-} \end{pmatrix} = [\emptyset],$$

(8.143)

respectively, as in the previous example no partial derivatives exist. Note, as well, that no left-hand side derivatives exist because decreasing a_3 makes the problem infeasible but right-hand side derivatives exist as it is shown in Fig. 8.7c: $\frac{\partial x}{\partial a_3^+} = 1$, $\frac{\partial y}{\partial a_3^+} = -1$, and $\frac{\partial z}{\partial a_3^+} = -1$. Figure 8.6c shows the new optimal point if a_3 is increased in one unit, where $x = 2$ and $y = 0$ increases and decreases one unit, respectively and $z = -4$ decreases in one unit (note that the unit increase is valid because we are dealing with an LP problem, but for NLPP only small perturbations are allowed). Note that the second inequality constraint ($y \leq a_4$) becomes inactive.

The same process is done for the last parameter $d\boldsymbol{a} = \begin{pmatrix} 0 & 0 & 0 & 1 \end{pmatrix}^T$ and $d\boldsymbol{a} = \begin{pmatrix} 0 & 0 & 0 & -1 \end{pmatrix}^T$, respectively, solving the two possible combinations of (8.84)–(8.85), the right- and left-hand side derivatives with respect to a_4 are obtained:

$$\begin{pmatrix} \frac{\partial x}{\partial a_4^+} \\ \frac{\partial y}{\partial a_4^+} \\ \frac{\partial \lambda}{\partial a_4^+} \\ \frac{\partial \mu_1}{\partial a_4^+} \\ \frac{\partial \mu_2}{\partial a_4^+} \\ \frac{\partial z}{\partial a_4^+} \\ \frac{\partial}{\partial a_4^+} \end{pmatrix} = \pi \begin{bmatrix} 0 \\ 0 \\ -1 \\ 1 \\ 1 \\ 0 \end{bmatrix}, \begin{pmatrix} \frac{\partial x}{\partial a_4^-} \\ \frac{\partial y}{\partial a_4^-} \\ \frac{\partial \lambda}{\partial a_4^-} \\ \frac{\partial \mu_1}{\partial a_4^-} \\ \frac{\partial \mu_2}{\partial a_4^-} \\ \frac{\partial z}{\partial a_4^-} \\ \frac{\partial}{\partial a_4^-} \end{pmatrix} = \pi \begin{bmatrix} 0 \\ 0 \\ -1 \\ 1 \\ 1 \\ 0 \end{bmatrix}, \begin{pmatrix} \frac{\partial x}{\partial a_4^+} \\ \frac{\partial y}{\partial a_4^+} \\ \frac{\partial \lambda}{\partial a_4^+} \\ \frac{\partial \mu_1}{\partial a_4^+} \\ \frac{\partial \mu_2}{\partial a_4^+} \\ \frac{\partial z}{\partial a_4^+} \\ \frac{\partial}{\partial a_4^+} \end{pmatrix} = \begin{bmatrix} 0 \\ 0 \\ 0 \\ 0 \\ 0 \\ 0 \end{bmatrix}, \begin{pmatrix} \frac{\partial x}{\partial a_4^-} \\ \frac{\partial y}{\partial a_4^-} \\ \frac{\partial \lambda}{\partial a_4^-} \\ \frac{\partial \mu_1}{\partial a_4^-} \\ \frac{\partial \mu_2}{\partial a_4^-} \\ \frac{\partial z}{\partial a_4^-} \\ \frac{\partial}{d a_4^-} \end{pmatrix} = [\emptyset],$$

(8.144)

respectively, which imply that no partial derivatives exist. Note that the vectors associated with the π-values, as in the previous cases, correspond to the possible changes in the multipliers that make the solution remain valid without changing the optimal values of $x, y,$ and z. This is possible because the dual problem has infinite solutions. Note, as well, that no left-hand side derivatives exist because decreasing a_4 makes the problem infeasible but the right-hand side derivatives exist as shown in Fig. 8.7d: $\frac{\partial x}{\partial a_4^-} = 0$, $\frac{\partial y}{\partial a_4^-} = -1$ and $\frac{\partial z}{\partial a_4^-} = 1$. Figure 8.6d shows that these derivatives are null because the optimal solution of the primal problem remains the same. Note that the second inequality constraint ($y \le a_4$) becomes inactive. □

8.6 Sensitivities of Active Constraints

Sometimes one can be interested in the sensitivities of a constraint with respect to data. In this case, we have the following theorem (see Castillo et al. [58]):

Theorem 8.3 (The sensitivities of active constraints). *If the problem*

$$\underset{x}{minimize} \quad z_P = f(\boldsymbol{x}) \quad (8.145)$$

subject to

$$\boldsymbol{h}(\boldsymbol{x}) = \boldsymbol{0} : \boldsymbol{\lambda} \quad (8.146)$$
$$\boldsymbol{g}(\boldsymbol{x}) \le \boldsymbol{0} : \boldsymbol{\mu} , \quad (8.147)$$

where $\boldsymbol{\lambda}$ and $\boldsymbol{\mu}$ are the corresponding dual variables, has an optimal regular solution $\boldsymbol{x}^*$ holding the sufficient conditions (4.4) for minimum, such that $z_P^* = f(\boldsymbol{x}^*)$ and that no degenerate (see Definition 4.3) inequality constraints exist, then

(a) *If $\lambda_s > 0$, the problem (8.145)–(8.147) is equivalent to the problem*

$$\underset{x}{minimize} \quad z_P = h_s(\boldsymbol{x}) \quad (8.148)$$

subject to

$$f(\boldsymbol{x}) = z_P^* : 1/\lambda_s \quad (8.149)$$
$$h_k(\boldsymbol{x}) = 0 : \lambda_k/\lambda_s, \quad k \ne s \quad (8.150)$$
$$g_j(\boldsymbol{x}) \le 0 : \mu_j/\lambda_s, \quad j = 1, 2, \ldots, q. \quad (8.151)$$

(b) *If $\lambda_s < 0$, the problem (8.145)–(8.147) is equivalent to the problem*

$$\underset{x}{maximize} \quad z_P = h_s(\boldsymbol{x}) \quad (8.152)$$

subject to

$$f(\boldsymbol{x}) = z_P^* : 1/\lambda_s \quad (8.153)$$
$$h_k(\boldsymbol{x}) = 0 : \lambda_k/\lambda_s, \quad k \ne s \quad (8.154)$$
$$g_j(\boldsymbol{x}) \le 0 j : \mu_j/\lambda_s \quad j = 1, 2, \ldots, q. \quad (8.155)$$

340   8 Local Sensitivity Analysis

(c) If $\mu_s > 0$, the constraint $g_s(x) \le 0$ becomes active, i.e., $g_s(x) = 0$ and then it can be treated as an equality with positive $\lambda_s = \mu_s$.

The multipliers or dual variables in (8.149), (8.150), and (8.151) or (8.153), (8.154), and (8.155) give the corresponding sensitivities of $h_s(x)$ to changes in z_P^*, h_k, and g_j, respectively. □

Proof. We only give the proof for the case $\lambda_s > 0$, because in the other cases it is similar. The KKT necessary optimality conditions for the NLPP (8.145)–(8.147) are

$$\nabla_x f(x^*) + \lambda^{*T} \nabla_x h(x^*) + \mu^{*T} \nabla_x g(x^*) = 0$$
$$h(x^*) = 0$$
$$g(x^*) \le 0$$
$$\mu^{*T} g(x^*) = 0$$
$$\mu^* \ge 0.$$

If no redundant constraints exist and the NLPP has an optimal solution x^* such that $z_P^* = f(x^*)$, then, if $\lambda_s > 0$, we can write

$$\nabla_x h_s(x^*) + \frac{1}{\lambda_s^*}\nabla_x f(x^*) + \sum_{k \ne s}^{\ell} \frac{\lambda_k^*}{\lambda_s^*}\nabla_x h_k(x^*) + \sum_{j=1}^{m} \frac{\mu_j^*}{\lambda_s^*}\nabla_x g_j(x^*) = 0$$

$$f(x^*) = z_P^*$$
$$h_k(x^*) = 0; \quad k \ne s$$
$$g_j(x^*) \le 0; \quad \forall j$$
$$\frac{\mu_j^*}{\lambda_s^*} g_j(x^*) = 0; \quad \forall j$$
$$\frac{\mu_j^*}{\lambda_s^*} \ge 0; \quad \forall j,$$

which are the KKTC for the problem (8.148)–(8.151) and lead to the indicated multipliers.

Since the problem (8.145)–(8.147) satisfies the second-order necessary conditions (Definition 4.5) and $f, h, g \in C^2$, then because of the nondegenerate inequalities, it also satisfies the Second-Order sufficient conditions (Definition 4.4). □

The results in this section, apart from giving the sensitivities of constraints, allow us to solve the initial problem in an alternative way. This has advantages in some cases in which one looks for optimal solutions of the initial problem (8.145)–(8.147) with a given optimal value and wants to fix the right-hand side of a constraint for that to be possible. In this case, the alternative statements (8.148)–(8.151) or (8.152)–(8.155) allows us to solve the problem in one step, while the initial problem requires iterations.

Illustrative Example 8.12 (Example 8.5 revisited). Consider the optimization problem in Example 8.5

$$\text{minimize} \quad z = x^2 + y^2 \quad (8.156)$$
$$x, y$$

subject to

$$\begin{aligned} x + y &= 5 \\ (x-4)^2 + (y-2)^2 &\leq 1, \end{aligned} \quad (8.157)$$

whose optimal solution is $z^* = 13$ at the point $(x^*, y^*) = (3, 2)$ and the corresponding values of the multipliers are $\lambda = -4$ and $\mu = 1$.

According to Theorem 8.3 and since $\lambda < 0$, the problem (8.156)–(8.157) is equivalent to the problem

$$\text{maximize} \quad z = x + y \quad (8.158)$$
$$x, y$$

subject to

$$\begin{aligned} x^2 + y^2 &= 13 \; : \; 1/\lambda = -1/4 \\ (x-4)^2 + (y-2)^2 &\leq 1 \; : \; \mu/\lambda = -1/4 \, , \end{aligned} \quad (8.159)$$

where the corresponding multipliers are indicated.

Similarly, according to Theorem 8.3 and since $\mu > 0$, the problem (8.156)–(8.157) is equivalent to the problem

$$\text{minimize} \quad z = (x-4)^2 + (y-2)^2 - 1 \quad (8.160)$$
$$x, y$$

subject to

$$\begin{aligned} x + y &= 5 \; : \; \lambda/\mu = -4 \\ x^2 + y^2 &= 13 \; : \; 1/\mu = 1 \, , \end{aligned} \quad (8.161)$$

where the corresponding multipliers are also indicated. □

8.7 Exercises

Exercise 8.1. Consider the following optimization problem:

$$\text{minimize} \quad -(x_1 + 1)^2 - (x_2 + 1)^2 \quad (8.162)$$
$$x_1, x_2$$

subject to

$$x_1^2 + x_2^2 \leq 1. \quad (8.163)$$

1. Obtain the Lagrangian function.
2. Solve the problem graphically.

3. Solve the problem using the KKT conditions and show that the optimal value of the objective function, attained at the point $(1/\sqrt{2}, 1/\sqrt{2})$, is $-(1+\sqrt{2})^2$.
4. Show that the dual function is $\phi(\mu) = \dfrac{\mu(1+\mu)}{1-\mu}$ and discuss its range.
5. State the dual problem.
6. Solve the dual problem and show that the optimal value of the objective function, attained at the point $\mu = 1 + \sqrt{2}$, is $-(1+\sqrt{2})^2$.
7. Check that there is no duality gap.
8. Replace 1 in the right-hand side of the constraint by parameter a and discuss the existence of derivatives with respect to a. Finally, obtain the derivatives of the optimal objective function value and the coordinates of the solution point with respect to a.

Exercise 8.2. A circle of minimum radius must be found such that it contains a given set of points. To this end:

1. Simulate 30 points whose random coordinates are $N(0, 1)$.
2. State the problem as one optimization problem. Note that the variables are the coordinates of its center (x_0, y_0) and its radius r.
3. Solve the problem using an appropriate solver.
4. Use the technique of converting the data to artificial variables and incorporating constraints stating that these variables are equal to their corresponding data values, to obtain the sensitivities of the resulting minimum radius with respect to the point coordinates.
5. Identify the three points that define the circle based on the above sensitivities.
6. Generalize the problem to the case of n dimensions.

Exercise 8.3. Consider the following optimization problem:

$$\underset{x_1, x_2}{\text{minimize}} \quad (x_1+1)^2 + (x_2+1)^2 \tag{8.164}$$

subject to

$$x_1^2 + x_2^2 \leq 1. \tag{8.165}$$

1. Obtain the Lagrangian function.
2. Solve the problem graphically.
3. Solve the problem using the KKT conditions and show that the optimal value of the objective function, attained at the point $(-1/\sqrt{2}, -1/\sqrt{2})$, is $(\sqrt{2}-1)^2$.
4. Show that the dual function is $\phi(\mu) = \dfrac{\mu(1-\mu)}{1+\mu}$ and discuss its range.
5. State the dual problem.
6. Solve the dual problem and show that the optimal value of the objective function, attained at the point $\mu = \sqrt{2} - 1$, is $(\sqrt{2}-1)^2$.
7. Check that there is no duality gap.

8. Replace 1 in the right-hand side of the constraint by parameter a and discuss the existence of derivatives with respect to a. Finally, obtain the derivatives of the optimal objective function value and the coordinates of the solution point with respect to a.

Exercise 8.4. Consider the water supply problem in Sect. 1.4.3 with the data in Fig. 1.12. Modify the optimization problem for the sensitivities of the objective function with respect to the capacities r_i and flows q_j to be directly obtained as the values of dual variables.

Exercise 8.5. Consider the problem

$$\underset{x,y}{\text{minimize}} \quad z = (x-x_0)^2 + (y-y_0)^2 \qquad (8.166)$$

subject to

$$(x-a)^2 + (y-b)^2 = r^2, \qquad (8.167)$$

where x_0, y_0, a, b and r are given constants.

1. Give a geometrical interpretation to this problem and plot a graph to explain it.
2. Solve the problem analytically.
3. Obtain the sensitivity of z with respect to x_0, y_0, a, b, and r based on the analytical solution.
4. Modify problem (8.166)–(8.167) to obtain the above sensitivities using the techniques developed in this chapter.

Exercise 8.6. Consider the following parametric optimization problem:

$$\underset{x_1,x_2}{\text{minimize}} \quad z = -x_1^2 - x_2^2 \qquad (8.168)$$

subject to

$$\begin{array}{rcl} x_1 + x_2 & \leq & a_1 \\ x_1 & \geq & a_2 \\ x_2 & \geq & a_3, \end{array} \qquad (8.169)$$

which for $a_1 = 1, a_2 = 0, a_3 = 0$ leads to the following optimal solution:

$$x_1^* = 0, \quad x_2^* = 1, \quad \mu_1^* = 2, \quad \mu_2^* = 2, \quad \mu_3^* = 0, \quad z^* = -1 .$$

Calculate all sensitivities at once using the technique developed in Sect. 8.4.

Exercise 8.7. Consider the problem

$$\underset{x,y}{\text{maximize}} \quad z = xy \qquad (8.170)$$

subject to

$$xy \le r^2 \qquad (8.171)$$
$$y = bx \qquad (8.172)$$
$$x \le c \qquad (8.173)$$
$$y \le d \qquad (8.174)$$
$$x \ge 0 \qquad (8.175)$$
$$y \ge 0, \qquad (8.176)$$

where $a, b, c,$ and d are given constants.

1. Give a geometrical interpretation to this problem and plot a graph to explain it.
2. Solve the problem analytically.
3. Obtain the sensitivity of z with respect to $a, b, c,$ and d based on the analytical solution.
4. Modify problem (8.171)–(8.176) to obtain the above sensitivities using the techniques developed in this chapter.
5. Discuss the existence of partial derivatives of z with respect to $a, b, c,$ and d, for the case $a = 4, b = 1, c = 2, d = 2$.

Exercise 8.8. Use the technique developed in Sect. 8.4 to calculate all sensitivities at once of the problem in Example 8.5, and check that the sensitivities you obtain for the objective function coincide with those given in Example 8.5.

Exercise 8.9. Consider the operation of a multiarea electricity network example in Chap. 1 and the numerical data in Tables 1.14, 1.15, and 1.16. Modify the problem to obtain the sensitivities of the objective function with respect to all data values.

Exercise 8.10. Write a GAMS program to check Example 8.12. Proceed as follows:

1. Write the code for the initial program (8.156)–(8.157).
2. Write the code for the modified program (8.158)–(8.159).
3. Write the code for the modified program (8.160)–(8.161).
4. Solve the above three problems and print the results in a file.
5. Check that the results are those expected by the theory.

Exercise 8.11. Consider the uniform random variable family with densities of the form
$$f(y; a, b) = \frac{1}{b-a}, \quad a \le y \le b,$$
with mean $\dfrac{a+b}{2}$ and variance $\dfrac{(b-a)^2}{12}$.

To estimate the parameters a and b based on a random sample, use the constrained method of moments, which consists of solving the optimization problem:

$$\underset{a,b}{\text{minimize}} \quad z = \left(\frac{a+b}{2} - \bar{y}\right)^2 + \left(\frac{(b-a)^2}{12} - \sigma^2\right)^2 \tag{8.177}$$

subject to

$$a - y_{\min} \leq 0 : \mu_1 \tag{8.178}$$
$$y_{\max} - b \leq 0 : \mu_2, \tag{8.179}$$

where $\bar{y}$ and σ^2 are the sample mean and variance, respectively, μ_1 and μ_2 are the corresponding dual variables, and $y_{\min}$ and $y_{\max}$ are the minimum and maximum values of the sample, respectively.

1. Simulate five points with random coordinates $U(0,1)$.
2. Calculate the sensitivities of the objective function z and the Lagrange multipliers μ_1 and μ_2 with respect to the sample values.

Exercise 8.12. Consider the following simple NLPP

$$\underset{x_1, x_2}{\text{minimize}} \quad f(\boldsymbol{x}) = a_1 x_1^2 + x_2^2 \tag{8.180}$$

subject to

$$h(\boldsymbol{x}) = x_1 x_2^2 - a_2 = 0 : \lambda \tag{8.181}$$
$$g(\boldsymbol{x}) = -x_1 + a_3 \leq 0 : \mu, \tag{8.182}$$

where λ and μ are the corresponding multipliers.

Using the techniques developed in Sect. 8.4 discuss and study the sensitivities with respect to the parameters for the particular case $a_1 = a_3 = 1$ and $a_2 = 2$.

Exercise 8.13. Consider the diet problem in Murty [62]. Let x_1, x_2, and x_3 be the amounts of greens, potatoes, and corn (foods) included in the diet, respectively. The amounts vitamins A, C, and D, respectively, in each nutrient and the minimum daily vitamin requirements are given in Table 8.3.

Then, the well-known diet problem becomes

$$\underset{x_1, x_2, x_3}{\text{minimize}} \quad z = 50x_1 + 100x_2 + 51x_3 \tag{8.183}$$

Table 8.3. Amounts of vitamins A, C, and D in each food and the minimum daily vitamin requirements

	Vitamin contents			Daily requirement
Food	Greens	Potatoes	Corn	of vitamin
Vitamin A	10	1	9	5
Vitamin C	10	10	10	50
Vitamin D	10	11	11	10
Cost ($/kg)	50	100	51	

subject to

$$10x_1 + x_2 + 9x_3 \leq 5 \quad : \quad \lambda_1 \quad (8.184)$$
$$10x_1 + 10x_2 + 10x_3 \leq 50 \quad : \quad \lambda_2 \quad (8.185)$$
$$10x_1 + 11x_2 + 11x_3 \leq 10 \quad : \quad \lambda_3 \quad (8.186)$$
$$-x_1 \leq 0 \quad : \quad \lambda_4 \quad (8.187)$$
$$-x_2 \leq 0 \quad : \quad \lambda_5 \quad (8.188)$$
$$-x_3 \leq 0 \quad : \quad \lambda_6 \ . \quad (8.189)$$

1. Solve the problem (8.183)–(8.189) and detect the active constraints.
2. Using the expressions (8.110) given in the Illustrative Example 8.7 obtain the sensitivities with respect to the parameters.
3. Which is the most important parameter to be controlled or modified to decrease the cost?

Exercise 8.14. Consider the following simple nonlinear programming problem:

$$\underset{x_1, x_2}{\text{minimize}} \quad f(\boldsymbol{x}) = x_1^2 + x_2^2$$

subject to

$$h(\boldsymbol{x}) = -x_1 + a_1 \quad = \quad 0 : \lambda \quad (8.190)$$
$$g_1(\boldsymbol{x}) = -x_1 - x_2 + 2a_1 \quad \leq \quad 0 : \mu_1 \quad (8.191)$$
$$g_2(\boldsymbol{x}) = a_2 x_1 - x_2 \quad \leq \quad 0 : \mu_2 \ , \quad (8.192)$$

where λ, μ_1, and μ_2 are the multipliers corresponding to the constraints (8.190)–(8.192).

Using the techniques developed in Sect. 8.4 discuss and study the sensitivities with respect to the parameters for the particular case $a_1 = a_2 = 1$.

Part IV

Applications

9

Applications

9.1 The Wall Design

Engineering design consists of selecting the dimensions and materials for an engineering work to satisfy the desired requirements and to become a reliable construction. Engineering design is a complicated and highly iterative process that usually requires a long experience. Iterations consist of a trial-and-error selection of the design variables or parameters, together with a check of the associated safety and functionality constraints, until reasonable designs, in terms of cost and safety, are obtained.

Since safety of structures is the fundamental criterion for design, the engineer first identifies all failure modes of the work being designed and then establishes the safety constraints to be satisfied by the design variables. To ensure satisfaction of the safety constraints, two approaches are normally used: (a) the classical approach based on safety factors, and (b) the modern approach based on failure probabilities or reliability indices.

In the design and reliability analysis of an engineering work, there are some random variables $(X_1, \ldots, X_n)$ involved. They include geometric variables, material properties, loads, etc.

To illustrate these concepts, we consider again the wall problem but including three modes of failure. The wall design has already been analyzed, but in a simple version, in Chap. 1, page 45 and in Chap. 7, p. 276. The wall is depicted in Fig. 9.1, where a and b are the width and the height of the wall, γ is the unit weight of the wall, w is its weight per unit length, t is the horizontal force acting on its right-hand side, h is the corresponding offset with respect to the soil level, σ_{mean} is the mean stress at the foundation level (it is used instead of the maximum stress in order to simplify the problem), s is the soil strength, and k is the friction coefficient between soil and wall. These are the actual values of the corresponding random variables that are denoted using the corresponding capital letters.

In this section we assume that $a, b, \gamma, t, h, s,$ and k are independent normal random variables. Note that from the random character point of view there

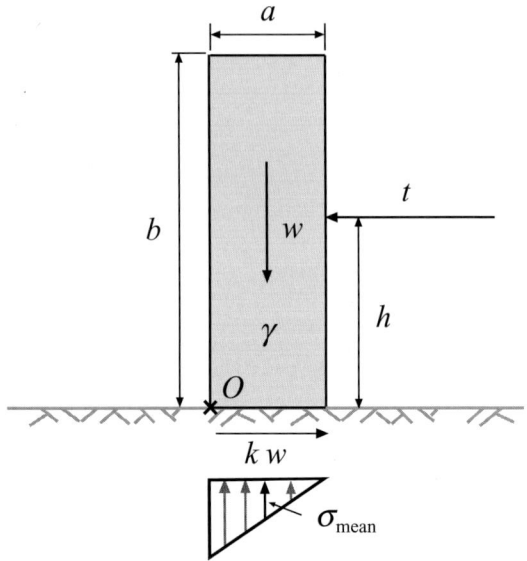

Fig. 9.1. Wall and acting forces

are two sets of variables $\{t, h, s, k\}$, which have a large importance because of its large dispersion, and $\{a, b, \gamma\}$ which are random but with small variability. In fact they could be considered as deterministic. Note also that a and b are random because in the real wall they will not take the exact values desired by the engineer due to constructions errors and imprecisions.

This set of variables involved in the problem can be partitioned into four subsets:

d: *Optimization design variables.* They are the design variables whose values are to be chosen by the optimization program to optimize the objective function (minimize the cost). Normally, they define the dimensions of the work being design, as width, thickness, height, cross sections, etc. In our wall example these variables are

$$d = \{a, b\} \ .$$

η: *Nonoptimization design variables.* They are the set of variables (deterministic or random) whose mean or characteristic values are fixed by the engineer or the code and must be given as data to the optimization program. Some examples are costs, material properties (unit weights, strength, Young modula, etc.), and other geometric dimensions of the work being designed. In our wall example

$$\eta = \{\gamma, t, h, s, k\} \ .$$

9.1 The Wall Design

κ: *Random model parameters.* They are the set of parameters used in the probabilistic design, defining the random variability, and dependence structure of the variables involved. In our wall example

$$\kappa = \{\sigma_a, \sigma_b, \sigma_\gamma, \sigma_t, \sigma_h, \sigma_s, \sigma_k\},$$

where σ refers to the standard deviation of the corresponding variable.

ψ: *Auxiliary or nonbasic variables.* They are auxiliary variables whose values can be obtained from the basic variables d and η, using some formulas. They are used to facilitate the calculations and the statement of the problem constraints. In the wall example

$$\psi = \{w, \sigma_{\text{mean}}\}.$$

The corresponding mean of d, and the mean or characteristic values of η will be denoted by $\bar{d}$ and $\tilde{\eta}$, respectively.

In the classical approach the safety factors are used as constraints and the variables are assumed to be deterministic, i.e., the mean or characteristic (extreme percentiles) values of the variables are considered.

Assume that the following three failure modes are considered (see Fig. 9.2):

1. *Overturning failure mode.* The overturning safety factor F_o is defined as the ratio of the stabilizing to the overturning moments with respect to some point (O in Fig. 9.1), as

$$F_o = g_o(\bar{d}, \tilde{\eta}) = \frac{\text{Stabilizing moment}}{\text{Overturning moment}} = \frac{w\bar{a}/2}{\tilde{h}\tilde{t}} = \frac{\bar{a}^2 \bar{b} \tilde{\gamma}}{2\tilde{h}\tilde{t}} \geq F_o^0, \quad (9.1)$$

where F_o^0 is the corresponding lower bound associated with the overturning failure, and the bars and tildes refer to the means and the characteristic values of the corresponding variables, respectively.

2. *Sliding failure mode.* The sliding safety factor F_s is the ratio of the stabilizing to the sliding forces as

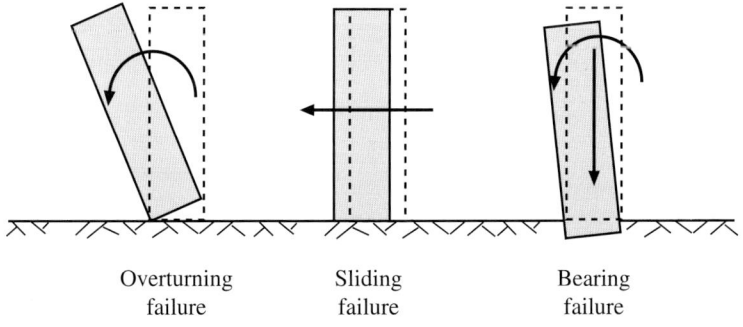

Overturning failure   Sliding failure   Bearing failure

Fig. 9.2. Illustration of the wall three modes of failure

$$F_{\rm s} = g_{\rm s}(\bar{d}, \tilde{\eta}) = \frac{\text{Stabilizing force}}{\text{Sliding force}} = \frac{\tilde{k}w}{\tilde{t}} = \frac{\bar{a}\bar{b}\tilde{k}\tilde{\gamma}}{\tilde{t}} \geq F_{\rm s}^0 , \qquad (9.2)$$

where $F_{\rm s}^0$ is the corresponding lower bound associated with the sliding failure.

3. *Bearing capacity failure mode.* The bearing capacity safety factor $F_{\rm b}$ is the ratio of the bearing capacity to the maximum stress at the bottom of the wall,

$$F_{\rm b} = g_{\rm b}(\bar{d}, \tilde{\eta}) = \frac{\text{Bearing capacity}}{\text{Maximum stress}} = \frac{\tilde{S}}{\sigma_{\rm mean}} \geq F_{\rm b}^0 , \qquad (9.3)$$

where $F_{\rm b}^0$ is the corresponding lower bound associated with a foundation failure.

The wall is safe if and only if $F_{\rm o}, F_{\rm s}, F_{\rm b} \geq 1$.

Three different design alternatives can be used.

1. **Classical design.** In a classical design the engineer minimizes the cost of building the engineering work subject to safety factor constraints (9.1)–(9.3), i.e.,

$$\underset{\bar{d}}{\text{minimize}} \quad c(\bar{d}, \tilde{\eta}) \qquad (9.4)$$

subject to

$$\begin{aligned}
g_i(\bar{d}, \tilde{\eta}) &\geq F_i^0; & \forall i \in I & \qquad (9.5) \\
h(\bar{d}, \tilde{\eta}) &= \psi & & \qquad (9.6) \\
r_j(\bar{d}, \tilde{\eta}) &\leq 0; & \forall j \in J , & \qquad (9.7)
\end{aligned}$$

where $c(\bar{d}, \tilde{\eta})$ is the objective function to be optimized (cost function), $I = \{o, s, b\}$ is the set of failure modes, $g_i(\bar{d}, \tilde{\eta})$ ($i \in I$) are the actual safety factor functions associated with all failure modes, respectively, constraints (9.6) are the equations that allow obtaining the auxiliary variables ψ from the basic variables d and η, and $r_j(\bar{d}, \tilde{\eta}) \leq 0$ ($j \in J$) are the geometric or code constraints.

2. **Modern design.** Alternatively, the modern design minimizes the cost subject to reliability constraints, i.e.,

$$\underset{\bar{d}}{\text{minimize}} \quad c(\bar{d}, \tilde{\eta}) \qquad (9.8)$$

subject to

$$\begin{aligned}
\beta_i(\bar{d}, \tilde{\eta}, \kappa) &\leq \beta_i^0; \forall i \in I & \qquad (9.9) \\
h(\bar{d}, \tilde{\eta}) &= \psi & \qquad (9.10) \\
r_j(\bar{d}, \tilde{\eta}) &\leq 0; \forall j \in J , & \qquad (9.11)
\end{aligned}$$

where β_i is the reliability index function associated with failure mode i, and β_i^0 the corresponding upper bound.

3. **Mixed design.** There exists another design, the mixed alternative, which combines safety factors and reliability indices (see Castillo et al. [15, 16, 17, 18, 19, 13]) and can be stated as

$$\underset{\bar{d}}{\text{minimize}} \quad c(\bar{d}, \tilde{\eta}) \tag{9.12}$$

subject to

$$g_i(\bar{d}, \tilde{\eta}) \geq F_i^0; \; \forall i \in I \tag{9.13}$$
$$\beta_i(\bar{d}, \tilde{\eta}, \kappa) \leq \beta_i^0; \; \forall i \in I \tag{9.14}$$
$$h(\bar{d}, \tilde{\eta}) = \psi \tag{9.15}$$
$$r_j(\bar{d}, \tilde{\eta}) \leq 0; \; \forall j \in J. \tag{9.16}$$

Unfortunately, the previous two alternatives cannot be solved directly, because evaluation of the constraints (9.9) and (9.14) involve additional optimization problems,

$$\beta_i(\bar{d}, \tilde{\eta}, \kappa) = \underset{d_i, \eta_i}{\text{minimum}} \quad \beta_i(d_i, \eta_i, \bar{d}, \tilde{\eta}, \kappa) \tag{9.17}$$

subject to

$$g_i(d_i, \eta_i) = 1 \tag{9.18}$$
$$h(d_i, \eta_i) = \psi. \tag{9.19}$$

Therefore, constraints (9.9) and (9.14) are the complicating constraints of our problem. Consequently, these two bilevel problems cannot be solved by standard techniques and decomposition methods, some of them given in Sect. 7.2, p. 276 are required.

To obtain the reliability index $\beta(d, \eta, \bar{d}, \tilde{\eta}, \kappa)$ one proceeds as follows:

1. The set of random variables $(a, b, k, t, \gamma, h, s)$ is transformed, using the Rosenblatt [63] transformation, into a set of independent standard normal random variables $(z_1, z_2, \ldots, z_n)$:

$$z_k = z_k(d, \eta, \bar{d}, \tilde{\eta}, \kappa); \; k = 1, 2, \ldots, n, \tag{9.20}$$

where n is the number of random variables involved in the problem ($n = 7$ in the wall problem).

2. The transformed failure region is replaced by the halfspace limited by the hyperplane tangent to the failure region boundary, at the point $z^* = (z_1^*, z_2^*, \ldots, z_n^*)$ in the failure region which is closest to the origin (see Fig. 9.3). This point z^* is known as the *design point* or *point of maximum*

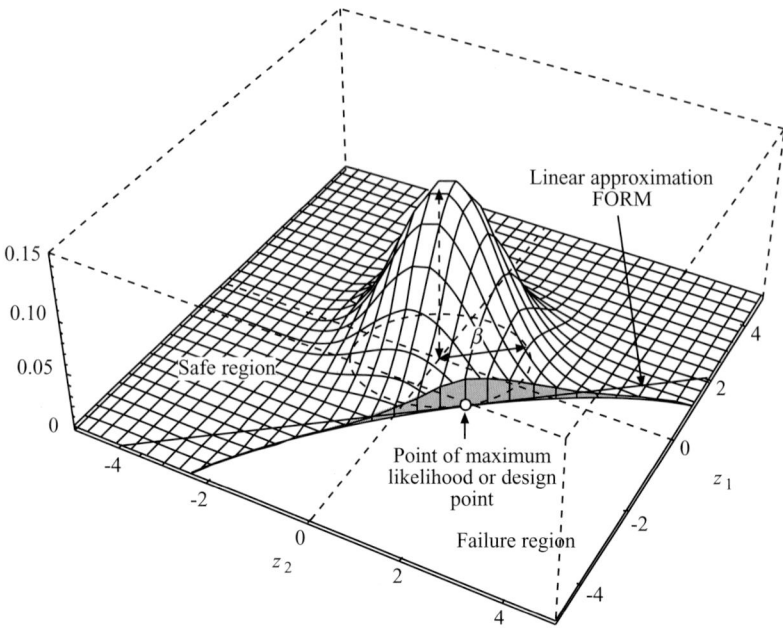

Fig. 9.3. Graphical illustration of the standard normal distribution function in the u-space, the β-value, and the *point of maximum likelihood* in the bidimensional case

likelihood. If this hyperplane has equation $\sum_{i=1}^{n} a_i z_i = c$, then its distance β to the origin is

$$\beta = \frac{c}{\sqrt{\sum_{i=1}^{n} a_i^2}}$$

and, since $\sum_{i=1}^{n} a_i z_i$ probability distribution is $N(0, \sum_{i=1}^{n} a_i^2)$, the failure probability becomes

$$\begin{aligned} P_{\mathrm{f}} &= P(\sum_{i=1}^{n} a_i z_i \leq c) = F_{N(0, \sum_{i=1}^{n} a_i^2)}(c) \\ &= \Phi\left(\frac{-c}{\sqrt{\sum_{i=1}^{n} a_i^2}}\right) = \Phi(-\beta) \\ &= \Phi\left(-\sum_{k=1}^{n} z_k^2(\boldsymbol{d}, \boldsymbol{\eta}, \bar{\boldsymbol{d}}, \tilde{\boldsymbol{\eta}}, \kappa)\right), \end{aligned} \qquad (9.21)$$

where $F_{N(0,\sigma^2)}(x)$ is the cumulative distribution function of $N(0,\sigma^2)$, and $\Phi(\cdot)$ is the cumulative distribution function of the standard normal random variable.

3. Then, since the design point is the point in the failure region that is closest to the origin, which implies a maximum failure probability, this is calculated solving the problem

$$\beta = \underset{\bar{d},\eta}{\text{minimum}} \sum_{k=1}^{n} z_k^2(\bar{d},\eta,\bar{d},\tilde{\eta},\kappa) \qquad (9.22)$$

subject to

$$z_k = z_k(\bar{d},\eta,\bar{d},\tilde{\eta},\kappa); \quad k=1,2,\ldots,n \qquad (9.23)$$
$$g(\bar{d},\eta) = 1 \qquad (9.24)$$
$$h(\bar{d},\eta) = \psi. \qquad (9.25)$$

Next, two alternative methods are given for solving the engineering design problem (9.12)–(9.16).

9.1.1 Method 1: Updating Safety Factor Bounds

The method presented in this section uses an iterative procedure that consists of repeating a sequence of three steps (see Castillo et al. [15, 16, 17]): (1) an optimal (in the sense of optimizing an objective function) classic design, based on given safety factors, is done; (2) reliability indices or bounds for all failures modes are determined; and (3) all mode safety factor bounds are adjusted. The three steps are repeated until convergence, i.e., until the safety factors lower bounds and the failure mode probability upper bounds are satisfied. More precisely, the method proceeds as follows:

Algorithm 9.1 (Updating safety factor bounds).

Input. The nonlinear programming problem (9.12)–(9.16).
Output. The solution of the problem obtained by iteratively updating the safety factor bounds.

Step 1: Solving the optimal classic design. An optimal classic design based on the actual safety factors, which are fixed initially to their corresponding lower bounds, is done. In other words, the following problem is solved

$$\underset{\bar{d}}{\text{minimize}} \quad c(\bar{d},\tilde{\eta}) \qquad (9.26)$$

subject to

$$g_i(\bar{d},\tilde{\eta}) \geq F_i^0; \quad \forall i \in I \qquad (9.27)$$
$$h(\bar{d},\tilde{\eta}) = \psi \qquad (9.28)$$
$$r_j(\bar{d},\tilde{\eta}) \leq 0; \quad \forall j \in J. \qquad (9.29)$$

The result of this process is a set of values of the design variables (their means) that satisfy the safety factor constraints (9.27) and the geometric and code ones (9.29).

Step 2: Evaluating the reliability indices β_i. The actual reliability indices β_i associated with all modes of failure are evaluated, based on the design values of Step 1, solving the problem

$$\beta_i = \underset{d_i, \eta_i}{\text{minimize}} \sum_{k=1}^{n} z_k^2(d_i, \eta_i, \bar{d}, \tilde{\eta}, \kappa) \qquad (9.30)$$

subject to

$$z_k = z_k(d_i, \eta_i, \bar{d}, \tilde{\eta}, \kappa); \quad k = 1, 2, \ldots, n \qquad (9.31)$$
$$h(d_i, \eta_i) = \psi \qquad (9.32)$$
$$g_i(d_i, \eta_i) = 1. \qquad (9.33)$$

At this step, as many optimization problems as the number of modes of failure are solved.

Step 3: Check convergence. If $||(\beta^{(\nu)} - \beta^{(\nu-1)})/\beta^{(\nu)}|| < \epsilon$, then stop. Otherwise, go to Step 4.

Step 4: Updating safety factor values. The safety factors bounds are adequately updated. To this end, the safety factors are modified using the increments

$$F_i^{(\nu+1)} = \max\left(F_i^{(\nu)} + \rho(\beta_i^0 - \beta^{(\nu)}), F_i^0\right),$$

where ρ is a small positive constant (a relaxation factor). Next, the iteration counter is updated $\nu \leftarrow \nu + 1$ and the process continues with Step 1.

Note that values of the actual reliability indices $\beta_i^{(\nu+1)}$ below the desired bound levels β_i^0, lead to an increase of the associated safety factor bound.

In addition, if, using this formula, any safety factor $F_i^{(\nu+1)}$ becomes smaller than the associated lower bound, it is kept equal to F_i^0. □

To perform a probabilistic design in the wall example, the joint probability density of all variables is required. Assume for example that all the variables involved are independent normal random variables, i.e.,

$$a \sim N(\bar{a}, \sigma_a), \quad b \sim N(\bar{b}, \sigma_b), \quad k \sim N(\tilde{k}, \sigma_k),$$
$$t \sim N(\tilde{t}, \sigma_t), \quad \gamma \sim N(\tilde{\gamma}, \sigma_\gamma), \quad h \sim N(\tilde{h}, \sigma_h), \quad s \sim N(\tilde{s}, \sigma_s),$$

where $\sim$ indicates probability distribution, $\bar{a}, \bar{b}, \tilde{k}, \tilde{t}, \tilde{\gamma}, \tilde{h}$, and $\tilde{s}$ are the mean values and $\sigma_a, \sigma_b, \sigma_k, \sigma_t, \sigma_\gamma, \sigma_h$, and σ_s are the standard deviations of a, b, k, t, γ, h, and s, respectively. The numerical values are given in Table 9.1.

9.1 The Wall Design

Table 9.1. Data for the wall example

	a (m)	b (m)	k	t (kN)	γ (kN/m^3)	h (m)	s (kN/m^2)
$\bar{x}$ o $\tilde{x}$	$\bar{a}$	$\bar{b}$	0.3	50	23	3	220
σ_x	0.01	0.01	0.05	15	0.46	0.2	16

Using the Rosenblatt [63] transformation, this set is transformed into a set of standard normal random variables $z_1, z_2, \ldots, z_7$ by

$$z_1 = \frac{a - \bar{a}}{\sigma_a}, \quad z_2 = \frac{b - \bar{b}}{\sigma_b}, \quad z_3 = \frac{k - \tilde{k}}{\sigma_k}, \quad z_4 = \frac{t - \tilde{t}}{\sigma_t},$$

$$z_5 = \frac{\gamma - \tilde{\gamma}}{\sigma_\gamma}, \quad z_6 = \frac{h - \tilde{h}}{\sigma_h}, \quad z_7 = \frac{s - \tilde{s}}{\sigma_s}.$$

(9.34)

Assume that the required safety factors and reliability bounds are

$$F_o^0 = 1.5, \quad F_s^0 = 1.6, \quad F_b^0 = 1.5, \quad \beta_o^0 = 3, \quad \beta_s^0 = 3, \quad \beta_b^0 = 3 \ .$$

For these numerical data, Algorithm 7.1 in this case consists of the following steps:

Step 0: Initialization. Let

$$\nu = 1, \quad F_o^{(1)} = 1.5, \quad F_s^{(1)} = 1.6, \quad F_b^{(1)} = 1.5 \ .$$

Step 1: Solve the classical problem.

$$\underset{\bar{a}, \bar{b}}{\text{minimize}} \quad \bar{a}\bar{b} \tag{9.35}$$

subject to

$$\frac{\bar{a}^2 \bar{b} \tilde{\gamma}}{2 \tilde{h} \tilde{t}} \geq F_o^{(\nu)} \tag{9.36}$$

$$\frac{\bar{a}\bar{b}\tilde{k}\tilde{\gamma}}{\tilde{t}} \geq F_s^{(\nu)} \tag{9.37}$$

$$\frac{\tilde{s}}{\sigma_{\text{mean}}} \geq F_b^{(\nu)} \tag{9.38}$$

$$\bar{b} = 2\bar{a} \ . \tag{9.39}$$

Step 2: Solve the subproblems. For $i = 1, 2, 3$:

$$\beta_i^{(\nu)} = \underset{a, b, k, t, \gamma, h, s}{\text{minimum}} \sum_{j=1}^{7} z_j^2 \tag{9.40}$$

subject to constraints (9.34) and

$$\frac{a^2 b\gamma}{2ht} = 1 \tag{9.41}$$

or

$$\frac{abk\gamma}{t} = 1 \tag{9.42}$$

or

$$\frac{s}{\sigma_{\text{mean}}} = 1, \tag{9.43}$$

depending on $i = 1, 2$, or 3, i.e., overturning, sliding, or foundation failures, respectively, are considered.

Step 3: Check convergence. If $||(\boldsymbol{\beta}^{(\nu)} - \boldsymbol{\beta}^{(\nu-1)})/\boldsymbol{\beta}^{(\nu)}|| < \epsilon$, then stop. Otherwise, go to Step 4.

Step 4: Update safety factors. Using

$$F_i^{(\nu+1)} = \max\left(F_i^{(\nu)} + \rho(\beta_i^0 - \beta^{(\nu)}), F_i^0\right),$$

the safety factors are updated. Next, the iteration counter is updated $\nu \leftarrow \nu+1$ and the process continues with Step 1.

Using this algorithm, the results shown in Table 9.2 are obtained. The solution is $\bar{a} = 3.053$ m and $\bar{b} = 6.107$ m. This table shows the progress and convergence of the process that requires only six iterations, and where only β_s is active and F_o, F_s, F_b, β_o, and β_b are inactive. This is illustrated in the last row of the table, where the active value has been boldfaced. This means that the reliability index for sliding is more restrictive than the corresponding safety factor.

Since the final overturning actual safety factor (4.364) and beta value (8.877) are very high compared with the design bounds (1.5 and 3, respectively), this implies that overturning is almost impossible.

Table 9.2. Illustration of the iterative process of the original algorithm

ν	Cost	a	b	Actual bounds F_o^0	F_s^0	F_b^0	Actual values F_o	F_s	F_b	Actual values β_o	β_s	β_b
1	11.594	2.408	4.816	1.500	1.600	1.500	2.140	1.600	1.986	3.456	1.491	6.763
2	20.332	3.189	6.377	1.500	1.600	1.500	4.970	2.806	1.500	10.132	3.245	4.508
3	18.918	3.076	6.152	1.500	1.600	1.500	4.461	2.611	1.555	9.084	3.042	4.832
4	18.672	3.056	6.112	1.500	1.600	1.500	4.374	2.577	1.565	8.900	3.005	4.890
5	18.645	3.054	6.107	1.500	1.600	1.500	4.365	2.573	1.566	8.879	3.000	4.897
6	18.642	3.053	6.107	1.500	1.600	1.500	4.364	2.573	1.566	8.877	**3.000**	4.897

9.1.2 Method 2: Using Cutting Planes

The iterative method presented in Sect. 9.1.1 requires a relaxation factor ρ that needs to be fixed by trial and error. An adequate selection leads to a fast convergence of the process, but an inadequate selection can lead to lack of convergence. In this section an alternative method (see Castillo et al. [51]) is given that solves this shortcoming, and in addition exhibits a better convergence. The method is as follows.

The problem (9.26)–(9.33) can be organized in the following steps:

Step 1: Solve the classical problem.

$$\underset{\bar{d}}{\text{minimize}} \quad c(\bar{d}, \tilde{\eta}) \tag{9.44}$$

subject to

$$g_i(\bar{d}, \tilde{\eta}) \geq F_i^0; \quad \forall i \in I \tag{9.45}$$
$$\beta_i^{(s)} + \boldsymbol{\lambda}_i^{(s)^T}(\bar{d} - \bar{d}^{(s)}) \geq \beta_i^0; \quad \forall i \in I; \quad s = 1, 2, \cdots, \nu - 1 \tag{9.46}$$
$$h(\bar{d}, \tilde{\eta}) = \psi \tag{9.47}$$
$$r_j(\bar{d}, \tilde{\eta}) \leq 0; \quad \forall j \in J. \tag{9.48}$$

Step 2: Solve the subproblems.

$$\beta_i^{(\nu)} = \underset{d_i, \eta_i, \bar{d}}{\text{minimum}} \sum_{k=1}^{n} z_k^2(d_i, \eta_i, \bar{d}, \tilde{\eta}, \kappa) \tag{9.49}$$

subject to

$$z_k = z_k(d_i, \eta_i, \bar{d}, \tilde{\eta}, \kappa); \quad k = 1, 2, \ldots, n \tag{9.50}$$
$$h(d_i, \eta_i) = \psi \tag{9.51}$$
$$g_i(d_i, \eta_i) = 1 \tag{9.52}$$
$$\bar{d} = \bar{d}^{(\nu)}: \boldsymbol{\lambda}_i^{(\nu)}. \tag{9.53}$$

Step 3: Check convergence. If $||(\boldsymbol{\beta}^{(\nu)} - \boldsymbol{\beta}^{(\nu-1)})/\boldsymbol{\beta}^{(\nu)}|| < \epsilon$, then stop. Otherwise, the iteration counter is updated $\nu \leftarrow \nu + 1$ and the process continues with Step 1.

It should be noted that problem (9.44)–(9.53) is a relaxation of problem (9.12)–(9.16) in the sense that functions $\beta_i^{(s)}(\cdot)$ are approximated using cutting hyperplanes. Functions $\beta_i^{(s)}(\cdot)$ become more precisely approximated as the iterative procedure progresses, which implies that problem (9.44)–(9.53) reproduces more exactly problem (9.12)–(9.16) (see Kelly [45]). Observe, additionally, that cutting hyperplanes are constructed using the dual variable vector associated with constraint (9.53) in problems (9.49)–(9.53) (the sub-problems). Constraints (9.53) in problems (9.49)–(9.53) fix to given values the optimization variable vector of problem (9.44)–(9.48) (the master problem).

For the wall example, this process consists of the following steps:

Step 0: Initialization. Let

$$\nu = 1, \quad F_o^{(1)} = 1.5, \quad F_s^{(1)} = 1.6, \quad F_b^{(1)} = 1.5 .$$

Step 1: Solve the classical problem.

$$\underset{\bar{a},\bar{b}}{\text{minimize}} \quad \bar{a}\bar{b} \tag{9.54}$$

subject to

$$\frac{\bar{a}^2 \bar{b} \tilde{\gamma}}{2 \tilde{h} \tilde{t}} \geq F_o^0 \tag{9.55}$$

$$\frac{\bar{a}\bar{b}\tilde{k}\tilde{\gamma}}{\tilde{t}} \geq F_s^0 \tag{9.56}$$

$$\frac{\tilde{s}}{\sigma_{\text{mean}}} \geq F_b^0 \tag{9.57}$$

$$\beta_i^{(s)} + \boldsymbol{\lambda}_i^{(s)T} \left(\begin{pmatrix} a \\ b \end{pmatrix} - \begin{pmatrix} a^{(s)} \\ b^{(s)} \end{pmatrix} \right) \geq \beta_i^0; \quad \forall i \in I; \quad s = 1, \cdots, \nu - 1 \tag{9.58}$$

$$\bar{b} = 2\bar{a} , \tag{9.59}$$

obtaining $\bar{a}^{(\nu)}$ and $\bar{b}^{(\nu)}$.

Step 2: Solve the subproblems. For $i = 1, 2, 3$:

$$\beta_i^{(\nu)} = \underset{a, b, k, t, \gamma, h, s}{\text{minimum}} \sum_{j=1}^{7} z_j^2 \tag{9.60}$$

subject to constraint (9.34),

$$a = \bar{a}^{(\nu)} : \lambda_{1i}^{(\nu)}$$

$$b = \bar{b}^{(\nu)} : \lambda_{2i}^{(\nu)}$$

and

$$\frac{a^2 b \gamma}{2ht} = 1 \tag{9.61}$$

or

$$\frac{abk\gamma}{t} = 1 \tag{9.62}$$

or

$$\frac{s}{\sigma_{\text{mean}}} = 1, \tag{9.63}$$

depending on $i = 1, 2$, or 3, i.e., overturning, sliding, or foundation failures, respectively, are considered.

Step 3: Check convergence. If $||(\beta^{(\nu)} - \beta^{(\nu-1)})/\beta^{(\nu)}|| < \epsilon$, then stop. Otherwise, the iteration counter is updated $\nu \leftarrow \nu + 1$, and the process continues with Step 1.

The iterative procedure leads to the results shown in Table 9.3, that provides the same information as Table 9.2 using the alternative procedure. In this case the process converges after five iterations.

Table 9.3. Illustration of the iterative process for the alternative algorithm

			Actual bounds			Actual values			Actual values		
ν	cost	a	b	F_o^0	F_s^0	F_b^0	F_o	F_s	F_b	β_o	β_s	β_b
1	11.594	2.408	4.816	1.500	1.600	1.500	2.140	1.600	1.986	3.456	1.491	6.763
2	17.463	2.955	5.910	1.500	1.600	1.500	3.956	2.410	1.619	7.989	2.807	5.180
3	18.603	3.050	6.100	1.500	1.600	1.500	4.350	2.567	1.568	8.848	2.994	4.906
4	18.642	3.053	6.107	1.500	1.600	1.500	4.363	2.573	1.567	8.877	3.000	4.897
5	18.642	3.053	6.107	1.500	1.600	1.500	4.363	2.573	1.567	8.877	**3.000**	4.897

9.2 The Bridge Crane Design

Modern industrial requirements need the application of equipment for handling large, heavy, or bulky objects. That is why there are engineers specializing in overhead material handling: bridge cranes, hoists, and monorails.

Focussing on crane bridges, an under running overhead crane with single girder is shown in Fig. 9.4. All its structural elements must be manufactured in accordance with current mandatory requirements of the National Safety and Health Act, OSHA Section 1910.179 and 1910.309 as applicable. Additionally, all American Crane & Equipment Corporation (ACECO) cranes are manufactured in accordance with the appropriate standard of the American National Standards Institute (ANSI) specifications, the National Electric Code, and the Crane Manufacturers Association of America (CMAA) specifications. Crane girders are designed using structural steel beams (reinforced as necessary) or fabricated plate box sections. Bridge girder to end truck connections are designed for loadings, stresses, and stability in accordance with current CMAA design specifications.

In this section, we apply the Engineering Design Method illustrated in Sect. 1.5.3, p. 45, for designing an overhead crane for an industrial nave (see

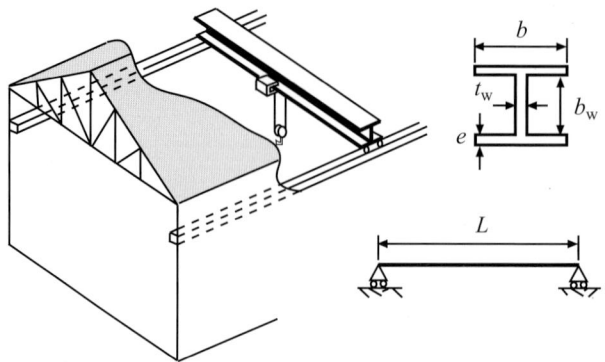

Fig. 9.4. Graphical illustration of the bridge crane design and its cross section

Fig. 9.4). In particular, we calculate the bridge girder dimensions that allow trolley traveling horizontally. It should be a box section fabricated from plate of structural steel, for the web, top plate and bottom plate, so as to provide for maximum strength at minimum dead weight. Maximum allowable vertical girder deflection should be a function of span.

Consider the girder and the cross section shown in Fig. 9.4, where L is the span or distance from centerline to centerline of runway rails, b and e are the flange width and thickness, respectively, and h_w and t_w are the web height and thickness, respectively.

The set of variables involved in this problem can be partitioned into four subsets:

d: *Optimization design variables.* They are the design variables whose values are to be chosen by the optimization program to optimize the objective function (minimize the cost). In this crane example these variables are (see Fig. 9.4):
$$d = \{b, e, t_w, h_w\} \ .$$

η: *Nonoptimization design variables.* They are the set of variables whose mean or characteristic values are fixed by the engineer or the code and must be given as data to the optimization program. In this bridge girder example,
$$\eta = \{f_y, P, L, \gamma_y, E, \nu, c_y\},$$
where f_y is the elastic limit of structural steel, P is the maximum load supported by the girder, L is the length of the span, γ_y is the steel unit weight, E is the Young modulus of the steel, ν is the Poisson modulus, and c_y is the steel cost.

κ: *Random model parameters.* They are the set of parameters defining the random variability, and dependence structure of the variables involved. In this example,

9.2 The Bridge Crane Design

$$\kappa = \{\sigma_{f_y}, \sigma_P, \sigma_L, \sigma_{\gamma_y}\} ,$$

where σ refers to the standard deviation of the corresponding variable.

ψ: *Auxiliary or nonbasic variables.* They are auxiliary variables whose values can be obtained from the basic variables d and η, using some formulas. In this example,

$$\psi = \{W, I_{xx}, I_{yy}, I_t, G, \sigma, \tau, M_{cr}, \delta\} ,$$

whose meaning is described below.

In the classical approach the safety factors are used as constraints and the variables are assumed to be deterministic, i.e., either the mean or characteristic (extreme percentiles) values of the variables are used.

Assume that the following four failure modes are considered (see Fig. 9.5):

1. *Maximum allowed deflection.* The maximum deflection safety factor F_d is defined (see Fig. 9.5a) as

$$F_d = \frac{\delta_{max}}{\delta} , \qquad (9.64)$$

where δ is the maximum deflection on the center of the girder, and δ_{max} is the maximum deflection allowed by codes.

2. *Damage limit state of the steel upper and lower flanges.* We consider the ratio of the actual strength to actual stresses (see Fig. 9.5b) as

$$F_u = \frac{f_y}{\sqrt{\sigma^2 + 3\tau^2}} , \qquad (9.65)$$

where F_u is the corresponding safety factor, and σ and τ are the normal and tangential stresses at the center of the beam, respectively.

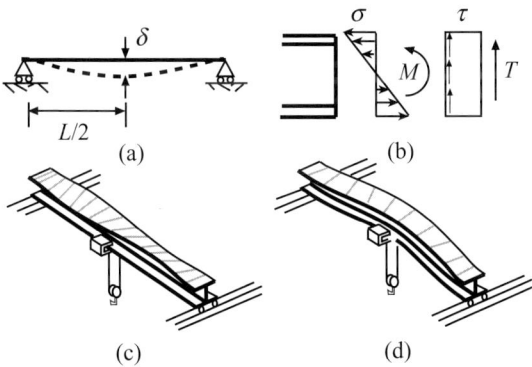

Fig. 9.5. Illustration of the bridge girder modes of failure: (**a**) maximum allowed deflection, (**b**) damage limit state of the steel upper and lower flanges and the steel web, (**c**) local buckling, and (**d**) global buckling

3. *Damage limit state of the steel web.* The bearing capacity safety factor F_w is the ratio of the shear strength capacity to actual shear stress (see Fig. 9.5b) at the center of the beam

$$F_w = \frac{f_y}{\sqrt{3}\tau} \ . \tag{9.66}$$

4. *Global buckling.* The global buckling safety factor F_b is the ratio of the critical moment against buckling (see Fig. 9.5d) of the cross section to the maximum moment applied at the center of the beam

$$F_b = \frac{M_{cr}}{M} \ . \tag{9.67}$$

The gilder bridge is safe if and only if $F_d, F_u, F_w,$ and $F_b \geq 1$.

9.2.1 Obtaining Relevant Constraints

In the following subsection we give full detail of how to obtain the constraints for the gelder bridge.

Geometrical and Mechanical Properties of the Girder

The moments of inertia I_{xx} and I_{yy} are obtained as

$$I_{xx} = \frac{1}{12}\left[b(h_w + 2e)^3 - (b - t_w)h_w^3\right] \tag{9.68}$$

$$I_{yy} = \frac{1}{12}\left(2eb^3 + h_w t_w^3\right) , \tag{9.69}$$

whereas the torsional moment of inertia is obtained using

$$I_t = \frac{1}{3}\left(2be^3 - h_w t_w^3\right) . \tag{9.70}$$

The deflection at the center of the beam is calculated using

$$\delta = \frac{PL^3}{48EI_{xx}} + \frac{5WL^4}{384EI_{xx}} , \tag{9.71}$$

where W is the girder bridge weight per unit length

$$W = \gamma_s(2eb + t_w h_w) . \tag{9.72}$$

The stresses at the center of the beam are calculated considering

$$T = P/2 \tag{9.73}$$

$$M = PL/4, \tag{9.74}$$

where T and M are the shear force and moment, respectively. Thus,

$$\sigma = \frac{M(h_\mathrm{w}+e)}{2I_\mathrm{xx}} \tag{9.75}$$

$$\tau = \frac{T}{h_\mathrm{w} t_\mathrm{w}}. \tag{9.76}$$

The critical moment for global buckling is

$$M_\mathrm{cr} = \frac{\pi}{L}\sqrt{EGI_\mathrm{yy} I_\mathrm{t}} \tag{9.77}$$

with the auxiliary parameter

$$G = \frac{E}{2(1+\nu)}.$$

Code and Other Requirements

The following constraints are established by the codes.
   The steel thickness must satisfy

$$0.008 \leq e \leq 0.15 \tag{9.78}$$
$$0.008 \leq t_\mathrm{w} \leq 0.15, \tag{9.79}$$

and the maximum deflection allowed is

$$\delta_\mathrm{max} = L/888.$$

To avoid local buckling (see Fig. 9.5c) the design satisfy the following restriction:

$$\frac{b}{2e} \leq 15\sqrt{\frac{276}{f_\mathrm{y}}}, \tag{9.80}$$

where f_y is the steel strength in MPa.
   The trolley is the unit which travels on the bottom flange of the bridge girder and carries the hoist. To support the trolley the minimum flange width is 0.30 m.

9.2.2 A Numerical Example

To perform a probabilistic design in the bridge girder example, the joint probability density of all variables is required. Assume for simplicity that all the variables involved are independent normal random variables, i.e.,

$$f_\mathrm{y} \sim N(\mu_{f_\mathrm{y}}, \sigma_{f_\mathrm{y}}), \quad P \sim N(\mu_P, \sigma_P), \quad L \sim N(\mu_L, \sigma_L), \quad \gamma_\mathrm{y} \sim N(\mu_{\gamma_\mathrm{y}}, \sigma_{\gamma_\mathrm{y}}).$$

The means are

$$\mu_{f_y} = 400 \text{ MPa}, \quad \mu_P = 600 \text{ kN}, \quad \mu_L = 6 \text{ m}, \quad \mu_{\gamma_y} = 78.5 \text{ kN/m}^3,$$

and the standard deviations

$$\sigma_{f_y} = 251 \text{ MPa}, \quad \sigma_P = 90 \text{ kN}, \quad \sigma_L = 0.05 \text{ m}, \quad \sigma_{\gamma_y} = 0.785 \text{ kN/m}^3.$$

The constant parameters are

$$E = 210{,}000 \text{ MPa}, \quad \nu = 0.3, \quad c_y = 0.24 \text{ \$/kN}.$$

Assume also that the required safety factors and reliability bounds are

$$F_d^0 = 1.15, \quad F_u^0 = 1.5, \quad F_w^0 = 1.5, \quad F_b^0 = 1.3,$$

and

$$\beta_d^0 = 1.5, \quad \beta_u^0 = 3.7, \quad \beta_w^0 = 3.7, \quad \beta_b^0 = 3.2.$$

Note that the "violation" of limit states with more serious consequences are associated with higher reliability indices.

Using the Rosenblatt [63] transformation, the above set of random variables is transformed into a set of standard normal random variables $z_1, z_2, \ldots, z_4$ by

$$z_1 = \frac{f_y - \mu_{f_y}}{\sigma_{f_y}}, \quad z_2 = \frac{P - \mu_P}{\sigma_P}, \quad z_3 = \frac{L - \mu_L}{\sigma_L}, \quad z_4 = \frac{\gamma_y - \mu_{\gamma_y}}{\sigma_{\gamma_y}}. \quad (9.81)$$

Using the proposed method, the results shown in Table 9.4 are obtained. This table shows the progress and convergence of the procedure that attains the solution after seven iterations. The first column shows the values of the design variables, and the actual safety factors and failure β-values, associated with the optimal classical design for the safety factors. Note that no lower safety factor bound is active. Note also that the β_b and β_d constraints (boldfaced in Table 9.4) are active. Then, the associated safety margins are increased until all constraints hold.

The last column of the table shows the design values of the design variables

$$b, e, t_w, h_w,$$

together with the safety factors and associated β-values.

The active values appear boldfaced in Table 9.4, from which the following conclusions can be drawn:

1. The process converges in only seven iterations.
2. The list of actual safety factors and β-reliability indices is provided.
3. No safety factor is active.
4. Due to the strict constraints imposed by the serviceability limit state constraint (maximum deflection) and global buckling, the probability bounds β_b and β_d are active.

Table 9.4. Illustration of the iterative process

ν	Units	1	2	3	4	5	6	7
$\text{Cost}^{(\nu)}$	(\$)	2,245.0	2,243.2	2,325.6	2,354.8	2,361.2	2,362.2	2,362.3
$b^{(\nu)}$	(cm)	41.81	30.00	36.14	39.25	40.43	40.67	40.68
$e^{(\nu)}$	(mm)	16.78	23.65	20.44	19.13	18.64	18.54	18.53
$t_w^{(\nu)}$	(mm)	8.00	8.00	8.00	8.00	8.00	8.00	8.00
$h_w^{(\nu)}$	(cm)	72.91	70.70	72.49	72.64	72.72	72.74	72.74
$F_u^{(\nu)}$	–	2.23	2.17	2.30	2.33	2.33	2.34	2.34
$F_t^{(\nu)}$	–	4.49	4.35	4.46	4.47	4.48	4.48	4.48
$F_b^{(\nu)}$	–	1.30	1.30	1.42	1.47	1.48	1.49	1.49
$F_d^{(\nu)}$	–	1.15	1.11	1.20	1.22	1.22	1.22	1.22
$\beta_u^{(\nu)}$	–	6.014	5.795	6.235	6.342	6.370	6.375	6.375
$\beta_t^{(\nu)}$	–	10.968	10.794	10.935	10.948	10.954	10.955	10.955
$\beta_b^{(\nu)}$	–	1.980	1.980	2.761	3.106	3.186	3.199	**3.200**
$\beta_d^{(\nu)}$	–	1.001	0.725	1.333	1.461	1.494	1.500	**1.500**
$\text{Error}^{(\nu)}$	–	0.6627	0.3815	0.4563	0.1111	0.0253	0.0041	0.0002

5. The final design (iteration 7) is more expensive than the initial design (iteration 1), because the initial design does not satisfy the β_b and β_d constraints.

The sensitivities for this bridge girder example are given in Table 9.5, which gives the cost sensitivities associated with the optimal classical design. It allows us to know how much a small change in a single data value changes the total cost of the bridge girder. This information is useful during the construction process to control the cost, and for analyzing how the changes in the safety factors required by the codes influence the total cost of engineering works. For example, a change of \$1 in the unit cost c_y of the steel leads to a cost increase of \$9843 (see the corresponding entry in Table 9.5). Similarly, an increase in the safety factor lower bound F_d does not change the cost, and an increase of one unit in the bridge span leads to an increase of the cost of \$747.

Additionally, Table 9.5 gives the sensitivities associated with the β-values. It is useful to know how much a small change in a single data value changes the corresponding β-value; for example, the means, standard deviations, etc. In this table the designer can easily analyze how the quality of the material (reduced standard deviations in f_y) or precision in the applied loads (reduced standard deviations in P) influence the safety of the beam. Note that an increase in the dispersion (standard deviations or coefficients of variation) leads to a decrease of the β indices.

Table 9.5. Sensitivities

x	$\partial \text{Cost}/\partial x$	$\partial \beta_u/\partial x$	$\partial \beta_t/\partial x$	$/\partial x \partial \beta_b$	$/\partial x \partial \beta_d$
b	—	12.851	0.000	46.864	17.717
e	—	280.902	0.000	993.835	408.587
t_w	—	352.458	698.939	122.267	108.587
h_w	—	11.974	7.687	0.448	23.088
μ_{f_y}	0.000	0.000	0.000	0.000	0.000
μ_P	1.268	−0.008	−0.005	−0.011	−0.011
μ_L	746.662	−0.975	0.000	−3.218	−2.722
μ_{γ_y}	30.125	0.000	0.000	0.000	−0.001
σ_{f_y}	0.000	0.000	0.000	0.000	0.000
σ_P	3.312	−0.036	−0.027	−0.035	−0.016
σ_L	149.935	−0.303	0.000	−1.657	−0.556
σ_{γ_y}	0.000	0.000	0.000	0.000	0.000
E	0.000	0.000	0.000	0.000	0.000
ν	290.378	0.000	0.000	−3.730	0.000
c_y	9,842.876	0.000	0.000	0.000	0.000
F_u	0.000	—	—	—	—
F_t	0.000	—	—	—	—
F_b	0.000	—	—	—	—
F_d	0.000	—	—	—	—
β_u	0.000	—	—	—	—
β_t	0.000	—	—	—	—
β_b	77.858	—	—	—	—
β_d	37.611	—	—	—	—

9.3 Network Constrained Unit Commitment

This application demonstrates the practical use of the Benders decomposition procedure, which was explained in Chap. 3.

9.3.1 Introduction

This section addresses the problem of supplying electric energy through an electric network during the hours of one day or one week. The sources of energy are power plants that can be started up and shut down as needed throughout the week. Particular emphasis is given to the modeling of the network and the possibility of starting-up and shutting-down plants.

Specifically, this application addresses the so-called multiperiod optimal power flow (OPF), properly modeling the start-up and shut-down of thermal units, and the transmission network in terms of line transmission capacity limits and line losses.

The OPF [64, 65, 66] considers a single time period (a snap shot in time) and determines the output of every on-line power plant so that a specified objective function is optimized. The transmission network is modeled in detail. Generating plants are considered either online or off-line.

The unit commitment problem [67, 32, 68, 69, 70] considers a multiperiod time horizon and determines the start-up and the shut-down schedules of thermal plants, as well as their productions. Thermal plants are modeled precisely, however, the transmission network is not considered.

This section addresses simultaneously the unit commitment problem and the OPF problem. This results in what is called a multiperiod OPF.

References [69, 71, 72, 73] address, with different levels of detail, the multiperiod OPF using Lagrangian relaxation. In this section the Benders decomposition [24, 25, 26] is used to solve this multiperiod problem. The Benders decomposition is a natural way to decompose the problem because the binary variable decisions are decoupled from continuous variable decisions. The master problem of this decomposition scheme fixes the start-up and shut-down schedules of thermal units, while the Benders subproblem solves a multiperiod OPF. The subproblem sends to the master problem marginal information on the "goodness" of the proposed start-up and shut-down schedule, which allows the master problem to suggest an improved start-up and shut-down schedule. The procedure continues until some cost tolerance is met.

The Benders decomposition shows good convergence properties for this application, being typically low the number of iterations required to attain convergence.

Readily available solution data provided by the model include power output and production cost per generator and time period.

Detailed numerical simulation results and other technical details can be found in [74].

9.3.2 Notation

The notation used is stated below.

Constants.
- A_j: start-up cost of power plant j in \$
- B_{np}: subsceptance (constructive parameter) of line np in $1/\Omega$
- C_{np}: transmission capacity limit of line np in MW
- D_{nk}: load demand at node n during period k in MW
- $E_j(t_{jk})$: nonlinear function representing the operating cost of power plant j as a function of its power output in period k in \$/MWh
- F_j: fixed cost of power plant j in \$/h
- K_{np}: loss constant (conductance) of line np in $1/\Omega$
- R_k: security (spinning reserve) requirement during period k in MW
- R_{np}: resistance of line np in Ω
- $T_j^{\max}$: maximum power output of thermal plant j in MW

$T_j^{\min}$: minimum power output of thermal plant j in MW
T: number of time periods
X_{np}: reactance of line np in Ω.

Variables.

t_{jk}: power output of plant j in period k in MW
v_{jk}: binary variable which is equal to 1 when plant j is committed in period k
$V_{jk}^{(\nu)}$: constant values of variables v_{jk} fixed by The Benders master problem at iteration ν
y_{jk}: binary variable which is equal to 1 when plant j is started up at the beginning of period k
$z^{(\nu)}$: total operating cost at iteration ν
$\alpha^{(\nu)}$: continuous variable which approximates operating costs in the Benders master problem at iteration ν in \$
δ_{nk}: angle of node n in period k in radians
$\lambda_{jk}^{(\nu)}$: dual variables provided by the subproblem which are associated to the decisions of fixing variables v_{jk} at constant values
ϕ_{nk}: locational marginal price of node n in period k in \$/MWh.

Sets.

$\mathcal{J}$: set of indices of all thermal plants
$\mathcal{K}$: set of period indices
$\mathcal{N}$: set of indices of all nodes
Λ_n: set of indices of the power plants at node n
Ω_n: set of indices of nodes connected and adjacent to node n
Υ: set of iteration indices.

9.3.3 Problem Formulation

The multiperiod OPF problem is stated below,

$$\underset{v_{jk}, y_{jk}, t_{jk}; k \in \mathcal{K}, j \in \mathcal{J}}{\text{minimize}} \quad \sum_{k \in \mathcal{K}} \sum_{j \in \mathcal{J}} [F_j v_{jk} + A_j y_{jk} + E_j(t_{jk})] \quad (9.82)$$

subject to

$$\sum_{j \in \Lambda_n} t_{jk} + \sum_{p \in \Omega_n} [B_{np}(\delta_{pk} - \delta_{nk}) - K_{np}(1 - \cos(\delta_{pk} - \delta_{nk}))]$$
$$= D_{nk} \quad : \phi_{nk}; \quad \forall n \in \mathcal{N}, \forall k \in \mathcal{K} \quad (9.83)$$

$$\sum_{j \in \mathcal{J}} T_j^{\max} v_{jk} \geq \sum_{n \in \mathcal{N}} D_{nk} + R_k; \quad \forall k \in \mathcal{K} \quad (9.84)$$

9.3 Network Constrained Unit Commitment

$$T_j^{\min} v_{jk} \leq t_{jk} \leq T_j^{\max} v_{jk}; \quad \forall j \in \mathcal{J}, \forall k \in \mathcal{K} \tag{9.85}$$

$$-C_{np} \leq B_{np}(\delta_{pk} - \delta_{nk}) \leq C_{np}; \quad \forall n \in \mathcal{N}, \forall p \in \Omega_n, \forall k \in \mathcal{K} \tag{9.86}$$

$$y_{jk} \geq v_{jk} - v_{j,k-1}; \quad \forall j \in \mathcal{J}, \forall k \in \mathcal{K} \tag{9.87}$$

$$v_{jk}, y_{jk} \in \{0,1\}; \quad \forall j \in \mathcal{J}, \forall k \in \mathcal{K} \,. \tag{9.88}$$

Variables ϕ_{nk} are the dual variables associated with constraints (9.83). These dual variables are denominated locational marginal prices. Variable ϕ_{nk} provides the marginal cost of supplying one additional megawatt hour during period k in bus (location) n.

The objective function (9.82) includes fixed cost, start-up cost, and operating cost.

A power balance constraint is written per node and period, as stated in (9.83). The first term is the summation of energies injected in the node by production plants, and the second term is the summation of the energies reaching the node through lines connected to it. Within this second term, the subtracting block corresponds to energy losses. The right-hand side term is the demand at the node.

A security (spinning reserve) constraint per period is written, (9.84). For every period, it states that online power should excess demand by a pre-specified margin.

Plant power output is limited above and below as stated in (9.85).

Transmission capacity limits of lines are stated in (9.86).

The two sets of (9.87) and (9.88) preserve the logic of running, start-up, and shut-down (e.g., a running plant cannot be started-up).

It should be noted that minimum up- and down-time constraints are not considered. However, they can be easily incorporated in the formulation as additional linear constraints.

9.3.4 Solution Approach

The multiperiod OPF problem formulated in the preceding section is a large-scale mixed-integer nonlinear optimization problem.

This problem is solved using the Benders decomposition method [24, 25]. This technique decomposes the original problem into a 0/1 mixed-integer linear master problem and a nonlinear subproblem. The subproblem is a multi-period OPF with the 0/1 variables fixed to given values. The master problem determines the running, start-up, and shut-down schedule of the plants. For the schedule fixed by the master problem, the Benders subproblem determines total operating cost, properly enforcing transmission capacity limits and taking into account line losses.

The solutions of the subproblem provide marginal information on the goodness of the scheduling decisions made at the master problem. This information enables the master problem to propose a refined running, start-up, and shut-down schedule. This iterative procedure continues until some cost tolerance is met.

Subproblem

The subproblem is a multiperiod operation problem that is continuous and in which the on-line power plants are specified. Its objective is to minimize cost while supplying the demand and using solely the plants that are on-line. This subproblem at iteration ν is formulated below.

$$\underset{t_{jk};\, j \in \mathcal{J},\, k \in \mathcal{K}}{\text{minimize}} \quad z^{(\nu)} = \sum_{k \in \mathcal{K}} \sum_{j \in \mathcal{J}} E_j(t_{jk}) \tag{9.89}$$

subject to

$$\sum_{j \in \Lambda_n} t_{jk} + \sum_{p \in \Omega_n} [B_{np}(\delta_{pk} - \delta_{nk}) - K_{np}(1 - \cos(\delta_{pk} - \delta_{nk}))]$$
$$= D_{nk}; \qquad \forall n \in \mathcal{N}, \quad \forall k \in \mathcal{K} \tag{9.90}$$

$$T_j^{\min} v_{jk} \leq t_{jk} \leq T_j^{\max} v_{jk}; \quad \forall j \in \mathcal{J}, \quad \forall k \in \mathcal{K} \tag{9.91}$$

$$-C_{np} \leq B_{np}(\delta_{pk} - \delta_{nk}) \leq C_{np}; \qquad \forall n \in \mathcal{N}, \quad \forall p \in \Omega_n, \forall k \in \mathcal{K} \tag{9.92}$$

$$v_{jk} = V_{jk}^{(\nu-1)} : \lambda_{jk}^{(\nu)}; \qquad \forall j \in \mathcal{J}, \quad \forall k \in \mathcal{K}. \tag{9.93}$$

The above subproblem formulation is similar to the original problem formulation once the 0/1 variables are fixed to given values. Therefore the description of equations is not repeated below.

The last block of constraints (9.93), which enforces the running, start-up, and shut-down schedule, fixed in the master problem, deserves special mention. The dual variables $\lambda_{jk}^{(\nu)}$ associated with this block of constraints provide the master problem with relevant dual information to improve the current schedule.

Master Problem

The multiperiod master problem allows us to decide which plants are online and which ones are off-line throughout the considered multiperiod decision horizon. This master problem is formulated below.

$$\underset{\alpha^{(\nu)},\, v_{jk},\, y_{jk};\, j \in \mathcal{J},\, k \in \mathcal{K}}{\text{minimize}} \quad \alpha^{(\nu)} + \sum_{k \in \mathcal{K}} \sum_{j \in \mathcal{J}} [F_j v_{jk} + A_j y_{jk}] \tag{9.94}$$

subject to

$$\alpha^{(\nu)} \geq z^{(\nu)} + \sum_{k \in \mathcal{K}} \sum_{j \in \mathcal{J}} \lambda_{jk}^{(\nu)} [v_{jk} - V_{jk}^{(\nu)}]; \qquad \forall \nu \in \Upsilon \tag{9.95}$$

$$y_{jk} \geq v_{jk} - v_{j,k-1}; \qquad \forall j \in \mathcal{J}, \quad \forall k \in \mathcal{K} \tag{9.96}$$

$$v_{jk}, y_{jk} \in \{0,1\}; \qquad \forall j \in \mathcal{J}, \quad \forall k \in \mathcal{K} \qquad (9.97)$$

$$\sum_{j \in \mathcal{J}} T_j^{\max} v_{jk} \geq \sum_{n \in \mathcal{N}} D_{nk} + R_k; \qquad \forall k \in \mathcal{K} \qquad (9.98)$$

$$\sum_{j \in \mathcal{J}} T_j^{\min} v_{jk} \leq \sum_{n \in \mathcal{N}} D_{nk}; \qquad \forall k \in \mathcal{K}. \qquad (9.99)$$

The objective function (9.94) includes an underestimate of total operating costs in all periods (variable α), fixed costs, and start-up costs.

The constraints (9.95) are the Benders cuts. These cuts provide a lower estimate of total operating costs in the Benders subproblem (as a function of the scheduling variables which are the variables of the master problem). An additional cut is added in every iteration.

The two sets of (9.96) and (9.97), as in the original problem, enforce the logic of running, start-up, and shut-down of the power plants.

Equations (9.98) and (9.99) are called feasibility cuts and force the master problem to generate solutions that satisfy the load and reserve requirements. These constraints enforce the feasibility of the subproblem.

It should be noted that the only real variable in the above problem is $\alpha^{(\nu)}$, all other variables are binary integer.

Stopping Criterion

The iterative Benders procedure stops if the operating cost computed through the master problem and the operating cost computed through the subproblem are close enough, i.e., if the equation below holds

$$\frac{|\alpha^{(\nu)} - z^{(\nu)}|}{|z^{(\nu)}|} \leq \varepsilon, \qquad (9.100)$$

where ε is a per unit cost tolerance.

Convergence behavior is appropiate as the total operating costs over all periods as a function of the scheduling variables (the optimization variables of the master problem) is a nonincreasing function. This is so because the larger the number of committed units is, the lower the total variable operating costs are over all periods.

Iterative Algorithm

In summary, the iterative Benders algorithm is described below.

Step 1 : Solve the initial or current master problem.
Step 2 : Update scheduling binary variables.
Step 3 : Solve the subproblems.
Step 4 : If the stopping criterion is met, stop; otherwise formulate a new Benders cut, add it to the master problem and go to Step 1.

9.4 Production Costing

This application expands the motivating Example in Sect. 1.3.4 in p. 23, and demonstrates the practical use of the Dantzig-Wolfe decomposition technique, which is explained in Chap. 2.

9.4.1 Introduction

The production costing problem consists of supplying a set of energy demands, expressed through power versus time curves, using production plants. This problem is multiperiod, i.e., an energy demand curve exists for every time period of the production medium- or long-term planning horizon. Production plants are related through intraperiod constraints within a given time period and through interperiod constraints across time periods. Each production plant is characterized by its production cost and its links with other production plants. The load curve in every period can be modeling using a probabilistic description, and analogously, plants unavailabilities can be described using probabilistic models. However, for the sake of clarity, no probability description is used in this application, although its inclusion is straightforward.

Production cost models [75, 76, 77] are widely used tools to estimate electric energy productions and costs in a long- or medium-term horizon. A typical time horizon is a year divided in monthly or weekly periods. The load in every period is modeled using a load duration curve, as can be seen in Fig. 1.6, p. 23. The load duration curve expresses the percentage of the load that is equal or exceed a fixed power value. Section 1.3.4 in p. 23 illustrates the basic blocks that constitute a production cost model.

To perform energy and cost calculations, plants are ordered from lower to higher operating cost. This order is referred to as merit order. This merit or loading order is used to load plants at maximum capacity (see Sect. 1.3.4 in p. 23). This plant loading rule is enough to carry out a production cost calculation so that total production costs are minimized. However, this rule holds only if neither intra- nor inter-period constraints, denominated dispatch constraints, are taken into account. The merit order is illustrated in Fig. 1.7, p. 24.

Dispatch constraints are of two types: intraperiod constraints, i.e., constraints that couple together decisions within a given period, and interperiod dispatch constraints, i.e., constraints that couple together decisions across time periods. Intraperiod constraints include among others those involved in: (i) energy storage plants, (ii) multiple block plants, and (iii) limited energy plants. Interperiod constraints include the ones involved in: (i) emission caps over multiple plants and (ii) complex cascaded hydroelectric plants. Interperiod constraints are particularly useful to model dispatch restrictions of the cascaded hydroelectric power plants of a complex river system.

For the formulation below, it should be noted that the unserved energy in a time period is the energy not supplied once a given set of production plants is loaded in the considered time period but not the remaining plants.

This section addresses a multiperiod production cost model including both intraperiod and interperiod constraints.

The work described in this section, originally proposed in [9], is built upon the facet LP formulation presented in [6]. Detailed numerical simulations and other technical details can be found in [9].

9.4.2 Notation

The notation used is stated below.

Indices.
- i: plant index
- j: intraperiod constraint index
- m: interperiod constraint index
- k: time period index
- s: index for the initial solutions
- l: iteration index
- ν: current iteration index.

Sets.
- Ω: set of plants
- Ψ_0^k: set of initial solutions of the subproblem associated with period k
- Ξ_ν^k: set of positive reduced cost solutions of the subproblem associated with period k from iteration 1 to iteration $\nu - 1$
- $\emptyset$: the empty set.

Numbers.
- I: number of plants
- J^k: number of intraperiod constraints in period k
- M: number of interperiod constraints
- K: number of periods.

Variables.
- e_i^k: energy produced by plant i in period k in MWh
- u_l^k: dimensionless weighting factor for the solution of the subproblem associated with period k at iteration l
- e_{il}^k: energy produced by plant i in the subproblem associated with period k at iteration l in MWh
- z: objective function of the original problem in \$
- z_ν: objective function of the master problem at iteration ν in \$
- z_ν^k: objective function of the subproblem associated with period k at iteration ν in \$.

Marginal values.

d_l^k: reduced costs of the subproblem of period k at iteration l

$\lambda_{m\nu}$: marginal value of the interperiod constraint m at iteration ν

$\mu_{j\nu}^k$: marginal value of the intraperiod constraint j of period k at iteration ν

σ_ν^k: marginal value of the convexity constraint for period k at iteration ν.

Constants.

C_i: running cost of plant i in \$/MWh

D_{mi}^k: coefficient of variable e_i^k in the interperiod constraint m in per unit; it relates the production of plant i with the productions of other plants across time periods

A_{ij}^k: coefficient of variable e_i^k in the intraperiod constraint j of period k in per unit; it relates the production of plant i with the productions of other plants within period k

F_m: right-hand side of the interperiod constraint m in MWh

B_j^k: right-hand side of the intraperiod constraint j of period k in MWh

$W^k(\Omega)$: unserved energy after loading plants of set Ω in period k in MWh

P_l^k: cost of the subproblem associated with period k at iteration l in \$

Q_{ml}^k: contribution to the right-hand side of the interperiod constraint m of solution l of the subproblem associated with period k in MWh

R_{jl}^k: contribution to the right-hand side of the intraperiod constraint j of solution l of the subproblem related to period k in MWh.

9.4.3 Problem Formulation

The multiperiod production cost model including dispatch constraints is formulated below.

$$\underset{e_i^k; \forall i, \forall k}{\text{minimize}} \quad z = \sum_{k=1}^{K} \sum_{i=1}^{I} C_i e_i^k \qquad (9.101)$$

subject to

$$\sum_{k=1}^{K} \sum_{i=1}^{I} D_{mi}^k e_i^k = F_m; \quad m = 1, \ldots, M \qquad (9.102)$$

$$\sum_{i=1}^{I} A_{ij}^k e_i^k = B_j^k; \quad j = 1, \ldots, J^k, \; k = 1, \ldots, K \qquad (9.103)$$

$$\sum_{i \in \Omega} e_i^k \leq W^k(\emptyset) - W^k(\Omega); \quad \forall \Omega \subset \{1, \ldots, I\}, \; k = 1, \ldots, K \qquad (9.104)$$

$$\sum_{i=1}^{I} e_i^k = W^k(\emptyset); \quad k = 1, \ldots, K \qquad (9.105)$$

$$e_i^k \geq 0; \quad i = 1, \ldots, I, \; k = 1, \ldots, K. \qquad (9.106)$$

Decision variables are the energies produced by every plant in every period.

The objective function in (9.101) represents generation costs over plants and periods.

The block of (9.102) represents interperiod linear constraints, i.e., constraints that couple together the production of a set of plants over different time periods. These constraints are particularly useful to model hydro system constraints. They allow the optimal allocation of hydro generation among periods.

The block of (9.103) represents intraperiod linear constraints, i.e., constraints that couple together the production of a set of plants within a given time period. These constraints are used to model environmental and different types of dispatch constraints.

Equations (9.104) and (9.105) are the facet constraints [6] used to express the merit order loading rule as a linear programming problem (See Sect. 1.3.4 in p. 23).

Finally, constraints (9.106) enforce the positiveness of energy values.

It should be noted that $W^k(\Omega)$ is the unserved energy value of period k after loading the plants in set Ω.

An example of the formulation (9.101)–(9.106) is (1.34)–(1.35), stated in Sect. 1.3.4, p. 27. Note the way in which the unserved energy value $W^k(\Omega)$ is calculated.

This problem cannot be solved directly. The number of constraints grows exponentially with the number of plants and, for systems of realistic sizes, it reaches extremely high values. A 12-period case study including 100 generating plants requires $12 \times (2^{(100+1)} - 1) = 3.04 \times 10^{31}$ facet constraints. Therefore, this problem has to be addressed using a decomposition procedure.

9.4.4 Solution Approach

Blocks of constraints (9.102) and (9.103) are the complicating constraints, as they couple together the productions of the plants across time periods and within any time period. They make problem (9.101)–(9.106) hard to solve. If these blocks of constraints are ignored, the resulting problem decomposes by time period, and every subproblem attains such a structure that it can be solved in a straightforward manner by direct application of the aforementioned merit order rule.

The Dantzig-Wolfe decomposition technique was developed to efficiently solve problems with the structure of problem (9.101)–(9.106). Through the Dantzig-Wolfe decomposition procedure the original problem is reformulated becoming the so-called master problem. In this master problem complicating constraints are explicitly considered, while the remaining constraints are implicitly considered. The master problem typically has a low number of constraints but a high number of variables. The variables (columns) to add to the master problem at every iteration are determined through the solution of the subproblems. Every subproblem is associated with a time period and includes only noncomplicating constraints. Therefore, the solution of every subproblem is independently obtained by straightforward application of the merit order rule.

The Master Problem

The master problem carries out a convex combination of subproblem solutions with the objective to meet complicating constraints while achieving minimum cost. The solution of the master problem provides marginal cost signals to be used by the subproblems to implicitly take into account the complicating constraints. The master problem at iteration ν has the form

$$\underset{u_s^k, u_l^k; \forall s, \forall l, \forall k}{\text{minimize}} \quad z_\nu = \sum_{k=1}^{K} \left(\sum_{s \in \Psi_0^k} P_s^k u_s^k + \sum_{l \in \Xi_\nu^k} P_l^k u_l^k \right) \quad (9.107)$$

subject to

$$\sum_{k=1}^{K} \left(\sum_{s \in \Psi_0^k} Q_{ms}^k u_s^k + \sum_{l \in \Xi_\nu^k} Q_{ml}^k u_l^k \right) = F_m : \lambda_{m\nu}; \quad m = 1, \ldots, M \quad (9.108)$$

$$\sum_{s \in \Psi_0^k} R_{js}^k u_s^k + \sum_{l \in \Xi_\nu^k} R_{jl}^k u_l^k = B_j^k : \mu_{j\nu}^k; \quad j = 1, \ldots, J^k, \quad k = 1, \ldots, K \quad (9.109)$$

$$\sum_{s \in \Psi_0^k} u_s^k + \sum_{l \in \Xi_\nu^k} u_l^k = 1 : \sigma_\nu^k; \quad k = 1, \ldots, K \quad (9.110)$$

$$u_l^k \geq 0 \; ; \quad l = 1, \ldots, \nu - 1, \quad k = 1, \ldots, K \quad (9.111)$$

$$u_s^k \geq 0 \; ; \quad s \in \Psi_0^k, \quad k = 1, \ldots, K, \quad (9.112)$$

where

$$\Xi_\nu^k = \{l \in \{1, 2, \ldots, \nu - 1\} : d_l^k > 0\}; \quad k = 1, \ldots, K \quad (9.113)$$

is the set of positive reduced cost solutions of the subproblem associated with period k from iteration 1 to iteration $\nu - 1$,

$$d_l^k = \sigma_l^k - z_l^k \quad (9.114)$$

are reduced costs,

$$P_l^k = \sum_{i=1}^{I} C_i e_{il}^k \quad (9.115)$$

is the contribution to the total cost of solution l of the subproblem associated with period k,

$$Q_{ml}^k = \sum_{i=1}^{I} D_{mi}^k e_{il}^k \quad (9.116)$$

is the contribution to the right-hand side of the interperiod constraint m of solution l of the subproblem associated with period k, and

$$R_{jl}^k = \sum_{i=1}^{I} A_{ij}^k e_{il}^k \qquad (9.117)$$

is the contribution to the right-hand side of the intraperiod constraint j of solution l of the subproblem associated with period k.

The variables of the master problem are the weighting factors for the solutions of the subproblems. The objective function (9.107) is a convex combination of the available subproblem solutions. Equations (9.108) enforce interperiod constraints while (9.109) enforce intraperiod constraints. Equations (9.110)–(9.112) enforce convex combination conditions for every period.

It should be noted that the master problem above is a small size linear programming problem whose number of constraints is constant and equal to the number of complicating constraints plus one convexity constraint for each period, and whose variable number grows with the number of iterations.

The Subproblems

Each subproblem represents the production cost problem corresponding to a single time period. The effect on intra- and interperiod constraints is taken into account through a modified objective function.

The subproblem at iteration ν, associated with period k has the following form:

$$\underset{e_{i\nu}^k; \forall i}{\text{minimize}} \quad z_\nu^k = \sum_{i=1}^{I} \left(C_i - \sum_{m=1}^{M} \lambda_{m\nu} D_{mi}^k - \sum_{j=1}^{J^k} \mu_{j\nu}^k A_{ij}^k \right) e_{i\nu}^k \qquad (9.118)$$

subject to

$$\sum_{i \in \Omega} e_{i\nu}^k \leq W^k(\emptyset) - W^k(\Omega); \qquad \forall \Omega \subset \{1, \ldots I\} \qquad (9.119)$$

$$\sum_{i=1}^{I} e_{i\nu}^k = W^k(\emptyset) \qquad (9.120)$$

$$e_{i\nu}^k \geq 0; \quad i = 1, \ldots, I. \qquad (9.121)$$

The variables of the subproblems are the energies produced by every plant. The cost associated with every variable in the objective function consists of three terms: (i) running cost, (ii) cost incurred for contributing to meet interperiod constraints, and (iii) cost incurred for contributing to meet intraperiod constraints.

It should be noted that the number of subproblems at iteration ν is equal to the number of periods K. It should also be noted that the subproblems are solved independently.

The optimal solution of every subproblem is obtained from a merit order criterion that is independent of the number of facet constraints.

Units affected by active dispatch constraints are forced to change their location in the load duration curve and to lie in the position determined by their respective equivalent costs. The equivalent cost of one unit in one period is the cost coefficient of that unit in the objective function of the subproblem associated with that period. The equivalent cost depends on the unit running cost, the dual value of any active dispatch (complicating) constraint and the linear coefficient of that unit in the (complicating) dispatch constraint.

Iterative Procedure

The solution at iteration ν of subproblem k is a useful solution if it has a positive reduced cost, i.e., if $\sigma_\nu^k - z_\nu^k$ is positive. Any useful subproblem solution can be incorporated into the master problem as a new variable to improve the current master problem solution [5].

At iteration ν the master problem carries out a convex combination of the useful subproblem solutions of the first $\nu - 1$ iterations with the objective to meet complicating constraints while achieving the minimum cost. The solution of the master problem provides cost signals (λs and μs) to be used by the subproblems to implicitly take into account the complicating constraints.

Once every plant production cost has been updated using master problem price signals, the subproblems are solved independently. The solutions of the subproblems provide the master problem with information on the usage of the "resources" (right-hand sides) associated with the complicating constraints.

The master problem and the subproblems are solved iteratively until no useful subproblem solution is found, i.e., until no master problem variable with positive reduced cost exists.

Once the iterative procedure is completed, energy values of plants in every period are computed as

$$e_i^k = \sum_{s \in \Psi_0^k} e_{is}^k u_s^k + \sum_{l \in \Xi_\nu^k} e_{il}^k u_l^k; \quad i = 1, \ldots, I, \quad k = 1, \ldots, K. \quad (9.122)$$

It should be noted that the convergence and robustness of the above procedure are guaranteed because all subproblems are bounded [5].

Iterative Algorithm

In summary, the iterative Dantzig-Wolfe algorithm is stated below.

Step 1: Solve an initial master problem.
Step 1: Solve independently the subproblems that incorporate the price signals from the master problem.
Step 1: Solve an updated master problem that incorporates the useful solutions of the last subproblems.

Step 1: If the stopping criterion is met, stop; otherwise continue with Step 2.

9.5 Hydrothermal Coordination

This application demonstrates the practical use of the Lagrangian relaxation (LR) technique that is explained in Chap. 5.

9.5.1 Introduction

The objective of the hydrothermal coordination problem is to serve the demand for electricity at minimum cost throughout a short-term multiperiod horizon, typically one week hour by hour. Available production plants include thermal ones, and hydroelectric plants embedded in river systems. Thermal plants can be both started-up and shut-down as needed. More specifically, this problem is solved to determine the start-up and shut-down schedule of thermal plants, as well as the power output of thermal and hydro plants during a short-term planning horizon. The objective is to meet customer demand with appropriate levels of security (measured through the so-called spinning reserve) so that total operating costs are minimized.

This is a large-scale mixed-integer nonlinear problem. As recognized in the technical literature, the Lagrangian Relaxation (LR) technique considered in Chap. 5 is the most appropriate procedure to solve this problem [78, 69, 79, 32].

The LR procedure decomposes the original problem in one subproblem per thermal plant and one subproblem per hydroelectric system.

Instead of solving the original problem, the LR technique solves its dual problem. As a spinoff of the solution of the dual problem a solution for the primal problem is obtained. However, more often than not, this primal solution is slightly infeasible. Heuristic procedures are required to get a feasible primal solution. Finally, plants should be dispatched to supply exactly the demand. Therefore, the LR technique consists of the three phases below:

Phase 1. To solve the dual problem.
Phase 2. To obtain a primal feasible solution.
Phase 3. To exactly dispatch committed generation to meet the demand.

Phase 1 requires the solution of a nondifferentiable maximization problem, and the most commonly used technique to address this phase is the subgradient technique [67, 79, 32]. However, the technique used in this application is a cutting plane method as explained in Sect. 5.3.4 of Chap. 5.

Phase 2 is easily accomplished using a subgradient algorithm as described in Chap. 5.

Phase 3 is a multiperiod economic dispatch, whose solution is well stated in the technical literature [80].

Detailed numerical simulation results and additional technical information for the hydrothermal coordination problem can be found in [81].

9.5.2 Notation

The notation used in this application is stated below:

Variables.
x_i: variable vector related to thermal plant i
y_j: variable vector related to the hydroelectric system j
(x, y): vector defined as $(x_i, y_j;\ \forall i, \forall j)$
$d(\lambda, \mu)$: solution of the decomposed primal problem.

Functions.
$f(x)$: total cost
$f_i(x_i)$: cost related to thermal plant i
$g_i(x_i)$: inequality constraint vector related to thermal plant i of dimension equal to the number of subperiods in the planning horizon; it represents the contribution of thermal plant i to meet security constraints
$g_j(y_j)$: inequality constraint vector related to hydroelectric system j of dimension equal to the number of subperiods in the planning horizon; it represents the contribution of hydroelectric system j to meet security constraints
$h_i(x_i)$: equality constraint vector related to thermal plant i of dimension equal to the number of subperiods in the planning horizon; it represents the contribution of thermal plant i to satisfy demand constraints
$h_i(y_j)$: equality constraint vector related to hydroelectric system j of dimension equal to the number of subperiods in the planning horizon; it represents the contribution of hydroelectric system j to satisfy demand constraints
$s_i(x_j)$: constraints pertaining to thermal plant i
$s_j(x_j)$: constraints pertaining to hydroelectric system j.

Constants.
G: vector of dimension equal to the number of subperiods in the planning horizon, related to inequality constraint; it represents the right-hand side of the security constraints
H: vector of dimension equal to the number of subperiods in the planning horizon, related to equality constraints; it represents the right-hand side of the demand constraints.

Dual variables.
λ: dual vector of dimension equal to the number of subperiods in the planning horizon; it is related to demand supply constraints

$\boldsymbol{\mu}$: dual vector of dimension equal to the number of subperiods in the planning horizon; it is related to security constraints

$\boldsymbol{\theta}$: vector defined as $(\boldsymbol{\lambda}, \boldsymbol{\mu})$.

9.5.3 Problem Formulation

Formulation

The short-term hydrothermal coordination problem can be formulated as follows:

$$\underset{\boldsymbol{x},\boldsymbol{y}}{\text{minimize}} \quad f(\boldsymbol{x}) = \sum_i f_i(\boldsymbol{x}_i) \tag{9.123}$$

subject to

$$s_i(\boldsymbol{x}_i) \leq \mathbf{0}; \quad \forall i \tag{9.124}$$

$$s_j(\boldsymbol{y}_j) \leq \mathbf{0}; \quad \forall j \tag{9.125}$$

$$g(\boldsymbol{x}, \boldsymbol{y}) = \boldsymbol{G} - \sum_i g_i(\boldsymbol{x}_i) - \sum_j g_j(\boldsymbol{y}_j) \leq \mathbf{0} \tag{9.126}$$

$$h(\boldsymbol{x}, \boldsymbol{y}) = \boldsymbol{H} - \sum_i h_i(\boldsymbol{x}_i) - \sum_j h_j(\boldsymbol{y}_j) = \mathbf{0}. \tag{9.127}$$

Equation (9.123) is the production cost to be minimized. Note that this equation depends solely of thermal plant variables, because there is no significant cost associated to produce with hydroelectric plants. Equation (9.124) are the constraints related to every thermal unit, (9.125) are the constraints related to every hydroelectric system, (9.126) enforces spinning reserve constraints, and (9.127) enforces demand constraints. It should be noted that time is embedded in the above formulation.

Equations (9.126) and (9.127) are global constraints which couple together thermal-related and hydro-related variables.

Problem (9.123)–(9.127) is denominated the primal problem (PP).

Global constraints are incorporated in the objective function of the primal problem through Lagrange multipliers to obtain the Lagrangian function:

$$\mathcal{L}(\boldsymbol{x}, \boldsymbol{y}, \boldsymbol{\lambda}, \boldsymbol{\mu}) = \sum_i f_i(\boldsymbol{x}_i) + \boldsymbol{\lambda}^T \left[\boldsymbol{H} - \sum_i h_i(\boldsymbol{x}_i) - \sum_j h_j(\boldsymbol{y}_j) \right]$$

$$+ \boldsymbol{\mu}^T \left[\boldsymbol{G} - \sum_i g_i(\boldsymbol{x}_i) - \sum_j g_j(\boldsymbol{y}_j) \right]. \tag{9.128}$$

The dual problem (DP) of the original primal problem (9.123)–(9.127) has the form

$$\underset{\boldsymbol{\lambda}, \boldsymbol{\mu}}{\text{maximize}} \quad \phi(\boldsymbol{\lambda}, \boldsymbol{\mu}) \tag{9.129}$$

subject to
$$\mu \geq 0, \tag{9.130}$$
where
$$\phi(\theta) = \phi(\lambda, \mu) = \lambda^T H + \mu^T G + d(\lambda, \mu), \tag{9.131}$$
where $\theta = (\lambda, \mu)$, and $d(\lambda, \mu)$ is the solution of the decomposed primal problem (DPP) stated below

$$\underset{x,y}{\text{minimize}} \quad \sum_i \left[f_i(x_i) - \lambda^T h_i(x_i) - \mu^T g_i(x_i) \right] - \sum_j \left[\lambda^T h_j(y_j) + \mu^T g_j(y_j) \right] \tag{9.132}$$

subject to
$$s_i(x_i) \leq 0; \quad \forall i \tag{9.133}$$
$$s_j(y_j) \leq 0; \quad \forall j. \tag{9.134}$$

The above problem decomposes in one subproblem per thermal unit and one subproblem per hydroelectric system.

The subproblem associated with thermal unit i is

$$\underset{x_i}{\text{minimize}} \quad f_i(x_i) - \lambda^T h_i(x_i) - \mu^T g_i(x_i) \tag{9.135}$$

subject to
$$s_i(x_i) \leq 0, \tag{9.136}$$

and the subproblem associated with hydroelectric system j is

$$\underset{y_j}{\text{minimize}} \quad \lambda^T h_j(y_j) + \mu^T g_j(y_j) \tag{9.137}$$

subject to
$$s_j(y_j) \leq 0. \tag{9.138}$$

9.5.4 Solution Approach

Iterative Algorithm

The LR procedure to solve the dual problem works as follows:

Step 1: Initialize multiplier vector $\theta = (\lambda, \mu)$.
Step 2: Solve the decomposed primal problem by solving 1 subproblem per thermal plant (9.135)–(9.136) and 1 subproblem per hydroelectric system (9.137)–(9.138).
Step 3: Update the multiplier vector using any of the procedures stated in Chap. 5.
Step 4: If the convergence criterion is satisfied, stop. Otherwise continue with Step 2. Convergence criteria depend on the multiplier updating procedure and are explained in Chap. 5.

Duality Gap

If the dual problem is solved using a piecewise linear reconstruction of the dual function, the objective function value of the current linearly reconstructed dual problem constitutes an upper bound of the optimal dual cost value. On the other hand, the objective function value of the dual problem (evaluated through the decomposed primal problem) provides at every iteration a lower bound of the optimal dual cost value. This can be mathematically stated as follows:

$$z^{(\nu)} \geq \phi^* \geq \phi^{(\nu)} ,\qquad(9.139)$$

where $z^{(\nu)}$ is the objective function value of the linear reconstruction of the dual problem at iteration ν, ϕ^* is the optimal dual cost value, and $\phi^{(\nu)} = \boldsymbol{\lambda}^{(\nu)T}\boldsymbol{H} + \boldsymbol{\mu}^{(\nu)T}\boldsymbol{G} + d(\boldsymbol{\lambda}^{(\nu)}, \boldsymbol{\mu}^{(\nu)})$ is the objective function value of the dual problem at iteration ν.

The size of the per unit gap $g^{(\nu)} = (z^{(\nu)} - \phi^{(\nu)})/\phi^{(\nu)}$ is an appropriate per unit cost criterion to stop the search for the dual optimum.

9.6 Multiarea Optimal Power Flow

This application demonstrates the practical use of the optimality condition decomposition technique that is explained in Sect. 5.5, p. 210.

9.6.1 Introduction

The objective of this problem is to determine the energy generated by production plants to supply the demand for electricity while minimizing the total production cost in a single time period. These production plants are located in an electric network that is composed by several areas that are interconnected by tie-lines. It is required to enforce the constraints imposed by the electric network and to determine the flows of energy through lines, and particularly through tie-lines.

The optimality condition decomposition is applied to a multiarea optimal power flow (OPF) problem in the context of an electric energy system that spans several interconnected areas. It is often desirable to preserve the autonomy of each area in these systems. A decentralized operation can be preserved while still attaining overall optimality by applying decomposition techniques to a centralized operation problem.

The multiarea OPF problem is an important problem for the secure and economic operation of an interconnected power system. The multiarea OPF determines, in a precise way, the active and reactive power that each generation unit in the system must generate. This is done to ensure that all demand and security constraints for the system are satisfied at a minimal cost for all interconnected areas. The resulting multiarea OPF problem is a large-scale optimization problem [80].

The decomposition procedure used allows the company in each area to operate its system independently of the other areas, while obtaining an optimally coordinated but decentralized solution. A central agent in the model is necessary to collect and distribute information for the whole system. This agent ensures the coordination of the global system and therefore, the proposed methodology is appropriate for an Independent System Operator (ISO) in charge of the technical operation of an electric energy system.

Detailed numerical simulations and further technical details of the multi-area OPF problem can be found in [82].

9.6.2 Notation

The notation used in the model is state below:

Numbers.
A: total number of areas
B: total number of nodes
G: total number of generators
L: total number of transmission lines.

Sets.
Λ_j: set of indices of generators in node j
Ω_j: set of indices of nodes connected to node j
Θ: set of indices of transmission lines.

Constants.
Y_{jk}: element jk of the admittance magnitude matrix in $1/\Omega$ (network constructive data)
δ_{jk}: element jk of the admittance phase matrix in radians (network constructive data)
$P_{G_i}^{\max}$: maximum active power production capacity of generator i in MW
$P_{G_i}^{\min}$: minimum active power production capacity of generator i in MW
$Q_{G_i}^{\max}$: maximum reactive power production capacity of generator i in MVAr
$Q_{G_i}^{\min}$: minimum reactive power production capacity of generator i in MVAr
$V_j^{\max}$: maximum voltage magnitude in node j in per unit V
$V_j^{\min}$: minimum voltage magnitude in node j in per unit V
P_{D_j}: active power demand in node j in MW
Q_{D_j}: reactive power demand in node j in MVAr
$S_{jk}^{\max}$: maximum transmission capacity of line jk in MVA.

Variables.
p_{G_i}: active power produced by generator i in MW
q_{G_i}: reactive power produced by generator i in MVAr
v_j voltage magnitude in node j in per unit V

θ_j: voltage phase in node j in radians
$\boldsymbol{p}_G$: vector of produced active powers
$\boldsymbol{q}_G$: vector of produced reactive powers
$\boldsymbol{v}$: voltage magnitude vector
$\boldsymbol{\theta}$: voltage phase vector.

9.6.3 Problem Formulation

The multiarea OPF model can be formulated as

$$\underset{\boldsymbol{p}_G, \boldsymbol{q}_G, \boldsymbol{v}, \boldsymbol{\theta}}{\text{minimize}} \quad f(\boldsymbol{p}_G, \boldsymbol{q}_G, \boldsymbol{v}, \boldsymbol{\theta}) \tag{9.140}$$

subject to

$$a_j(\boldsymbol{p}_G, \boldsymbol{v}, \boldsymbol{\theta}) = 0; \quad j = 1, \ldots, B \tag{9.141}$$
$$r_j(\boldsymbol{q}_G, \boldsymbol{v}, \boldsymbol{\theta}) = 0; \quad j = 1, \ldots, B \tag{9.142}$$
$$t_j(\boldsymbol{v}, \boldsymbol{\theta}) \leq 0; \quad j = 1, \ldots, L \tag{9.143}$$
$$P_{G_i}^{\min} \leq p_{G_i} \leq P_{G_i}^{\max}; \quad i = 1, \ldots, G \tag{9.144}$$
$$Q_{G_i}^{\min} \leq q_{G_i} \leq Q_{G_i}^{\max}; \quad i = 1, \ldots, G \tag{9.145}$$
$$V_j^{\min} \leq v_j \leq V_j^{\max}; \quad j = 1, \ldots, B \tag{9.146}$$
$$-\pi \leq \theta_j \leq \pi; \quad j = 1, \ldots, B. \tag{9.147}$$

Function (9.140) is the objective function that typically represents production cost.

The power flow equations are included in the model as constraints (9.141)–(9.142). There are two equations for each node of the global system, representing the active and reactive power balance in each node,

$$\sum_{i \in \Lambda_j^a} p_{G_i}^a - P_{D_j}^a = v_j^a \sum_{k \in \Omega_j^a} Y_{jk}^a v_k^a \cos(\theta_j^a - \theta_k^a - \delta_{jk}^a); \quad j = 1, \ldots, N^a \tag{9.148}$$

$$\sum_{i \in \Lambda_j^a} q_{G_i}^a - Q_{D_j}^a = v_j^a \sum_{k \in \Omega_j^a} Y_{jk}^a v_k^a \sin(\theta_j^a - \theta_k^a - \delta_{jk}^a); \quad j = 1, \ldots, N^a \tag{9.149}$$

where $a = 1, \ldots, A$, and the superscripts a indicate the area for each constant and variable.

Constraints (9.143) are the transmission capacity limits for each line of the global system,

$$\left(v_j^a v_k^a Y_{jk}^a \cos(\theta_j^a - \theta_k^a - \delta_{jk}^a)\right)^2 + \left(v_j^a v_k^a Y_{jk}^a \sin(\theta_j^a - \theta_k^a - \delta_{jk}^a)\right)^2 \leq$$
$$(S_{jk}^{\max})^2; \quad (j,k) \in \Theta^a, \quad a = 1, \ldots, A. \tag{9.150}$$

Constraints (9.144)–(9.147) represent technical limits over variables. Model (9.140)–(9.147) can be written in compact form as

$$\underset{\boldsymbol{x}_a}{\text{minimize}} \quad \sum_{a=1}^{A} f_a(\boldsymbol{x}_a) \tag{9.151}$$

subject to

$$\boldsymbol{h}_a(\boldsymbol{x}_1, \ldots, \boldsymbol{x}_A) \leq \boldsymbol{0}; \quad a = 1, \ldots, A \tag{9.152}$$
$$\boldsymbol{g}_a(\boldsymbol{x}_a) \leq \boldsymbol{0}; \quad a = 1, \ldots, A, \tag{9.153}$$

where $\boldsymbol{x}_a$ are the state variables for each area a of the global system that contain node voltage magnitudes and node phase angles. Equations (9.152) represent the power flow equations and transmission capacity limits (9.148)–(9.150) for those lines and nodes interconnecting different areas. Constraints (9.153) include the power flow equations and transmission capacity limits (9.148)–(9.150), only for those lines and nodes lying within a given area, and limits over dependent and control variables (9.144)–(9.147). It should be noted that the sets of (9.152) and (9.153) represent both equality and inequality constraints.

The multiarea OPF model (9.151)–(9.153) is a large-scale optimization problem. Equations (9.152) are the complicating constraints. These equations contain variables from different areas and prevent each system from operating independently from the others. If these equations are removed from problem (9.151)–(9.153), the resulting problem can be trivially decomposed into one subproblem for each area.

The complicating constraints (9.152) include the power balance equations at the interconnecting nodes of area a (the nodes in area a connected to nodes in areas b different than area a). Also, the transmission capacity limits for the interconnecting lines of the global system are complicating constraints. It should be noted that the only variables appearing in the complicating constraints are those corresponding to the interconnecting nodes of the global system.

Equation (9.153) contain only variables belonging to area a for $a = 1, \ldots, A$. These constraints represent balance equations, transmission limits, and technical constraints for area a.

The proposed decomposition is as follows. Problem (9.151)–(9.153) is equivalent to the problem below

$$\underset{\boldsymbol{x}_a; a=1,\ldots,A}{\text{minimize}} \quad \sum_{a=1}^{A} f_a(\boldsymbol{x}_a) + \sum_{a=1}^{A} \boldsymbol{\lambda}_a^T \, \boldsymbol{h}_a(\boldsymbol{x}_1, \ldots, \boldsymbol{x}_A) \tag{9.154}$$

subject to

$$\boldsymbol{h}_a(\boldsymbol{x}_1, \ldots, \boldsymbol{x}_A) \leq \boldsymbol{0}; \quad a = 1, \ldots, A \tag{9.155}$$
$$\boldsymbol{g}_a(\boldsymbol{x}_a) \leq \boldsymbol{0}; \quad a = 1, \ldots, A \,. \tag{9.156}$$

Given trial values to all variables and multipliers (indicated by overlining) different than those in area a, problem (9.154)–(9.156) reduces to

$$\underset{\boldsymbol{x}_a}{\text{minimize}} \quad k + f_a(\boldsymbol{x}_a) + \sum_{b=1, b \neq a}^{A} \overline{\boldsymbol{\lambda}}_b^T \, \boldsymbol{h}_b(\overline{\boldsymbol{x}}_1, \ldots, \overline{\boldsymbol{x}}_{a-1}, \boldsymbol{x}_a, \overline{\boldsymbol{x}}_{a+1}, \ldots, \overline{\boldsymbol{x}}_A) \tag{9.157}$$

$$\boldsymbol{h}_a(\overline{\boldsymbol{x}}_1, \ldots, \overline{\boldsymbol{x}}_{a-1}, \boldsymbol{x}_a, \overline{\boldsymbol{x}}_{a+1}, \ldots, \overline{\boldsymbol{x}}_A) \leq \boldsymbol{0} \tag{9.158}$$

$$\boldsymbol{g}_a(\boldsymbol{x}_a) \leq \boldsymbol{0}, \tag{9.159}$$

where $k = \sum_{b=1, b \neq a}^{A} f_b(\overline{\boldsymbol{x}}_b)$ is a constant.

The dual variable vector corresponding to constraint (9.158) is denoted by λ_a.

The reduced problem (9.157)–(9.159) can be obtained for every area. The proposed decomposition technique is actually based on the solutions of these reduced area problems, as stated in Chap. 5.

9.6.4 Solution Approach

An outline of the optimality condition decomposition algorithm is as follows:

Step 1: Each area initializes its variables and parameters.
Step 2: Each area carries out one iteration for its corresponding subproblem and obtains search directions.
Step 3: Each area updates its variables and parameters.
The central agent distributes updated information of border nodes and lines.
Step 4: The algorithm stops if variables do not change significantly in two consecutive iterations. Otherwise, it continues with Step 2.

9.7 Sensitivity in Regression Models

In this application we develop a sensitivity analysis associated with the standard linear regression model

$$y_i = \alpha + \beta x_i + \varepsilon_i, \tag{9.160}$$

where $y_i, x_i : i = 1, 2, \ldots, n$ are the values of the response and predictor variables, respectively, α and β are the regression coefficients or parameters, and $\varepsilon_i : i = 1, 2, \ldots, n$ are the values of the random errors.

In the Minimax regression problem (MM), the maximum of the distances between observed and predicted values is minimized, i.e.,

$$\underset{\alpha, \beta, \varepsilon}{\text{minimize}} \; Z_{\text{MM}} = \varepsilon \tag{9.161}$$

Table 9.6. Data set

i	x_i	y_i
1	0.99624	1.88400
2	0.15747	1.03862
3	0.98227	1.68746
4	0.52450	1.24458
5	−0.73862	0.33268
6	0.27944	1.54415
7	−0.68096	0.10369
8	−0.49984	0.63275
9	0.33786	1.55854
10	−0.12929	0.71364

subject to

$$y_i - \alpha - \beta x_i \leq \varepsilon, \quad i = 1, \ldots, n \quad (9.162)$$
$$\alpha + \beta x_i - y_i \leq \varepsilon, \quad i = 1, \ldots, n, \quad (9.163)$$

where we note that the usual extra constraint $\varepsilon \geq 0$ is not required because it is implied by (9.162) and (9.163).

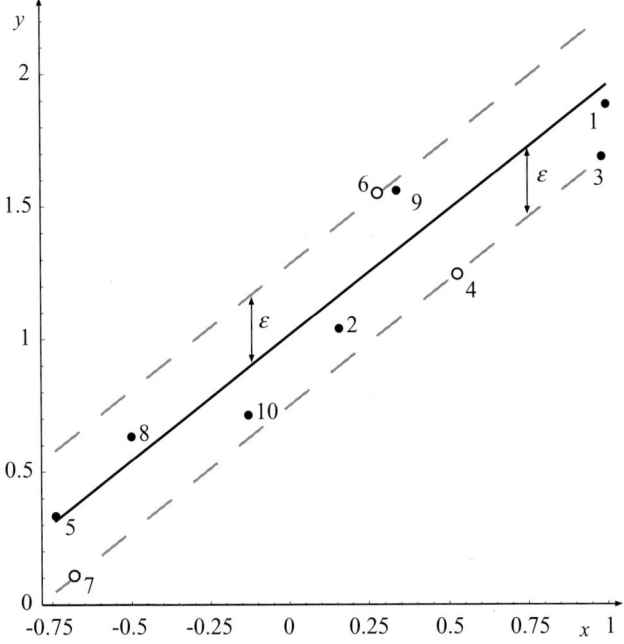

Fig. 9.6. Data set, fitted MM regression line, and error lines (dotted)

9.7 Sensitivity in Regression Models

Assume that we have the data in Table 9.6. Then, the resulting MM regression model leads to the regression line shown in Fig. 9.6, where two parallel dashed lines have been added at distances $\pm\varepsilon$ above and below such a line. One important property of these two lines is that they pass through two and one data point, and that these are the points giving the maximum distance to the MM regression line. These three points (4, 6, and 7) are shown in Fig. 9.6.

Since (9.162) and (9.163) cannot simultaneously hold for the same i, unless $\varepsilon = 0$, the Minimax problem (9.161)–(9.163), assuming a nondegenerated case, i.e., a case in which $\varepsilon \neq 0$, is equivalent to the problem

$$\underset{\alpha,\beta,\varepsilon}{\text{minimize}}\ Z_{\text{MM}} = \varepsilon \qquad (9.164)$$

subject to

$$y_i - \alpha - \beta x_i = \varepsilon : \lambda_i^{(1)}, \quad i \in I_1, \qquad (9.165)$$
$$\alpha + \beta x_i - y_i = \varepsilon : \lambda_i^{(2)}, \quad i \in I_2, \qquad (9.166)$$

where I_1 and I_2 are disjoint subsets of $\{1, 2, \ldots, n\}$, such that $|I_1| + |I_2| \geq 3$. In this section we assume that the solution of the nonlinear MM regression problem (9.161)–(9.163) is regular and nondegenerate. Then, $|I_1| + |I_2| = 3$ and $\varepsilon > 0$.

The set I_1 contains the data points above the MM regression line at distance ε from that line. Similarly, the set I_2 contains the data points below the MM regression line at distance ε from that line. In the example of Fig. 9.6 the set I_1 contains one point and the set I_2 contains two points, but in other cases the reverse can occur. For the sake of simplicity, only this case is discussed:

$$|I_1| = \{(x_r, y_r)\}; \quad |I_2| = \{(x_s, y_s), (x_t, y_t)\}\ .$$

Note that these points correspond to the points 6, 4, and 7, respectively, in the numerical example.

Then, problem (9.164)–(9.166) becomes

$$\underset{\alpha,\beta,\varepsilon}{\text{minimize}}\ Z_{MM} = \varepsilon \qquad (9.167)$$

subject to

$$y_r - \alpha - \beta x_r = \varepsilon : \lambda_r \qquad (9.168)$$
$$\alpha + \beta x_s - y_s = \varepsilon : \lambda_s \qquad (9.169)$$
$$\alpha + \beta x_t - y_t = \varepsilon : \lambda_t, \qquad (9.170)$$

where (9.168)–(9.170) can be written in matrix form as

$$\begin{pmatrix} -1 & -x_r & -1 \\ 1 & x_s & -1 \\ 1 & x_t & -1 \end{pmatrix} \begin{pmatrix} \alpha \\ \beta \\ \varepsilon \end{pmatrix} = \begin{pmatrix} -y_r \\ y_s \\ y_t \end{pmatrix}, \qquad (9.171)$$

392   9 Applications

and then we have

$$\begin{pmatrix} \hat{\alpha} \\ \hat{\beta} \\ \hat{\varepsilon} \end{pmatrix} = \begin{pmatrix} -1 & -x_r & -1 \\ 1 & x_s & -1 \\ 1 & x_t & -1 \end{pmatrix}^{-1} \begin{pmatrix} -y_r \\ y_s \\ y_t \end{pmatrix} \quad (9.172)$$

$$= \begin{pmatrix} -\frac{1}{2} & \frac{x_r + x_t}{2x_t - 2x_s} & \frac{x_r + x_s}{2x_s - 2x_t} \\ 0 & \frac{1}{x_s - x_t} & \frac{1}{x_t + x_s} \\ -\frac{1}{2} & \frac{x_r - x_t}{2x_t - 2x_s} & \frac{x_r - x_s}{2x_s - 2x_t} \end{pmatrix} \begin{pmatrix} -y_r \\ y_s \\ y_t \end{pmatrix} \quad (9.173)$$

$$= \begin{pmatrix} \frac{y_r}{2} + \frac{(x_r + x_t)y_s}{2x_t - 2x_s} + \frac{(x_r + x_s)y_t}{2x_s - 2x_t} \\ \frac{y_s - y_t}{x_s - x_t} \\ \frac{y_r}{2} + \frac{(x_r - x_t)y_s}{2x_t - 2x_s} + \frac{(x_r - x_s)y_t}{2x_s - 2x_t} \end{pmatrix} = \begin{pmatrix} 1.01393 \\ 0.94643 \\ 0.26575 \end{pmatrix}, \quad (9.174)$$

which are closed formulas for the parameter estimates and the optimal objective function value, and the corresponding numerical values for our example. The dual variables in the numerical example are

$$\lambda_r = 0.5, \quad \lambda_s = 0.39835, \quad \lambda_t = 0.10165 .$$

Finally, using the sensitivity formulas for the linear programming case (8.110), one gets

$$\begin{pmatrix} \frac{\partial \alpha}{\partial x_r} & \frac{\partial \alpha}{\partial x_s} & \frac{\partial \alpha}{\partial x_t} \\ \frac{\partial \beta}{\partial x_r} & \frac{\partial \beta}{\partial x_s} & \frac{\partial \beta}{\partial x_t} \\ \frac{\partial \varepsilon}{\partial x_r} & \frac{\partial \varepsilon}{\partial x_s} & \frac{\partial \varepsilon}{\partial x_t} \end{pmatrix} = \beta \begin{pmatrix} -\frac{1}{2} & \frac{x_r + x_t}{2x_s - 2x_t} & \frac{x_r + x_s}{2x_t - 2x_s} \\ 0 & \frac{1}{x_t - x_s} & \frac{1}{x_s - x_t} \\ -\frac{1}{2} & \frac{x_r - x_t}{2x_s - 2x_t} & \frac{x_s - x_t}{x_s - x_r} \end{pmatrix}$$

$$= \left(\frac{y_s - y_t}{x_s - x_t} \right) \begin{pmatrix} -\frac{1}{2} & \frac{x_r + x_t}{2x_s - 2x_t} & \frac{x_r + x_s}{2x_t - 2x_s} \\ 0 & \frac{1}{x_t - x_s} & \frac{1}{x_s - x_t} \\ -\frac{1}{2} & \frac{x_r - x_t}{2x_s - 2x_t} & \frac{x_s - x_t}{x_s - x_r} \end{pmatrix}$$

$$= \begin{pmatrix} -0.473217 & -0.157621 & -0.315596 \\ 0 & -0.785123 & 0.785123 \\ -0.473217 & 0.377016 & 0.0962011 \end{pmatrix}$$

and

$$\begin{pmatrix} \dfrac{\partial \alpha}{\partial y_r} & \dfrac{\partial \alpha}{\partial y_s} & \dfrac{\partial \alpha}{\partial y_t} \\ \dfrac{\partial \beta}{\partial y_r} & \dfrac{\partial \beta}{\partial y_s} & \dfrac{\partial \beta}{\partial y_t} \\ \dfrac{\partial \varepsilon}{\partial y_r} & \dfrac{\partial \varepsilon}{\partial y_s} & \dfrac{\partial \varepsilon}{\partial y_t} \end{pmatrix} = \begin{pmatrix} \dfrac{1}{2} & \dfrac{x_r + x_t}{2x_t - 2x_s} & -\dfrac{x_r + x_s}{2x_t - 2x_s} \\ 0 & \dfrac{1}{x_s - x_t} & \dfrac{1}{x_t - x_s} \\ \dfrac{1}{2} & \dfrac{x_s - x_t}{x_r - x_t} & \dfrac{x_t - x_s}{x_s - x_r} \\ & \dfrac{}{2x_t - 2x_s} & \dfrac{}{2x_t - 2x_s} \end{pmatrix}$$

$$= \begin{pmatrix} 0.5 & 0.166542 & 0.333458 \\ 0 & 0.829559 & -0.829559 \\ 0.5 & -0.398354 & -0.101646 \end{pmatrix},$$

which are closed formulas for the sensitivities of all the primal variables with respect to the data points, and the corresponding values for our numerical example.

Note that the sensitivity $\dfrac{\partial \varepsilon}{\partial y_r}$ is not $-\lambda_r$ but λ_r due to the minus sign preceding y_r in (9.171).

Part V

Computer Codes

A
Some GAMS Implementations

A.1 GAMS Implementation of the Dantzig-Wolfe Decomposition Algorithm

A GAMS general implementation of the Dantzig-Wolfe decomposition algorithm is given below. The aim of this code is simply to clarify how the algorithm works. Note that the code is not optimized because it solves each of the linear programming problems from scratch, without using the previous results.

```
* This program solves linear programming problems with complicating
* constraints using the Dantzig-Wolfe decomposition.
* It is considered the Computational Example 1.2.
$title dw1_2

* A file to keep the results of the problem is declared.
* The internal file name is 'out1' and it refers to
* an external file named 'dw1_2.out'.
file out1/dw1_2.out/; put out1;

* First, single scalar data are declared and defined.
SCALARS
    IP number of available solutions of the relaxed problem
    ncomp number of complicating constraints /4/
    sigma normalized constraint dual variable value in master problem
    ninitsol number of initial basic solutions of the relaxed problem /2/
    objectf objective function value of the relaxed problem /0/
    lowerB original problem lower bound of the objective function /1E+08/
    upperB original problem upper bound of the objective function /-1E+08/
    new to control new basic solution
    MMM master problem large enough positive penalizing constant /20/
    err to control error
    toler tolerance /0.0001/
    step steps of the algorithm /0/;

* Now, indices are declared and defined.
SETS
    N number of variables /1*2/
    P maximum number of feasible solutions of the relaxed problem /1*10/
    K number of subproblems of the relaxed problem /1*2/
    M number of constraints including complicating constraints /1*6/
* Complicating constraints are the 'ncomp' last constraints
    NN(N) subset of variables
```

```
         MM(M) subset of constraints
         COUNT(P) counter of calculated solutions of the relaxed problem;

* Initialize this counter to the number of initial solutions
         COUNT(P)$(ord(P) le ninitsol)=yes;

* Vectors of data are defined as parameters.
PARAMETER
         Z(P) optimal value of the objective function for solution P
         R(M,P) value of the complicating constraint M for solution P
         XX(N,P) value of the basic solution P of the relaxed problem
* The parameters LV and UV indicate which variables belong to
* each subproblem of the relaxed problem.
         LV(K) first variable in subproblem K /1 1, 2 2/
         UV(K) last variable in subproblem K /1 1, 2 2/
* The parameters LC and UC indicate which constraints belong to
* each subproblem of the relaxed problem.
         LC(K) first constraint in subproblem K /1 1, 2 2/
         UC(K) last constraint in subproblem K /1 1, 2 2/
         C(N) cost coefficients of the original objective function /1 2, 2 1/
         B(M) known terms of each constraint /1 5, 2 5, 3 9, 4 4, 5 -2, 6 -3/
         C1(N) cost coefficients for each subproblem of the relaxed problem
         lambda(M) values of dual variables associated with each master problem...
         ...constraint without considering the normalized constraint;
* Initialize the value of the each solution of the relaxed problem.
XX(N,P)=0;

* The data matrices are defined as a table.
         TABLE E(M,N) constraints coefficients including complicating constraints
               1  2
         1     1
         2        1
         3     1  1
         4     1 -1
         5    -1 -1
         6    -3 -1
         TABLE CC(P,N) cost coefficients (relaxed problem)
               1  2
         1    -1 -1
         2    -2  1;

VARIABLES z1 objective function variable of each problem solved;

POSITIVE VARIABLES
         U(P) variables of the master problem
         X(N) variables of the relaxed problem
         W(M) artificial variable to avoid infeasibility in master problem
         W1 artificial variable to avoid infeasibility  in master problem;
* Initialize the value of the solutions of the relaxed problem.
X.l(N)=1;

EQUATIONS
* Equations are declared.
         zglobaldef original objective function equation
         subproblemdef objective function equation of the subproblems constraints
         constraints equations of the original problem
* or the relaxed problem depending of the subset MM
         zmasterdef objective function of the master problem
         masterc(M) constraint equations of the master problem
         normalize normalized equation of the master problem;

* Once the equations have been declared, they need to be defined.
* The following sentences formulate the above declared equations.
         zglobaldef..z1=e=sum(NN,C(NN)*X(NN));
         subproblemdef..z1=e=sum(NN,C1(NN)*X(NN));
         constraints(MM)..sum(NN,E(MM,NN)*X(NN))=l=B(MM);
         zmasterdef..z1=e=sum(COUNT,Z(COUNT)*U(COUNT))+MMM*(sum(MM,W(MM))+W1);
```

A.1 Dantzig-Wolfe Algorithm

```
        masterc(MM)..sum(COUNT,R(MM,COUNT)*U(COUNT))+W(MM)-W1=l=B(MM);
        normalize..sum(COUNT,U(COUNT))=e=1;

* The next sentences name each model and list its constraints.
* The model 'global' represents the original problem.
* The model 'subprob' represents a subproblem of the relaxed problem.
* The model 'master' represents the master problem.
MODEL global/zglobaldef,constraints/;
MODEL subprob/subproblemdef,constraints/;
MODEL master/zmasterdef,masterc,normalize/;

*** The GLOBAL SOLUTION of the problem is calculated ***
put "starts the global solution"/;

* All variables and constraints are considered in original problem.
NN(N)=yes; MM(M)=yes;

* The next sentence directs GAMS to solve the global model using
* a linear programming solver (lp) to minimize the objective.
SOLVE global USING lp MINIMIZING z1;

* Next sentences write the results in the external file. Omit this
* part of the program if you are only interested in the algorithm.
* The values of the objective function of the original problem,
* the model status and the solver status are shown.
put "z1=",z1.l:12:8," mdst=",global.modelstat,", svst=",global.solvestat//;
put "Global solution"/; put "---------------"/;
* The value of the variables is shown.
loop(N,put X.l(N):8:4," ";); put ""/;

*** The solution is calculated using the DANTZIG-WOLFE DECOMPOSITION ***
put "***%***%***%***%***%***%***%***%***%***%***%***%*"/;
put "Starts the Dantzig-Wolfe decomposition technique"/;

** INITIALIZATION
* Initialize the number of available solutions of the relaxed problem.
IP=0;
* The initial solutions of the relaxed problem are calculated.
loop(COUNT,
* To obtain the initial solution, each subproblem is solved.
    loop(K,
* In each subproblem, the associated variables are assigned by subset NN.
* The same is done with the constraints, which are assigned by subset MM.
        NN(N)=no; NN(N)$((ord(N) ge LV(K)) and (ord(N) le UV(K)))=yes;
        MM(M)=no; MM(M)$((ord(M) ge LC(K)) and (ord(M) le UC(K)))=yes;
* The cost coefficients for each subproblem are assigned.
        C1(NN)=CC(COUNT,NN);

* Next sentences write the results in the external file.
* This part of the program can be omitted.
        put "***%***%***%***%***%***%***%***%***%***%***%***%*"/;
        put "Subproblem ",(IP+1):2," in block ",K.tl:2," is solved minimizing"/;
* The cost coefficients of the subproblems are shown.
        loop(NN,put C1(NN):5:2," ";);put ""/; put "and constraints:"/;
* The coefficients of the subproblem constraints are shown.
        loop(MM,loop(NN,put E(MM,NN):5:2," ");put " <=",B(MM):5:2/;);

* The next sentence directs GAMS to solve the the subprob model using a
* linear programming solver (lp) to minimize the objective.
        SOLVE subprob USING lp MINIMIZING z1;

* Next sentences write the results in the external file. Omit this
* part of the program if you are only interested in the algorithm.
* The objective function values of the subproblem, the model status
* and the solver status are shown.
        put "z1=",z1.l:12:8," mdst=",global.modelstat,",svst=",global.solvestat//;
        put "with solution"/;
```

```
* The variables values of the subproblem are shown.
      loop(NN,put X.1(NN):8:1," ";);put ""/; );

* Checks if the obtained solution already exists.
* Initialize err.
      err=1;
* the counter for identifying new solutions is initialized.
      new=0;
* The new solution is compared with existing solutions.
      loop(P$(ord(P) le IP),
* err measures the difference between actual and calculated solutions
* solution P already available.
            err=sum(N,abs(XX(N,P)-X.1(N)));
* If err is bigger than a tolerance, then the actual solution is new.
            if(err>toler,new=new+1);
      );

      if(new=IP,
* If the actual solution is new, it steps forward and added to the list.
            step=step+1; IP=IP+1; XX(N,P)$(ord(P) = IP)=X.1(N);
* The objective function value is calculated for the new solution.
            Z(P)$(ord(P) = IP)=sum(N,C(N)*X.1(N));
* The complicating constraints values are calculated for the new solution.
            loop(M$(ord(M)>card(M)-ncomp),
                  R(M,P)$(ord(P)=IP)=sum(N,E(M,N)*X.1(N));
            );
      else
* If the solution calculated isn't new, it is rejected.
            put "THIS SOLUTION ALREADY EXISTS"/;
      );
);

* The maximum value of the counter is the number of available solutions.
   COUNT(P)=no; COUNT(P)$(ord(P)le IP)=yes;

* Next sentences write the results in the external file.
* You can omit this part of the program.
* The variable values of each initial solution are shown.
   put "Values of X"/; put "-----------"/;
   loop(COUNT,
         loop(N,put XX(N,COUNT):8:1;);put""/;
   );
* The objective function values of each initial solution are shown.
   put "Values of Z"/; put "-----------"/;
   loop(COUNT,put Z(COUNT):8:1/;);
* The complicating constraints values for each initial solution are shown.
put "Values of R"/; put "-----------"/;
loop(COUNT,
      loop(M$(ord(M)>card(M)-ncomp),put R(M,COUNT):8:1;); put""/;
);

* The master problem and the relaxed problem are solved while
*  the objective function value of the relaxed problem is smaller than
* the dual variable value of the normalized constraint in the
* master problem.
sigma=objectf+1;
while(objectf<sigma,

** MASTER PROBLEM
      put "MASTER PROBLEM"/;
* The original problem constraints are the complicating constraints.
      MM(M)=no; MM(M)$(ord(M) > card(M)-ncomp)=yes;

* Next sentences write the results in the external file.
* You can omit this part of the program.
      put "***%***%***%***%***%***%***%***%***%***%*"/;
      put "Master problem is solved with objective function "/;
```

A.1 Dantzig-Wolfe Algorithm

```
* The cost coefficients of the master problem are shown.
    loop(COUNT,put Z(COUNT):9:3," ";);put ""/;
    put "with normalizing constraint and constraints:"/;
* The coefficients of the constraints of the master problem are shown.
    loop(MM,loop(COUNT,put R(MM,COUNT):5:2," ");put " <= ",B(MM):5:2/;);

* The next sentence directs GAMS to solve the master model using a
* linear programming solver (lp) to minimize the objective.
    SOLVE master USING lp MINIMIZING z1;
* The original problem objective function upper bound is the
* objective function value of the master problem.
    upperB=z1.l;

* Next sentences write the results in the external file.
* You can omit this part of the program.
* The value of the upper bound is shown.
    put "UPPER BOUND=",upperB/;
* The objective function value, the model, and the solver status are shown.
    put "z1=",z1.l:12:8," mdst=",master.modelstat,", svst=",master.solvestat//;
    put "with solution"/;
* The variable values of the master problem are shown.
    loop(COUNT,put U.l(COUNT):8:1, " ";);put " W: ";
* The artificial variable values of the master problem are shown.
    loop(MM,put W.l(MM):8:1," ";);put " W1: ",W1.l:8:1/;

* The lambda value is the marginal value of each complicating constraint.
    lambda(MM)=masterc.m(MM);
* The sigma value is the marginal value of the normalized constraint.
    sigma=normalize.m;

* Next sentences write the results in the external file.
* You can omit this part of the program.
* The values of lambda and sigma are shown.
    put "values of lambda"/; loop(MM,put lambda(MM):8:1," ");put ""/;
    put "and sigma"/; put sigma:8:1/;

** RELAXED PROBLEM
    put "RELAXED PROBLEM"/;
* The cost coefficients for each subproblem are assigned.
    C1(N)=C(N)-sum(MM,lambda(MM)*E(MM,N));
* The relaxed problem solution is obtained solving the subproblems.
    loop(K,
* In each subproblem, the associated variables are assigned by subset NN.
* The same is done with the constraints, which are assigned by subset MM.
        NN(N)=no; NN(N)$((ord(N) ge LV(K)) and (ord(N) le UV(K)))=yes;
        MM(M)=no; MM(M)$((ord(M) ge LC(K)) and (ord(M) le UC(K)))=yes;

* Next sentences write the results in the external file.
* You can omit this part of the program.
        put "***%***%***%***%***%***%***%***%***%***%***%***%*"/;
        put "Subproblem ",IP:2," in block ",K.tl:2," is solved minimizing"/;

* The cost coefficients of the subproblems are shown.
        loop(NN,put C1(NN):5:2," ";);put ""/; put "and constraints:"/;
* The coefficients of the subproblem constraints are shown.
        loop(MM,loop(NN,put E(MM,NN):5:2," ");put " <= ",B(MM):5:2/;);

* The next sentence directs GAMS to solve the subprob model using a
* linear programming solver (lp) to minimize the objective function.
        SOLVE subprob USING lp MINIMIZING z1;

* Next sentences write the results in the external file.
* You can omit this part of the program.
* The objective function values of the subproblem, the model status
* and the solver status are shown.
        put "z1=",z1.l:12:8," mdst=",global.modelstat,", svst=",global.solvestat//;
        put "Generating new solution", IP," Block ",K.tl:2/;
```

```
              put "----------------"/;
* The subproblem variable values are shown.
          loop(NN,put X.l(NN):8:1;);put ""/;
       );

* The optimal objective function lower bound of the problem is calculated.
       lowerB=sum(N,C1(N)*X.1(N))+sum(M$(ord(M)>card(M)-ncomp),lambda(M)*B(M));
* The objective function of the relaxed problem is calculated.
       objectf=sum(N,C1(N)*X.1(N));

* Next sentences write the results in the external file.
* You can omit this part of the program.
* The lowerB and objectf values are shown.
         put "LOWER BOUND=",lowerB; put "objectf=",objectf/;

* Once, the relaxed problem solution (the solution of all subproblems)
* is obtained, it checks if a new solution has been obtained
* Initialize err.
      err=1;
* the counter for identifying new solutions is initialized.
      new=0;
* The new solution is compared with existing solutions.
      loop(P$(ord(P) le IP),
* err measures the difference between actual and calculated solutions
           err=sum(N,abs(XX(N,P)-X.1(N)));
* If err is bigger than the tolerance, then the actual solution is new.
           if(err>toler,new=new+1);
      );
      if(new=IP,
* If the actual solution is new, it steps forward and added to the list.
           step=step+1; IP=IP+1;
           COUNT(P)$(ord(P) eq IP)=yes; XX(N,P)$(ord(P)= IP)=X.1(N);
* The objective function value is calculated for the new solution.
           Z(P)$(ord(P) = IP)=sum(N,C(N)*X.1(N));
* The complicating constraints values are calculated for the new solution.
           loop(M$(ord(M)>card(M)-ncomp),
               R(M,P)$(ord(P) =IP)=sum(N,E(M,N)*X.1(N));
           );
       else
* If the solution calculated isn't new, it is rejected.
           put "THIS SOLUTION ALREADY EXISTS"/;
      );
);
* Next sentences write the results in the external file.
* You can omit this part of the program.
* The variable values of each initial solution are shown.
put "Values of X"/; put "-----------"/;
loop(COUNT, loop(N,put XX(N,COUNT):8:1;); put""/;);
* The objective function values of each initial solution are shown.
put "Values of Z"/; put "-----------"/;
loop(COUNT,put Z(COUNT):8:1/;);
* The complicating constraints values for each initial solution are shown.
put "Values of R"/; put "-----------"/;
loop(COUNT,
     loop(M$(ord(M)>card(M)-ncomp),put R(M,COUNT):8:1;);
     put""/;
);


* The optimal value of the variables of the master problem is shown.
put "values of U"/; put "------------"/;
loop(COUNT,put U.1(COUNT):8:1, " ";);put""/; put "SOLUTION:\\"/;
* The optimal objective function value of the original problem is shown.
put "z=",(sum(P$(ord(P) <= IP),U.1(P)*Z(P))):8:1;
* The optimal variable values of the original problem are shown.
loop(N, put (sum(P$(ord(P) <= IP),U.1(P)*XX(N,P))):8:1," "; );
put ""/; put "------------------ END ------------------"/;
```

A.2 GAMS Implementation of the Benders Decomposition Algorithm

A GAMS implementation of the Benders decomposition algorithm is given below. The aim of this code is simply to clarify how the algorithm works. Note that the code is not optimized because it solves each of the linear programming problems from scratch, without using the previous results.

```
* This program solves linear programming problems with complicating
* variables using the Benders decomposition.
* We consider the computational example 1.7.
$title b1_7

* We define a file to store the results of the problem.
* The internal file name is 'out1' and it is referred to as external
* name 'b1_7.out'.
file out1/b1_7.out/; put out1;

* First, single scalar data are declared and defined.
SCALARS
    IP counter of iterations
    ncomp number of complicating variables /1/
    ncmast number of constraints of the master original problem /1/
    toler tolerance to control convergence of the algorithm /1E-8/
    lowerB lower bound of the original problem objective function /-1E+08/
    upperB upper bound of the original problem objective function /1E+08/
    alphadown lower bound of alpha /-25/
    MMM large enough positive constant to penalize the artificial variables...
    ....value of the subproblems /20/;

* Indices are declared and defined.
SETS
    N number of variables including complicating variables /1*2/
    K number of subproblems /1*1/
    M number of constraints /1*5/
* The master problem constraints are the 'ncmast' last constraints
    NN(N) subset of variables
    NNC(N) subset of complicating variables
    MM(M) subset of constraints
    P maximum number of solutions of the master problem /1*15/
    COUNT(P) number of calculated feasible solutions of the master problem;
* The complicating variables are the 'ncomp' last variables.
    NNC(N)=no; NNC(N)$(ord(N) gt card(N)-ncomp)=yes;
* Initialize the counter of solutions of the master problem.
    COUNT(P)=no;

* Vectors of data are defined as parameters.
PARAMETER
* LV and UV store which variables belong to each subproblem.
    LV(K) first variable in subproblem K /1 1/
    UV(K) last variable in subproblem K /1 1/
* LC and UC store which constraints belong to each subproblem.
    LC(K) first constraint in subproblem K /1 1/
    UC(K) last constraint in subproblem K /1 4/
    C(N) cost coefficients of the original objective function /1 -1, 2 -0.25/
    B(M) known terms of each constraint /1 5, 2 7.5, 3 17.5, 4 10, 5 16/
    OBJSUB(P) objective function value of the subproblem
    lambda(P,N) value of the dual variables associated with the constraints...
    ...that fix the value of the complicating variables in the subproblem
    XC_fix(P,N) value of the complicating variables of the solution P;

* The data matrices are defined as a table.
    TABLE E(M,N) coefficients of the constraints
           1    2
```

```
1      1   -1
2      1   -0.5
3      1    0.5
4     -1    1
5           1;

VARIABLES
    alpha variable alpha of the master problem
    z1 objective function variable of each problem solved;

POSITIVE VARIABLES
    X(N) variables of the subproblems to avoid infeasibility
    W(M) artificial variable used in the subproblems to avoid infeasibility
    W1 artificial variable used in the subproblems to avoid infeasibility;

EQUATIONS
* Equations are declared.
    zglobaldef original objective function equation constraints
    constraints equations of the original or the master problem
    subproblemdef objective function equation of each subproblem
    subproblemc constraint equations of each subproblem
    frozenc frozing constraint of the complicating variables
    zmasterdef objective function of the master problem
    benders_cut Benders' cut constraint
    lowerb_alpha lower bound constraint of alpha;

* Once the equations have been declared, they need to be defined.
* The following sentences formulate the above declared equations.
zglobaldef..z1=e=sum(NN,C(NN)*X(NN));
constraints(MM)..sum(NN,E(MM,NN)*X(NN))=l=B(MM);
subproblemdef..z1=e=sum(NN,C(NN)*X(NN))+MMM*(sum(MM,W(MM))+W1);
subproblemc(MM)..sum(N,E(MM,N)*X(N))+W(MM)-W1=l=B(MM);
frozenc(COUNT,NNC)..X(NNC)=e=XC_fix(COUNT,NNC);
zmasterdef..z1=e=sum(NN,C(NN)*X(NN))+alpha;
benders_cut(COUNT)..alpha=g=OBJSUB(COUNT)
                    +sum(NN,lambda(COUNT,NN)*(X(NN)-XC_fix(COUNT,NN)));
lowerb_alpha..alpha=g=alphadown;

* The next sentences name each model and list its constraints.
* The first model 'global' represents the original problem.
* The second mode 'subprob' represents a subproblem.
* The third model 'master' represents the master problem.
MODEL global/zglobaldef,constraints/;
MODEL subprob/subproblemdef,subproblemc,frozenc/;
MODEL master/zmasterdef,constraints,benders_cut,lowerb_alpha/;

*** The GLOBAL SOLUTION of the problem is calculated ***
put "starts the global solution"/;

* The original problem contains all variables and constraints.
NN(N)=yes; MM(M)=yes;

* The next sentence directs GAMS to solve the global model using
* a linear programming solver (lp) to minimize the objective.
SOLVE global USING lp MINIMIZING z1;

* Next sentences write the results in the external file.
* You can omit this part of the program.
* The values of the objective function of the original problem,
* the model status, and the solver status are shown.
put "z1=",z1.1:12:8," mst= ",global.modelstat," sst= ",global.solvestat//;
put "Global solution"/; put "---------------"/;
* The value of the variables is shown.
loop(N,put X.1(N):8:4," ";); put ""/;

*** The problem is solved using the BENDERS DECOMPOSITION ***
put "***%***%***%***%***%***%***%***%***%***%***%*"/;
```

A.2 Benders Decomposition Algorithm

```
put "Starts the Benders decomposition technique"/;

* Initialize the counter of iteration.
IP=0;

* The master problem and the subproblem or subproblems are solved
* while the objective function upper bound is bigger than its lower bound.
while((upperB-lowerB)>toler,
* Increase the counter of iterations.
    IP=IP+1;

** MASTER PROBLEM
* The variables considered are the 'ncomp' last variables.
    NN(N)=no; NN(N)$(ord(N) gt card(N)-ncomp)=yes;
* The original problem constraints are the 'ncmast' last constraints.
    MM(M)=no; MM(M)$(ord(M) gt card(M)-ncmast)=yes;
* In the first iteration there is not Benders' cuts (Initialization)
    if (IP=1,
        lambda(COUNT,NN)=0; XC_fix(COUNT,NN)=0; OBJSUB(COUNT(P))=alphadown;

* Next sentences write the results in the external file.
* You can omit this part of the program.
        put "***%***%***%***%***%***%***%***%***%***%***%***%*"/; put
"Initialization is solved with objective function "/;
* The cost coefficients of the master problem are shown.
        loop(NN,put C(NN):9:3," ";);put "+ alpha";put ""/;
        put "with constraints:"/;
* The constraints coefficients of the master problem are printed.
        loop(MM,loop(NN,put E(MM,NN):5:2," ");put " <= ",B(MM):6:2/;);
        put "alpha >=",alphadown/;

* In the rest of the iterations Benders' cut constraints appear.
    else

* Next sentences write the results in the external file.
* You can omit this part of the program.
        put "***%***%***%***%***%***%***%***%***%***%***%***%*"/; put
"Master problem is solved with objective function "/;
* The cost coefficients of the master problem are shown.
        loop(NN,put C(NN):9:3," ";);put "+ alpha";put ""/;
        put "with constraints:"/;
* The constraints coefficients of the master problem are shown.
        loop(MM,loop(NN,put E(MM,NN):5:2," ");put " <= ",B(MM):6:2/;);
        put "alpha >=",alphadown/;
        loop(COUNT,
            put "alpha >=",OBJSUB(COUNT):8:2," ";
            loop(NN,put lambda(COUNT,NN):10:2;);put ""/;);
    );

* The next sentence directs GAMS to solve the master model using
* a linear programming solver (lp) to minimize the objective function.
    SOLVE master USING lp MINIMIZING z1;
* The counter of solutions only shows the actual solution.
    COUNT(P)=no; COUNT(P)$(ord(P)eq IP)=yes;
* The values of the master problem complicating variables are stored.
    XC_fix(COUNT,NN)=X.l(NN);
* The lower bound of the optimal objective function of the original
* problem  is the objective function value of the master problem.
    lowerB=z1.l;

* Next sentences write the results in the external file.
* You can omit this part of the program.
* The value of the lower bound is shown.
        put "LOWER BOUND=",lowerB/;
* The values of the objective function of the master problem,
* the model status, and the solver status are shown.
        put "z1=",z1.l:12:8," mst= ",master.modelstat,", sst= ",master.solvestat//;
```

```
        put "with solution"/;
* The values of the variables of the master problem are shown.
        loop(NN,put X.l(NN):8:1, " "/;);
* The value of alpha is shown.
        put "value of alpha "/; put alpha.l:8/;

** SUBPROBLEM
* Initialize the subproblem objective function value and the lambda value
* If there are several subproblems, the final objective function value
* of the subproblem is the sum of the subproblem objective function values.
* The lambda final value is the sum of the subproblem lambdas.
        OBJSUB(COUNT)=0; lambda(COUNT,NN)=0;
* The subproblem or subproblems are solved.
        loop(K,
* For each subproblem, the associated variables are assigned (subset NN).
* The same with the constraints that are assigned to the subset MM.
            NN(N)=no; NN(N)$((ord(N) ge LV(K)) and (ord(N) le UV(K)))=yes;
            MM(M)=no; MM(M)$((ord(M) ge LC(K)) and (ord(M) le UC(K)))=yes;

* Next sentences write the results in the external file.
* You can omit this part of the program.
            put "***%***%***%***%***%***%***%***%***%***%***%***%*"/;
            put "Subproblem ",IP:2," in block ",K.tl:2," is solved minimizing"/;
* The cost coefficients of the subproblem considered are shown.
            loop(NN,put C(NN):5:2," ";);put ""/; put "and constraints:"/;
* The subproblem constraints coefficients are shown.
            loop(MM,
                loop(NN,put E(MM,NN):5:2;);
                loop(NNC,put E(MM,NNC):10:2, " ");
                put " <= ",B(MM):6:2/;
            );

* The next sentence directs GAMS to solve the subprob model
* using a linear programming solver (lp) to minimize the objective function.
            SOLVE subprob USING lp MINIMIZING z1;
* The value of the objective function of the subproblem is assigned.
            OBJSUB(COUNT)=OBJSUB(COUNT)+z1.l;
* Assign lambda value (marginal value of the frozen subproblem constraints).
            lambda(COUNT,NNC)=lambda(COUNT,NNC)+frozenc.m(COUNT,NNC);

* Next sentences write the results in the external file.
* You can omit this part of the program.
* The values of the objective function of the subproblem considered,
* the model status, and the solver status are shown.
            put "z1=",z1.l:12:8," mst= ",global.modelstat,", sst= ",global.solvestat//;
            put "with solution in block ",K.tl:2/; put "----------------"/;
* The values of the variables of the subproblem solved are shown.
            loop(NN,put X.l(NN):8:1, " ";);put " W: ";
* The values of the artificial variables of the subproblem are shown.
            loop(MM,put W.l(MM):8:1," ";);put " W1: ",W1.l:8:1/;
* The value of lambda is shown.
            put "values of lamda "/;
            loop(NNC,loop(COUNT,put frozenc.m(COUNT,NNC):8:3/;););
        );

* The original problem optimal objective function upper bound is calculated.
        upperB=sum(COUNT,OBJSUB(COUNT))+sum(NNC,C(NNC)*X.l(NNC));
        put "UPPER BOUND=",upperB/;
* The counter of solutions shows the subset of solutions calculated.
        COUNT(P)=no; COUNT(P)$(ord(P)le IP)=yes;
);

* Next sentences write the results in the external file.
* You can omit this part of the program.
* The optimal value of the variables is shown.
put "OPTIMAL SOLUTION:\\"/; loop(N,put X.l(N):8:1;); put ""/;
put "------------------ END ------------------"/;
```

A.3 GAMS Code for the Rubblemound Breakwater Example

A GAMS implementation of the Benders decomposition algorithm for solving the reliability-based optimization of a rubblemound breakwater, introduced in Sect. 1.5.4 and in the Computational Example 5.1, p. 228 is given below.

```
$title rubblemound breakwater example

* Initializing the output file
file out /rebasebenders.out/; put out;

SETS
    I number of random variables /1*2/
    IT iterations /1*20/
    ITER(IT) dinamic set to control Benders cuts;

ITER(IT)=no;

SCALARS
    pi /3.1415926535898/
    gra /9.81/
    error /1/
    Csup upper bound of the total cost /INF/
    Cinf lower bound of the total cost /5000/
    Toler admissible tolerance/1e-3/;

SCALARS
* Non-optimization design variables
    Dwl design water level (m) /20/
    Au experimental parameter for runup /1.05/
    Bu experimental parameter for runup /-0.67/
    cc concrete cost (dollars) /60/
    ca armor cost (dollars) /2.4/
* Random model parameters, sea state descriptors
    Hs significant wave height (m) /5/
    Tz mean period (s) /10/
    dst sea state duration (s) /3600/
* Auxiliary scalars
    pf probability of failure for one wave
    pfD probability of failure for the design sea state;

PARAMETERS
    Ctotal(IT) vector of total costs for each iteration
    ValorFc(IT) vector of freeboards for each iteration
    LambdaFc(IT) vector of partial derivatives of the total cost function...
    ...with respect to the freeboard for each iteration
    ValorTan(IT) vector of seaside slopes for each iteration
    LambdaTan(IT) vector of partial derivatives of the total cost...
    ...function with respect to seaside slopes for each iteration;

VARIABLES
    cd construction cost
    ci insurance cost
    beta reliability index
    alfacost auxiliary variable for the master problem
    z(I) normalized random variables;

POSITIVE VARIABLES
* Random variables
    H wave height
    T wave period
* Optimization design variables, complicating variables
    tan seaside slope
```

```
        Fc freeboard level
        auxtan auxiliary seaside slope for the master problem
        auxFc auxiliary freeboard level for the master problem
*Auxiliary or non-basic variables
        Ir Iribarren number
        L wavelength
        Ru run up
        d breakwater height
        vc concrete volume
        va armor volume;

* Limits for the design variables,
* very important to achieve convergence
auxFc.up=20; auxFc.lo=5; auxtan.lo=1/5; auxtan.up=1/2;
L.lo=10; H.up=2.2*Hs; T.up=2.2*Tz;

EQUATIONS
* Overall optimization
        cddf construction cost definition
        ddf breakwater height definition
        vcdf concrete volume definition
        vadf armor volume definition
* Sub-level optimization problem
        betadef reliability index definition
        Zdef1 definition of the standard normal random variables z1
        Zdef2 definition of the standard normal random variables z2
        Iridf Irribarren number definition
        Ldf dispersion equation
        rudf runup definition
        verdf definition of the verification equation runup equal to Fc
* Master problem
        Restric(IT) Benders cuts
        auxmaster;

* Overall optimization
        cddf..cd=e=cc*vc+ca*va; ddf..Fc=e=2+d; vcdf..vc=e=10*d;
        vadf..va=e=0.5*(Dwl+2)*(Dwl+46+(Dwl+2)/tan);

* Sub-level optimization problem
        betadef..beta=e=sqrt(sum(I,sqr(z(I))));
        Zdef1..errorf(z('1'))=e=1-exp(-2*sqr(H/Hs));
        Zdef2..errorf(z('2'))=e=1-exp(-0.675*power((T/Tz),4));
        Iridf..Ir*sqrt(H/L)=e=tan;
        Ldf..2*pi*L*(exp(2*pi*Dwl/L)+exp(-2*pi*Dwl/L))=e=
        T*T*gra*(exp(2*pi*Dwl/L)-exp(-2*pi*Dwl/L));
        rudf..Ru=e=H*Au*(1-exp(Bu*Ir)); verdf..Fc=e=Ru;

* Master
        Restric(ITER)..alfacost=g=Ctotal(ITER)+LambdaFc(ITER)*(auxFc-ValorFc(ITER))+
        LambdaTan(ITER)*(auxtan-ValorTan(ITER));
        auxmaster..alfacost=g=5000;

MODEL sublevel/betadef,Zdef1,Zdef2,Iridf,Ldf,rudf,verdf/;
MODEL cdirect/cddf,ddf,vcdf,vadf/;
MODEL Master/Restric,auxmaster/;

Ctotal(IT)=0.0; LambdaFc(IT)=0.0; ValorFc(IT)=0.0;
LambdaTan(IT)=0.0; ValorTan(IT)=0.0;

* Initial values for the complicating variables
Fc.fx=7; tan.fx=1/3;

loop(IT$(error gt TOLER), put "        Iteration= ",ord(IT):12:8//;

    if(ORD(IT)>1,
* Solving the master problem for obtaining new values of
* the complicating variables
```

A.3 GAMS Code for the Rubblemound Breakwater Example

```
        SOLVE Master USING lp MINIMIZING alfacost;
        put "alfacost= ",alfacost.l:12:4,", modelstat= ",Master.modelstat,
        ", solvestat= ",Master.solvestat/;
* New values for the complicating variables
        Fc.fx=auxFc.l; tan.fx=auxtan.l;
* Lower bound of solution at iteration IT
        Cinf=alfacost.l;
    );
* Saving the values of the complicating variables for iteration IT
    ValorFc(IT)=Fc.l; ValorTan(IT)=tan.l;
* Activating Benders cut
    ITER(IT)=yes;

put "Complicating variables: Fc=",Fc.l:6:3,", tan=",tan.l:6:3/;

* Initial values, very important to achieve convergence
    H.l=1.5*Hs; T.l=1.1*Tz; L.l=136.931; Ir.l=tan.l/sqrt(H.l/L.l);
    Ru.l=H.l*Au*(1-exp(Bu*Ir.l)); z.l('1')=2.28; z.l('2')=0.32;
    beta.l=sqrt(sum(I,sqr(z.l(I))));

* Solve the reliability problem for fixed values
* of the complicating variables
    SOLVE sublevel USING nlp MINIMIZING beta;

    put "pf= ",(errorf(-beta.l)):12:8,", modelstat= ",
    sublevel.modelstat,", solvestat= ",sublevel.solvestat/;

* Probabilities of failure
    pf=errorf(-beta.l); pfD=1-(1-pf)**(dst/Tz);
* Partial derivatives of pfD with respect to complicating variables
    LambdaFc(IT)=-dst*(1-pf)**(dst/Tz-1)*exp(-beta.l*beta.l/2)*
    (Fc.m)/(Tz*sqrt(2*pi));
    LambdaTan(IT)=-dst*(1-pf)**(dst/Tz-1)*exp(-beta.l*beta.l/2)*
    (tan.m)/(Tz*sqrt(2*pi));
* Insurance cost as a function of pfD
    ci.l=5000+125000000*pfD**2;

    put "pfD= ",pfD:12:8/; put "insurance cost= ",(ci.l):12:4/;
    put "LambdaFc1(",ord(IT):2," )=",LambdaFc(IT):12:4,
    "LambdaTn1(",ord(IT):2," )=",LambdaTan(IT):12:4/;

    Ctotal(IT)=ci.l;
* Partial derivatives of insurance cost with respect to
* complicating variables
    LambdaFc(IT)=LambdaFc(IT)*(2*125000000*pfD);
    LambdaTan(IT)=LambdaTan(IT)*(2*125000000*pfD);

    put "LambdaFc2(",ord(IT):2," )=",LambdaFc(IT):12:4,
    "LambdaTn2(",ord(IT):2," )=",LambdaTan(IT):12:4/;

* Auxiliary model for calculating the construction cost and their
* partial derivatives with respect to complicating variables
    SOLVE cdirect USING nlp MINIMIZING cd;

    put "direct cost= ",cd.l:12:4,", modelstat= ",cdirect.modelstat,",
    solvestat= ",cdirect.solvestat/;

* Total cost for fixed values of the complicating variables
    Ctotal(IT)=Ctotal(IT)+cd.l;
* Partial derivatives of total cost with respect to complicating variables
    LambdaFc(IT)=LambdaFc(IT)+Fc.m; LambdaTan(IT)=LambdaTan(IT)+tan.m;

    put "Ctotal(",ord(IT):2," )=",Ctotal(IT):12:4/; put
    "LambdaFc(",ord(IT):2," )=",LambdaFc(IT):12:4,
    "LambdaTn(",ord(IT):2," )=",LambdaTan(IT):12:4/;

* Upper bound of solution
```

```
        Csup=Ctotal(IT);

*  Calculating error
      error=(abs(Csup-Cinf)/Cinf);

      put "Upper bound= ",Csup:12:4/; put "Lower bound= ",Cinf:12:4/;
      put "error= ",error:15:10//;
);
```

A.4 GAMS Code for the Wall Problem

A.4.1 The Relaxation Method

This version of the GAMS code contains commands to perform a sensitivity analysis with respect to the cost (classic design) and reliability indices β_o, β_s, and β_b (probabilistic design). Note that this program solves the wall problem introduced in Sect. 9.1, p. 349 using the relaxation method explained in Subsection 9.1.1.

```
$title wall example method 1 with sensitivity analysis

file out1 /wall1.out/;

SETS
    V set of variables /a,b,nu,T,gamma,H,S/
    D(V) set of optimized design variables     /a,b/
    A(V) set of non-optimized design variables /nu,T,gamma,H,S/
    M failure modes /turn,slid,bear/
    IT iterations /1*15/;

ALIAS(M,Maux);

SCALARS

*  convergence control

      epsilon maximum allowable tolerance /1e-5/
      error error in actual iteration/1/
      iteration iteration number /0/
      rho relaxation factor /0.8/;

PARAMETERS
*  Safety factors lower bounds
    Flo(M) /
    turn    1.5
    slid    1.6
    bear    1.5/
    Fr(M) Real safety factors values
    Fpar(M)
    betalo(M) Beta lower bounds /
    turn    3.0
    slid    3.0
    bear    3.0/
    betaa(M) Actual beta values
    betaux(M) Auxiliar beta values for error checking
    Faux(M)
    mean(V) mean values for classic model/
    a       3
    b       6
    nu      0.3
    T       50.0
```

```
        gamma   23.0
        H        3.0
        S      220.0
        /
        sigma(V) standard deviations /
        a       0.01
        b       0.01
        nu      0.05
        T      15.0
        gamma   0.46
        H       0.2
        S      16.0/
        PML(M,V)  Points of maximum likelihood;

        PARAMETERS
* Auxiliar parameters for sensitivity analysis
        sensFclas(M), sensAclas(A), sensBclas(V), sensMclas(M), sensF(M,M)
        sens(M,V), sensB(M,V);

VARIABLES
        beta(M) Actual beta values
        cost Master objective function
        ZC
        VarD(V) Variables
        VarR(V) Random variables
        VarB(V) Statistical data variables
        Fa(M) Safety factors
        Z(V) Auxiliary variables for subproblems
        Zeta(M,V);

EQUATIONS
* Equations for primal problem WITHOUT sensitivity analysis
ZClassic, turn, slid, bear, geometric
* Equations for failure problems WITHOUT sensitivity analysis
Zbeta, Zdef, Fturn, Fslid, Fbear
* Equations for first primal problem WITH sensitivity analysis
turnS, slidS, bearS, Zsens, betasens
* Equations for failure problems WITH sensitivity analysis
ZdefS;

* Equations for primal problem WITHOUT sensitivity analysis
ZClassic..cost=e=VarD('a')*VarD('b');
turn..VarD('a')*VarD('a')*VarD('b')*mean('gamma')=g=2*mean('H')
        *mean('T')*Fpar('turn');
slid..VarD('a')*VarD('b')*mean('nu')*mean('gamma')=g=mean('T')
        *Fpar('slid');
bear..mean('S')=g=VarD('b')*mean('gamma')*Fpar('bear');
geometric..VarD('a')*2=e=VarD('b');

* Equations for failure problems WITHOUT sensitivity analysis
Zbeta..ZC=e=sqrt(sum(V,sqr(Z(V))));
Zdef(V)..Z(V)=e=(VarR(V)-mean(V))/sigma(V);
Fturn..VarR('a')*VarR('a')*VarR('b')*VarR('gamma')=e=2*VarR('H')
        *VarR('T');
Fslid..VarR('a')*VarR('b')*VarR('nu')*VarR('gamma')=e=VarR('T');
Fbear..VarR('b')*VarR('gamma')=e=VarR('S');

* Equations for primal problem WITH sensitivity analysis
turnS..VarD('a')*VarD('a')*VarD('b')*VarD('gamma')=g=2*VarD('H')
        *VarD('T')*Fa('turn');
slidS..VarD('a')*VarD('b')*VarD('nu')*VarD('gamma')=g=VarD('T')
        *Fa('slid');
bearS..VarD('S')=g=VarD('b')*VarD('gamma')*Fa('bear');
Zsens(M,V)..Zeta(M,V)=e=(PML(M,V)-VarD(V))/(VarB(V));
betasens(M)..sqrt(sum(V,Zeta(M,V)*Zeta(M,V)))=g=betalo(M);

* Equations for failure problems WITH sensitivity analysis
```

412   A Some GAMS Implementations

```
    ZdefS(V)..Z(V)=e=(VarR(V)-VarD(V))/VarB(V);

***%***% INITIAL MODEL WITHOUT SENSITIVITY ANALYSIS ***%***%

    MODEL classic /ZClassic,turn,slid,bear,geometric/;
    MODEL mturn /Zbeta,Zdef,Fturn/;
    MODEL mslid /Zbeta,Zdef,Fslid/;
    MODEL mbear /Zbeta,Zdef,Fbear/;

***%***% MODELS WITH SENSITIVITY ANALYSIS***%***%

    MODEL classicS /ZClassic,turnS,slidS,bearS,geometric,Zsens,betasens/;
    MODEL mturnS /Zbeta,ZdefS,Fturn/;
    MODEL mslidS /Zbeta,ZdefS,Fslid/;
    MODEL mbearS /Zbeta,ZdefS,Fbear/;

    Fpar(M)=Flo(M); betaa(M)=betalo(M); PML(M,V)=0;

    put out1;
    put "-----------------------------------------------------------------"/;
    put " n  Cost     a      b      Fo     Fs     Fb     Bo     Bs     Bb     Error"/;
    put "-----------------------------------------------------------------"/;
    loop(IT$(error>epsilon),
        iteration=iteration+1;
        betaux(M)=betaa(M);
*   Initialize design variables

        VarD.l(D)=mean(D);
        SOLVE classic USING nlp MINIMIZING cost;

*   Actual safety factors

        Fr('turn')=(VarD.l('a')*VarD.l('a')*VarD.l('b')*mean('gamma')/
            (2*mean('H')*mean('T')));
        Fr('slid')=(VarD.l('a')*VarD.l('b')*mean('nu')*mean('gamma'))
            /mean('T'));
        Fr('bear')=mean('S')/(VarD.l('b')*mean('gamma'));
        mean(D)=VarD.l(D);
        loop(M,
            VarR.l(V)=mean(V);
                if(ORD(M) eq 1,
                    SOLVE mturn USING nlp MINIMIZING zc;
                else if(ORD(M) eq 2,
                    SOLVE mslid USING nlp MINIMIZING zc;
                else
                    SOLVE mbear USING nlp MINIMIZING zc;
                );
        );
*   Save beta values and points of maximum likelihood
        betaa(M)=zc.l;
        PML(M,V)=VarR.l(V);
    );
*   update safety factor bounds

    Fpar(M)=Fpar(M)+rho*(betalo(M)-betaa(M));
    loop(M,if(Fpar(M)<Flo(M),Fpar(M)=Flo(M);););
*   error evaluation

    error=0.0;
    loop(M,
        if(abs((betaa(M)-betaux(M))/betaa(M))>error,
            error=abs((betaa(M)-betaux(M))/betaa(M));
        );
    );
*------------------------Printing Table ---------------------*
    put iteration:2:0;
    put cost.l:6:2;
```

A.4 GAMS Code for the Wall Problem

```
        loop(D,put mean(D):6:2;);
        loop(M,put Fr(M):6:2;);
        loop(M, put betaa(M):6:2;);
        put error:9:5/;
*------------------- End Printing Table End ------------------*
); put
"---------------------------------------------------------------"//;

VarD.fx(A)=mean(A); VarB.fx(V)=sigma(V); Fa.fx(M)=Flo(M);

loop((M,D,V)$(ORD(D)=ORD(V)),
    Zeta.l(M,V)=(PML(M,D)-VarD.l(D))/(VarB.l(V));
);
loop((M,A,V)$(ORD(A)+CARD(D)=ORD(V)),
    Zeta.l(M,V)=(PML(M,A)-VarD.l(A))/(VarB.l(V));
);

* Final loop for calculating  sensitivities

SOLVE classicS USING nlp MINIMIZING cost;

sensFclas(M)=Fa.m(M); sensAclas(A)=VarD.m(A);
sensBclas(V)=VarB.m(V); sensMclas(M)=betasens.m(M);
VarD.fx(D)=mean(D);

loop(M,
    VarR.l(D)=mean(D);
    VarR.l(A)=mean(A);
    if(ORD(M) eq 1,

        SOLVE mturnS USING nlp MINIMIZING zc;
    else if(ORD(M) eq 2,
        SOLVE mslidS USING nlp MINIMIZING zc;
    else
        SOLVE mbearS USING nlp MINIMIZING zc;);
    );
    sens(M,V)=VarD.m(V);
    sensB(M,V)=VarB.m(V);
);

*---------- Print table of sensitivities ------------*
put "----------------------------------------"/;
put " x       c       Bo      Bs      Bb"/;
put "-------------------MEAN------------------"/;
loop(D,
    put D.tl:7,"    --   ";
    loop(M,put sens(M,D):9:3;); put ""/;
);
loop(A,
    put A.tl:7,sensAclas(A):9:3;
    loop(M,put sens(M,A):9:3;); put ""/;
);
put "-------------------SIGMA----------------- "/;
loop(V,
    put V.tl:7,sensBclas(V):9:3;
    loop(M,put sensB(M,V):9:3;); put ""/;
);
put "----------------------------------------"/;
loop(M,
    put "F",M.tl:6,sensFclas(M):9:3;
    loop(Maux,put "       -- ";); put ""/;
);
put "----------------------------------------"/;
loop(M,
    put "B",M.tl:6,sensMclas(M):9:3;
    loop(Maux,put "       -- ";);put ""/;
```

```
);
put "----------------------------------------"/;
```

A.4.2 The Cutting Hyperplanes Method

This version of the GAMS code contains commands to perform a sensitivity analysis with respect to the cost (classic design) and reliability indices β_o, β_s, and β_b (probabilistic design). Analogously to the previous code this program solves the wall problem introduced in Sect. 9.1, p. 349 but using the cutting plane (CP) method explained in detail in SubSection 9.1.2.

```
$title wall example method 2 with sensitivity analysis

file out1 /Wall2.out/; put out1;

SETS
    V set of variables /a,b,nu,T,gamma,H,S/
    D(V) set of optimized design variables   /a,b/
    A(V) set of non-optimized design variables  /nu,T,gamma,H,S/
    M failure modes /turn,slid,bear/
    IT iterations /1*15/
    ITER(IT) index to control hyperplane cuts
    FIRST(M);

ALIAS(M,Maux);

FIRST(M)=no; ITER(IT)=no;

SCALARS
    cost objective function values
    epsilon maximum allowable tolerance /1e-5/
    error error in actual iteration /1/
    iteration iteration number /0/;

PARAMETERS
* Safety factors lower bounds
    Flo(M)/
    turn    1.5
    slid    1.6
    bear    1.5/
    Fr(M) Real safety factors values
* Beta lower bounds
    betalo(M) Beta lower bounds/
    turn    3.0
    slid    3.0
    bear    3.0/
    betaa(M) Actual beta values
    betaux(M) Auxiliar beta values for error checking
* Auxiliar parameters for cuts
    betaK(IT,M), lambdaK(IT,M,D), muK(M,A), deltaK(M,V), XesK(IT,D)
* Design variables
    mean(V) mean values for classic model/
    a       3
    b       6
    nu      0.3
    T       50.0
    gamma   23.0
    H       3.0
    S       220.0
    /
    sigma(V) standard deviations /
    a       0.01
    b       0.01
```

A.4 GAMS Code for the Wall Problem

```
         nu    0.05
         T     15.0
         gamma 0.46
         H     0.2
         S     16.0/
         PML(M,V)  Points of maximum likelihood;

PARAMETERS
errors(IT)
* Auxiliar parameters for sensitivity analysis
sensFclas(M),sensAclas(A),sensBclas(V),sensMclas(M)
sensX(M,D),sensA(M,A),sensB(M,V);

VARIABLES
* Actual beta values
     beta(M)
* Master objective function
     zc
* Design variables
     VarD(V) variables
     VarR(V) Random variables
     VarB(V) Statistical data variables
     Fa(M) Safety factors
     Z(V) Auxiliary variables for subproblems
     Zeta(M,V);

EQUATIONS
* Equations for primal problem without sensitivity analysis
Zclass,turn,slid,bear,geometric,betadef,auxbeta
* Equations for failure problems without sensitivity analysis
Zbeta,ZDdef,ZAdef,Fturn,Fslid,Fbear,fixedX,fixedFa
* Equations for primal problem with sensitivity analysis
turnS,slidS,bearS,betadefS,fixedA(A),fixedB(V)
* Equations for failure problems with sensitivity analysis
ZdefS;

* Equations for primal problem without sensitivity analysis
Zclass..zc=e=VarD('a')*VarD('b');
turn..VarD('a')*VarD('a')*VarD('b')*mean('gamma')=g=
     2*mean('H')*mean('T')*Flo('turn');
slid..VarD('a')*VarD('b')*mean('nu')*mean('gamma')=g=
     mean('T')*Flo('slid');
bear..mean('S')=g=VarD('b')*mean('gamma')*Flo('bear');
geometric..VarD('a')*2=e=VarD('b');
betadef(ITER,M)$(FIRST(M))..beta(M)=l=betaK(ITER,M)
     +sum(D,lambdaK(ITER,M,D)*(VarD(D)-XesK(ITER,D)));
auxbeta(M)$(FIRST(M))..beta(M)=g=betalo(M);

* Equations for failure problems without sensitivity analysis
Zbeta..zc=e=sqrt(sum(V,sqr(Z(V))));
ZDdef(D)..Z(D)=e=(VarR(D)-VarD(D))/(sigma(D));
ZAdef(A)..Z(A)=e=(VarR(A)-mean(A))/(sigma(A));
Fturn..VarR('a')*VarR('a')*VarR('b')*VarR('gamma')=e=
      2*VarR('H')*VarR('T');
Fslid..VarR('a')*VarR('b')*VarR('nu')*VarR('gamma')=e=VarR('T');
Fbear..VarR('b')*VarR('gamma')=e=VarR('S');
fixedX(D)..VarD(D)=e=mean(D);

* Equations for first primal problem with sensitivity analysis
turnS..VarD('a')*VarD('a')*VarD('b')*VarD('gamma')=g=
     2*VarD('H')*VarD('T')*Fa('turn');
slidS..VarD('a')*VarD('b')*VarD('nu')*VarD('gamma')=g=
     VarD('T')*Fa('slid');
bearS..VarD('S')=g=VarD('b')*VarD('gamma')*Fa('bear');
betadefS(ITER,M)$(FIRST(M))..beta(M)=l=betaK(ITER,M)
     +sum(D,lambdaK(ITER,M,D)*(VarD(D)-XesK(ITER,D)))
     +sum(A,muK(M,A)*(VarD(A)-mean(A)))
```

```
          +sum(V,deltaK(M,V)*(VarB(V)-sigma(V)));

* Equations for failure problems with sensitivity analysis
ZdefS(V)..Z(V)=e=(VarR(V)-VarD(V))/(VarB(V));

***%***% INITIAL MODELS WITHOUT SENSITIVITY ANALYSIS ***%***%

MODEL classic /Zclass,turn,slid,bear,geometric,betadef,auxbeta/;
MODEL mturn/Zbeta,ZDdef,ZAdef,Fturn,fixedX/;
MODEL mslid/Zbeta,ZDdef,ZAdef,Fslid,fixedX/;
MODEL mbear/Zbeta,ZDdef,ZAdef,Fbear,fixedX/;

***%***% MODELS WITH SENSITIVITY ANALYSIS ***%***%

MODEL mturnS /Zbeta,ZdefS,Fturn,fixedX/;
MODEL mslidS /Zbeta,ZdefS,Fslid,fixedX/;
MODEL mbearS /Zbeta,ZdefS,Fbear,fixedX/;
MODEL classicS /Zclass,turnS,slidS,bearS,geometric,betadefS,auxbeta/;

betaa(M)=betalo(M); lambdaK(IT,M,D)=0.0; XesK(IT,D)=0.0;
betaK(IT,M)=0.0; PML(M,V)=0;

put "---------------------------------------------------------------"/;
put " n  Cost    a      b      Fo    Fs    Fb    Bo    Bs    Bb    Error"/;
put "---------------------------------------------------------------"/;
loop(IT$(error>epsilon),
    iteration=iteration+1;
    betaux(M)=betaa(M);

* Initialize the variables

    VarD.l(D)=mean(D);
    SOLVE classic USING nlp MINIMIZING zc;
    cost=zc.l;
    Fr('turn')=(VarD.l('a')*VarD.l('a')*VarD.l('b')*
        mean('gamma')/(2*mean('H')*mean('T')));
    Fr('slid')=(VarD.l('a')*VarD.l('b')*mean('nu')*
        mean('gamma')/mean('T'));
    Fr('bear')=mean('S')/(VarD.l('b')*mean('gamma'));
    mean(D)=VarD.l(D);
    XesK(IT,D)=mean(D);
    loop(M,
        VarR.l(V)=mean(V);
        if(ORD(M) eq 1,SOLVE mturn USING nlp MINIMIZING zc;
        else if(ORD(M) eq 2,SOLVE mslid USING nlp MINIMIZING zc;
        else SOLVE mbear USING nlp MINIMIZING zc;
            );
        );
* Save beta values and maximum likelihood points

        betaa(M)=zc.l;
        lambdaK(IT,M,D)=fixedX.m(D);
        betaK(IT,M)=zc.l;
        PML(M,V)=VarR.l(V);
    );
    error=0.0;
    loop(M,
        if((abs(betaa(M)-betaux(M))/betaa(M))>error and betaa(M)>0,
            error=(abs(betaa(M)-betaux(M))/betaa(M));
            errors(IT)=error;
        );
    );
    FIRST(M)=yes;
    ITER(IT)=yes;
*------------------------Printing Table ---------------------*
    put iteration:2:0;
    put cost:6:2;
```

```
    loop(D,put mean(D):6:2;);
    loop(M,put Fr(M):6:2;);
    loop(M, put betaa(M):6:2;);
    put error:9:5/;
*------------------ End Printing Table End ------------------*
);
put "----------------------------------------------------------"//;

**%%-------- LAST ITERATION FOR SENSITIVITY ANALYSIS ----------%%**

muK(M,A)=0.0; deltaK(M,V)=0.0; VarD.fx(A)=mean(A);
VarB.fx(V)=sigma(V);
loop(M,
    VarR.l(V)=mean(V);
    if(ORD(M) eq 1,
        SOLVE mturnS USING nlp MINIMIZING zc;
    elseif(ORD(M) eq 2),
        SOLVE mslidS USING nlp MINIMIZING zc;
    else
        SOLVE mbearS USING nlp MINIMIZING zc;
    );
    muK(M,A)=VarD.m(A);
    deltaK(M,V)=VarB.m(V);
    sensX(M,D)=fixedX.m(D);
    sensA(M,A)=VarD.m(A);
    sensB(M,V)=VarB.m(V);
);

Fa.fx(M)=Flo(M);

SOLVE classicS USING nlp MINIMIZING zc;
sensFclas(M)=Fa.m(M);
sensAclas(A)=VarD.m(A);
sensBclas(V)=VarB.m(V);
sensMclas(M)=auxbeta.m(M);

put "----------------------------------------"/;
put "x    Cost     Bo      Bs      Bb    "/;
put "-------------------- MEAN ----------------"/;
loop(D,put D.tl:8:0,"    -- ";
    loop(M,put sensX(M,D):9:3;); put ""/;
);
loop(A,
    put A.tl:8:0,sensAclas(A):9:3;
    loop(M,put sensA(M,A):9:3;); put ""/;
);
put "-------------------- SIGMA ----------------"/;
loop(V,
    put V.tl:8:0,sensBclas(V):9:3;
    loop(M,put sensB(M,V):9:3;); put ""/;
);
put "----------------------------------------"/;
loop(M,
    put "F",M.tl:7,sensFclas(M):9:3;
    loop(Maux,put "     -- ";);put ""/;
);
put "----------------------------------------"/;
loop(M,
    put "B",M.tl:7,sensMclas(M):9:3;
    loop(Maux,put "     -- ";);put ""/;
);
put "----------------------------------------"/;
```

Part VI

Solution to Selected Exercises

B

Exercise Solutions

B.1 Exercises from Chapter 1

Solution to Exercise 1.2. Considering the wall shape shown in Fig. 1.20, the overturning safety factor is

$$F_o = \frac{w_1(d-a/2) + w_2(d-a)/2}{\tilde{h}\tilde{t}} = \frac{ab\gamma(2d-a) + c(d-a)^2\gamma}{2\tilde{h}\tilde{t}} \geq F_o^0.$$

Since the objective of this problem is minimizing the cost of building the wall, the formulation of the wall design problem is

$$\begin{array}{c} \text{minimize} \\ a,b,c,d \end{array} \quad ab + c(d-a)$$

subject to safety factor, reliability, and geometric constraints

$$\begin{aligned} \frac{ab\gamma(2d-a) + c(d-a)^2\gamma}{2\tilde{h}\tilde{t}} &\geq F_o^0 \\ \beta(a,b,c,d) &\geq \beta^0 \\ b &\geq b_0 \\ c &\geq a \\ d &\geq a, \end{aligned}$$

where

$$\beta = \begin{array}{c} \text{minimum} \\ h,t \end{array} \quad z_1^2 + z_2^2$$

subject to

$$z_1 = \frac{t - \mu_t}{\sigma_t}$$

$$z_2 = \frac{h - \mu_h}{\sigma_h}$$

$$\frac{ab\gamma(2d-a) + c(d-a)^2\gamma}{2ht} = 1.$$

This bilevel problem can be solved considering the second constraint of the first problem as a complicating constraint.

Solution to Exercise 1.4. Considering the notation used in Sect. 1.3.3, the formulation of the river basin operation problem, if storage facilities are not available, is shown below. Note that the objective of the problem is to maximize the expected benefit. That is,

$$\underset{d_{ti}, r_{ti}; \forall i; \forall t}{\text{maximize}} \quad z = \sum_{t=1}^{m} \lambda_t \left(\sum_{i=1}^{n} k_i d_{ti} - e_t \right)$$

subject to water balance constraints

$$d_{ti} = w_{ti} + \sum_{j \in \Omega_i} d_{tj}; \quad t = 1, \ldots, m; \quad i = 1, \ldots, n$$

$$\sum_{t=1}^{m} \sum_{i=1}^{n} w_{ti} = \sum_{t=1}^{m} d_{tn},$$

demand constraints

$$\sum_{i=1}^{n} k_i d_{ti} \geq e_t; \quad t = 1, \ldots, m,$$

and discharge bounds

$$0 \leq d_{ti} \leq d_i^{\max}; \quad t = 1, \ldots, m; \quad i = 1, \ldots, n.$$

The structure of this problem is illustrated considering two periods of time and two reservoirs. Arranging variables in order $d_{11}, d_{21}, d_{12}, d_{22}$, the matrix corresponding to the constraints (without considering) bounds is

$$\begin{pmatrix} 1 & & & \\ & 1 & & \\ -1 & & 1 & \\ & -1 & & 1 \\ & & -1 & -1 \\ k_1 & & k_2 & \\ & k_1 & & k_2 \end{pmatrix}.$$

It should be noted that the last two constraints are complicating constraints. If they are relaxed, the resulting matrix exhibits a structure that can be computationally exploited.

Solution to Exercise 1.6. Considering the notation used in Sect. 1.4.1, the formulation of the 2-year coal, oil, and gas procurement problem, for five demand scenarios in the second year is

$$\underset{c_0, c_s, g_0, g_s}{\text{minimize}} \quad a_0 c_0 + b_0 g_0 + \sum_{s=1}^{5} p_s \left(a_s c_s + b_s g_s \right)$$

subject to the first year demand constraint

$$c_0 + g_0 \geq d_0 \, ,$$

supply total demand (first and second year) constraints for all scenarios

$$c_0 + g_0 + c_s + g_s = d_0 + d_s; \quad s = 1, \cdots, 5 \, ,$$

maximum and minimum bounds on coal consumption

$$c_0 + c_s \leq \frac{2}{3}(d_0 + d_s); \quad s = 1, \cdots, 5$$

$$-c_0 - c_s \leq -\frac{1}{3}(d_0 + d_s); \quad s = 1, \cdots, 5 \, ,$$

and maximum and minimum bounds on gas consumption

$$g_0 + g_s \leq \frac{2}{3}(d_0 + d_s); \quad s = 1, \cdots, 5$$

$$-g_0 - g_s \leq -\frac{1}{3}(d_0 + d_s); \quad s = 1, \cdots, 5 \, .$$

If the order of variables is $c_1, g_1, c_2, g_2, c_3, g_3, c_4, g_4, c_5, g_5, c_0, g_0$, the constraint matrix of the problem above is

$$\begin{pmatrix}
 & & & & & & & & & & 1 & 1 \\
1 & 1 & & & & & & & & & 1 & 1 \\
1 & & & & & & & & & & 1 & \\
-1 & & & & & & & & & & -1 & \\
 & 1 & & & & & & & & & & 1 \\
 & -1 & & & & & & & & & & -1 \\
\hline
 & & 1 & 1 & & & & & & & 1 & 1 \\
 & & 1 & & & & & & & & 1 & \\
 & & -1 & & & & & & & & -1 & \\
 & & & 1 & & & & & & & & 1 \\
 & & & -1 & & & & & & & & -1 \\
\hline
 & & & & 1 & 1 & & & & & 1 & 1 \\
 & & & & 1 & & & & & & 1 & \\
 & & & & -1 & & & & & & -1 & \\
 & & & & & 1 & & & & & & 1 \\
 & & & & & -1 & & & & & & -1 \\
\hline
 & & & & & & 1 & 1 & & & 1 & 1 \\
 & & & & & & 1 & & & & 1 & \\
 & & & & & & -1 & & & & -1 & \\
 & & & & & & & 1 & & & & 1 \\
 & & & & & & & -1 & & & & -1 \\
\hline
 & & & & & & & & 1 & 1 & 1 & 1 \\
 & & & & & & & & 1 & & 1 & \\
 & & & & & & & & -1 & & -1 & \\
 & & & & & & & & & 1 & & 1 \\
 & & & & & & & & & -1 & & -1 \\
\end{pmatrix}$$

Variables c_0 and g_0 are complicating variables that prevent a distributed solution of the problem. If these variables are fixed to given values, the problem decomposes by blocks.

Solution to Exercise 1.8. Considering the example presented in Sect. 1.4.2, the formulation of the multiperiod capacity expansion planning problem including nonlinear investment and operation costs, and discrete investment variables is shown below. The objective of the production company is to minimize both investment and operation costs, i.e.,

$$\underset{x_{it},\, y_{it},\, f_{ij,t}}{\text{minimize}} \quad \sum_{t=1}^{T} \left[\sum_{i=1}^{2} c_{it}(x_{it} - x_{i,t-1})^2 + \sum_{(i,j) \in \mathcal{P}} e_{ij} f_{ij,t}^2 \right]$$

subject to balance constraints at production nodes

$$\begin{aligned} y_{1t} &= f_{13,t} + f_{12,t} - f_{21,t}; & t &= 1, \ldots, T \\ y_{2t} &= f_{23,t} + f_{21,t} - f_{12,t}; & t &= 1, \ldots, T \,, \end{aligned}$$

balance constraint at the consumption node

$$d_t = f_{13,t} + f_{23,t}; \quad t = 1, \ldots, T \,,$$

production bounds

$$0 \leq y_{it} \leq x_{it}; \quad i = 1, 2; \quad t = 1, \ldots, T \,,$$

constraints on maximum capacity expansion

$$x_{it} \leq x_{i,t+1}; \quad i = 1, 2; \quad t = 1, \ldots, T-1 \,,$$

expansion capacity bounds

$$0 \leq x_{it} \leq x_i^{\max}; \quad i = 1, 2; \quad t = 1, \ldots, T \,,$$

constraints on transportation capacity

$$0 \leq f_{ij,t} \leq f_{ij}^{\max}; \quad (i,j) \in \mathcal{P}; \quad t = 1, \ldots, T \,,$$

and discrete investment variables

$$x_{it} \in \mathbb{N}; \quad i = 1, 2; \quad t = 1, \ldots, T \,.$$

Considering two periods of time, the constraint matrix of the above problem is

$$\left(\begin{array}{cccc|cccc|cccc}
1 & -1 & -1 & 1 & & & & & & & & \\
 & 1 & -1 & 1 & -1 & & & & & & & \\
 & & 1 & 1 & & & & & & & & \\
1 & & & & & & & & -1 & & & \\
1 & & & & & & & & & -1 & & \\
 & & & & 1 & -1 & -1 & 1 & & & & \\
 & & & & & 1 & -1 & 1 & -1 & & & \\
 & & & & & & 1 & 1 & & & & \\
 & & & & 1 & & & & & & -1 & \\
 & & & & & 1 & & & & & & -1 \\
 & & & & & & & & 1 & & -1 & \\
 & & & & & & & & & 1 & & -1
\end{array}\right)$$

Note that the variable order is y_{11}, y_{21}, $f_{13,1}$, $f_{23,1}$, $f_{12,1}$, $f_{21,1}$; y_{12}, y_{22}, $f_{13,2}$, $f_{23,2}$, $f_{12,2}$, $f_{21,2}$; x_{11}, x_{21}, x_{12}, x_{22}.

It should be noted that discrete variables x_{it} ($i = 1, 2; t = 1, 2$) are complicating variables. If they are fixed to given values, the problem above decomposes by time period.

Solution to Exercise 1.10. Considering the example presented in Sect. 1.7.1, the formulation of the 24-h unit commitment of production units is shown below. The objective of this problem is to minimize cost, i.e.,

$$\operatorname*{minimize}_{P_{Git}} \sum_{t=1}^{24} \sum_{i=1}^{2} c_{it} P_{Git}$$

subject to production capacity limits for the facilities

$$u_{it} P_{Gi}^{\min} \leq P_{Git} \leq u_{it} P_{Gi}^{\max}; \quad i = 1, 2; \quad t = 1, \ldots, 24,$$

production balance at node 1

$$P_{G1t} + [G(1 - \cos \delta_{1t}) - B \sin \delta_{1t}] = P_{D1t},$$

production balance at node 2

$$P_{G2t} + [G(1 - \cos \delta_{1t}) + B \sin \delta_{1t}] = P_{D2t},$$

security of supply

$$u_{1t} P_{G1}^{\max} + u_{2t} P_{G2}^{\max} \geq P_{D1t} + P_{D2t},$$

and ramping limits

$$P_{Gi,t-1} - P_{Git} \leq R_i^{\mathrm{down}}; \quad i = 1, 2; \quad t = 1, \ldots, 24,$$

$$P_{Git} - P_{Gi,t-1} \leq R_i^{\mathrm{up}}; \quad i = 1, 2; \quad t = 1, \ldots, 24,$$

where R_i^{down} is the ramp-down limit for unit i and R_i^{up} is the ramp-up limit for unit i.

It should be noted that the problem above is mixed-integer and nonlinear. The balance constraints are complicating constraints. If these constraints are linearized, the problem becomes mixed-integer and linear.

The minimum up time constraints are formulated as follows:

$$[x_{i,t-1} - \text{MUT}_i][u_{i,t-1} - u_{it}] \geq 0 \qquad \forall i; \quad t = 1, \ldots, 24,$$

where $x_{i,t}$ is the number of hours that unit i has been on at the end of hour t, and MUT_i is the minimum up time of unit i (minimum number of hours that the unit should be on line once started up).

This equation is nonlinear. However, linear constraints to enforce minimum up time for unit i is shown below.

$$\sum_{t=1}^{L_i} [1 - u_{i,t}] = 0; \qquad \forall i$$

$$\sum_{\tau=t}^{t+\text{MUT}_i - 1} u_{i,\tau} \geq \text{MUT}_i y_{it}; \qquad \forall i; \quad t = L_i + 1, \ldots, 24 - \text{MUT}_i + 1$$

$$\sum_{\tau=t}^{24} [u_{i\tau} - y_{it}] \geq 0; \qquad \forall i; \quad t = 24 - \text{MUT}_i + 2, \ldots, 24,$$

where y_{it} is the start-up status for unit i in period t and $L_i = \text{Min}[24, (\text{MUT}_i - \text{UT}_i)u_{i,0}]$. Note that UT_i represents the number of time periods that unit i has been on-line at the beginning of the planning horizon.

B.2 Exercises from Chapter 2

Solution to Exercise 2.2.

1. If the original problem is solved directly, we obtain the following global solution:

$$x_1 = 1, \quad x_2 = 2, \quad x_3 = 1, \quad x_4 = 0, \quad z = -5 \ .$$

2. It should be noted that the last two constraints are complicating constraints. Table 2.5 shows two different feasible solutions (x_1, x_2, x_3, x_4) of the relaxed problem and the associated values of r_1, r_2, and z, obtained by minimizing the objective functions:

$$\begin{aligned}
z_1 &= -x_1 - x_2 & &+ x_4 \\
z_2 &= x_1 + x_2 & -x_3 & \\
z_3 &= x_1 & -x_3 & + x_4 \\
z_4 &= 2x_1 + x_2 & & + 3x_4 \ .
\end{aligned}$$

3. The problem is solved using the Dantzig-Wolfe decomposition as follows:

   Step 1: Master problem solution. The master problem below is solved.

   $$\underset{u_1, u_2}{\text{minimize}} \quad -10u_1 - 3u_2$$

   subject to

   $$\begin{array}{rcll}
   5u_1 + 3u_2 & \leq & 2 : & \lambda_1 \\
   5u_1 + 0u_2 & \leq & 3 : & \lambda_2 \\
   u_1 + u_2 & = & 1 : & \sigma \\
   u_1, u_2 & \geq & 0. &
   \end{array}$$

   Its solution is $u_1^{(1)} = 0$ and $u_2^{(1)} = 1$ with dual variable values $\lambda_1^{(1)} = -20$, $\lambda_2^{(1)} = 0$, and $\sigma_1 = 57$.

   Step 2: Relaxed problem solution. The subproblems are solved below to obtain a solution for the current relaxed problem.
   The objective function of the first subproblem is

   $$\left(c_1 - \lambda_1^{(1)} a_{11} - \lambda_2^{(1)} a_{21}\right) x_1 + \left(c_2 - \lambda_1^{(1)} a_{12} - \lambda_2^{(1)} a_{22}\right) x_2 =$$

   $$(-2 + 20)x_1 + (-1)x_2 = 18x_1 - x_2,$$

   and its solution, obtained by inspection, is $x_1 = 0$ and $x_2 = 2.5$.
   The objective function of the second subproblem is

   $$\left(c_3 - \lambda_1^{(1)} a_{13} - \lambda_2^{(1)} a_{23}\right) x_3 + \left(c_4 - \lambda_1^{(1)} a_{14} - \lambda_2^{(1)} a_{24}\right) x_4$$

   $$= (-1 + 20)x_3 + (1)x_4 = 19x_3 + x_4,$$

   and its solution is $x_3 = 0$ and $x_4 = 0$.
   For this relaxed problem solution, the objective function value of the original problem is $z = -2.5$ and the values of the complicating constraints $r_1 = 0$ and $r_2 = 2.5$, respectively.

   Step 3: Convergence checking. The objective function value of the current relaxed problem is

   $$v_1 = 18x_1 - x_2 + 19x_3 + x_4 = -2.5.$$

   Note that $v_1 < \sigma_1$ ($-2.5 < 57$) and therefore the current solution of the relaxed problem can be used to improve the solution of the master problem.
   The iteration counter is updated, $\nu = 1 + 1 = 2$, and the number of available solutions of the relaxed problem is also updated, $p^{(2)} = 2 + 1 = 3$. The algorithm continues with Step 1.

Table 2.5 shows the new solutions of the relaxed problem obtained through the Dantzig-Wolfe decomposition algorithm, together with the corresponding upper and lower bounds associated with each step. This table also shows the results (u_1, u_2, u_3, u_4), λ_1, λ_2, and σ for the master problem for different iterations with the corresponding lower and upper bounds. The solution obtained by decomposition is

$$x_1 = 1.6, \quad x_2 = 1.4, \quad x_3 = 0.4, \quad x_4 = 0, \quad z = -5.$$

4. Note that the solution obtained by decomposition is different from the global solution obtained by solving the global problem. However, the values of the objective function coincide. This means that the solution of this problem is not unique.

Solution to Exercise 2.4. If the original problem is solved directly, the solution is

$$x_1 = 2, \ x_2 = 1, \ x_3 = 0, \ x_4 = 1.5, \ x_5 = 0.5,$$
$$x_6 = 0.5, \ x_7 = 2, \ x_8 = 1, \ x_9 = 0, \ x_{10} = 0, \ z = -21.5.$$

It should be noted that the last constraint prevents a decomposed solution of the problem; therefore, it is a complicating constraint.

Minimizing the following objective functions we obtain five feasible initial solutions of the relaxed problem that are shown in Table B.1,

$$
\begin{aligned}
z_1 &= -x_3 - x_6 - x_9 \\
z_2 &= -x_3 - x_6 - x_7 \\
z_3 &= -x_3 - x_4 \\
z_4 &= -x_3 - x_4 - x_7 \\
z_5 &= -x_1 - x_6.
\end{aligned}
$$

The solution resulting from the Dantzig-Wolfe decomposition algorithm is given in Table B.1. It should be noted that the solution obtained by decomposition is the same than the one obtained without decomposition.

Solution to Exercise 2.6. The solution of the original problem is

$$x_1 = 1, \ x_2 = 1, \ x_3 = 1, \ x_4 = 0, \ z = -4.$$

It should be noted that the last two constraints prevent a decomposed solution of the problem; therefore, they are complicating constraints.

We can obtain two feasible initial solutions of the relaxed problem minimizing the objective functions,

$$
\begin{aligned}
z_1 &= -x_1 - x_2 + x_3 \\
z_2 &= x_1 + x_2 - x_3.
\end{aligned}
$$

Table B.1. Initial solutions for the subproblems and additional solutions obtained through the Dantzig-Wolfe decomposition algorithm in Exercise 2.4

Iteration ν	Bounds Lower	Bounds Upper	x_1	x_2	x_3	x_4	x_5	x_6	x_7	x_8	x_9	x_{10}	r_1	z
0–1	$-\infty$	∞	1.0	0.0	1.0	1.0	0.0	1.0	1.0	0.0	1.0	0.0	9.0	-11.0
0–2	$-\infty$	∞	1.0	0.0	1.0	1.0	0.0	1.0	2.0	1.0	0.0	0.0	13.0	-17.0
0–3	$-\infty$	∞	1.0	0.0	1.0	2.0	1.0	0.0	1.0	0.0	1.0	0.0	11.0	-12.0
0–4	$-\infty$	∞	1.0	0.0	1.0	2.0	1.0	0.0	2.0	1.0	0.0	0.0	15.0	-18.0
0–5	$-\infty$	∞	2.0	1.0	0.0	1.0	0.0	1.0	1.0	0.0	1.0	0.0	12.0	-15.0
Solutions for the subproblem														
1	-22.0	-18.0	2.0	1.0	0.0	2.0	1.0	0.0	2.0	1.0	0.0	0.0	18.0	-22.0
2	-22.0	-21.0	2.0	1.0	0.0	1.0	0.0	1.0	2.0	1.0	0.0	0.0	16.0	-21.0
3	-21.5	-21.5	–	–	–	–	–	–	–	–	–	–	–	–

			Master solutions									
			u_1	u_2	u_3	u_4	u_5	u_6	u_7	λ_1	σ	Feasible
1	$-\infty$	-18.0	0.0	0.0	0.0	1.0	0.0	0.0	0.0	0.0	-18.0	Yes
2	-22.0	-21.0	0.0	0.2	0.0	0.0	0.0	0.8	0.0	-1.0	-4.0	Yes
3	-22.0	-21.5	0.0	0.0	0.0	0.0	0.0	0.5	0.5	-0.5	-13.0	Yes

The solution resulting from the Dantzig-Wolfe decomposition algorithm is provided in Table B.2. Note that the solution obtained by decomposition is identical to the one obtained without decomposition.

Table B.2. Initial solutions for the subproblems and additional solutions obtained through the Dantzig-Wolfe decomposition algorithm in Exercise 2.6

Iteration ν	Bounds Lower	Bounds Upper	x_1	x_2	x_3	x_4	r_1	r_2	z
0–1	$-\infty$	∞	1.0	1.0	0.0	0.0	1.0	5.0	-3.0
0–2	$-\infty$	∞	0.0	0.0	1.3	0.0	1.3	0.0	-1.3
Solutions for the subproblem									
1	-4.33	-3.0	1.0	1.0	1.3	0.0	2.3	5.0	-4.3
2	-4.0	-4.0	–	–	–	–	–	–	–

			Master solutions						
			u_1	u_2	u_3	λ_1	λ_2	σ	Feasible
1	$-\infty$	-3.0	1.0	0.0	0.0	0.0	0.0	-3.0	Yes
2	-4.33	-4.0	0.2	0.0	0.8	-1.0	0.0	-2.0	Yes

Solution to Exercise 2.8. The formulation of the stochastic hydro scheduling problem is

$$\underset{d_{ts}, r_{ts}; \forall t; \forall s}{\text{minimize}} \quad z = (-30 \ -45 \ -20 \ -30 \ -20 \ -30 \ -30 \ -45) \begin{pmatrix} d_{11} \\ d_{21} \\ d_{12} \\ d_{22} \\ d_{13} \\ d_{23} \\ d_{14} \\ d_{24} \end{pmatrix}$$

subject to water balance constraints

$$\begin{aligned} r_{11} &= 50 - d_{11} + 20 \\ r_{21} &= r_{11} - d_{21} + 25 \\ r_{12} &= 50 - d_{12} + 20 \\ r_{22} &= r_{12} - d_{22} + 35 \\ r_{13} &= 50 - d_{13} + 30 \\ r_{23} &= r_{13} - d_{23} + 25 \\ r_{14} &= 50 - d_{14} + 30 \\ r_{24} &= r_{14} - d_{24} + 35 \,, \end{aligned}$$

reservoir level limits

$$20 \le r_{11} \le 140, \quad 20 \le r_{12} \le 140, \quad 20 \le r_{13} \le 140, \quad 20 \le r_{14} \le 140,$$
$$20 \le r_{21} \le 140, \quad 20 \le r_{22} \le 140, \quad 20 \le r_{23} \le 140, \quad 20 \le r_{24} \le 140 \,,$$

discharge limits

$$d_{11} \le 60, \quad d_{12} \le 60, \quad d_{13} \le 60, \quad d_{14} \le 60,$$
$$d_{21} \le 60, \quad d_{22} \le 60, \quad d_{23} \le 60, \quad d_{24} \le 60 \,,$$

and nonanticipativity constraints

$$d_{11} = d_{12}, \quad r_{11} = r_{12}, \quad d_{13} = d_{14}, \quad r_{13} = r_{14} \,.$$

The nonanticipativity constraints are complicating constraints, the Dantzig-Wolfe procedure is therefore used to solve this problem. The solution of the original problem not using decomposition is

$$z = -11250, \quad d_{11} = d_{12} = 25, \quad d_{13} = d_{14} = 35,$$
$$d_{21} = 50, \quad d_{22} = 60, \quad d_{23} = 503, \quad d_{24} = 60 \,.$$

B.2 Exercises from Chapter 2   431

We obtain two feasible initial solutions of the relaxed problem minimizing the following objective functions:

$$z_1 = 25d_{11} + 20d_{21} + 35d_{12} + 30d_{22} + 30d_{13} + 25d_{23} + 35d_{14} + 30d_{24}$$
$$z_2 = -10d_{11} - 25d_{21} - 35d_{12} - 25d_{22} - 20d_{13} - 35d_{23} - 35d_{14} - 45d_{24}.$$

The solution resulting from the Dantzig-Wolfe decomposition algorithm is provided in Table B.3.

Table B.3. Initial solutions for the subproblems and additional solutions obtained through the Dantzig-Wolfe decomposition algorithm in Exercise 2.8

Iteration	Bounds		Initial solutions for the subproblems												
ν	Lower	Upper	d_{11}	d_{21}	d_{12}	d_{22}	d_{13}	d_{23}	d_{14}	d_{24}	r_1	r_2	r_3	r_4	z
0-1	$-\infty$	∞	0	0	0	0	0	0	0	0	0	0	0	0	0
0-2	$-\infty$	∞	60	20	50	35	25	60	35	60	-35	35	-10	10	$-11{,}250$
			Solutions for the subproblem												
1	$-15{,}475$	$-7{,}750$	50	25	0	60	25	60	35	60	50	-50	-10	10	$-10{,}475$
2	$-15{,}461$	$-10{,}022$	15	60	25	60	60	25	0	60	-10	10	60	-60	$-10{,}100$
3	$-11{,}388$	$-10{,}799$	15	60	25	60	60	25	35	60	-10	10	25	-25	$-11{,}150$
4	$-11{,}445$	$-10{,}967$	15	60	25	60	25	60	35	60	-10	10	-10	10	$-11{,}500$
5	$-11{,}302$	$-11{,}229$	50	25	25	60	60	25	35	60	25	-25	25	-25	$-10{,}625$
6	$-11{,}322$	$-11{,}250$	50	25	25	60	25	60	35	60	25	-25	-10	10	$-10{,}975$
7	$-11{,}250$	$-11{,}250$	–	–	–	–	–	–	–	–	–	–	–	–	–
									Master solutions						
			u_1	u_2	u_3	u_4	u_5	u_6	u_7	u_8	λ_1	λ_2	λ_3	λ_4	σ	Feasible
1	$-\infty$	$-7{,}750$	0	1	0	0	0	0	0	0	20	-80	20	20	$-7{,}750$	No
2	$-15{,}475$	$-10{,}022$	0	0.7	0.3	0	0	0	0	0	20	11	20	-71	$-10{,}022$	No
3	$-15{,}461$	$-10{,}799$	0	0.5	0.4	0.1	0	0	0	0	20	11	20	7	$-10{,}799$	Yes
4	$-11{,}388$	$-10{,}967$	0	0.4	0.3	0	0.3	0	0	0	20	11	16	20	$-10{,}967$	Yes
5	$-11{,}445$	$-11{,}229$	0	0	0.2	0	0.3	0.5	0	0	20	3	20	10	$-11{,}229$	Yes
6	$-11{,}302$	$-11{,}250$	0	0	0	0	0	0.7	0.3	0	20	3	20	12	$-11{,}250$	Yes
7	$-11{,}323$	$-11{,}250$	0	0	0	0	0.3	0.4	0	0.3	20	5	20	10	$-11{,}250$	Yes

The solution obtained by decomposition is the same as one obtained without decomposition:

$$z = -11250, \quad d_{11} = d_{12} = 25, \quad d_{13} = d_{14} = 35,$$
$$d_{21} = 50, \quad d_{22} = 60, \quad d_{23} = 503, \quad d_{24} = 60.$$

Solution to Exercise 2.10. The problem to be solved consists of supplying the energy demand at minimum cost. In order to formulate this problem, the definition of the maximum energy produced by a set of production devices was introduced in Sect. 1.3.4. This cost minimization problem can be stated as

$$\underset{x_i, \forall i}{\text{minimize}} \quad z = x_1 + 2x_2 + 3x_3 + 4x_4 + 5x_5$$

subject to

$$\begin{array}{lllllr}
x_1 & & & & & \leq 0.97 \\
& x_2 & & & & \leq 1.90 \\
& & x_3 & & & \leq 2.77 \\
& & & x_4 & & \leq 2.77 \\
& & & & x_5 & \leq 4.37 \\
x_1 & +x_2 & & & & \leq 2.77 \\
x_1 & & +x_3 & & & \leq 3.60 \\
x_1 & & & +x_4 & & \leq 3.60 \\
x_1 & & & & +x_5 & \leq 5.10 \\
& x_2 & +x_3 & & & \leq 4.37 \\
& x_2 & & +x_4 & & \leq 4.37 \\
& x_2 & & & +x_5 & \leq 5.77 \\
& & x_3 & +x_4 & & \leq 5.10 \\
& & x_3 & & +x_5 & \leq 6.40 \\
& & & x_4 & +x_5 & \leq 6.40 \\
x_1 & +x_2 & +x_3 & & & \leq 5.10 \\
x_1 & +x_2 & & +x_4 & & \leq 5.10 \\
x_1 & +x_2 & & & +x_5 & \leq 6.40 \\
x_1 & & +x_3 & +x_4 & & \leq 5.77 \\
x_1 & & +x_3 & & +x_5 & \leq 6.97 \\
x_1 & & & +x_4 & +x_5 & \leq 6.97 \\
& x_2 & +x_3 & +x_4 & & \leq 6.40 \\
& x_2 & +x_3 & & +x_5 & \leq 7.50 \\
& x_2 & & +x_4 & +x_5 & \leq 7.50 \\
& & x_3 & +x_4 & +x_5 & \leq 7.50 \\
x_1 & +x_2 & +x_3 & +x_4 & & \leq 6.97 \\
x_1 & +x_2 & +x_3 & & +x_5 & \leq 7.50 \\
x_1 & +x_2 & & +x_4 & +x_5 & \leq 7.50 \\
x_1 & & +x_3 & +x_4 & +x_5 & \leq 7.50 \\
& x_2 & +x_3 & +x_4 & +x_5 & \leq 7.50 \\
x_1 & +x_2 & +x_3 & +x_4 & +x_5 & \leq 7.50 \\
-x_1 & -x_2 & -x_3 & -x_4 & -x_5 & \leq -7.50 \\
x_1 & & +x_3 & & & \leq 3.00 \\
& & & x_4 & +x_5 & \leq 4.00,
\end{array}$$

where x_i is the energy produced by device i.

If the last two constraints are relaxed, the solution of the problem can be obtained using a merit order rule; therefore these equations are complicating constraints.

The solution of this problem is

$$z = 21.98, \quad x_1 = 0.88, \quad x_2 = 1.90, \quad x_3 = 2.13, \quad x_4 = 2.07, \quad x_5 = 0.53\ .$$

We can obtain two feasible initial solutions of the relaxed problem minimizing the following objective functions:

$$z_1 = 2x_1 - 3x_2 + 5x_3 + 6x_4 + x_5$$
$$z_2 = x_1 + 2.5x_2 + 3x_3 + 4x_4 - x_5 .$$

The solution resulting from the Dantzig-Wolfe decomposition algorithm is provided in Table B.4.

Table B.4. Initial solutions for the subproblems and additional solutions obtained through the Dantzig-Wolfe decomposition algorithm in Exercise 2.10

Iteration	Bounds		Initial solutions for the subproblems							
ν	Lower	Upper	x_1	x_2	x_3	x_4	x_5	r_1	r_2	z
0–1	$-\infty$	∞	0.62	1.90	1.10	0.00	3.88	1.72	3.88	27.10
0–2	$-\infty$	∞	0.72	1.30	1.10	0.00	4.38	1.82	4.38	28.50
			Solutions for the subproblem							
1	21.68	27.10	0.98	1.80	2.33	1.88	0.53	3.30	2.40	21.68
2	18.59	22.71	0.72	1.90	0.00	2.48	2.40	0.72	4.88	26.43
3	21.76	22.23	0.88	1.90	1.87	2.32	0.53	2.75	2.85	22.22
4	21.97	21.98	0.88	1.90	2.32	1.87	0.53	3.20	2.40	21.78
			Master solutions							
			u_1	u_2	u_3	u_4	u_5	λ_1	λ_2	σ	Feasible
1	$-\infty$	27.10	1.00	0.00	0.00	0.00	0.00	0.00	0.00	27.16	Yes
2	21.68	22.71	0.19	0.00	0.81	0.00	0.00	-3.44	0.00	33.04	Yes
3	18.59	22.23	0.00	0.00	0.88	0.12	0.00	-1.84	0.00	27.76	Yes
4	21.76	21.98	0.00	0.00	0.45	0.00	0.55	-1.00	0.00	24.97	Yes

The solution obtained by decomposition is

$$z = 21.98, \quad x_1 = 0.92, \quad x_2 = 1.85, \quad x_3 = 2.08, \quad x_4 = 2.12, \quad x_5 = 0.53 .$$

Solution to Exercise 2.12.

1. We are interested in maximizing the number of planes that can be manufactured. This problem is formulated as a linear programming problem. The objective function can be expressed as

$$\underset{x_i; \forall i}{\text{minimize}} \quad z = -\sum_{i=1}^{3} x_i,$$

where x_i is the number of planes manufactured at location i.

For technical reasons, the available labor time and the fuselage material are limited to a maximum amount at each location. These constraints are expressed as

Available labor time constraints

$$10x_1 \leq 100$$
$$10x_2 \leq 120$$
$$10x_3 \leq 60 .$$

Fuselage material constraints

$$15x_1 \leq 50$$
$$15x_2 \leq 40$$
$$15x_3 \leq 55\,.$$

Finally, the number of engines manufactured centrally are limited to a maximum amount. This limit is enforced through the constraint below

$$x_1 + x_2 + x_3 \leq 9\,.$$

Consequently, the problem has the form

$$\text{minimize}_{x_i;\forall i} \quad z = -\sum_{i=1}^{3} x_i$$

subject to

$$
\begin{aligned}
10x_1 & & & \leq 100 \\
15x_1 & & & \leq 50 \\
& 10x_2 & & \leq 120 \\
& 15x_2 & & \leq 40 \\
& & 10x_3 & \leq 60 \\
& & 15x_3 & \leq 55 \\
x_1 & +x_2 & +x_3 & \leq 9\,.
\end{aligned}
$$

2. Considering the number of planes a real variable, it can be noted that the last constraint prevents a decomposed solution of the problem. Therefore, the problem can be solved using the Dantzig-Wolfe decomposition.

We can obtain two feasible initial solutions of the relaxed problem minimizing the following objective functions:

$$
\begin{aligned}
z_1 &= -x_1 - x_2 + x_3 \\
z_2 &= x_1 + x_2 - x_3\,.
\end{aligned}
$$

The solution resulting from the Dantzig-Wolfe decomposition algorithm is provided in Table B.5.

The solution obtained by decomposition is

$$z = -9.00, \quad x_1 = 3.33, \quad x_2 = 2.67, \quad x_3 = 3.00\,.$$

Table B.5. Initial solutions for the subproblems and additional solutions obtained through the Dantzig-Wolfe decomposition algorithm in Exercise 2.12

Iteration	Bounds		Initial solutions for the subproblems				
ν	Lower	Upper	x_1	x_2	x_3	r_1	z
0–1	$-\infty$	∞	3.33	2.67	0.00	6.00	−6.00
0–2	$-\infty$	∞	0.00	0.00	3.67	3.67	−3.67
			Solutions for the subproblem				
1	−9.67	−6.00	3.33	2.67	3.67	9.67	−9.67
2	−9.00	−9.00	–	–	–	–	–
					Master solutions		
			u_1	u_2	u_3	λ_1	σ	Feasible
1	$-\infty$	−6.00	1.00	0.00	0.00	0.00	−6.00	Yes
2	−9.67	−9.00	0.18	0.00	0.82	−1.00	0.00	Yes

B.3 Exercises from Chapter 3

Solution to Exercise 3.2.

1. If the original problem is solved directly, the solution obtained is

$$x_1 = 1, \ x_2 = 0, \ x_3 = 2.2, \ x_4 = 3.4, \ x_5 = 0.5, \ z = 18.2 \ .$$

2. If variable x_5 is considered to be a complicating variable, the above problem can be solved using the Benders decomposition as follows.

Step 0: Initialization. The iteration counter is initialized, $\nu = 1$. The initial master problem is solved:

$$\underset{x_5, \alpha}{\text{minimize}} \quad z = 3x_5 + \alpha$$

subject to

$$-100 \ \leq \ \alpha \ .$$

The solution of this problem is $x_5^{(1)} = 0$ and $\alpha^{(1)} = -100$. The value for the objective function is $z^{(1)} = -100$.

Step 1: Subproblem solution. The subproblem is infeasible, then artificial variables w_1 and w_2 are included in the subproblem,

$$\underset{x_1, x_2, x_3, x_4}{\text{minimize}} \quad z = 2x_1 + 2.5x_2 + 0.5x_3 + 4x_4 + 20(w_1 + w_2)$$

subject to

$$\begin{array}{rrrrrrr}
-2x_1 & +3x_2 & & & -4x_5 & -w_1 & \le -4 \\
2x_1 & +4x_2 & & & +x_5 & -w_1 & \le 2.5 \\
& & 2x_3 & -x_4 & -x_5 & -w_2 & \le 0.5 \\
& & -0.5x_3 & -x_4 & +3x_5 & -w_2 & \le -3 \\
& & & & x_5 & & = 0.
\end{array}$$

The solution of this problem is $x_1^{(1)} = 1.6$, $x_2^{(1)} = 0$, $x_3^{(1)} = 1.4$, $x_4^{(1)} = 2.3$, $w_1^{(1)} = 0.8$, and $w_2^{(1)} = 0$. The objective function optimal value is $z^{(1)} = 28.1$. Note that the optimal value of the dual variable associated with the constraint $x_5 = 0$ is $\lambda^{(1)} = -32.5$.

Step 2: Convergence checking. An upper bound of the objective function optimal value is computed as

$$z_{\text{up}}^{(1)} = 28.1 + (-100) - (-100) = 28.15 \ .$$

A lower bound of the objective function optimal value is

$$z_{\text{down}}^{(1)} = -100 \ .$$

Since the difference $z_{\text{up}}^{(1)} - z_{\text{down}}^{(1)} = 128.1 > \varepsilon$, the procedure continues.

Step 3: Master problem solution. The iteration counter is updated, $\nu = 1 + 1 = 2$. The master problem below is solved.

$$\underset{x_5, \alpha}{\text{minimize}} \quad z = 3x_5 + \alpha$$

subject to

$$\begin{array}{rl}
28.15 - 32.5(x_5 - 0) & \le \alpha \\
-100 & \le \alpha \ .
\end{array}$$

The solution of this problem is $x_5^{(2)} = 3.9$ and $\alpha^{(2)} = -100$. The value for the objective function is $z^{(2)} = -88.2$.

The procedure continues with Step 1.

The solution resulting from the Benders decomposition algorithm is given in Table B.6.
The solution of this problem is $x_1 = 1$, $x_2 = 0$, $x_3 = 2.2$, $x_4 = 3.4$, $x_5 = 0.5$ with an objective function value $z = 18.2$.

Solution to Exercise 3.4. If variable x_1 is considered a complicating variable, the considered problem can be solved using the Benders decomposition as follows.

Table B.6. Evolution of the values of the master and subproblem variables using the Benders decomposition in Exercise 3.2

ν	$x_1^{(\nu)}$	$x_2^{(\nu)}$	$x_3^{(\nu)}$	$x_4^{(\nu)}$	$x_5^{(\nu)}$	$\alpha^{(\nu)}$	$\lambda^{(\nu)}$	$z_{up}^{(\nu)}$	$z_{down}^{(\nu)}$	$w_1^{(\nu)}$	$w_2^{(\nu)}$
1	1.6	0	1.4	2.3	0	-100	-32.5	28.1	-100	0.8	0
2	0	0	7.7	11	3.9	-100	29.6	88.4	-88.1	1.4	0
3	0	0	3.2	4.7	1.1	-7.6	9.6	23.7	-4.3	0	0
4	1.1	0	2.1	3.3	0.4	14.1	-22.9	19.5	15.4	0.1	0
5	0.9	0	2.3	3.5	0.6	15.3	5.6	18.7	17	0	0
6	1	0	2.2	3.4	0.5	16.7	5.6	18.2	18.2	0	0

Step 0: Initialization. The iteration counter is initialized, $\nu = 1$. The initial master problem is solved:

$$\underset{x_1, \alpha}{\text{minimize}} \quad -4x_1 + \alpha$$

subject to

$$\begin{aligned}
x_1 &\leq 4 \\
2x_1 &\leq 6 \\
-x_1 &\leq -1 \\
-100 &\leq \alpha \ .
\end{aligned}$$

The solution of this problem is $x_1^{(1)} = 3$ and $\alpha^{(1)} = -100$. The value for the objective function is $z^{(1)} = -112$.

Step 1: Subproblem solution. The subproblem is infeasible, then an artificial variable w_1 is included in the subproblem:

$$\underset{y_1, y_2, y_3}{\text{minimize}} \quad -y_1 - 3y_2 - y_3 - 4x_1 + 20w_1$$

subject to

$$\begin{aligned}
-y_1 +x_1 - w_1 &\leq 1 \\
2y_2 +2x_1 - w_1 &\leq 4 \\
2y_1 + y_2 + 2y_3 + 2x_1 - w_1 &\leq 9 \\
x_1 &= 3.
\end{aligned}$$

The solution of this problem is $y_1^{(1)} = 2.5$, $y_2^{(1)} = 0$, $y_3^{(1)} = 0$, and $w_1^{(1)} = 2$. The objective function optimal value is $z^{(1)} = 37.5$. Note that the optimal value of the dual variable associated with the constraint $x_1 = 3$ is $\lambda^{(1)} = 40$.

Step 2: Convergence checking. An upper bound of the objective function optimal value is computed as

$$z_{up}^{(1)} = 37.5 + (-112) - (-100) = 25.5 \ .$$

A lower bound of the objective function optimal value is

$$z_{\text{down}}^{(1)} = -112 \ .$$

Since the difference $z_{\text{up}}^{(1)} - z_{\text{down}}^{(1)} = 137.5 > \varepsilon$, the procedure continues.

Step 3: Master problem solution. The iteration counter is updated, $\nu = 1 + 1 = 2$. The master problem below is solved

$$\begin{array}{cc} \text{minimize} & -4x_1 + \alpha \\ x_1, \alpha & \end{array}$$

subject to

$$\begin{aligned} x_1 &\leq 4 \\ 2x_1 &\leq 6 \\ -x_1 &\leq -1 \\ 37.5 + 40(x_1 - 3) &\leq \alpha \\ -100 &\leq \alpha \ . \end{aligned}$$

The solution of this problem is $x_1^{(2)} = 1$ and $\alpha^{(2)} = -42.5$. The value for the objective function is $z^{(2)} = -46.5$.

The procedure continues with Step 1.

The solution resulting from the Benders decomposition algorithm is provided in Table B.7.

Table B.7. Evolution of the values of the master and subproblem variables using the Benders decomposition in Exercise 3.4

ν	$y_1^{(\nu)}$	$y_2^{(\nu)}$	$y_3^{(\nu)}$	$x_1^{(\nu)}$	$\alpha^{(\nu)}$	$\lambda^{(\nu)}$	$z_{(\nu)}^{\text{up}}$	$z_{(\nu)}^{\text{down}}$	$w_1^{(\nu)}$
1	2.5	0	0	3	-100	40	25.5	-112	2
2	0	1	3	1	-42.5	3.5	-10	-46.5	0
3	2.5	0	0	2	-2.5	40	-10.5	-10.5	0

The solution of this problem is $y_1 = 2.5$, $y_2 = 0$, $y_3 = 0$, $x_1 = 2$ with an objective function value $z = 10.5$.

Solution to Exercise 3.6. The formulation of the multiperiod investment problem is

$$\begin{array}{c} \text{minimize} \\ x_{it}, y_{it}, f_{ij,t}; i = 1, 2; t = 1, 2; (i,j) \in \mathcal{P} \end{array}$$

$$2x_{11} + 3.5x_{21} + 2.5(x_{12} - x_{11}) + 3.0(x_{22} - x_{21}) +$$

$$(0.7f_{13,1} + 0.8f_{23,1} + 0.5f_{12,1} + 0.6f_{21,1}) + (0.7f_{13,2} + 0.8f_{23,2} + 0.5f_{12,2} + 0.6f_{21,2})$$

subject to product balance constraints in period 1

$$y_{11} = f_{13,1} + f_{12,1} - f_{21,1}$$
$$y_{21} = f_{23,1} + f_{21,1} - f_{12,1}$$
$$19 = f_{13,1} + f_{23,1},$$

and in period 2

$$y_{12} = f_{13,2} + f_{12,2} - f_{21,2}$$
$$y_{22} = f_{23,2} + f_{21,2} - f_{12,2}$$
$$15 = f_{13,2} + f_{23,2},$$

production bounds

$$0 \leq y_{it} \leq x_{it}; \quad i = 1, 2; \quad t = 1, 2,$$

expansion limits

$$x_{it} \leq x_{i,t+1}; \quad i = 1, 2; \quad t = 1,$$

expansion capacity bounds

$$0 \leq x_{1t} \leq 10; \quad t = 1, 2,$$
$$0 \leq x_{2t} \leq 12; \quad t = 1, 2,$$

and transportation capacity limits

$$0 \leq f_{13,t} \leq 11; \quad t = 1, 2$$
$$0 \leq f_{23,t} \leq 9; \quad t = 1, 2$$
$$0 \leq f_{12,t} \leq 5; \quad t = 1, 2$$
$$0 \leq f_{21,t} \leq 5; \quad t = 1, 2.$$

Note that variables x_{it} are complicating variables. If they are fixed to given values, the problem decomposes by time periods. The solution resulting from the Benders decomposition algorithm is provided in Table B.8. Some of the subproblems are infeasible; therefore, artificial variables w_1 and w_2 are included to achieve feasibility.

The solution of this problem corresponds with the results provided in Sect. 1.4.2.

Solution to Exercise 3.8.

1. The formulation of the optimal scheduling problem is

$$\operatorname*{minimize}_{q_{it},\, y_{it},\, x_{it};\, i=1,2;\, t=1,2,3} \sum_{t=1}^{3} (10q_{1t} + 12q_{2t} + 4y_{1t} + 2y_{2t})$$

Table B.8. Evolution of the values of the master and subproblem variables using the Benders decomposition in Exercise 3.6

	Subproblem solutions											
ν	$y_{11}^{(\nu)}$	$y_{21}^{(\nu)}$	$f_{13,1}^{(\nu)}$	$f_{23,1}^{(\nu)}$	$f_{12,1}^{(\nu)}$	$f_{21,1}^{(\nu)}$	$y_{12}^{(\nu)}$	$y_{22}^{(\nu)}$	$f_{13,2}^{(\nu)}$	$f_{23,2}^{(\nu)}$	$f_{12,2}^{(\nu)}$	$f_{21,2}^{(\nu)}$
1	3.8	3.8	7.6	7.6	0	0	3	3	6	6	0	0
2	10	9	10	9	0	0	10	5	10	5	0	0
3	10.4	7.4	10.8	7.8	0	0	10	5	10	5	0	0
4	8.5	9.3	9.6	9	0	0.7	8.1	6.9	8.1	6.9	0	0
5	10	9	10	9	0	0	10	5	10	5	0	0

	Master problem solutions											
ν	$x_{11}^{(\nu)}$	$x_{21}^{(\nu)}$	$x_{12}^{(\nu)}$	$x_{22}^{(\nu)}$	$\alpha^{(\nu)}$	$\lambda_{11}^{(\nu)}$	$\lambda_{21}^{(\nu)}$	$\lambda_{12}^{(\nu)}$	$\lambda_{22}^{(\nu)}$	$z_{(\nu)}^{\text{up}}$	$z_{(\nu)}^{\text{down}}$	$w_1^{(\nu)}$	$w_2^{(\nu)}$
1	0	0	0	0	−100	−3.9	−3.8	−3.9	−3.8	−100	156.4	3.8	3
2	10	12	10	12	−12.8	−0.1	0	0	0	49.2	87.2	0	0
3	10	7	10	7	25.2	0	−3.8	0	0	69.7	77.3	0.4	0
4	8.1	8.9	8.1	8.9	25.4	−4.1	−3.5	−0.1	0	72.9	80.9	0.4	0
5	10	9	10	9	25.2	0	0	0	0	76.7	76.7	0	0

subject to

production balances

$$q_{11} + q_{21} = 100$$
$$q_{12} + q_{22} = 140$$
$$q_{13} + q_{23} = 200\,,$$

ramping limits

$$q_{1,t+1} - q_{1t} \leq 60; \quad t = 1, 2$$
$$q_{1t} - q_{1,t+1} \leq 60; \quad t = 1, 2$$
$$q_{2,t+1} - q_{2t} \leq 60; \quad t = 1, 2$$
$$q_{2t} - q_{2,t+1} \leq 60; \quad t = 1, 2\,,$$

production capacity limits

$$10x_{1t} \leq q_{1t}; \quad t = 1, 2, 3$$
$$50x_{2t} \leq q_{2t}; \quad t = 1, 2, 3$$
$$q_{1t} \leq 150x_{1t}; \quad t = 1, 2, 3$$
$$q_{2t} \leq 180x_{2t}; \quad t = 1, 2, 3\,,$$

logic of running and start-up status for production devices

$$
\begin{aligned}
y_{11} &\geq x_{11} \\
y_{1t} &\geq x_{1t} - x_{1,t-1}; & t = 2,3 \\
y_{21} &\geq x_{21} \\
y_{2t} &\geq x_{2t} - x_{2,t-1}; & t = 2,3 ,
\end{aligned}
$$

feasibility constraints (these constraints force the master problem to generate feasible solutions for the subproblem)

$$
\begin{aligned}
10x_{11} + 50x_{21} &\leq 100 \\
150x_{11} + 180x_{21} &\geq 100 \\
10x_{12} + 50x_{22} &\leq 140 \\
150x_{12} + 180x_{22} &\geq 140 \\
10x_{13} + 50x_{23} &\leq 200 \\
10x_{13} + 50x_{23} &\leq 200 \\
150x_{13} + 180x_{23} &\geq 200 ,
\end{aligned}
$$

and discrete and binary variable declarations

$$
\begin{aligned}
q_{it} &\in \mathbb{N}; & i = 1,2; & \quad t = 1,2,3 \\
x_{it} &\in \{0,1\}; & i = 1,2; & \quad t = 1,2,3 \\
y_{it} &\in \{0,1\}; & i = 1,2; & \quad t = 1,2,3 .
\end{aligned}
$$

Variables q_{it} represent the production of device i in period t, x_{it} represent the status variables of device i in period t, and variables y_{it} represent the start-up variable of device i in period t.

2. It should be noted that variables x_{it} and y_{it} are complicating variables. The solution resulting from the Benders decomposition algorithm is provided in Table B.9. Note that dual variables λ are associated with the constraints that fix the values of the complicating variables in the subproblem, therefore there are 12 dual variables. In Table B.9 we only provide the values of the dual variables which are different to zero. Dual variables λ_3, λ_4, and λ_5 correspond with constraints that fix the values of x_{13}, x_{21}, and x_{22}, respectively.

B.4 Exercises from Chapter 4

Solution to Exercise 4.2. Using the rules in Sect. 4.3.2 we obtain the dual problem

$$
\underset{y_1, y_2, y_3}{\text{maximize}} \quad z = 4y_1 + y_3
$$

Table B.9. Evolution of the values of the master and subproblem variables using the Benders decomposition in Exercise 3.8

	Subproblem solutions											
ν	$q_{11}^{(\nu)}$	$q_{12}^{(\nu)}$	$q_{13}^{(\nu)}$	$q_{21}^{(\nu)}$	$q_{22}^{(\nu)}$	$q_{23}^{(\nu)}$	$x_{11}^{(\nu)}$	$x_{12}^{(\nu)}$	$x_{13}^{(\nu)}$	$x_{21}^{(\nu)}$	$x_{22}^{(\nu)}$	$x_{23}^{(\nu)}$
1	50	90	150	50	50	50	1	1	1	1	1	1
2	100	140	150	0	0	50	1	1	1	0	0	1
3	100	140	150	0	0	50	1	1	1	0	0	1

	Master problem solution											
ν	$y_{11}^{(\nu)}$	$y_{12}^{(\nu)}$	$y_{13}^{(\nu)}$	$y_{21}^{(\nu)}$	$y_{22}^{(\nu)}$	$y_{23}^{(\nu)}$	$\alpha^{(\nu)}$	$\lambda_3^{(\nu)}$	$\lambda_4^{(\nu)}$	$\lambda_5^{(\nu)}$	$z_{\text{up}}^{(\nu)}$	$z_{\text{down}}^{(\nu)}$
1	1	0	0	1	0	0	−1,000	0	100	200	−994	4,706
2	1	0	0	0	0	1	4,400	−300	0	0	4,406	4,506
3	1	0	0	0	0	1	4,500	−300	0	0	4,506	4,506

subject to

$$\begin{array}{rcl}
y_1 + 2y_2 & \leq & 3 \\
y_1 - y_2 + 3y_3 & \leq & 1 \\
-y_1 + y_3 & = & 0 \\
-y_1 + y_2 - 2y_3 & \geq & -1 \\
y_2 & \leq & 0 \\
y_3 & \geq & 0 \, .
\end{array}$$

Note that

1. Since the primal problem is a minimization problem, the dual is a maximization problem.
2. The coefficients of the dual objective function are the right-hand side terms of the primal constraints.
3. The right-hand side terms of the dual problem constraints are the coefficients of the primal objective function.
4. The matrix of the coefficients of the dual constraints is the transpose of the corresponding primal matrix.
5. The equal, less-equal, or greater-equal signs are derived from the rules in Sect. 4.3.2.

The solution of the primal is

$$x_1^* = 0, \quad x_2^* = 5/4, \quad x_3^* = -11/4, \quad x_4^* = 0, \quad z_P^* = 5/4 \, ,$$

and the solution of the dual

$$y_1^* = 1/4, \quad y_2^* = 0, \quad y_3^* = 1/4, \quad z_D^* = 5/4 \, .$$

Note that the optimal objective function values of the primal and dual problems coincide.

Solution to Exercise 4.4. The Lagrangian function is

$$\mathcal{L}(x, y, \lambda, \mu) = x^2 + y^2 + \lambda(x - 5) + \mu(3 - xy)$$

and the KKT conditions

$$\begin{aligned} 2x + \lambda - \mu y &= 0 \\ 2y - \mu x &= 0 \\ x &= 5 \\ xy &\geq 3 \\ \mu(3 - xy) &= 0 \\ \mu &\geq 0 \, , \end{aligned}$$

which lead to the primal and dual solutions

$$x^* = 5, \quad y^* = 3/5, \quad \lambda^* = -1{,}232/125, \quad \mu^* = 6/25 \, .$$

For obtaining the dual problem we first calculate the dual function

$$\phi(\lambda, \mu) = \inf_{x,y} \mathcal{L}(x, y, \lambda, \mu) = \inf_{x,y} \left[x^2 + y^2 + \lambda(x - 5) + \mu(3 - xy) \right] \, ,$$

and since

$$\begin{aligned} \frac{\partial \mathcal{L}(x, y, \lambda, \mu)}{\partial x} &= 2x + \lambda - \mu y = 0 \\ \frac{\partial \mathcal{L}(x, y, \lambda, \mu)}{\partial y} &= 2y - \mu x = 0 \, , \end{aligned}$$

it leads to

$$x = \frac{2\lambda}{\mu^2 - 4}, \quad y = \frac{\lambda \mu}{\mu^2 - 4} \, .$$

The dual function becomes

$$\phi(\lambda, \mu) = \frac{\lambda^2 - 5\lambda \left(\mu^2 - 4\right) + 3\mu \left(\mu^2 - 4\right)}{\mu^2 - 4}$$

and then the dual problem is

$$\underset{\lambda, \mu}{\text{maximize}} \quad \phi(\lambda, \mu) = \frac{\lambda^2 - 5\lambda \left(\mu^2 - 4\right) + 3\mu \left(\mu^2 - 4\right)}{\mu^2 - 4}$$

subject to $\mu \geq 0$.
Finally, since

$$\begin{aligned} \frac{\partial \phi(\lambda, \mu)}{\partial \lambda} &= \frac{20 + 2\lambda - 5\mu^2}{\mu^2 - 4} = 0 \\ \frac{\partial \phi(\lambda, \mu)}{\partial \mu} &= \frac{-2\lambda^2 \mu + 3(\mu^2 - 4)^2}{(\mu^2 - 4)^2} = 0 \, , \end{aligned}$$

one obtains
$$\lambda^* = -1232/125, \quad \mu^* = 6/25,$$
which is the solution of the dual problem, and obviously coincides with that obtained from the KKT conditions above.

Solution to Exercise 4.6. The problem is illustrated graphically in Fig. B.1.
The Lagrangian function is
$$\begin{aligned}\mathcal{L}(x_1, x_2, \lambda, \mu_1, \mu_2, \mu_3, \mu_4) &= 2x_1^2 + x_2^2 - 2x_1 x_2 - 6x_2 - 4x_1 \\ &+ \lambda(x_1^2 + x_2^2 - 1) + \mu_1(-x_1 + 2x_2) \\ &+ \mu_2(x_1 + x_2 - 8) - \mu_3 x_1 - \mu_4 x_2,\end{aligned}$$

and the KKT conditions

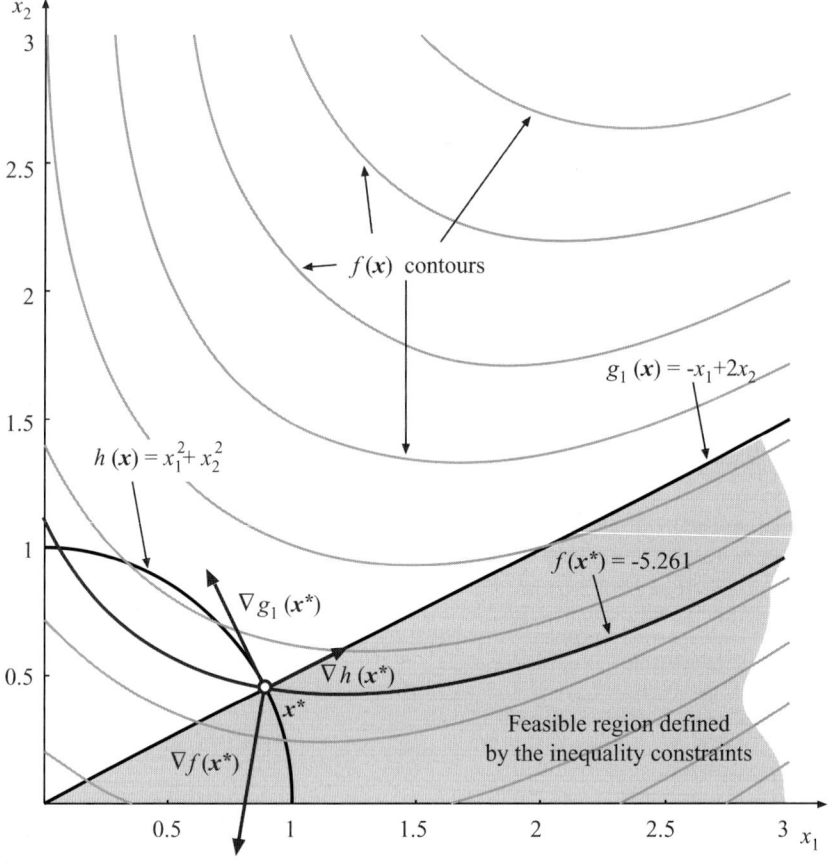

Fig. B.1. Illustration of the problem in Exercise 4.6

and the dual problem becomes

$$\underset{\lambda,\mu_1,\mu_2,\mu_3,\mu_4}{\text{maximize}} \quad z_D = \phi(\lambda,\mu_1,\mu_2,\mu_3,\mu_4)$$

subject to $\mu_1, \mu_2, \mu_3, \mu_4 \geq 0$, whose solution is

$$\lambda^* = \frac{7}{\sqrt{5}} - 1, \quad \mu_1^* = \frac{2}{5}(4+\sqrt{5}), \quad \mu_2^* = \mu_3^* = \mu_4^* = 0,$$

which coincides with the one given above.

The optimal objective function value of the dual problem is $z_D = 1 - \dfrac{14}{\sqrt{5}}$ that coincides with the optimal objective function value of the primal problem, therefore no duality gap exists.

Solution to Exercise 4.8. Consider the scheme of the problem in Fig. 4.13. Since the equation of the straight line passing through the points $(x, 0)$ and $(0, y)$ is

$$v = -\frac{y}{x}(u - x)$$

and the critical case occurs where the point (a, b) touches the ladder, the problem can be stated as

$$\underset{x,y}{\text{minimize}} \quad x^2 + y^2$$

subject to

$$xb + y(a - x) \leq 0.$$

The Lagrangian function is

$$\mathcal{L}(x, y, \mu) = x^2 + y^2 + \mu\left[xb + y(a - x)\right]$$

and the KKT conditions

$$\begin{aligned}
2x + \mu(b - y) &= 0 \\
2y + \mu(a - x) &= 0 \\
xb + y(a - x) &\leq 0 \\
\mu\left[xb + y(a - x)\right] &= 0,
\end{aligned}$$

which for $\mu \neq 0$ leads to

$$\begin{aligned}
x^* &= a + a^{1/3}b^{2/3} \\
y^* &= b + a^{2/3}b^{1/3} \\
\mu^* &= 2\left(\frac{a^{2/3} + b^{2/3}}{a^{1/3}b^{1/3}}\right).
\end{aligned}$$

B.4 Exercises from Chapter 4   445

$$
\begin{aligned}
4x_1 - 2x_2 - 4 + 2\lambda x_1 - \mu_1 + \mu_2 - \mu_3 &= 0 \\
2x_2 - 2x_1 - 6 + 2\lambda x_2 + 2\mu_1 + \mu_2 - \mu_4 &= 0 \\
x_1^2 + x_2^2 &= 1 \\
-x_1 + 2x_2 &\leq 0 \\
x_1 + x_8 &\leq 8 \\
\mu_1(-x_1 + 2x_2) + \mu_2(x_1 + x_2 - 8) - \mu_3 x_1 - \mu_4 x_2 &= 0 \\
\mu_1, \mu_2, \mu_3, \mu_4 &\geq 0,
\end{aligned}
$$

which leads to

$$x_1^* = \frac{2}{\sqrt{5}}, \quad x_2^* = \frac{1}{\sqrt{5}}, \quad \lambda^* = \frac{7}{\sqrt{5}} - 1, \quad \mu_1^* = \frac{2}{5}(4+\sqrt{5}), \quad \mu_2^* = \mu_3^* = \mu_4^* = 0,$$

with an objective function value $z_P = 1 - \dfrac{14}{\sqrt{5}}$.

The dual function is

$$\phi(\lambda, \mu_1, \mu_2, \mu_3, \mu_4) = \inf_{x_1, x_2} \mathcal{L}(x_1, x_2, \lambda, \mu_1, \mu_2, \mu_3, \mu_4)$$

and since

$$\frac{\partial \mathcal{L}(x_1, x_2, \lambda, \mu_1, \mu_2, \mu_3, \mu_4)}{\partial x_1} = 4x_1 - 2x_2 - 4 + 2\lambda x_1 - \mu_1 + \mu_2 - \mu_3 = 0$$

$$\frac{\partial \mathcal{L}(x_1, x_2, \lambda, \mu_1, \mu_2, \mu_3, \mu_4)}{\partial x_2} = 2x_2 - 2x_1 - 6 + 2\lambda x_2 + 2\mu_1 + \mu_2 - \mu_4 = 0$$

it leads to

$$x_1 = \frac{10 - \mu_1 - 2\mu_2 + \mu_3 + \lambda(4 + \mu_1 - \mu_2 + \mu_3) + \mu_4}{2(1 + 3\lambda + \lambda^2)}$$

$$x_2 = \frac{16 - 3\mu_1 - 3\mu_2 + \mu_3 + 2\mu_4 + \lambda(6 - 2\mu_1 - \mu_2 + \mu_4)}{2(1 + 3\lambda + \lambda^2)},$$

the dual function becomes

$$
\begin{aligned}
\phi(\lambda, \mu_1, \mu_2, \mu_3, \mu_4) =\ & -\frac{136 + 4\lambda^3 + 5\mu_1^2 - 20\mu_2 + 5\mu_2^2 + 4\lambda^2(3 + 8\mu_2)}{4(1 + 3\lambda + \lambda^2)} \\
& - \frac{20\mu_3 - 4\mu_2\mu_3 + \mu_3^2 + 2\mu_1(-22 + 4\mu_2 - \mu_3 - 3\mu_4)}{4(1 + 3\lambda + \lambda^2)} \\
& - \frac{32\mu_4 - 6\mu_2\mu_4 + 2\mu_3\mu_4 + 2\mu_4^2 + \lambda(56 + 5\mu_1^2)}{4(1 + 3\lambda + \lambda^2)} \\
& - \frac{\lambda(2\mu_2^2 + 8\mu_3 + \mu_3^2 + 12\mu_4 + \mu_4^2 + 76\mu_2)}{4(1 + 3\lambda + \lambda^2)} \\
& - \frac{\lambda(-2\mu_2\lambda(\mu_3 + \mu_4) + 2\mu_1(\mu_2 + \mu_3 - 2(4 + \mu_4)))}{4(1 + 3\lambda + \lambda^2)},
\end{aligned}
$$

The dual function is

$$\phi(\mu) = \inf_{x,y} \mathcal{L}(x,y,\mu) = \inf_{x,y} \left\{ x^2 + y^2 + \mu \left[xb + y(a-x) \right] \right\} .$$

Since

$$\frac{\partial \mathcal{L}(x,y,\mu)}{\partial x} = 2x + \mu(b-y) = 0$$

and

$$\frac{\partial \mathcal{L}(x,y,\mu)}{\partial y} = 2y + \mu(a-x) = 0$$

it leads to

$$x = \frac{2b\mu + a\mu^2}{\mu^2 - 4}, \quad y = \frac{2a\mu + b\mu^2}{\mu^2 - 4},$$

we have

$$\phi(\mu) = \frac{\mu^2(a^2 + b^2 + ab\mu)}{\mu^2 - 4} .$$

Then, the dual problem is

$$\text{maximize}_{\mu} \quad \frac{\mu^2(a^2 + b^2 + ab\mu)}{\mu^2 - 4}$$

subject to $\mu \geq 0$.
Since

$$\frac{\partial \phi(\mu)}{\partial \mu} = \frac{\mu \left(ab\mu \left(-12 + \mu^2\right) - 8a^2 - 8b^2 \right)}{(\mu^2 - 4)^2} = 0 ,$$

we have

$$\mu^* = 2 \left(\frac{a^{2/3} + b^{2/3}}{a^{1/3} b^{1/3}} \right) ,$$

i.e., the same value obtained from the KKT conditions.

Solution to Exercise 4.10. The problem can be stated as

$$\text{minimize}_{x,y,z} \quad x^2 + y^2 + z^2$$

subject to

$$xyz = 1; \quad x, y, z \geq 0 .$$

The Lagrangian function is

$$\mathcal{L}(x,y,\mu) = x^2 + y^2 + z^2 + \lambda(xyz - 1) - \mu_1 x - \mu_2 y - \mu_3 z$$

and the KKT conditions

$$\begin{aligned} 2x + \lambda yz - \mu_1 &= 0 \\ 2y + \lambda xz - \mu_2 &= 0 \\ 2z + \lambda xy - \mu_3 &= 0 \\ xyz &= 1 \\ x, y, z &\geq 0 \\ \mu_1 x + \mu_2 y + \mu_3 z &= 0 \\ \mu_1, \mu_2, \mu_3 &\geq 0 , \end{aligned}$$

which for $\mu_1 = \mu_2 = \mu_3 = 0$ leads to

$$x^* = y^* = z^* = 1, \quad \lambda^* = -2 . \tag{B.1}$$

The dual function is

$$\begin{aligned} \phi(\lambda, \mu_1, \mu_2, \mu_3) &= \inf_{x,y} \mathcal{L}(x, y, \lambda, \mu_1, \mu_2, \mu_3) \\ &= \inf_{x,y} \left\{ x^2 + y^2 + z^2 + \lambda(xyz - 1) - \mu_1 x - \mu_2 y - \mu_3 z \right\} . \end{aligned}$$

Since

$$\frac{\partial \mathcal{L}(x, y, \lambda, \mu_1, \mu_2, \mu_3)}{\partial x} = 2x + \lambda yz - \mu_1 = 0$$

$$\frac{\partial \mathcal{L}(x, y, \lambda, \mu_1, \mu_2, \mu_3)}{\partial y} = 2y + \lambda xz - \mu_2 = 0$$

$$\frac{\partial \mathcal{L}(x, y, \lambda, \mu_1, \mu_2, \mu_3)}{\partial z} = 2z + \lambda xy - \mu_3 = 0 ,$$

and we can use the symmetry property of the problem, we have

$$x = y = z = \frac{-1 + \sqrt{1 + \lambda \mu}}{\lambda}$$

and $\mu_1 = \mu_2 = \mu_3 = \mu$. Then, we have

$$\phi(\lambda, \mu) = \frac{2 - \lambda^3 - 2\sqrt{1 + \lambda \mu} + \lambda \mu \sqrt{1 + \lambda \mu}}{\lambda^2} .$$

Then, the dual problem is

$$\underset{\lambda, \mu}{\text{maximize}} \quad \phi(\lambda, \mu) = \frac{2 - \lambda^3 - 2\sqrt{1 + \lambda \mu} + \lambda \mu \sqrt{1 + \lambda \mu}}{\lambda^2}$$

subject to $\mu \geq 0$, whose solution is

$$\mu^* = 0, \quad \lambda^* = -2 ,$$

which implies $\mu_1^* = \mu_2^* = \mu_3^* = 0$, i.e., the same value obtained from the KKT conditions.

Solution to Exercise 4.12. We must introduce one multiplier for each constraint. We denote as λ the multiplier for the equality constraint, and μ_1, μ_2, and μ_3 the multipliers for the inequality constraints. Then, the Lagragian function is

$$\mathcal{L}(\lambda, \mu_1, \mu_2, \mu_3) = -x_1 + x_2 + \lambda(-x_1^2 + x_2) + \mu_1(x_1^2 + x_2^2 - 4) - \mu_2 x_1 - \mu_3 x_2. \quad \text{(B.2)}$$

The KKTC conditions are

1. The stationary condition of the Lagrangian function is stated as

$$\begin{pmatrix} -1 \\ 1 \end{pmatrix} + \lambda \begin{pmatrix} -2x_1 \\ 1 \end{pmatrix} + \mu_1 \begin{pmatrix} 2x_1 \\ 2x_2 \end{pmatrix} + \mu_2 \begin{pmatrix} -1 \\ 0 \end{pmatrix} + \mu_3 \begin{pmatrix} 0 \\ -1 \end{pmatrix} = \begin{pmatrix} 0 \\ 0 \end{pmatrix}. \quad \text{(B.3)}$$

2. The primal feasibility conditions are

$$\begin{aligned} x_1^2 + x_2^2 - 4 &\leq 0 \\ -x_1 &\leq 0 \\ -x_2 &\leq 0 \\ -x_1^2 + x_2 &= 0. \end{aligned} \quad \text{(B.4)}$$

3. The slackness conditions are

$$\mu_1(x_1^2 + x_2^2 - 4) = 0 \quad \text{(B.5)}$$
$$\mu_2(-x_1) = 0 \quad \text{(B.6)}$$
$$\mu_3(-x_2) = 0. \quad \text{(B.7)}$$

4. The dual feasibility conditions are

$$\mu_1, \mu_2, \mu_3 \geq 0. \quad \text{(B.8)}$$

To solve this system of equalities and inequalities, we consider the following cases (full enumeration of possibilities):

Case 1. $\mu_2 \neq 0$. If $\mu_2 \neq 0$, then using condition (B.6), $x_1 = 0$, and condition (B.3) implies

$$\begin{aligned} -1 - \mu_2 &= 0 \\ 1 + 2x_2\mu_1 - \mu_3 + \lambda &= 0. \end{aligned}$$

Since $\mu_2 = -1$ and the non-negativity condition (B.8) does not hold, then any KKT points must satisfy $\mu_2 = 0$.

Case 2. $\mu_3 \neq 0$ and $\mu_2 = 0$. If $\mu_3 \neq 0$ then using condition (B.7), $x_2 = 0$, and using the relationship $-x_1^2 + x_2 = 0$ of condition (B.4) we obtain $x_1 = 0$, and using condition (B.3), we obtain the contradiction

$$-1 = 0 \,.$$

This shows that any KKT point must have $\mu_3 = 0$.

Case 3. $\mu_1 \neq 0$ and $\mu_2 = \mu_3 = 0$. If $\mu_1 \neq 0$, then using condition (B.5), we obtain $x_1^2 + x_2^2 - 4 = 0$, and using the feasibility condition, we form the following system of equations

$$\begin{aligned} x_1^2 + x_2^2 - 4 &= 0 \\ -x_1^2 + x_2 &= 0 \,. \end{aligned} \tag{B.9}$$

The only solution satisfying condition (B.9) is $(x_1, x_2) = (\sqrt{\delta}, \delta)$, where $\delta = \dfrac{(-1+\sqrt{17})}{2}$ and, using condition (B.3), we get

$$\begin{aligned} -1 + 2\sqrt{\delta}\mu_1 - 2\sqrt{\delta}\lambda &= 0 \\ 1 + 2\delta\mu_1 + \lambda &= 0 \,, \end{aligned}$$

with solution

$$\begin{aligned} \mu_1 &= \frac{2\delta - \sqrt{\delta}}{2\delta(1 + 2\delta)} > 0 \\ \lambda &= \frac{-1}{2\sqrt{\delta}} + \frac{2\delta - \sqrt{\delta}}{2\delta(1 + 2\delta)} \,, \end{aligned}$$

which is a KKT point.

Case 4. The last case is $\mu_1 = \mu_2 = \mu_3 = 0$. Using condition (B.3), we obtain the system of equations

$$\begin{aligned} -1 - 2\lambda x_1 &= 0 \\ 1 + \lambda &= 0 \,, \end{aligned}$$

and we obtain the solution $\lambda = -1$ and $x_1 = 1/2$. Using now condition (B.4), we get $x_2 = x_1^2 = 1/4$. Since this point is feasible, it is also a KKT point. Thus, there are two candidates, shown in Cases 3 and 4:

$$(x_1, x_2, \lambda, \mu_1, \mu_2, \mu_3) = \left(\sqrt{\delta}, \delta, \frac{-1}{2\sqrt{\delta}} + \frac{2\delta - \sqrt{\delta}}{2\delta(1+2\delta)}, \frac{2\delta - \sqrt{\delta}}{2\delta(1+2\delta)}, 0, 0 \right)$$

with

$$\delta = \frac{(-1+\sqrt{17})}{2}$$

and

$$(x_1, x_2, \lambda, \mu_1, \mu_2, \mu_3) = (1/2, 1/4, -1, 0, 0, 0) \,.$$

B.5 Exercises from Chapter 5

Solution to Exercise 5.2. The solutions for the different items are as follows:

1. The Lagrangian problem is

$$\mathcal{L}(x, y, \lambda) = x^2 + y^2 + \lambda(x + y - 10)$$

subject to

$$\begin{aligned} x &\geq 0 \\ y &\geq 0. \end{aligned}$$

2. The LR decomposition reconstructs and solves the dual problem in a distributed fashion. Consider the dual problem

$$\underset{\lambda}{\text{maximize}} \quad \sum_{i=1}^{2} \phi_i(\lambda)$$

where

$$\phi_1(\lambda) = \underset{x}{\text{minimize}} \quad \mathcal{L}_1(\lambda) = x^2 + \lambda x - 5\lambda$$

subject to

$$x \geq 0,$$

and

$$\phi_2(\lambda) = \underset{y}{\text{minimize}} \quad \mathcal{L}_2(\lambda) = y^2 + \lambda y - 5\lambda$$

subject to

$$y \geq 0.$$

If, as stated in Sect. 5.3.4, a subgradient procedure with proportionality "constant" for iteration ν equal to $k^{(\nu)} = \dfrac{1}{a + b\,\nu}$ is used, then

$$\lambda \leftarrow \lambda + \frac{1}{a + b\nu} \frac{(x + y - 10)}{|(x + y - 10)|}.$$

With $a = 1$, $b = 0.1$, and an initial multiplier value $\lambda^{(0)} = -8$, we can use the following program in GAMS to solve the problem:

```
$Title Exercise 5.2.2

file out/Exercise5.2.2.out/; put out;

SCALARS
    a updating parameter /1/
    b updating parameter /0.1/
    error control eror parameter /1/
    epsilon maximum tolerable error /1e-3/
    itmax maximum iteration number /200/
    nu iteration counter /0/
    lambdaold;

PARAMETERS
    lambda Lagrange multiplier value /-8/;

POSITIVE VARIABLES x,y;

VARIABLES phi1,phi2;

EQUATIONS dual1,dual2;

dual1..phi1=e=x*x+lambda*x-5*lambda;
dual2..phi2=e=y*y+lambda*y-5*lambda;

MODEL dual1df /dual1/; MODEL dual2df /dual2/;

x.l=0; y.l=0;
phi1.l=x.l*x.l+lambda*x.l-5*lambda;
phi2.l=y.l*y.l+lambda*y.l-5*lambda;

while(error>epsilon and nu<itmax,
    nu=nu+1;
    put "    Iteration ",nu:3:0//;
    SOLVE dual1df using nlp MINIMIZING phi1;
    put "Modelstat= ",dual1df.modelstat,"; Solvestat= ",...
    ...dual1df.solvestat/;
    put "Phi1= ",phi1.l:8:3,"; x= ",x.l:8:3/;
    SOLVE dual2df using nlp MINIMIZING phi2;
    put "Modelstat= ",dual2df.modelstat,"; Solvestat= ",...
    ...dual2df.solvestat/;
    put "Phi2= ",phi2.l:8:3,"; y= ",y.l:8:3/;
    put "z= ",(phi1.l+phi2.l):8:3/;
*   Updating multiplier
    lambdaold=lambda;
    lambda=lambdaold+(x.l+y.l-10)/(abs(x.l+y.l-10))/(b+a*nu);
    put "New multiplier lambda= ",lambda:8:3//;
*   Updating error
    if(lambda ne 0,
        error=abs((lambda-lambdaold)/lambda);
    else
        error=abs((lambda-lambdaold));
    );
    put "Error= ",error:12:9//;
);
```

The optimal solution is $x^* = 5.005$, $y^* = 5.005$, $z^* = 50.000$ that is attained after 100 iterations.

3. Using the bundle (BD) method for updating the multipliers the quadratic programming problem below must be solved (see Sect. 5.3.4)

$$\begin{array}{c} \text{maximize} \\ z, \lambda \in C \end{array} \quad z - \alpha^{(\nu)}|\lambda - M^{(\nu)}|^2$$

subject to

$$z \leq \phi^{(k)} + s^{(k)}\left(\lambda - \lambda^{(k)}\right); \qquad k = 1, \ldots, \nu, \tag{B.10}$$

where $C = \{\lambda, -10 \leq \lambda \leq 0\}$, $\phi^{(k)}$ and $s^{(k)}$ are the dual function and the subgradient at iteration k, respectively. Then, we have

$$\phi^{(k)} = \left(\left(x^{(k)}\right)^2 + \left(y^{(k)}\right)^2 + \lambda^{(k)}\left(x^{(k)} + y^{(k)} - 10\right)\right)$$
$$s^{(k)} = \left(x^{(k)} + y^{(k)} - 10\right).$$

The penalty parameter can be calculated as $\alpha^{(\nu)} = d\ \nu$, where d is a constant scalar. If $\nu = 1$, the center of gravity is $M^{(1)} = \lambda^0$. Otherwise, the center of gravity is computed as

$$\begin{aligned} &\text{if} \quad \phi\left(\lambda^{(\nu)}\right) - \phi\left(M^{(\nu-1)}\right) \geq m\delta^{(\nu-1)} \\ &\text{then} \qquad M^{(\nu)} = \lambda^{(\nu)} \\ &\text{else} \qquad M^{(\nu)} = M^{(\nu-1)}, \end{aligned}$$

where

$$\phi\left(\mu^{(\nu)}\right) = \left((x^{(\nu)})^2 + (y^{(\nu)})^2 + \lambda^{(\nu)}(x^{(\nu)} + y^{(\nu)} - 10)\right)$$
$$\phi\left(M^{(\nu-1)}\right) = \left((x^{(\nu-1)})^2 + (y^{(\nu-1)})^2 + M^{(\nu-1)}(x^{(\nu-1)} + y^{(\nu-1)} - 10)\right).$$

The objective function gap is

$$\delta^{(\nu-1)} = z^{(\nu-1)} - d(\nu-1)|\lambda^{(\nu-1)} - M^{(\nu-1)}|^2 - \phi\left(\lambda^{(\nu-1)}\right).$$

With $C = \{\lambda, -20 \leq \lambda \leq 0\}$, $m = 0.5$, $d = 0.02$, and an initial multiplier value $\lambda^0 = -8$, the following GAMS program can be used for solving this problem:

```
$Title Exercise 5.2.3

file out/Exercise5.2.3.out/; put out;

Set
    IT maximum number of iterations /1*30/
    ITER(IT) dynamic set for activating cutting hyperplanes
    ACT(IT) dynamic set for the actual iteration;

ALIAS(IT,IT1); ITER(IT)=no;

SCALARS
    error control eror parameter /1/
    epsilon maximum tolerable error /1e-3/
    phiup dual function upper bound /INF/
    philo dual function lower bound /-INF/
    alpha penalty parameter
    d constant /0.02/
    me /0.5/;
```

```
PARAMETERS
    lambdas(IT) lagrange multipliers values for each iteration
    zs(IT) dual function values for each iteration
    s(IT) subgradient values for each iteration
    M(IT)
    phiM(IT)
    delta(IT);

POSITIVE VARIABLES x,y;

VARIABLES
    z,phi1,phi2,objcut
    lambda Lagrange multipliers value;

EQUATIONS dual1,dual2,cutobj,cutting,lambdaup,lambdalo;

dual1..phi1=e=x*x+lambda*x-5*lambda;
dual2..phi2=e=y*y+lambda*y-5*lambda;
cutobj(ACT)..objcut=e=z-alpha*sqr(lambda-M(ACT));
cutting(ITER)..z=l=zs(ITER)+s(ITER)*(lambda-lambdas(ITER));
lambdaup..lambda=g=-20; lambdalo..lambda=l=0;

MODEL dual1df /dual1/;
MODEL dual2df /dual2/;
MODEL updating updating /cutobj,cutting,...
...lambdaup,lambdalo/;

x.l=0; y.l=0;

lambdas(IT)=0;
zs(IT)=0;
s(IT)=0;
M(IT)=0;
phiM(IT)=0;
delta(IT)=0;

lambda.fx=-8;

loop(IT$(error>epsilon),

    put "    Iteration ",ORD(IT):3:0//;

    alpha=d*ORD(IT);

    phi1.l=x.l*x.l+lambda.l*x.l-5*lambda.l;
    phi2.l=y.l*y.l+lambda.l*y.l-5*lambda.l;
    SOLVE dual1df using nlp MINIMIZING phi1;
    put "Modelstat= ",dual1df.modelstat,"; Solvestat= ",...
    ...dual1df.solvestat/;
    put "Phi1= ",phi1.l:8:3,"; x= ",x.l:8:3/;
    SOLVE dual2df using nlp MINIMIZING phi2;
    put "Modelstat= ",dual2df.modelstat,"; Solvestat= ",...
    ...dual2df.solvestat/;
    put "Phi2= ",phi2.l:8:3,"; y= ",y.l:8:3/;

    lambdas(IT)=lambda.l;
    zs(IT)=phi1.l+phi2.l;
    s(IT)=x.l+y.l-10;
    philo=zs(IT);
    put "z= ",zs(IT):8:3,"; lambda= ",lambdas(IT):8:3,";"...
    ...subgradient= ",s(IT):8:3,"; Dual lower bound= ",philo:8:3/;

    if(ORD(IT)=1,
        M(IT)=lambda.l;
    else
```

```
            if(zs(IT)-sum(IT1$(ORD(IT1)+1=ORD(IT)),phiM(IT1))...
            ...ge me*sum(IT1$(ORD(IT1)+1=ORD(IT)),delta(IT1)),
                M(IT)=lambda.l;
            else
                M(IT)=sum(IT1$(ORD(IT1)+1=ORD(IT)),M(IT1));
            );
        );
        phiM(IT)=x.l**2+y.l**2+M(IT)*s(IT);

        ACT(IT1)=no;
        ACT(IT)=yes;
        ITER(IT)=yes;

        lambda.up=Inf;
        lambda.lo=-Inf;
        SOLVE updating using nlp MAXIMIZING objcut;
        put "Modelstat= ",updating.modelstat,"; Solvestat= ",...
        ...updating.solvestat/;
        put "objcut= ",objcut.l:8:3,"z= ",z.l:8:3,"; lambda= ",...
        ...lambda.l:8:3/;

        delta(IT)=z.l-alpha*sqr(lambdas(IT)-M(IT))-zs(IT);
        lambda.fx=lambda.l;
        phiup=objcut.l;
        put "Dual function upper bound= ",phiup:8:3/;

*            Updating error
        if(lambda.l ne 0,
            error=abs((lambda.l-lambdas(IT))/lambda.l);
        else
            error=abs(lambda.l-lambdas(IT));
        );
        put "Error= ",error:12:9//;
    );
```

The optimal solution is $x^* = 5.027$, $y^* = 5.027$, $z^* = 49.999$ that is attained after nine iterations.

4. Note that both methods converge to the same solution but the bundle method (9 iterations) is much faster than the subgradient method (100 iterations) with the same prespecific error tolerance ($\varepsilon = 0.001$).

Solution to Exercise 5.4. The production-scheduling problem formulated in Sect. 1.5.1, p. 39, can be solved using the Lagrangian relaxation (LR) method. If the demand constraints (1.51) for the periods ($m = 2$) are considered as complicating constraints, the nonlinear production-scheduling problem decomposes by production device ($n = 2$). The solution of this problem is the same as the solution of the following dual problem:

$$\underset{\boldsymbol{\lambda}}{\text{maximize}} \quad \phi(\boldsymbol{\lambda})$$

subject to

$$\boldsymbol{\lambda}^{\text{down}} \leq \boldsymbol{\lambda} \leq \boldsymbol{\lambda}^{\text{up}},$$

where $\boldsymbol{\lambda}$ is the Lagrange multipliers vector related to the demand constraints (1.51) for all periods.

Note that the dual function can be decomposed by production device, thus $\phi(\boldsymbol{\lambda}) = \sum_{i=1}^{n} \phi_i(\boldsymbol{\lambda})$ where the dual functions $\phi_i(\boldsymbol{\lambda})$ are

$$\phi_i(\boldsymbol{\lambda}) = \underset{x_{it}}{\text{minimize}} \sum_{t=1}^{m}\left(a_i x_{it} + \frac{1}{2}b_i x_{it}^2 + \lambda_t\left(x_{it} - \frac{d_t}{n}\right)\right)$$

subject to

$$\begin{aligned}
x_{it} &\geq 0; & t &= 1,\ldots,m \\
x_{it} &\leq x_i^{\max}; & t &= 1,\ldots,m \\
x_{i1} - x_i^0 &\leq r_i^{\max} & & \\
x_{it} - x_{i,t-1} &\leq r_i^{\max}; & t &= 2,\ldots,m \\
x_i^0 - x_{i1} &\leq r_i^{\max} & & \\
x_{i,t-1} - x_{it} &\leq r_i^{\max}; & t &= 2,\ldots,m \,,
\end{aligned} \qquad (\text{B.11})$$

where the demand for each period d_t has been equally distributed among all devices.

For the Lagrange multipliers updating the cutting plane method (CP) is used. To this aim, the following problem is solved:

$$\underset{z,\lambda_t;\, t=1,\ldots,m}{\text{maximize}} \quad z$$

subject to

$$z \leq \phi^{(k)} + \sum_{i=1}^{t} s_t^{(k)}\left(\lambda_t - \lambda_t^{(k)}\right); \qquad k = 1,\ldots,\nu$$

$$\boldsymbol{\lambda}^{\text{down}} \leq \boldsymbol{\lambda} \leq \boldsymbol{\lambda}^{\text{up}},$$

where $s_t = \sum_{i=1}^{n} x_{it} - d_t$ is the subgradient of the dual function.

Assuming $\boldsymbol{\lambda}^{\text{up}} = 10$, $\boldsymbol{\lambda}^{\text{down}} = -10$, $\boldsymbol{\lambda}^{(0)} = 0$, and $x_{it}^{(0)} = 0$ where $i = 1,2;\; t = 1,2$, the following GAMS program is used to solve this problem:

```
$Title Exercise 5.4.1

file out/Exercise5.4.1.out/; put out;

Set
      m the number of time periods /1*2/
      n the number of production devices /1*2/
      ACT(n) dynamic set for the actual device;

Alias(m,m1); Alias(n,n1);

PARAMETERS
      xmax(n) the output capacity of device n/
      1       6
      2       8/
      rmax(n) the ramping (up and down) limit of device n/
      1       1.5
      2       3/
      x0(n) initial output level of device n/
      1       2
      2       2.5/
      a(n) linear coefficients in the cost function of device n/
      1       2
      2       2.5/
      b(n) nonlinear coefficients in the cost function of device n/
```

```
        1       0.6
        2       0.5/
    d(m) demand for period m/
        1       9
        2       12/;

POSITIVE VARIABLES x(n,m) the output of device n during period m;

Equation maxout maximum output capacity;
    maxout(ACT,m)..x(ACT,m)=l=xmax(ACT);

Equation rampup0,rampup ramping up limits;
    rampup0(ACT)..x(ACT,'1')-x0(ACT)=l=rmax(ACT);
    rampup(ACT,m,m1)$(ORD(m)+1=ORD(m1))..x(ACT,m1)-x(ACT,m)=l=...
    ...rmax(ACT);

Equation rampdown0,rampdown ramping down limits;
    rampdown0(ACT)..x0(ACT)-x(ACT,'1')=l=rmax(ACT);
    rampdown(ACT,m,m1)$(ORD(m)+1=ORD(m1))..x(ACT,m)-x(ACT,m1)=l=...
    ...rmax(ACT);

Set
    IT iteration number /1*50/
    ITER(IT) dynamic set for activating cutting hyperplanes;

SCALARS
    error control error parameter /0/
    maxerror /1/
    epsilon maximum tolerable error /1e-3/
    phiup dual function upper bound /INF/
    philo dual function lower bound /-INF/;

PARAMETERS
    lambdas(IT,m) Lagrange multipliers values associated with demand...
    ...constraint m for iteration IT
    zs(IT) dual function values for iteration IT
    s(IT,m) subgradient values associated with demand constraint m ...
    ...for iteration IT;

VARIABLES z, phi, lambda(m) lagrange multipliers value;

EQUATION dual associated with each device;
    dual(ACT)..phi=e=sum(m,a(ACT)*x(ACT,m)+0.5*b(ACT)*x(ACT,m)**2+...
    ...lambda(m)*(x(ACT,m)-d(m)/card(n))));

EQUATION cutting cutting planes;
    cutting(ITER)..z=l=zs(ITER)+sum(m,s(ITER,m)*(lambda(m)-...
    ...lambdas(ITER,m)));

EQUATIONS lambdaup,lambdalo upper and lower Lagrange multiplier
limits;
    lambdaup(m)..lambda(m)=l=10;
    lambdalo(m)..lambda(m)=g=-10;

MODEL dualdf /dual,maxout,rampup0,rampup,rampdown0,rampdown/;
MODEL updating multiplier updating /cutting,lambdaup,lambdalo/;

x.l(n,m)=0;

ITER(IT)=no;

lambdas(IT,m)=0; zs(IT)=0; s(IT,m)=0;

lambda.fx(m)=0;

loop(IT$(maxerror>epsilon),
```

```
            put "    Iteration ",ORD(IT):3:0//;

            loop(n,
                ACT(n1)=no;
                ACT(n)=yes;
                phi.l=sum(m,a(n)*x.l(n,m)+0.5*b(n)*x.l(n,m)**2+...
                ...lambda.l(m)*(x.l(n,m)-d(m)/card(n)));
                SOLVE dualdf using nlp MINIMIZING phi;
                put "Modelstat= ",dualdf.modelstat,"; Solvestat= ",...
                ...dualdf.solvestat/;
                put "phi(",ORD(n):3:0,")= ",phi.l:8:3/;
                loop(m,
                      put "x(",ORD(n):3:0,", ",ORD(m):3:0,")= ",x.l(n,m):8:3/;
                );
                zs(IT)=zs(IT)+phi.l;
            );

            s(IT,m)=sum(n,x.l(n,m))-d(m);
            lambdas(IT,m)=lambda.l(m);
            philo=zs(IT);
            put "z= ",zs(IT):8:3,"; Dual function lower bound= ",philo:8:3/;
            loop(m,
                 put "subgrad(",ORD(IT):3:0,", ",ORD(m):3:0,")= ",s(IT,m):8:3/;
            );

            ITER(IT)=yes;

            lambda.up(m)=Inf;
            lambda.lo(m)=-Inf;
            SOLVE updating using nlp MAXIMIZING z;
            put "Modelstat= ",updating.modelstat,"; Solvestat= ",...
            ...updating.solvestat/;
            put "z= ",z.l:8:3/;
            loop(m,
                 put "lambda(",ORD(m):3:0,")= ",lambda.l(m):8:3/;
            );

            lambda.fx(m)=lambda.l(m);
            phiup=z.l;
            put "Dual function upper bound= ",phiup:8:3/;

*           Updating error

            maxerror=0;
            loop(m,

                 if(lambda.l(m) ne 0,
                     error=abs((lambda.l(m)-lambdas(IT,m))/lambda.l(m));
                 else
                     error=abs((lambda.l(m)-lambdas(IT,m)));
                 );
                 if(error>maxerror, maxerror=error;);
            );
            put "Error= ",maxerror:12:9//;
);
```

The optimal solution is

$$x^*_{11} = 3.500, \quad x^*_{12} = 5.000, \quad x^*_{21} = 5.500, \quad x^*_{22} = 7.003,$$

which is attained after 22 iterations with a relative error tolerance of $\varepsilon = 0.001$. Note that it is the same solution provided in Sect. 1.5.1.

For solving this problem using the augmented Lagrangian (AL) decomposition (see Sect. 5.4) the following augmented Lagrangian function is defined:

$$\mathcal{A}(\boldsymbol{x},\boldsymbol{\lambda})=\sum_{i=1}^{n}\sum_{t=1}^{m}\left(a_i x_{it}+\frac{1}{2}b_i x_{it}^2\right)+\sum_{t=1}^{m}\lambda_t\left(\sum_{i=1}^{n} x_{it}-d_t\right)+\frac{\alpha}{2}\sum_{t=1}^{m}\lambda_t\left(\sum_{i=1}^{n} x_{it}-d_t\right)^2,$$

which can be decomposed by production device as

$$\mathcal{A}_i(\boldsymbol{x},\boldsymbol{\lambda})=\sum_{t=1}^{m}\left[a_i x_{it}+\frac{1}{2}b_i x_{it}^2 + \lambda_t\left(x_{it}-\frac{d_t}{n}\right)\right]+\frac{\alpha}{2n}\sum_{t=1}^{m}\lambda_t\left(x_{it}+\sum_{\substack{j\neq i \\ j=1}}^{n} x_{jt}^{(\nu-1)}-d_t\right)^2.$$

Note that, as in the LR procedure, the demand for each period d_t has been equally distributed among all devices. To achieve separability the variables associated with other devices are fixed to the values of the previous iteration ($x_{jt}^{(\nu-1)}$) and the quadratic term is also equally distributed among all devices. These functions together with (B.11) constitute the main subproblems. The multipliers are updated using a gradient of the augmented Lagrangian function (see Sect. 5.4.4)

$$\lambda_t^{(\nu+1)} = \lambda_t^{(\nu)} + \alpha\left(\sum_{i=1}^{n} x_{it}-d_t\right).$$

Assuming $\alpha = 1$, $x_{it}^{(0)} = 0$ ($i = 1, 2;\ t = 1, 2$), and $\lambda_t^{(0)} = 0$ ($t = 1, 2$), the following GAMS program is used to solve the problem (the first part of the code until the ramp-down equation definition is the same as the previous GAMS program):

```
SCALARS
    error control error parameter /0/
    maxerror /1/
    itmax maximum iterations number /50/
    nu iteration counter /0/
    epsilon maximum tolerable error /1e-3/
    alpha /1/;

PARAMETERS
    z
    lambda(m) Lagrange multipliers value/
    1       0
    2       0/
    lambdaold(m)
    xaux(n,m) auxiliar the output of device n during period m;

VARIABLES phi;

EQUATION dual associated with each device;
    dual(ACT)..phi=e=sum(m,a(ACT)*x(ACT,m)+...
    ...0.5*b(ACT)*sqr(x(ACT,m))+lambda(m)*(x(ACT,m)-...
    ...d(m)/card(n)))+0.5*alpha*sum(m,sqr(x(ACT,m)+...
    ...sum(n1$(not ACT(n1)),xaux(n1,m))-d(m)))/card(n);
EQUATION xfix;
    xfix(n1,m)$(not ACT(n1))..xaux(n1,m)=e=x(n1,m);

MODEL dualdf /dual,maxout,rampup0,rampup,rampdown0,rampdown/;

x.l(n,m)=0; xaux(n,m)=x.l(n,m);

put "Initial multipliers"/; loop(m,
```

```
            put "lambda(",ORD(m):3:0,")= ",lambda(m):8:3/;
        );put /;

        while(maxerror>epsilon and nu<itmax,
            nu=nu+1;
            put "    Iteration ",nu:3:0//;
            z=0;
            loop(n,
                ACT(n1)=no;
                ACT(n)=yes;
                phi.l=sum(m,a(n)*x.l(n,m)+0.5*b(n)*sqr(x.l(n,m))+...
                ...lambda(m)*(x.l(n,m)-d(m)/card(n)))+...
                ...0.5*alpha*sum(m,sqr(sum(n1,x.l(n1,m))-d(m)))/card(n);
                SOLVE dualdf using nlp MINIMIZING phi;
                put "Modelstat= ",dualdf.modelstat,"; Solvestat= ",...
                ...dualdf.solvestat/;
                put "phi(",ORD(n):3:0,")= ",phi.l:8:3/;
                loop(m,
                    put "x(",ORD(n):3:0,", ",ORD(m):3:0,")= ",x.l(n,m):8:3/;
                );
                z=z+phi.l;
            );
            xaux(n,m)=x.l(n,m);
            put "Augmented Lagrangian function= ",z:8:3/;
*           Updating multiplier
            lambdaold(m)=lambda(m);
            lambda(m)=lambdaold(m)+alpha*(sum(n,x.l(n,m))-d(m));
            put "New multipliers"/;
            loop(m,
                put "lambda(",ORD(m):3:0,")= ",lambda(m):8:3/;
            );put /;
*           Updating error
            maxerror=0;
            loop(m,
                if(lambda(m) ne 0,
                    error=abs((lambda(m)-lambdaold(m))/lambda(m));
                else
                    error=abs((lambda(m)-lambdaold(m)));
                );
                if(error>maxerror, maxerror=error;);
            );
            put "Error= ",maxerror:15:13//;
        );
```

The optimal solution is

$$x_{11}^* = 3.500, \quad x_{12}^* = 5.000, \quad x_{21}^* = 5.500, \quad x_{22}^* = 7.000,$$

which is attained after four iterations with a relative error tolerance of $\varepsilon = 10^{-8}$.

Note that this example points out the slow and oscillating behavior of the LR procedure, that converges in 22 iterations, while the quadratic penalty term in the AL procedure corrects this anomaly, obtaining a more precise solution in just 4 iterations.

Solution to Exercise 5.6. The optimal operation of the multiarea electricity network addressed in Sect. 1.5.2, p. 42, can be solved using the optimality condition decomposition (OCD). Consider Table B.10 where all the variables (columns) and equations (rows) are specified. Each cell contains the value 1 if the corresponding variable (column) appears in the corresponding equation

Table B.10. Structure of optimal operation of a multiarea electricity network problem

	x_1	x_2	e_{12}	e_{21}	e_{13}	e_{31}	e_{23}	e_{32}	e_{34}	δ_1	δ_2	δ_3													
b_1	0	0	1	0	1	0	0	0	0	0	0	0													
b_2	1	0	0	1	0	0	1	0	0	0	0	0													
b_3	0	1	0	0	0	1	0	1	1	0	0	0													
e_{12}	0	0	1	0	0	0	0	0	0	1	1	0													
e_{21}	0	0	0	1	0	0	0	0	0	1	1	0													
e_{13}	0	0	0	0	1	0	0	0	0	1	0	1													
e_{31}	0	0	0	0	0	1	0	0	0	1	0	1													
e_{23}	0	0	0	0	0	0	1	0	0	0	1	1													
e_{32}	0	0	0	0	0	0	0	1	0	0	1	1													
e_{34}	0	0	0	0	0	0	0	0	1	0	0	1	1	0	0	0	0	0	0	0	0	0	0	0	e_{34}
e_{43}	0	0	0	0	0	0	0	0	0	0	0	1	1	0	0	1	0	0	0	0	0	0	0	0	e_{43}
													0	0	0	0	0	0	0	1	0	1	0	0	b_4
													0	0	0	0	0	1	1	0	0	0	1	0	b_5
													0	0	0	1	1	0	0	0	1	0	0	1	b_6
													1	1	0	0	1	0	0	0	0	0	0	0	e_{45}
													1	1	0	0	0	1	0	0	0	0	0	0	e_{54}
													0	1	1	0	0	0	1	0	0	0	0	0	e_{56}
													0	1	1	0	0	0	0	1	0	0	0	0	e_{65}
													1	0	1	0	0	0	0	0	1	0	0	0	e_{46}
													1	0	1	0	0	0	0	0	0	1	0	0	e_{64}
													δ_4	δ_5	δ_6	e_{43}	e_{45}	e_{54}	e_{56}	e_{65}	e_{46}	e_{64}	y_1	y_2	

(row) and the value 0 otherwise. Note that equations e_{ij}, $\forall (i,j)$, correspond to (1.60) and b_i ($i = 1, \ldots, 6$) are the balance equation constraints in all nodes, (1.54)–(1.59). If the constraints corresponding to the flow of energy between frontier nodes 3 and 4 (e_{34} and e_{43}) are relaxed, the problem decomposes by area. Therefore, the balance equations at the frontier nodes are the complicating constraints.

The subproblems to achieve the optimality condition decomposition are the following:

$$\text{minimize} \atop x_i; \ i = 1, 2$$

$$\sum_{i=1}^{2} \left(a_i^x x_i + \frac{1}{2} b_i^x x_i^2 \right) + \lambda^y (\bar{e}_{43} - G_{43} \cos(\bar{\delta}_4 - \delta_3) - B_{43} \sin(\bar{\delta}_4 - \delta_3) + G_{43})$$

subject to

$$\begin{aligned}
-d^x &= e_{12} + e_{13} \\
x_1 &= e_{21} + e_{23} \\
x_2 &= e_{31} + e_{32} + e_{34} \\
x_i &\leq x_i^{\max}; \ i = 1, 2 \\
e_{ij} &= G_{ij} \cos(\delta_i - \delta_j) + B_{ij} \sin(\delta_i - \delta_j) - G_{ij}; \\
& (i,j) \in \{(1,2),(2,1),(1,3),(3,1),(2,3),(3,2)\} \\
e_{34} &= G_{34} \cos(\delta_3 - \bar{\delta}_4) + B_{34} \sin(\delta_3 - \bar{\delta}_4) - G_{34} : \lambda^x \ ,
\end{aligned}$$

corresponding to system X, where the bar refers to fixed values and λ^x is the Lagrange multiplier associated with the last constraint, and

$$\text{minimize} \atop y_i; \ i=1,2$$

$$\sum_{i=1}^{2} \left(a_i^y y_i + \frac{1}{2} b_i^y y_i^2 \right) + \lambda^x (\bar{e}_{34} - G_{34} \cos(\bar{\delta}_3 - \delta_4) - B_{34} \sin(\bar{\delta}_3 - \delta_4) + G_{34})$$

subject to

$$\begin{aligned}
-d^y &= e_{65} + e_{64} \\
y_1 &= e_{56} + e_{54} \\
y_2 &= e_{45} + e_{46} + e_{43} \\
y_i &\leq y_i^{\max}; \ i = 1, 2 \\
e_{ij} &= G_{ij} \cos(\delta_i - \delta_j) + B_{ij} \sin(\delta_i - \delta_j) - G_{ij}; \\
&\quad (i,j) \in \{(4,5),(5,4),(4,6),(6,4),(5,6),(6,5)\} \\
e_{43} &= G_{43} \cos(\delta_4 - \bar{\delta}_3) + B_{43} \sin(\delta_4 - \bar{\delta}_3) - G_{43} : \lambda^y,
\end{aligned}$$

corresponding to system Y, where λ^y is the Lagrange multiplier related to the last constraint.

The following GAMS program can be used to solve this problem:

```
$Title Exercise 5.6

file out/Exercise5.6.out/; put out;

Set
    lo zone location /x,y/
    ge number of generators /1*2/
    n nodes /1*6/
    AL(n,n) dynamic set for the active lines
    FIXC(n,n) dynamic set for active mismatches
    ACL(lo) dynamic set for the active location;

Alias(n,n1);
ALIAS(lo,lo1);

PARAMETERS
    d(lo) hourly energy demands of systems x and y/
    x       8
    y       9.5/
    G(n,n) conductance of line ij
    Bs(n,n) susceptance of line ij;

AL(n,n)=no;
AL('1','2')=yes; G('1','2')=-1.0; Bs('1','2')=7.0;
AL('1','3')=yes; G('1','3')=-1.5; Bs('1','3')=6.0;
AL('2','3')=yes; G('2','3')=-0.5; Bs('2','3')=7.5;
AL('3','4')=yes; G('3','4')=-1.3; Bs('3','4')=5.5;
AL('4','5')=yes; G('4','5')=-1.0; Bs('4','5')=9.0;
AL('4','6')=yes; G('4','6')=-0.9; Bs('4','6')=7.0;
AL('5','6')=yes; G('5','6')=-0.3; Bs('5','6')=6.5;

loop((n,n1)$(ORD(n)>ORD(n1) and AL(n1,n)),
        AL(n,n1)=yes;
        G(n,n1)=G(n1,n);
```

```
            Bs(n,n1)=Bs(n1,n);
);

TABLE xmax(lo,ge) maximum production capacities...
...of the two generators of areas X and Y
            1       2
x           6       7
y           6       8;

TABLE a(lo,ge) linear cost coefficients
            1       2
x           2       2.5
y           3       2.5;

TABLE b(lo,ge) quadratic cost coefficients
            1       2
x           0.6     0.5
y           0.7     0.5;


POSITIVE VARIABLES x(lo,ge) energy productions of generators...
...of areas X and Y;

VARIABLES
        delta(n) relative heights or phases of nodes
        e(n,n) electric energy flowing through the line between...
...nodes
        cost total production cost
        c(n,n) complicating constraint mismatches;

delta.fx('1')=0;

Equations balance1,balance2,balance3,balance4,balance5,...
...balance6,auxc energy balance in every node;

balance1$(ACL('x'))..e('1','2')+e('1','3')+d('x')=e=0;
balance2$(ACL('x'))..e('2','1')+e('2','3')-x('x','1')=e=0;
balance3$(ACL('x'))..e('3','1')+e('3','2')+e('3','4')-x('x','2')=e=0;
balance4$(ACL('y'))..e('6','4')+e('6','5')+d('y')=e=0;
balance5$(ACL('y'))..e('5','4')+e('5','6')-x('y','1')=e=0;
balance6$(ACL('y'))..e('4','3')+e('4','5')+e('4','6')-x('y','2')=e=0;
auxc(n,n1)$(FIXC(n,n1) and AL(n,n1))..c(n,n1)=e=0;

Equation edf electric energy flowing through the line between nodes;
edf(n,n1)$(AL(n,n1))..c(n,n1)=e=e(n,n1)-G(n,n1)*cos(delta(n)-delta(n1))...
...-Bs(n,n1)*sin(delta(n)-delta(n1))+G(n,n1);

Equation prodpu maximum production limit per generator;
        prodpu(ACL,ge)..x(ACL,ge)=l=xmax(ACL,ge);

Parameter lambda(n,n),lambdaold(n,n) complicating constraint Lagrange...
...multipliers;

Equation costlo total production cost per location definition;
costlo..cost=e=sum((ge,ACL),a(ACL,ge)*x(ACL,ge)+0.5*b(ACL,ge)*sqr(x(ACL,ge)))...
...+sum((n,n1)$(not FIXC(n,n1)),lambda(n,n1)*c(n,n1));

Model localcost /costlo,balance1,balance2,balance3,balance4,balance5,...
...balance6,auxc,edf,prodpu/;

SCALARS
        error control error parameter /0/
        maxerror /1/
        epsilon maximum tolerable error /1e-5/
        itmax maximum iteration number /20/
        nu iteration counter /0/;
```

```
lambda(n,n1)=0;
x.l(lo,ge)=0;
delta.l(n)=0;
e.l(n,n1)$(AL(n,n1))=0;

while(maxerror>epsilon and nu<itmax,
    nu=nu+1;
    loop(lo,
        ACL(lo1)=no;
        ACL(lo)=yes;
        AL(n,n1)=no;
        FIXC(n,n1)=yes;
        if(ORD(lo)=1,
            AL('1','2')=yes;
            AL('2','1')=yes;
            AL('1','3')=yes;
            AL('3','1')=yes;
            AL('2','3')=yes;
            AL('3','2')=yes;
            AL('3','4')=yes;
            AL('4','3')=yes;
            FIXC('4','3')=no;
            e.fx('4','3')=e.l('4','3');
            delta.fx('4')=delta.l('4');
        );
        if(ORD(lo)=2,
            AL('4','5')=yes;
            AL('5','4')=yes;
            AL('4','6')=yes;
            AL('6','4')=yes;
            AL('5','6')=yes;
            AL('6','5')=yes;
            AL('3','4')=yes;
            AL('4','3')=yes;
            FIXC('3','4')=no;
            e.fx('3','4')=e.l('3','4');
            delta.fx('3')=delta.l('3');
        );

        SOLVE localcost using nlp MINIMIZING cost;
        put "Modelstat= ",localcost.modelstat,"; Solvestat= ",localcost.solvestat/;
        put "cost(",ORD(lo):3:0,")= ",cost.l:8:3/;

        if(ORD(lo)=1,
            lambdaold('3','4')=lambda('3','4');
            lambda('3','4')=edf.m('3','4');
        );
        if(ORD(lo)=2,
            lambdaold('4','3')=lambda('4','3');
            lambda('4','3')=edf.m('4','3');
        );

        e.lo('4','3')=-Inf;      e.up('4','3')=Inf;
        delta.lo('4')=-Inf;      delta.up('4')=Inf;
        e.lo('3','4')=-Inf;      e.up('3','4')=Inf;
        delta.lo('3')=-Inf;      delta.up('3')=Inf;
    );

*   Updating error

    maxerror=0;
    loop((n,n1)$((ORD(n)=3 and ORD(n1)=4) or (ORD(n)=4 and ORD(n1)=3)),
            if(lambda(n,n1) ne 0,
                    error=abs((lambda(n,n1)-lambdaold(n,n1))/lambda(n,n1));
            else
                    error=abs(lambda(n,n1)-lambdaold(n,n1));
            );
```

```
            if(error>maxerror, maxerror=error;);
        );
        put "Error= ",maxerror:12:9//;

);

loop((ge,lo),
    put "x(",lo.tl:2,",",ge.tl:2,")= ",x.l(lo,ge):8:3/;
);
loop((n,n1)$(AL(n,n1)),
    put "e(",n.tl:2,",",n1.tl:2,")= ",e.l(n,n1):8:3/;
);put /;
loop(n,
    put "delta(",n.tl:2,")= ",delta.l(n):8:3/;
);put /;
```

The optimal solution is

$$x_1^* = 5.26 \text{ MW}, \quad x_2^* = 4.87 \text{ MW}, \quad y_1^* = 3.82 \text{ MW}, \quad y_2^* = 5.65 \text{ MW}.$$

The flow of energy in the tie line is

$$e_{34}^* = 0.89 \text{ MW},$$

which is attained after seven iterations with a relative error tolerance of $\varepsilon = 10^{-5}$.

Note that in this example the subproblems are solved until the optimality conditions hold at each iteration. This strategy is valid but it is inefficient because the algorithm would converge just performing a single iteration for each subproblem, and then updating variable values. In this example, subproblems are solved to optimality for the sake of an easy implementation.

Solution to Exercise 5.8. The energy flow problem can be solved considering the constraint that fixes the total energy transmitted as a complicating constraint. Using a LR procedure with a CP method for updating the parameters, the algorithm proceeds as follows:

Step 0: Initialization. Set the counter $\nu = 1$ and initialize the Lagrange multiplier corresponding to the complicating constraint to its initial value $\lambda^{(1)} = 0$. This means that no energy is transmitted, therefore the optimal solution of the corresponding subproblem will be $x_1^{(1)} = 0, x_2^{(1)} = 0$, and $x_3^{(1)} = 0$ (check it in the running program).

Step 1: Solution of the relaxed primal problem. The relaxed primal problem decomposes into the three subproblems below (the total energy transmitted (10) is equally distributed between subproblems):

$$\phi_1^{(\nu)} = \underset{x_1}{\text{minimize}} \quad \frac{x_1^2}{2} + \lambda^{(\nu)}\left(x_1 - \frac{10}{3}\right)$$

subject to
$$x_1 \geq 0,$$

$$\phi_2^{(\nu)} = \underset{x_2}{\text{minimize}} \quad \frac{x_2^2}{2} + \lambda^{(\nu)}\left(x_2 - \frac{10}{3}\right)$$

subject to
$$x_2 \geq 0,$$

and

$$\phi_3^{(\nu)} = \underset{x_3}{\text{minimize}} \quad \frac{x_3^2}{20} + x_3 + \lambda^{(\nu)}\left(x_3 - \frac{10}{3}\right)$$

subject to
$$x_3 \geq 0,$$

whose solutions are denoted, respectively, by $x_1^{(\nu)}$, $x_2^{(\nu)}$, and $x_3^{(\nu)}$. Note that the Lagrange multiplier is fixed.

Update the values $\phi^{(\nu)} = \phi_1^{(\nu)} + \phi_2^{(\nu)} + \phi_3^{(\nu)}$, $s^{(\nu)} = x_1^{(\nu)} + x_2^{(\nu)} + x_3^{(\nu)} - 10$.

Step 2: Multiplier updating. The updated multiplier is obtained by solving the linear programming problem

$$\underset{z, \lambda \in C}{\text{maximize}} \quad z \qquad \qquad \text{(B.12)}$$

subject to

$$z \leq \phi^{(k)} + s^{(k)}\left(\lambda - \lambda^{(k)}\right); \qquad k = 1, \ldots, \nu. \qquad \text{(B.13)}$$

Step 3: Convergence checking. If multiplier λ has not changed sufficiently, stop; the optimal solution has been found. Otherwise, update the iteration counter $\nu = \nu + 1$ and continue the procedure with Step 1.

Assuming $C = \{-10, 10\}$ and $\lambda^{(0)} = 0$, the following GAMS program can be used to solve the problem:

```
$Title Exercise 5.8.1

file out/Exercise5.8.1.out/; put out;

Set
    l the number of parallel lines /1*3/
    IT iteration number /1*15/
    ITER(IT) dynamic set for activating cutting hyperplanes;

Alias(l,l1); Alias(IT,IT1);

PARAMETER e total demanded energy /10/;

POSITIVE VARIABLES x(l) volume of energy transmitted through line i;
```

```
SCALARS
    error control error parameter /1/
    epsilon maximum tolerable error /1e-3/
    phiup dual function upper bound /INF/
    philo dual function lower bound /-INF/;

PARAMETERS
    lambdas(IT) Lagrange multipliers values associated with demand...
    ...constraint for iteration IT
    zs(IT) dual function values for iteration IT
    s(IT) subgradient values associated with demand constraint m...
    ...for iteration IT;

VARIABLES z, phi, lambda Lagrange multipliers value;

EQUATION dual1,dual2,dual3 associated to each device;
    dual1..phi=e=0.5*sqr(x('1'))+lambda*(x('1')-e/card(1));
    dual2..phi=e=0.5*sqr(x('2'))+lambda*(x('2')-e/card(1));
    dual3..phi=e=0.05*sqr(x('3'))+x('3')+lambda*(x('3')-e/card(1));
EQUATION cutting cutting planes;
    cutting(ITER)..z=l=zs(ITER)+s(ITER)*(lambda-lambdas(ITER));
EQUATIONS lambdaup,lambdalo upper and lower Lagrange multiplier
limits;
    lambdaup..lambda=l=10;
    lambdalo..lambda=g=-10;

MODEL dual1df /dual1/; MODEL dual2df /dual2/; MODEL
dual3df/dual3/; MODEL updating multiplier
updating/cutting,lambdaup,lambdalo/;

x.l(1)=0;

ITER(IT)=no;

lambdas(IT)=0; zs(IT)=0; s(IT)=0;

lambda.fx=0;

loop(IT$(error>epsilon),
    put "   Iteration ",ORD(IT):3:0//;

    loop(1,
        if(ORD(1)=1,
            phi.l=0.5*sqr(x.l('1'))+lambda.l*(x.l('1')-e/card(1));
            SOLVE dual1df using nlp MINIMIZING phi;
            put "Modelstat= ",dual1df.modelstat,"; Solvestat= ",...
            ...dual1df.solvestat/;
        elseif (ORD(1)=2),
            phi.l=0.5*sqr(x.l('2'))+lambda.l*(x.l('2')-e/card(1));
            SOLVE dual2df using nlp MINIMIZING phi;
            put "Modelstat= ",dual2df.modelstat,"; Solvestat= ",...
            ...dual2df.solvestat/;
        else
            phi.l=0.05*sqr(x.l('3'))+x.l('3')+lambda.l*...
            ...(x.l('3')-e/card(1));
            SOLVE dual3df using nlp MINIMIZING phi;
            put "Modelstat= ",dual3df.modelstat,"; Solvestat= ",...
            ...dual3df.solvestat/;
        );

        put "phi(",ORD(1):3:0,")= ",phi.l:8:3/;
        put "x(",ORD(1):3:0,")= ",x.l(1):8:3/;
        zs(IT)=zs(IT)+phi.l;

    );

    s(IT)=sum(1,x.l(1))-e;
```

```
        lambdas(IT)=lambda.l;
        philo=zs(IT);
        put "z= ",zs(IT):8:3,"; Dual function lower bound= ",philo:8:3/;
        put "subgradient(",ORD(IT):3:0,")= ",s(IT):8:3/;

        ITER(IT)=yes;

        lambda.up=Inf;
        lambda.lo=-Inf;
        SOLVE updating using nlp MAXIMIZING z;
        put "Modelstat= ",updating.modelstat,"; Solvestat= ",...
        ...updating.solvestat/;
        put "z= ",z.l:8:3/;
        put "lambda= ",lambda.l:8:3/;

        lambda.fx=lambda.l;
        phiup=z.l;
        put "Dual function upper bound= ",phiup:8:3/;

*               Updating error
        if(lambda.l ne 0,
        error=abs((lambda.l-lambdas(IT))/lambda.l);
                else
        error=abs((lambda.l-lambdas(IT)));
        );
        put "Error= ",error:12:9//;
);
```

The optimal solution is

$$x_1^* = 1.667, \quad x_2^* = 1.667, \quad x_3^* = 6.674, \quad p^* = 11.667,$$

which is attained after 14 iterations with a relative tolerance error of $\varepsilon = 10^{-3}$.

Considering x_3 as a complicating variable the problem can be solved using the Benders decomposition. In this case, the solution algorithm proceeds as follows:

Step 0: Initialization. The iteration counter is initialized, $\nu = 1$. The initial value for the complicating variable x_3 is set to $x_3^{(1)} = 0$. The lower bound of the objective function is set to $z_{\text{down}}^{(1)} = 0$.

Step 1: Subproblem solution. The subproblem below is solved:

$$\begin{aligned} \underset{x_1, x_2, x_3}{\text{minimize}} \quad & z = \frac{1}{2}\left(x_1^2 + x_2^2 + \frac{x_3^2}{10}\right) + x_3 \end{aligned}$$

subject to

$$\begin{aligned} x_1 + x_2 &= 10 - x_3 \\ x_3 &= x_3^{(\nu)} : \lambda^{(\nu)}, \end{aligned}$$

whose solution is $x_1^{(\nu)}, x_2^{(\nu)}$, and $\lambda^{(\nu)}$. The upper bound of the objective function optimal value is $z_{\text{up}}^{(\nu)} = z^{(\nu)}$.

Step 2: Convergence checking. If $|z_{\text{up}}^{(\nu)} - z_{\text{down}}^{(\nu)}|/|z_{\text{down}}^{(\nu)}|$ is not small enough, the procedure continues with Step 2 and the iteration counter is updated $\nu = \nu + 1$; otherwise, the optimal solution has been found.

Step 2: Master problem solution. The master problem below is solved:

$$\underset{\alpha, x_3}{\text{minimize}} \quad \alpha$$

subject to

$$\begin{aligned}
z^{(k)} + \lambda^{(k)}(x_3 - x_3^{(k)}) &\leq \alpha; \quad k = 1, \ldots, \nu - 1 \\
\alpha &\geq 0 \\
x_3 &\geq 0,
\end{aligned}$$

the solution of which is the new value of the complicating variable $x_3^{(\nu)}$ and $\alpha^{(\nu)}$. The lower bound of the objective function optimal value is $z_{\text{down}}^{(\nu)} = \alpha^{(\nu)}$. The procedure continues with Step 1.

Assuming $x_3^{(1)} = 0$ and $\alpha_{\text{down}} = 0$, the following GAMS program can be used to solve the problem:

```
$Title Exercise 5.8.2

file out/Exercise5.8.2.out/; put out;

Set
    l the number of parallel lines /1*3/
    IT iteration number /1*15/
    ITER(IT) dynamic set for activating cutting hyperplanes;

Alias(l,l1); Alias(IT,IT1);

PARAMETERS
    e total demanded energy /10/;

POSITIVE VARIABLES
    x(l),auxx3 volume of energy transmitted through line i;

SCALARS
    error control error parameter /1/
    epsilon maximum tolerable error /1e-3/
    zup objective function upper bound /INF/
    zlo objective function lower bound /0/;

PARAMETERS
    x3(IT) fixed values of the complicating variables
    zs(IT) objective function value for iteration IT
    lambda(IT) dual variables associated with the complicating...
    ...variable;

VARIABLES alpha, loss Lagrange multipliers value;

EQUATION obj associated with each device;
    obj..loss=e=0.5*sqr(x('1'))+0.5*sqr(x('2'))+0.05*...
    ...sqr(x('3'))+x('3');
EQUATION demand;
    demand..sum(l,x(l))=e=e;
```

```
EQUATION cutting,alphalo cutting planes;
    cutting(ITER)..alpha=g=zs(ITER)+lambda(ITER)*(auxx3-x3(ITER));
    alphalo..alpha=g=0;

MODEL subproblem /obj,demand/;
MODEL master multiplier updating /cutting,alphalo/;

x.1(1)=0;

ITER(IT)=no;

auxx3.up=e;

lambda(IT)=0; zs(IT)=0; x3(IT)=0;

loop(IT$(error>epsilon),

    put "   Iteration ",ORD(IT):3:0//;

    if(ORD(IT)=1,
        x.fx('3')=0;
    else
        SOLVE master using lp MINIMIZING alpha;
        put "Modelstat= ",master.modelstat,"; Solvestat= ",...
        ...master.solvestat/;
        put "alpha= ",alpha.l:8:3/;
        put "auxx3= ",auxx3.l:8:3/;

        x.fx('3')=auxx3.l;
        zlo=alpha.l;
        put "Objective function lower bound= ",zlo:8:3/;
    );

    SOLVE subproblem using nlp MINIMIZING loss;
    put "Modelstat= ",subproblem.modelstat,"; Solvestat= ",...
    ...subproblem.solvestat/;

    loop(1,
        put "x(",ORD(1):3:0,")= ",x.1(1):8:3/;
    );
    zs(IT)=loss.l;
    lambda(IT)=x.m('3');
    x3(IT)=x.l('3');
    zup=zs(IT);
    put "loss= ",zs(IT):8:3,"; Objective function upper...
    ...bound= ",zup:8:3/;

*           Updating error

    if(zlo ne 0,
            error=abs((zup-zlo)/zlo);
    else
            error=abs((zup-zlo));
    );
    put "Error= ",error:12:9//;
    ITER(IT)=yes;
);
```

The optimal solution is

$$x_1^* = 1.699, \quad x_2^* = 1.699, \quad x_3^* = 6.602, \quad p^* = 11.668,$$

which is attained after eight iterations with a relative error tolerance of smaller than $\varepsilon = 10^{-3}$.

Solution to Exercise 5.10. The capacity expansion planning problem analyzed in Sect. 1.6.1 can be solved in a decomposed manner. If variables x_{it} ($i = 1, 2$; $t = 1, 2$) are considered complicating variables, then the problem can be decomposed by time period.

Using the Benders decomposition, the solution algorithm proceeds as follows:

Step 0: Initialization. The iteration counter is initialized, $\nu = 1$. Initial values for the complicating variables $x_{it}^{(1)}$ are set to the maximum production capacity that can be built at location i. The lower bound of the objective function is $z_{\text{down}}^{(1)} = 0$. Note that the minimum cost is zero.

Step 1: Subproblem solution. The subproblems associated with each time period are solved:

$$\underset{x_{i1};\ i=1,2}{\text{minimize}} \quad \sum_{i=1}^{2} [(c_{i1} - c_{i2})\, x_{i1}]$$

subject to

$$\begin{aligned}
y_{11} &= f_{13,1} + f_{12,1} - f_{21,1} \\
y_{21} &= f_{23,1} + f_{21,1} - f_{12,1} \\
d_1 &= f_{13,1} + f_{23,1} \\
0 &\leq y_{i1} \leq x_{i1}\ ; \quad i = 1, 2 \\
0 &\leq f_{ij,1} \leq f_{ij}^{\max}\ ; \quad (i,j) \in \mathcal{P} \\
f_{ij,1} &= G_{ij} \cos(\delta_{i1} - \delta_{j1}) + B_{ij} \sin(\delta_{i1} - \delta_{j1}) \\
&\quad - G_{ij},\ ;\ (i,j) \in \mathcal{P} \\
x_{i1} &= x_{i1}^{(\nu)} : \lambda_{i,1}^{(\nu)}\ ; \quad i = 1, 2\,,
\end{aligned}$$

and

$$\underset{x_{i2};\ i=1,2}{\text{minimize}} \quad \sum_{i=1}^{2} (c_{i2} x_{i2})$$

subject to

$$\begin{aligned}
y_{12} &= f_{13,2} + f_{12,2} - f_{21,2} \\
y_{22} &= f_{23,2} + f_{21,2} - f_{12,2} \\
d_2 &= f_{13,2} + f_{23,2} \\
0 &\leq y_{i2} \leq x_{i2}\ ; \quad i = 1, 2 \\
0 &\leq f_{ij,2} \leq f_{ij}^{\max}\ ; \quad (i,j) \in \mathcal{P} \\
f_{ij,2} &= G_{ij} \cos(\delta_{i2} - \delta_{j2}) + B_{ij} \sin(\delta_{i2} - \delta_{j2}) - G_{ij}; \quad (i,j) \in \mathcal{P} \\
x_{i2} &= x_{i2}^{(\nu)} : \lambda_{i,2}^{(\nu)}\ ; \quad i = 1, 2\,,
\end{aligned}$$

whose solutions are $x_{it}^{(\nu)}$. Note that as the terms of the objective functions in both subproblems are the complicating variables (fixed values), then the subproblems are indeed nonlinear systems of equations that can be solved without using optimization techniques, and the values of the dual variables related to the complicating constraints are the partial derivatives of the cost function $\lambda_{i1}^{(\nu)} = c_{i1} - c_{i2}$ and $\lambda_{i2}^{(\nu)} = c_{i2}$, respectively. The upper bound of the objective function optimal value is $z_{\text{up}}^{(\nu)} = z^{(\nu)} = \sum_{i=1}^{2}\left[c_{i1} x_{i1}^{(\nu)} + c_{i2}\left(x_{i2}^{(\nu)} - x_{i1}^{(\nu)}\right)\right]$. Note that if one of the first subproblems is infeasible, this means that the global problem is infeasible, either because the maximum production that can be built or the transmission line capacities are not enough for ensuring demand satisfaction.

Step 2: Convergence checking. If $|z_{\text{up}}^{(\nu)} - z_{\text{down}}^{(\nu)}|/|z_{\text{down}}^{(\nu)}|$ is not small enough, the procedure continues with Step 2 and the iteration counter is updated $\nu \leftarrow \nu + 1$; otherwise, the optimal solution has been found.

Step 3: Master problem solution. The master problem below is solved.

$$\underset{x_{it};\ i=1,2;\ t=1,2}{\text{minimize}} \quad \alpha$$

subject to

$$\alpha \geq z^{(k)} + \sum_{i=1}^{2}\sum_{t=1}^{2} \lambda_{it}^{(k)}(x_{it} - x_{it}^{(k)}); \qquad k = 1,\ldots,\nu - 1$$

$$\alpha \geq 0$$

$$x_{it} \leq x_{i,t+1}; \quad i = 1,2;\ t = 1$$

$$0 \leq x_{it} \leq x_i^{\max}; \quad i = 1,2;\ t = 1,2$$

$$\sum_{i=1}^{2} x_{it} = d_t;\ t = 1,2. \tag{B.14}$$

The solutions of this problem are the new values of the complicating variables $x_{it}^{(\nu)}$ and $\alpha^{(\nu)}$. The lower bound of the objective function optimal value is $z_{\text{down}}^{(\nu)} = \alpha^{(\nu)}$. Note that to ensure that the maximum production capacities are enough for supplying the demand in each time period, an additional constraint (B.14) must be added to the master problem with respect to the initial problem. The procedure continues with Step 1.

The following GAMS program can be used to solve this problem using the Benders decomposition:

```
$Title Exercise 5.10

file out/Exercise5.10.out/; put out;

Set
    t time period /1*2/
```

```
lo production facilities /1*2/
n line nodes /1*3/
AL(n,n) dynamic set for the active lines
ACL(lo) dynamic set for the active location
ACT(t) dynamic set for the active time period
IT iteration number /1*15/
ITER(IT) dynamic set for activating cutting hyperplanes;

Alias(n,n1); Alias(t,t1,t2); ALIAS(lo,lo1); Alias(IT,IT1);

SCALARS
    error control error parameter /1/
    epsilon maximum tolerable error /1e-3/
    zup objective function upper bound /INF/
    zlo objective function lower bound /0/;

PARAMETERS
    d(t) demand during time period/
    1      7
    2      5/
    G(n,n) conductance of line ij
    B(n,n) susceptance of line ij
    fmax(n,n) maximum capacity of the line
    xmax(lo) maximum production capacity that can be built...
    ...in location/
    1      5
    2      6/;

AL(n,n1)=no;
AL('1','2')=yes; G('1','2')=-0.5; B('1','2')= 9.0;
fmax('1','2')=2.5;
AL('1','3')=yes; G('1','3')=-0.4; B('1','3')=15.0;
fmax('1','3')=6.0;
AL('2','1')=yes; G('2','1')=-0.5; B('2','1')= 9.0;
fmax('2','1')=2.0;
AL('2','3')=yes; G('2','3')=-0.7; B('2','3')=18.0;
fmax('2','3')=4.0;

Table c(lo,t) building cost for location and time period
           1      2
1          2      2.5 2      3.5      3;

POSITIVE VARIABLES
    x(lo,t) production capacity already built at location at the...
    ...beginning of time period
    y(lo,t) actual production at location during time period $t$
    xaux(lo,t) auxiliary variable for the master problem;

PARAMETERS
    xs(IT,lo,t) fixed values of the complicating variables
    zs(IT) objective function value for iteration IT
    lambda(IT,lo,t) dual variables associated with the complicating...
    ...variable;

VARIABLES
    f(n,n,t) energy sent from one location to another during period
    delta(n,t) relative height of location with respect to the...
    reference location during period.
    cost total production cost
    alpha;

Equations balance1,balance2 energy balances at production
locations 1 and 2... ...respectively;
    balance1(t)$(ACT(t))..y('1',t)-f('1','3',t)-f('1','2',t)+...
    ...f('2','1',t)=e=0;
    balance2(t)$(ACT(t))..y('2',t)-f('2','3',t)-f('2','1',t)+...
    ...f('1','2',t)=e=0;
```

```
Equation demand energy balance in the city;
    demand(t)$(ACT(t))..d(t)=e=f('1','3',t)+f('2','3',t);

Equation prodlim production capacity limits;
    prodlim(lo,t)$(ACT(t))..y(lo,t)=l=x(lo,t);

Equation translim transmission capacity limits;
    translim(n,n1,t)$(AL(n,n1) and ACT(t))..f(n,n1,t)=l=fmax(n,n1);

Equation fdf transmitted commodity through lines;
    fdf(n,n1,t)$(AL(n,n1) and ACT(t))..f(n,n1,t)=e=G(n,n1)*...
    ...cos(delta(n,t)-delta(n1,t))+B(n,n1)*sin(delta(n,t)-...
    ...delta(n1,t))-G(n,n1);

Equation costdf total production cost definition;
    costdf..cost=e=sum((t,t1)$(ACT(t) and ORD(t)+1=ORD(t1) and...
    ...ORD(t)<CARD(t)),sum(lo,(c(lo,t)-c(lo,t1))*x(lo,t)))+...
    ...sum(t$(ACT(t) and ORD(t)=CARD(t)),sum(lo,c(lo,t)*x(lo,t)));

Equation expansion expansion constraints;
    expansion(lo,t,t1)$(ORD(t)+1=ORD(t1))..xaux(lo,t)=l=xaux(lo,t1);

Equation expbound expansion bounds;
    expbound(lo,t)..xaux(lo,t)=l=xmax(lo);

Equation capacity;
    capacity(t)..sum(lo,xaux(lo,t))=g=d(t);

EQUATION cutting,alphalo cutting planes;
    cutting(ITER)..alpha=g=zs(ITER)+sum((lo,t),lambda(ITER,lo,t)*...
    ...(xaux(lo,t)-xs(ITER,lo,t)));
    alphalo..alpha=g=0;

Option iterlim=1000;

MODEL primal
/costdf,balance1,balance2,demand,prodlim,translim,fdf/;
MODEL updating multiplier updating
/cutting,alphalo,expansion,expbound,capacity/;

lambda(IT,lo,t)=0; zs(IT)=0; xs(IT,lo,t)=0; ITER(IT)=no;

y.l(lo,t)=0; f.lo(n,n1,t)=0;

loop(IT$(error>epsilon),
    put "   Iteration ",ORD(IT):3:0//;

    if(ORD(IT)=1,
        x.fx(lo,t)=xmax(lo);
    else
        xaux.lo(lo,t)=y.l(lo,t);
        SOLVE updating using lp MINIMIZING alpha;
        put "Modelstat= ",updating.modelstat,"; Solvestat= ",...
        ...updating.solvestat/;
        put "alpha= ",alpha.l:8:3/;
        loop((lo,t),
            put "x(",lo.tl:2,",",t.tl:2,"= ",xaux.l(lo,t):8:3/;
        );

        x.fx(lo,t)=xaux.l(lo,t);
        zlo=alpha.l;
        put "Optimal objective function lower bound= ",zlo:8:3/;
    );

    loop(t2,
        ACT(t)=no;    ACT(t2)=yes;
        SOLVE primal using nlp MINIMIZING cost;
```

```
            put "Modelstat= ",primal.modelstat,"; Solvestat= ",...
            ...primal.solvestat/;
            zs(IT)=zs(IT)+cost.l;
            lambda(IT,lo,t2)=x.m(lo,t2);
            xs(IT,lo,t2)=x.l(lo,t2);
        );

        zup=zs(IT);
        put "cost= ",zs(IT):8:3,"; Objective function upper bound= "...
        ...,zup:8:3/;

*           Updating error
        if(ORD(IT)>1,
            if(zlo ne 0,
                    error=abs((zup-zlo)/zlo);
            else
                    error=abs((zup-zlo));
            );
        );
        put "Error= ",error:12:9//;
        ITER(IT)=yes;
    );

    loop((lo,t),
        put "x(",lo.tl:2,",",t.tl:2,")= ",x.l(lo,t):8:3/;
    );put /; loop((lo,t),
        put "y(",lo.tl:2,",",t.tl:2,")= ",y.l(lo,t):8:3/;
    );put /; loop((n,n1,t)$(AL(n,n1)),
        put "f(",n.tl:2,",",n1.tl:2,",",t.tl:2,")= ",f.l(n,n1,t):8:3/;
    );put /; loop((n,t),
        put "delta(",n.tl:2,",",t.tl:2,")= ",delta.l(n,t):8:3/;
    );put /;
```

The optimal solution is shown in Table B.11.

Table B.11. Optimal solution to Exercise 5.10

Period	Location 1		Location 2		Flows			
t	Cap.	Prod.	Cap.	Prod.	1–3	1–2	2–1	2–3
1	3.2	3.2	3.8	3.8	3.2	0.0	0.0	3.8
2	3.2	2.3	3.8	2.7	2.3	0.0	0.0	2.7

B.6 Exercises from Chapter 6

Solution to Exercise 6.2. The graphical description of the problem is shown in Fig. B.2, where it can be observed that the point $(3,2)$ is an optimal solution. Substituting this point into the constraints we get

$$\begin{aligned}
z = 2 \times 3 + 3 \times 2 &= 12 \\
2 \times 3 = 6 &\geq 6 \\
\exp(2) + 3 = 10.39 &\geq 2 \\
x = 3 &\leq 10 \\
y = 2 &\leq 3 \\
x = 3 &\in \mathbb{N} ,
\end{aligned}$$

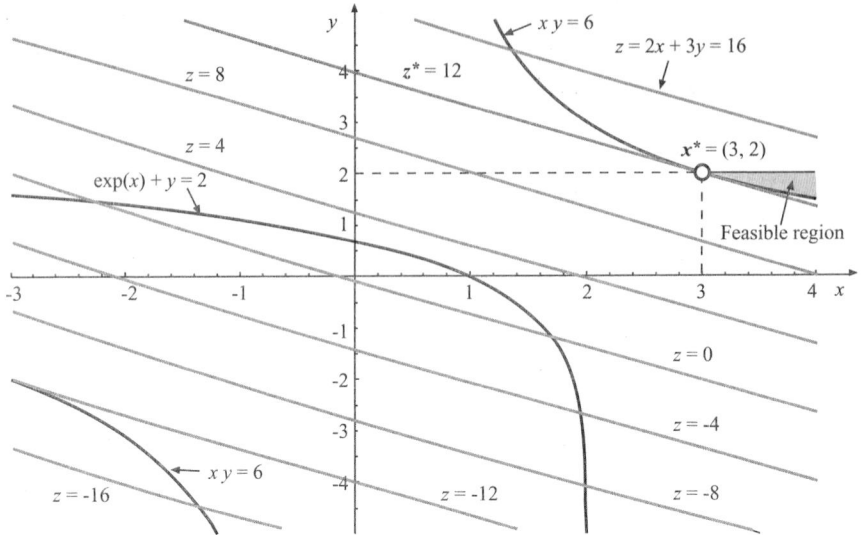

Fig. B.2. Illustration of the problem in Exercise 6.2

where we can see that, because all the constraints are satisfied, it is a solution of the problem.

This problem can be solved considering the integer variable x as a complicating variable using the Benders decomposition, then the solution algorithm proceeds as follows:

Step 0: Initialization. The iteration counter is initialized, $\nu = 1$. Initial values for the complicating variables are selected by setting $x_i^{(1)} = 1$. The lower bound of the objective function is $z_{\text{down}}^{(1)} = -\infty$.

Step 1: Subproblem solution. The following subproblem is solved:

$$\underset{x,y}{\text{minimize}} \quad 2x + 3y + 10w$$

subject to

$$\begin{aligned} xy + w &\geq 6 \\ \exp(y) + x &\geq 2 \\ y &\leq 3 \\ w &\geq 0 \\ x &= x^{(1)} = 1 : \lambda^{(1)}, \end{aligned}$$

whose solution is $y^{(1)} = 3$, $w^{(1)} = 3$, and $\lambda_1^{(1)} = -28$. The upper bound of the objective function optimal value is updated to $z_{\text{up}}^{(1)} = z^{(1)} = 41$. Note that an additional variable (w) has been added to the first constraint to avoid infeasibility.

Step 2: Convergence checking. Since $|z_{\text{up}}^{(1)} - z_{\text{down}}^{(1)}|/|z_{\text{down}}^{(1)}| = 41$ and is not small enough, the procedure continues with Step 2 and the iteration counter is updated $\nu \leftarrow \nu + 1 = 2$.

Step 2: Master problem solution. The master problem is solved.

$$\underset{\alpha, x}{\text{minimize}} \quad \alpha$$

subject to

$$\begin{aligned} 41 - 28(x-1) &\leq \alpha \\ \alpha &\geq 0 \\ x &\leq 10 \\ x &\in \mathbb{N}. \end{aligned}$$

The solution of this problem gives the new value of the complicating variable $x^{(2)} = 10$ and $\alpha^{(2)} = -211$. The lower bound of the objective function optimal value is updated to $z_{\text{down}}^{(2)} = \alpha^{(2)} = -211$. The procedure continues with Step 1.

Step 1: Subproblem solution. The subproblem is solved again fixing the complicating variable to the following value:

$$x = x^{(2)} = 10 : \lambda^{(2)},$$

whose solution is $y = 0.6$, $w = 0$, and $\lambda^{(2)} = 1.82$. The upper bound of the objective function optimal value is updated to $z_{\text{up}}^{(2)} = z^{(2)} = 21.8$.

Step 2: Convergence checking. Since $|z_{\text{up}}^{(2)} - z_{\text{down}}^{(2)}|/|z_{\text{down}}^{(2)}| = 1.1$ and is not small enough, the procedure continues with Step 2 and the iteration counter is updated $\nu \leftarrow \nu + 1 = 3$.

Step 2: Master problem solution. The master problem below is solved:

$$\underset{\alpha, x}{\text{minimize}} \quad \alpha$$

subject to

$$\begin{aligned} 41 - 28(x-1) &\leq \alpha \\ 21.8 + 1.82(x-10) &\leq \alpha \\ \alpha &\geq 0 \\ x &\leq 10 \\ x &\in \mathbb{N}. \end{aligned}$$

The solution of this problem is $x^{(3)} = 3$ and $\alpha^{(3)} = 9.06$. The lower bound of the objective function optimal value is updated to $z_{\text{down}}^{(3)} = \alpha^{(3)} = 9.06$. The procedure continues with Step 1.

Step 1: Subproblem solution. The subproblem is solved fixing the complicating variable to the following value:

$$x = x^{(3)} = 3 : \lambda^{(3)},$$

whose solution is $y = 2$, $w = 0$, and $\lambda^{(3)} = 0$. The upper bound of the objective function optimal value is updated to $z_{\text{up}}^{(3)} = z^{(3)} = 12$.

Step 2: Convergence checking. Since $|z_{\text{up}}^{(3)} - z_{\text{down}}^{(3)}|/|z_{\text{down}}^{(3)}| = 0.32$ and is not small enough, the procedure continues with Step 2 and the iteration counter is updated $\nu \leftarrow \nu + 1 = 4$.

Step 2: Master problem solution. The master problem is solved:

$$\underset{\alpha, x}{\text{minimize}} \quad \alpha$$

subject to

$$\begin{aligned}
41 - 28(x - 1) &\leq \alpha \\
21.8 + 1.82(x - 10) &\leq \alpha \\
12 + 0(x - 3) &\leq \alpha \\
\alpha &\geq 0 \\
x &\leq 10 \\
x &\in \mathbb{N}.
\end{aligned}$$

The solution of this problem gives the new value of the complicating variable $x^{(4)} = 3$ and $\alpha^{(4)} = 12$. The lower bound of the objective function optimal value is updated to $z_{\text{down}}^{(4)} = \alpha^{(4)} = 12$. Then, the procedure continues with Step 1. In Fig. B.3 the graphical description of the three Benders cuts in the last master problem and the reconstruction of the $\alpha(x)$ function using points, are shown.

Step 1: Subproblem solution. The following subproblem with this new value of the complicating variable is solved:

$$x = x^{(4)} = 3 : \lambda^{(4)},$$

whose solution is $y = 2$, $w = 0$, and $\lambda^{(4)} = 3$. The upper bound of the objective function optimal value is updated to $z_{\text{up}}^{(4)} = z^{(4)} = 12$.

Step 2: Convergence checking. Since $|z_{\text{up}}^{(4)} - z_{\text{down}}^{(4)}|/|z_{\text{down}}^{(4)}| = 0$ and is small enough, the optimal solution has been found. Note that it is the same as that obtained graphically: $x^* = 3$, $y^* = 2$, and $z^* = 12$.

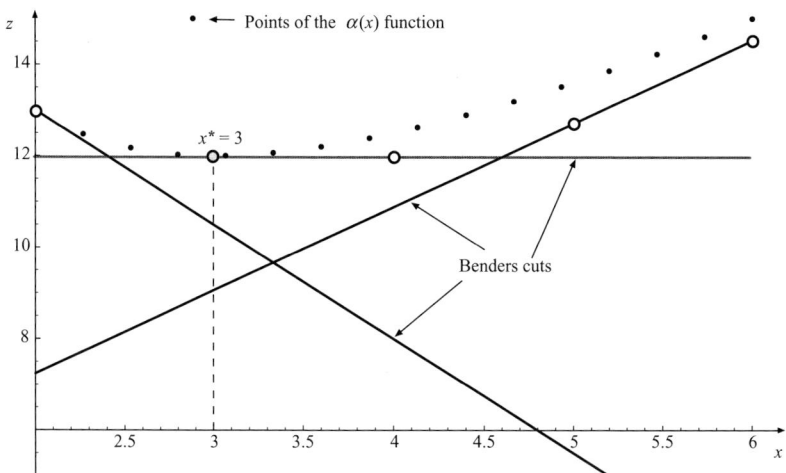

Fig. B.3. Illustration of the $\alpha(x)$ function and the Benders cuts in one of the master problems used in Exercise 6.2

The following GAMS program can be used to solve the problem:

```
$Title Exercise 6.2

file out/Exercise6.2.out/; put out;

Option mip=CPLEX;

Set
    IT iteration number /1*15/
    ITER(IT) dynamic set for activating cutting hyperplanes;

Alias(IT,IT1);

VARIABLES z, y, x;

SCALARS
    error control error parameter /1/
    epsilon maximum tolerable error /1e-3/
    zup objective function upper bound /INF/
    zlo objective function lower bound /0/;

PARAMETERS
    xs(IT) fixed values of the complicating variables
    zs(IT) objective function value for iteration IT
    lambda(IT) dual variables associated with the complicating...
    ...variable;

VARIABLES alpha;

INTEGER VARIABLE xaux; POSITIVE VARIABLE w;

EQUATION obj associated to each device;
    obj..z=e=2*x+3*y+10*w;
EQUATION cons1, cons2, cons3 constraints;
    cons1..x*y+w=g=6;
    cons2..x+exp(y)=g=2;
    cons3..y=l=3;
```

```
EQUATION cutting cutting planes;
    cutting(ITER)..alpha=g=zs(ITER)+lambda(ITER)*(xaux-xs(ITER));

MODEL subproblem /obj,cons1,cons2,cons3/; MODEL master multiplier
updating /cutting/;

ITER(IT)=no;

xaux.up=10;

lambda(IT)=0; zs(IT)=0; xs(IT)=0;

loop(IT$(error>epsilon),
    put "   Iteration ",ORD(IT):3:0//;

    if(ORD(IT)=1,
        x.fx=1;
    else
        SOLVE master using mip MINIMIZING alpha;
        put "Modelstat= ",master.modelstat,"; Solvestat= ",...
        ...master.solvestat/;
        put "alpha= ",alpha.l:8:3/;
        put "xaux= ",xaux.l:8:3/;

        x.fx=xaux.l;
        zlo=alpha.l;
        put "Objective function lower bound= ",zlo:8:3/;
    );

    SOLVE subproblem using nlp MINIMIZING z;
    put "Modelstat= ",subproblem.modelstat,"; Solvestat= ",...
    ...subproblem.solvestat/;
    put "x= ",x.l:8:3, "y= ",y.l:8:3," w= ",w.l:8:3,...
    ..."z= ",z.l:8:3/;
    put "lambda= ",x.m:16:8/;

    zs(IT)=z.l;
    lambda(IT)=x.m;
    xs(IT)=x.l;
    zup=zs(IT);
    put "Objective function upper bound= ",zup:8:3/;

*           Updating error

    if(zlo ne 0,
        error=abs((zup-zlo)/zlo);
    else
        error=abs((zup-zlo));
    );
    put "Error= ",error:12:9//;
    ITER(IT)=yes;
);
```

Solution to Exercise 6.4. Substituting the vector $(0,1)$ into the constraints we obtain
$$\begin{aligned} z = -7 \times 0 + 4 \times 01 &= 4 \\ 1^2 + 0^2 = 1 &\leq 1 \\ 1 &\leq 2 \\ 0 &\in \mathbb{N} \ . \end{aligned}$$

Since all the constraints hold it is a candidate to optimal solution, but we do not know if it is really optimal.

This problem can be solved using the previously stated outer linearization algorithm. A tolerance of 10^{-4} is considered. The solution procedure is as follows.

Step 0: Initialization. The iteration counter is initialized to $\nu = 1$.
The initial MILP problem is

$$\text{maximize}_{x,y} \quad z = -7x + 4y$$

subject to

$$y \leq 2$$
$$x \in \mathbb{N},$$

whose optimal solution is $x^{(1)} = 0$, $y^{(1)} = 2$, with an optimal objective function value $z^{(1)} = 8$.

Step 1: Determining the most violated constraint. Since this problem has a unique nonlinear constraint, it is the most violated one,

$$g(x^{(1)}, y^{(1)}) = 0^2 + 2^2 - 1 = 3 .$$

Step 2: Convergence check. Since $g(x^{(1)}, y^{(1)}) = 3$ and is not small enough, the algorithm continues.

Step 3: Linearization. The nonlinear constraint is linearized. Its gradient in the current solution is

$$\nabla g(x^{(1)}, y^{(1)}) = \begin{pmatrix} 2x^{(1)} & 2y^{(1)} \end{pmatrix}^T = \begin{pmatrix} 0 & 4 \end{pmatrix}^T .$$

The corresponding linear constraint is

$$l_1(x, y) = g(x^{(1)}, y^{(1)}) + \left(\nabla g(x^{(1)}, y^{(1)})\right)^T \begin{pmatrix} x - x^{(1)} \\ y - y^{(1)} \end{pmatrix}$$

or

$$l_1(x, y) = 3 + \begin{pmatrix} 0 & 4 \end{pmatrix} \begin{pmatrix} x - 0 \\ y - 2 \end{pmatrix}.$$

Step 4: Solution of the linearized problem. The current MILP problem is

$$\text{maximize}_{x,y} \quad z = -7x + 4y$$

subject to

$$l_1(x, y) \leq 0$$
$$y \leq 2$$
$$x \in \mathbb{N},$$

whose optimal solution is $x^{(2)} = 0$, $y^{(2)} = 1.25$, with an optimal objective function value $z^{(2)} = 5$.

Update iteration counter, $\nu = 1 + 1 = 2$, and continue with Step 1.

Step 1: Determining the most violated constraint. The violation of the nonlinear constraint is evaluated,

$$g(x^{(2)}, y^{(2)}) = 0^2 + 1.25^2 - 1 = 0.5625 \ .$$

Step 2: Convergence check. Since $g(x^{(2)}, y^{(2)}) = 0.5625$ and is not small enough, the algorithm continues.

Step 3: Linearization. The nonlinear constraint is linearized. Its gradient in the current solution is

$$\nabla g(x^{(2)}, y^{(2)}) = \begin{pmatrix} 2x^{(2)} & 2y^{(2)} \end{pmatrix}^T = \begin{pmatrix} 0 & 2.5 \end{pmatrix}^T \ .$$

The corresponding linear constraint is

$$l_2(x, y) = g(x^{(2)}, y^{(2)}) + \left(\nabla g(x^{(2)}, y^{(2)})\right)^T \begin{pmatrix} x - x^{(2)} \\ y - y^{(2)} \end{pmatrix}$$

or

$$l_2(x, y) = 0.5625 + \begin{pmatrix} 0 & 2.5 \end{pmatrix} \begin{pmatrix} x - 0 \\ y - 1.25 \end{pmatrix} \ .$$

Step 4: Solution to the linearized problem. The current MILP problem is

$$\begin{aligned} \text{maximize} \quad & z = -7x + 4y \\ x, y \end{aligned}$$

subject to

$$\begin{aligned} l_i(x, y) &\leq 0; \quad i = 1, 2 \\ y &\leq 2 \\ x &\in \mathbb{N} \ , \end{aligned}$$

whose optimal solution is $x^{(3)} = 0$, $y^{(3)} = 1.025$, and the optimal objective function value $z^{(3)} = 4.1$.

Update the iteration counter, $\nu = 2 + 1 = 3$, and continue with Step 1.

Step 1: Determining the most violated constraint. The new violation of the nonlinear constraint is

$$g(x^{(3)}, y^{(3)}) = 0^2 + 1.025^2 - 1 = 0.050625 \ .$$

Step 2: Convergence check. Since $g(x^{(3)}, y^{(3)}) = 0.050625$ and is not small enough, the algorithm continues.

Step 3: Linearization. The nonlinear constraint is linearized. Its gradient in the current solution is

$$\nabla g(x^{(3)}, y^{(3)}) = \begin{pmatrix} 2x^{(3)} & 2y^{(3)} \end{pmatrix}^T = \begin{pmatrix} 0 & 2.05 \end{pmatrix}^T.$$

The corresponding linear constraint is

$$l_3(x, y) = 0.050625 + \begin{pmatrix} 0 & 2.05 \end{pmatrix} \begin{pmatrix} x - 0 \\ y - 1.025 \end{pmatrix}.$$

Step 4: Solution to the linearized problem. The current MILP problem is

$$\underset{x, y}{\text{maximize}} \quad z = -7x + 4y$$

subject to

$$\begin{aligned} l_i(x, y) &\leq 0; \quad i = 1, 2, 3 \\ y &\leq 2 \\ x &\in \mathbb{N}, \end{aligned}$$

whose optimal solution is $x^{(4)} = 0$, $y^{(4)} = 1.0003$, and the optimal objective function value $z^{(4)} = 4.001$.

Update the iteration counter, $\nu = 2 + 1 = 3$, and continue with Step 1.

Step 1: Determining the most violated constraint. The new violation of the nonlinear constraint is

$$g(x^{(4)}, y^{(4)}) = 0^2 + 1.0003^2 - 1 = 0.00061.$$

Step 2: Convergence check. Since $g(x^{(4)}, y^{(4)}) = 0.00061$ and is not small enough, the algorithm continues.

Step 3: Linearization. The nonlinear constraint is linearized. Its gradient in the current solution is

$$\nabla g(x^{(4)}, y^{(4)}) = \begin{pmatrix} 2x^{(4)} & 2y^{(4)} \end{pmatrix}^T = \begin{pmatrix} 0 & 2.0006 \end{pmatrix}^T.$$

The corresponding linear constraint is

$$l_4(x, y) = 0.00061 + \begin{pmatrix} 0 & 2.0006 \end{pmatrix} \begin{pmatrix} x - 0 \\ y - 1.0003 \end{pmatrix}.$$

Step 4: Solution to the linearized problem. The current MILP problem is

$$\underset{x, y}{\text{maximize}} \quad z = -7x + 4y$$

subject to
$$l_i(x, y) \leq 0; \quad i = 1, 2, 3, 4$$
$$y \leq 2$$
$$x \in \mathbb{N},$$

whose optimal solution is $x^{(5)} = 0$, $y^{(5)} = 1$, and the optimal objective function value $z^{(5)} = 4$.

Update the iteration counter, $\nu = 2 + 1 = 3$, and continue with Step 1.

Step 1: Determining the most violated constraint. The new violation of the nonlinear constraint is

$$g(x^{(5)}, y^{(5)}) = 0^2 + 1^2 - 1 = 0.0 \,.$$

Step 2: Convergence check. Since $g(x^{(5)}, y^{(5)}) = 0.0$ and is small enough ($\leq 10^{-4}$), the algorithm terminates, and the optimal solution has been found to be $x^* = 0$, $y^* = 1$, with an optimal objective function value $z^* = 4$.

The following GAMS program can be used to solve this problem:

```
$Title Exercise 6.4

file out/Exercise6.4.out/; put out;

Option mip=CPLEX;

Set
    IT iteration number /1*10/
    ITER(IT) dynamic set for activating cutting hyperplanes;

Alias(IT,IT1);

VARIABLES z, y;
INTEGER VARIABLE x;

SCALARS
    error control error parameter /1/
    epsilon maximum tolerable error /1e-4/;

PARAMETERS
    xs(IT) value of the variable x for iteration IT
    ys(IT) value of the variable y for iteration IT
    g(IT) constraint for iteration IT
    gradx(IT) dual variable associated with the complicating...
    ...variable x
    grady(IT) dual variable associated with the complicating...
    ...variable x;

EQUATION obj associated with each device;
    obj..z=e=-7*x+4*y;
EQUATION cons1 constraints;
    cons1..y=l=2;

EQUATION approx constraint approximation;
    approx(ITER)..g(ITER)+gradx(ITER)*(x-xs(ITER))+...
    ...grady(ITER)*(y-ys(ITER))=l=0;
```

```
MODEL main /obj,cons1,approx/;

ITER(IT)=no;

xs(IT)=0;
ys(IT)=0;
g(IT)=0;
gradx(IT)=0;
grady(IT)=0;

loop(IT$(error>epsilon),
    put "    Iteration ",ORD(IT):3:0//;

    SOLVE main using mip MAXIMIZING z;
    put "Modelstat= ",main.modelstat,"; Solvestat= ",main.solvestat/;
    put "x= ",x.l:8:3, "y= ",y.l:8:3, "z= ",z.l:8:3/;

    g(IT)=x.l*x.l+y.l*y.l-1;
    gradx(IT)=2*x.l;
    grady(IT)=2*y.l;
    xs(IT)=x.l;
    ys(IT)=y.l;

    put "g(",IT.tl:2,")= ",g(IT):12:8/;
    put "gradx(",IT.tl:2,")= ",gradx(IT):12:8," x(",IT.tl:2,")=...
    ... ",xs(IT):12:8/;
    put "grady(",IT.tl:2,")= ",grady(IT):12:8," y(",IT.tl:2,")=...
    ... ",ys(IT):12:8/;

    ITER(IT)=yes;

*   Updating error
    error=g(IT);
    put "Error= ",g(IT):12:9//;
);
```

Solution to Exercise 6.6. The production cost c_i in dollars of each production plant i can be expressed as

$$c_i = \begin{cases} 0 & \text{if} \quad P_i = 0 \\ f_i + v_i P_i & \text{if} \quad 0 < P_i \leq P_i^{\max} \end{cases},$$

where $P_i^{\max}$ is the maximum output capacity.

Alternatively, this function can be replaced, using binary variables, by the following set of constraints:

$$\begin{aligned} c_i &= y_i f_i + v_i P_i \\ 0 &\leq P_i \\ P_i &\leq y_i P_i^{\max} \\ y_i &\in \{0, 1\}. \end{aligned}$$

Note that there are following two possibilities:

Case 1. If $y_i = 0$, then $0 \leq P_i \leq 0$, so that $P_i = 0$ and then $c_i = 0$.
Case 2. If $y_i = 1$, then $0 \leq P_i \leq P_i^{\max}$, so that $c_i = f_i + v_i P_i$.

Therefore, the single-period minimum production cost problem can be formulated as the following MILP problem

$$\text{minimize} \atop P_i;\ i=1,2 \quad \sum_{i=1}^{2} c_i P_i$$

subject to

$$\begin{aligned}
0 &= P_1 - e_{13} - e_{12} \\
0 &= P_2 + e_{12} - e_{23} \\
0 &= -d + e_{13} + e_{23} \\
e_{ij} &\leq e_{ij}^{\max};\ (i,j) \in \{(1,2),(1,3),(2,3)\} \\
e_{ij} &= B_{ij} \sin(\delta_i - \delta_j); \\
&\quad (i,j) \in \{(1,2),(1,3),(2,3)\} \\
c_i &= y_i f_i + v_i P_i \\
0 &\leq P_i \\
P_i &\leq y_i P_i^{\max} \\
y_i &\in \{0,1\};\ i=1,2,
\end{aligned}$$

where the first three constraints are the corresponding energy balance equations in nodes 1, 2, and 3, respectively, e_{ij}, $(i,j) \in \{(1,2),(1,3),(2,3)\}$ are the electric energy flows through the line between nodes i and j, B_{ij} is the susceptance (structural parameter) of line ij, δ_i is the relative "height" or phase of node i, and y_i are binary variables.

This problem can be solved using the Benders decomposition algorithm (6.1), considering the binary variables y_i ($i = 1, 2$) as complicating variables. Note that a physical interpretation of the solution can be done. The production facilities operation can be controlled by the binary variables, if the binary variable $y_i = 0$, then $P_i = 0$, therefore, the production facility i is not working.

The optimal solution of this problem is the set of productions that minimizes the costs satisfying the corresponding demand, and we would like to compare different situations with respect to working facilities, such as both production facilities are operating at the same time, only one, or none is working. Besides, government will penalize the production facilities if they are not able to satisfy the demand and, since the whole demand must be satisfied, this penalty will be high, $10u_d \sum_{i=1}^{2} f_i$, where u_d is the unsatisfied demand.

Using the Benders decomposition, the solution algorithm proceeds as follows:

Step 0: Initialization. The iteration counter is initialized to $\nu = 1$. Initial values for the complicating variables are set to $y_i^{(1)} = 1$, i.e., initially we consider that both production facilities are operating. The lower bound of the objective function is set to $z_{\text{down}}^{(1)} = 0$. Note that the minimum cost is zero.

Step 1: Subproblem solution. The following subproblem, where the unsatisfied demand is considered in the cost function, is solved:

$$\begin{aligned}\text{minimize} & \quad \sum_{i=1}^{2} c_i P_i + 10 u_d \sum_{i=1}^{2} f_i \\ P_i; \; i = 1, 2 & \end{aligned}$$

subject to

$$\begin{aligned}
0 &= P_1 - e_{13} - e_{12} \\
0 &= P_2 + e_{12} - e_{23} \\
0 &= -d + e_{13} + e_{23} + u_d \\
e_{ij} &\leq e_{ij}^{\max}; \; (i,j) \in \{(1,2),(1,3),(2,3)\} \\
e_{ij} &= B_{ij} \sin(\delta_i - \delta_j); \\
& \quad (i,j) \in \{(1,2),(1,3),(2,3)\} \\
c_i &= y_i f_i + v_i P_i \\
0 &\leq P_i \\
P_i &\leq y_i P_i^{\max} \\
y_i &= y_i^{(1)} = 1 : \lambda_i^{(1)}; \; i = 1, 2,
\end{aligned}$$

whose solution is $P_1^{(1)} = 0.265$, $P_2^{(1)} = 0.585$, $\lambda_1^{(1)} = 2.654$, and $\lambda_2^{(1)} = 2.923$. The upper bound of the objective function optimal value is updated to $z_{\text{up}}^{(1)} = z^{(1)} = 8.392$. Note that it corresponds to the optimal solution when both production facilities are operating and that the unsatisfied demand u_d is equal to zero.

Step 2: Convergence checking. Since $|z_{\text{up}}^{(1)} - z_{\text{down}}^{(1)}|/|z_{\text{down}}^{(1)}| = 8.392$ and is not small enough, the procedure continues with Step 2 and the iteration counter is updated $\nu \leftarrow \nu + 1 = 2$.

Step 2: Master problem solution. Based on the information obtained from the first subproblem, we try to know what is the cheapest option for decreasing the cost, thus the following master problem is solved:

$$\begin{aligned}\text{minimize} & \quad \alpha \\ \alpha, y_1, y_2 & \end{aligned}$$

subject to

$$\begin{aligned}
8.392 + 2.654(y_1 - 1) + 2.923(y_2 - 1) &\leq \alpha \\
\alpha &\geq 0 \\
y_i &\in \{0, 1\}; \quad i = 1, 2,
\end{aligned}$$

the solution of which is the new value of the complicating variables $y_i^{(2)} = 0$ $(i = 1, 2)$ i.e., the cheapest option is to stop the production of both facilities and, based on the actual information, a linear estimation of the cost is $\alpha^{(2)} = 2.815$. The lower bound of the objective function optimal value is updated to $z_{\text{down}}^{(2)} = \alpha^{(2)} = 2.815$. The procedure continues with Step 1.

Step 1: Subproblem solution. As the exact cost of this new situation (both production facilities are not operating) is sought, the subproblem is solved again fixing the complicating variables to the following values:
$$y_i = y_i^{(2)} = 0 : \lambda_i^{(2)}, \qquad i = 1, 2,$$
whose solution is $P_1^{(2)} = 0$, $P_2^{(2)} = 0$, $\lambda_1^{(2)} = -1{,}350$, and $\lambda_2^{(2)} = -1{,}350$. The upper bound of the objective function optimal value is updated to $z_{up}^{(2)} = z^{(2)} = 1{,}275$. Note that it is the optimal solution when both production facilities are not operating and that the unsatisfied demand u_d is equal to 0.85; this means that no demand is satisfied at all and because of the penalty, the cost is much higher than in the first situation.

Step 2: Convergence checking. Since $|z_{up}^{(2)} - z_{down}^{(2)}|/|z_{down}^{(2)}| = 451.9$ and is not small enough, the procedure continues with Step 2 and the iteration counter is updated $\nu = \nu + 1 = 3$.

Step 2: Master problem solution. Based on the new information obtained from the second subproblem, we try to know if there is another alternative cheaper than the previous ones, thus the following master problem is solved:

$$\underset{\alpha, y_1, y_2}{\text{minimize}} \quad \alpha$$

subject to
$$8.392 + 2.654(y_1 - 1) + 2.923(y_2 - 1) \leq \alpha$$
$$1{,}275 - 1{,}350(y_1 - 0) - 1{,}350(y_2 - 0) \leq \alpha$$
$$\alpha \geq 0$$
$$y_i \in \{0, 1\}; \qquad i = 1, 2,$$

the solution of which is the new value of the complicating variables $y_1^{(3)} = 1$ and $y_2^{(3)} = 0$, i.e., only the first production facility is operating, and based on the actual information, a linear estimation of the new cost is $\alpha^{(3)} = 5.469$. The lower bound of the objective function optimal value is updated to $z_{down}^{(3)} = \alpha^{(3)} = 5.469$. The procedure continues with Step 1.

Step 1: Subproblem solution. The exact cost of this new situation (only the first production facility is operating) is obtained, solving the subproblem again while fixing the complicating variables to the following values:

$$y_1 = y_1^{(3)} = 1 : \lambda_1^{(3)}$$
$$y_2 = y_2^{(3)} = 0 : \lambda_2^{(3)},$$

whose solution is $P_1^{(3)} = 0.85$, $P_2^{(3)} = 0$, $\lambda_1^{(3)} = 8.5$, and $\lambda_2^{(3)} = -18.18$. The upper bound of the objective function optimal value is updated to $z_{up}^{(3)} = z^{(3)} = 12.835$. Note that the unsatisfied demand u_d is now equal to 0.

Step 2: Convergence checking. Since $|z_{\text{up}}^{(3)} - z_{\text{down}}^{(3)}|/|z_{\text{down}}^{(3)}| = 1.347$ and is not small enough, the procedure continues with Step 2 and the iteration counter is updated $\nu \leftarrow \nu + 1 = 4$.

Step 2: Master problem solution. Based on the new information obtained from the previous subproblem, we try to know if there is another alternative cheaper than the previous ones, thus the master problem is solved:

$$\underset{\alpha, y_i;\ i=1,2}{\text{minimize}} \quad \alpha$$

subject to

$$\begin{aligned}
8.392 + 2.654(y_1 - 1) + 2.923(y_2 - 1) &\leq \alpha \\
1{,}275 - 1{,}350(y_1 - 0) - 1{,}350(y_2 - 0) &\leq \alpha \\
12.835 + 8.5(y_1 - 1) - 18.18(y_2 - 0) &\leq \alpha \\
\alpha &\geq 0 \\
y_i &\in \{0, 1\}; \quad i = 1, 2,
\end{aligned}$$

the solution of which is the new value of the complicating variables $y_1^{(4)} = 0$ and $y_2^{(4)} = 1$, i.e., only the second production facility is operating, and based on the actual information, a linear estimation of the new cost is $\alpha^{(4)} = 5.738$. The lower bound of the objective function optimal value is updated to $z_{\text{down}}^{(4)} = \alpha^{(4)} = 5.738$. The procedure continues with Step 1.

Step 1: Subproblem solution. The exact cost of this new situation (only the second production facility is operating) is obtained solving the subproblem again with the following values of the complicating variables:

$$\begin{aligned}
y_1 &= y_1^{(4)} = 0 : \lambda_1^{(4)} \\
y_2 &= y_2^{(4)} = 1 : \lambda_2^{(4)},
\end{aligned}$$

whose solution is $P_1^{(4)} = 0$, $P_2^{(4)} = 1$, $\lambda_1^{(4)} = -15.21$, and $\lambda_2^{(4)} = 4.25$. The upper bound of the objective function optimal value is updated to $z_{\text{up}}^{(4)} = z^{(4)} = 9.3075$. Note that the unsatisfied demand u_d is equal to 0.

Step 2: Convergence checking. Since $|z_{\text{up}}^{(4)} - z_{\text{down}}^{(4)}|/|z_{\text{down}}^{(4)}| = 0.622$ and is not small enough, the procedure continues with Step 2 and the iteration counter is updated $\nu \leftarrow \nu + 1 = 5$.

Step 2: Master problem solution. Note that we have already checked all the possibilities, now we should take the cheapest one that, based on the information obtained from the previous iterations, is the first one. Let us solve the master problem again to see what we get

$$\underset{\alpha, y_i;\ i=1,2}{\text{minimize}} \quad \alpha$$

subject to

$$\begin{aligned}
8.392 + 2.654(y_1 - 1) + 2.923(y_2 - 1) &\leq \alpha \\
1{,}275 - 1{,}350(y_1 - 0) - 1{,}350(y_2 - 0) &\leq \alpha \\
12.835 + 8.5(y_1 - 1) - 18.18(y_2 - 0) &\leq \alpha \\
9.3075 - 15.21(y_1 - 0) + 4.25(y_2 - 1) &\leq \alpha \\
\alpha &\geq 0 \\
y_i &\in \{0, 1\}; \quad i = 1, 2 \, .
\end{aligned}$$

The solution of this problem gives the new values of the complicating variables $y_1^{(5)} = 1$ and $y_2^{(5)} = 1$, that is the case where both production facilities are operating, and based on the actual information, a linear estimation of the new cost is $\alpha^{(5)} = 8.392$. The lower bound of the objective function optimal value is updated to $z_{\text{down}}^{(5)} = \alpha^{(5)} = 8.392$. Note that it is exactly the same as that obtained from the first subproblem. The procedure continues with Step 1.

Step 1: Subproblem solution. The exact cost of this new situation (the two production facilities are operating) is obtained solving the first subproblem again. The optimal solution is $P_1^{(5)} = 0.265$, $P_2^{(5)} = 0.585$. The upper bound of the objective function optimal value is updated to $z_{\text{up}}^{(5)} = z^{(5)} = 8.392$.

Step 2: Convergence checking. Since $|z_{\text{up}}^{(5)} - z_{\text{down}}^{(5)}|/|z_{\text{down}}^{(5)}| = 0$, the optimal solution has been found.

The following GAMS program can be used to solve the problem:

```
$Title Exercise 6.6

file out/Exercise6.6.out/; put out;

Set
       lo production facilities /1*2/
       n line nodes /1*3/
       AL(n,n) dynamic set for the active lines
       IT iteration number /1*15/
       ITER(IT) dynamic set for activating cutting hyperplanes;

Alias(n,n1); ALIAS(lo,lo1); Alias(IT,IT1);

SCALARS
       error control error parameter /1/
       epsilon maximum tolerable error /1e-3/
       zup objective function upper bound /INF/
       zlo objective function lower bound /0/;

PARAMETERS
       d demand /0.85/
       B(n,n) susceptance of line ij
       emax(n,n) maximum capacity of the line
       Pmax(lo) maximum production capacity/
       1        0.9
       2        0.9/
       f(lo) fixed cost for each production plant/
       1     10
       2      5/
```

```
            v(lo) variable cost for each production plant/
            1       6
            2       7/;

AL(n,n1)=no; AL('1','2')=yes; B('1','2')= 2.5; emax('1','2')=0.3;
AL('1','3')=yes; B('1','3')= 3.5; emax('1','3')=0.7;
AL('2','3')=yes; B('2','3')= 3.0; emax('2','3')=0.7;

POSITIVE VARIABLES
        P(lo) actual production at location
        c(lo) cost coefficients
        ud unsatisfied demand;

BINARY VARIABLES yaux(lo);

VARIABLES
        e(n,n) energy sent from one location to another
        delta(n) relative height of location
        cost total production cost
        y(lo)
        alpha;

PARAMETERS
        ys(IT,lo) fixed values of the complicating variables
        zs(IT) objective function value for iteration IT
        lambda(IT,lo) dual variables associated with the complicating...
        ...variable;

Equations balance1,balance2,balance3 energy balances in nodes...
        ...1 2 and 3 respectively;
        balance1..P('1')-e('1','3')-e('1','2')=e=0;
        balance2..P('2')+e('1','2')-e('2','3')=e=0;
        balance3..-d+e('1','3')+e('2','3')+ud=e=0;

Equation translim transmission capacity limits;
        translim(n,n1)$(AL(n,n1))..e(n,n1)=l=emax(n,n1);

Equation edf transmitted commodity through lines;
        edf(n,n1)$(AL(n,n1))..e(n,n1)=e=B(n,n1)*sin(delta(n)-delta(n1));

Equation copar cost coefficient definition;
        copar(lo)..c(lo)=e=f(lo)*y(lo)+v(lo)*P(lo);

Equation prodlim production capacity limits;
        prodlim(lo)..P(lo)=l=y(lo)*Pmax(lo);

Equation costdf total production cost definition;
        costdf..cost=e=sum(lo,c(lo)*P(lo))+10*sum(lo,10*f(lo))*ud;

EQUATION cutting,mincov,alphalo cutting planes;
        cutting(ITER)..alpha=g=zs(ITER)+sum(lo,lambda(ITER,lo)*...
        ...(yaux(lo)-ys(ITER,lo)));
        alphalo..alpha=g=0;

MODEL subproblem /costdf,balance1,balance2,balance3,translim,...
        ...edf,copar,prodlim/;
MODEL master multiplier updating /cutting,alphalo/;

lambda(IT,lo)=0; zs(IT)=0; ys(IT,lo)=0; ITER(IT)=no;

loop(IT$(error>epsilon),

        put "    Iteration ",ORD(IT):3:0//;

        if(ORD(IT)=1,
            y.fx(lo)=1;
        else
```

```
SOLVE master using mip MINIMIZING alpha;
put "Modelstat= ",master.modelstat,...
..."; Solvestat= ",master.solvestat/;
put "alpha= ",alpha.l:12:6/;
loop(lo,
      put "yaux(",lo.tl:2,")= ",yaux.1(lo):12:6/;
);

y.fx(lo)=yaux.1(lo);
zlo=alpha.1;
put "Optimal objective function lower bound= ",zlo:12:6/;
);

SOLVE subproblem using nlp MINIMIZING cost;
put "Modelstat= ",subproblem.modelstat,...
..."; Solvestat= ",subproblem.solvestat/;
zs(IT)=cost.1;
lambda(IT,lo)=y.m(lo);
ys(IT,lo)=y.1(lo);

zup=zs(IT);
put "cost= ",zs(IT):12:6,"; Objective function...
...upper bound= ",zup:12:6/;
loop(lo,
      put "P(",lo.tl:2,")= ",P.1(lo):12:6,"lambda(",lo.tl:2,...
      ...")= ",lambda(IT,lo):12:6/;
);put /;
put "Unsatisfied demand= ",ud.1:12:6/;

*         Updating error
if(zlo ne 0,
            error=abs((zup-zlo)/zlo);
else
            error=abs((zup-zlo));
);
put "Error= ",error:12:9//;
ITER(IT)=yes;
);
```

Solution to Exercise 6.8. If some connections are lacking and the nonlinear objective function is considered, the following cost coefficients could be used:

$$c_i = \begin{cases} 0 & \text{if } x_i = 0 \\ f_i + v_i x_i^2 & \text{if } 0 < x_i \leq x_i^{\max} \end{cases} \; ; i = 1, \ldots, 25 \; ,$$

where $x_i^{\max}$ is the maximum flow capacity for connection i, c_i is the connection cost in dollars, f_i is the fixed cost owing to the presence of the connection, and v_i is the variable cost depending on the equilibrium flow that goes through connection x_i.

These constraints can be replaced using binary variables by the following set of constraints for each connection i:

$$\begin{aligned} c_i &= y_i f_i + v_i x_i^2 \\ 0 \leq x_i &\leq y_i x_i^{\max} \\ y_i &\in \{0,1\} \; . \end{aligned}$$

Therefore, the nonlinear water supply problem considering the existence or nonexistence of connections can be stated as

$$\text{minimize} \quad \sum_{i=1}^{25} c_i$$
$$x_i, y_i; \; i = 1, 2, \ldots, 25$$

subject to

$$\sum_{j \in \mathcal{P}_i} x_j = \pm q_i; \quad i = 1, \ldots, 18,$$
$$c_i = y_i f_i + v_i x_i^2; \quad i = 1, \ldots, 25$$
$$0 \le x_i \le y_i x_i^{\max}; \quad i = 1, \ldots, 25$$
$$y_i \in \{0, 1\}; \quad i = 1, \ldots, 25,$$

where $\mathcal{P}_i$ is the set of connections connected to supply node i, x_j is positive for outgoing flow and negative for ingoing flow, whereas q_i is positive for supply nodes ($q_i \in J$) and negative for consumption nodes ($q_i \in I$).

This mixed-integer nonlinear programming (MILP) problem can be solved on decomposed by city using the Benders decomposition algorithm, considering the binary variables and the flow x_{13} that goes from one city to another as complicating variables. The following master problem and subproblems for iteration ν are considered:

Master Problem. The MILP problem for obtaining the new values of the complicating variables is

$$\text{minimize} \quad \alpha$$
$$x_{13}, y_i; \; i = 1, 2, \ldots, 25$$

subject to

$$\alpha \ge z_{C_1}^{(k)} + z_{C_2}^{(k)} + \left(\mu_{C_1}^{(k)} + \mu_{C_2}^{(k)}\right)\left(x_{13} - x_{13}^{(k)}\right)$$
$$+ \sum_{i=1}^{25} \lambda_i^{(k)} \left(y_i - y_i^{(k)}\right); \quad k = 1, \ldots, \nu - 1$$
$$0 \le x_{13} \le y_{13} x_{13}^{\max}$$
$$y_i \in \{0, 1\}; \quad i = 1, \ldots, 25,$$

where z_{C_1} and z_{C_2} are the optimal costs for each city, obtained from the subproblems. The optimal solution of this master problem provides the new values for the complicating variables $x_{13}^{(\nu)}$ and $y_i^{(\nu)}$ ($i = 1, \ldots, 25$). Note that the dual variable $\lambda_{13}^{(k)}$ is equal to $\lambda_{13_{C_1}}^{(k)} + \lambda_{13_{C_2}}^{(k)}$.

Subproblem 1. For the first city C_1, the following subproblem is considered:

$$\text{minimize} \quad \sum_{i=1}^{13} c_i$$
$$x_i, y_i; \; i = 1, 2, \ldots, 13$$

subject to

$$\sum_{j \in \mathcal{P}_i} x_j = \pm q_i; \quad i = 1, \ldots, 9$$

$$c_i = y_i f_i + v_i x_i^2; \quad i = 1, \ldots, 13$$

$$0 \leq x_i \leq y_i x_i^{\max}; \quad i = 1, \ldots, 13$$

$$y_i = y_i^{(\nu)} : \lambda_i^{(\nu)}; \quad i = 1, \ldots, 12$$

$$y_{13} = y_{13}^{(\nu)} : \lambda_{13_{C_1}}^{(\nu)}$$

$$x_{13} = x_{13}^{(\nu)} : \mu_{C_1}^{(\nu)},$$

the solution of which provides the optimal flows x_i ($i = 1, 2, \ldots, 12$), the optimal cost in the first city $z_{C_1}^{(\nu)}$ and its sensitivities $\lambda_i^{(\nu)}$ ($i = 1, \ldots, 12$), $\lambda_{13_{C_1}}^{(\nu)}$, and $\mu_{C_1}^{(\nu)}$ with respect to the fixed values of the complicating variables.

Subproblem 2. For the second city C_2, the following subproblem is considered:

$$\underset{x_i, y_i;\ i=13, 14, \ldots, 25}{\text{minimize}} \quad \sum_{i=14}^{25} c_i$$

subject to

$$\sum_{j \in \mathcal{P}_i} x_j = \pm q_i; \quad i = 10, \ldots, 18$$

$$c_i = y_i f_i + v_i x_i^2; \quad i = 14, \ldots, 25$$

$$0 \leq x_i \leq y_i x_i^{\max}; \quad i = 13, \ldots, 25$$

$$y_i = y_i^{(\nu)} : \lambda_i^{(\nu)}; \quad i = 14, \ldots, 25$$

$$y_{13} = y_{13}^{(\nu)} : \lambda_{13_{C_2}}^{(\nu)}$$

$$x_{13} = x_{13}^{(\nu)} : \mu_{C_2}^{(\nu)},$$

the solution of which provides the optimal flows x_i ($i = 14, \ldots, 25$), the optimal cost in the first city $z_{C_2}^{(\nu)}$ and its sensitivities $\lambda_i^{(\nu)}$ ($i = 14, \ldots, 25$), $\lambda_{13_{C_2}}^{(\nu)}$, and $\mu_{C_2}^{(\nu)}$ with respect to the fixed values of the complicating variables.

Depending on the values of the complicating variables x_{13} and y_{13}, the subproblems could be infeasible because the total consumption quantity could not coincide with the total supply in one of the cities or both.

Solution to Exercise 6.10. The multiperiod network-constrained production planning problem considering a 3-h planning horizon ($m = 3$) and two production plants ($n = 2$) can be formulated as follows:

$$\underset{x_{it}, u_{it};\ i=1, 2;\ t=1, 2, 3}{\text{minimize}} \quad \sum_{t=1}^{3} \sum_{i=1}^{2} (c_i x_{it} + s_{it}),$$

where the objective function includes the production and the start-up cost, subject to

$$\begin{aligned}
x_{it} &\geq 0; & t=1,2,3; \quad i=1,2 \\
x_{1t} &= e_{12,t} + e_{13,t}; & t=1,2,3 \\
x_{2t} &= -e_{12,t} + e_{24,t}; & t=1,2,3 \\
d_{3t} &= e_{13,t} - e_{34,t}; & t=1,2,3 \\
d_{4t} &= e_{24,t} + e_{34,t}; & t=1,2,3 \\
e_{ij,t} &= B_{ij}\sin(\delta_{i,t}-\delta_{j,t}); & (i,j)\in\mathcal{P}; \quad t=1,2,3 \\
e_{ij,t} &= e_{ij}^{\max}; & (i,j)\in\mathcal{P}; \quad t=1,2,3 \\
u_{it}x_i^{\min} &\leq x_{it} \leq u_{it}x_i^{\max}; & t=1,2,3; \quad i=1,2 \\
\sum_{i=1}^{2} u_{it}x_i^{\max} &\geq \sum_{i=1}^{2} d_{it}; & t=1,2,3 \\
u_{it} &\in \{0,1\}; & t=1,2,3; \quad i=1,2\,,
\end{aligned} \quad (\text{B.15})$$

where the first block of constraints ensures the production positiveness, the next four constraints are the balance equations in the nodes, the next block of constraints express the electric energy flow through the line between nodes i and j being $\mathcal{P}=\{(1,2),(1,3),(2,3),(3,4)\}$, the seventh block of constraints enforces the production capacity limits of the plants, and the next block establishes certain levels of supply security. Note that the binary variables u_{it} control the functioning of plant i during period t.

The production cost c_{it} in dollars of each production plant i in time period t can be expressed as

$$c_{it} = f_i + v_i x_{it}\,.$$

Additionally, the start-up costs (s_{it}) for each plant in every time period are obtained using the following equations:

$$s_{i1} = \begin{cases} s_i^{\text{up}} & \text{if } u_{i1}-u_i^0 > 0 \\ 0 & \text{if } u_{i1}-u_i^0 \leq 0 \end{cases}$$

and

$$s_{it} = \begin{cases} s_i^{\text{up}} & \text{if } u_{it}-u_{i,t-1} > 0 \\ 0 & \text{if } u_{it}-u_{i,t-1} \leq 0 \end{cases} ; t=2,3\,,$$

where s_i^{up} is the start-up cost for plant i, and u_i^0 is the status of plant i at the beginning of the time planning horizon.

It should be noted that binary variables u_{it} ($i=1,2; t=1,2,3$) are complicating variables whereas the last two constraints are complicating constraints. If binary variables are fixed to given values, the mixed-integer nonlinear multi-period network-constrained production planning problem can be solved using nonlinear programming methods.

If the Benders decomposition is used, two problems have to be solved iteratively.

Subproblem. The subproblem for given values of the binary variables $u_{it} = u_{it}^{(k)}$ for iteration k ($i=1,2; t=1,2,3$) without considering the security of supply constraint is solved. In this problem we minimize the production cost for given status (on- or off-line) of plants in every time period, and we obtain the dual variables $\lambda_{it}^{(k)}$ associated with the constraints that fix the values of the binary variables to their actual values. Note that once we have selected the actual values of the binary variables, the complicating constraint no longer complicate the solution of the problem.

Master Problem. The master problem is

$$\begin{array}{c} \text{minimize} \\ \alpha, u_{it}; \ i = 1, 2; \ t = 1, 2, 3 \end{array} \quad \alpha$$

subject to

$$z^{(k)} + \sum_{t=1}^{3} \sum_{i=1}^{2} \lambda_{it}^{(k)} \left(u_{it} - u_{it}^{(k)} \right) \leq \alpha; \quad k = 1, \ldots, \nu - 1$$

$$\sum_{i=1}^{2} u_{it} x_i^{\max} \geq \sum_{i=1}^{2} d_{it}; \quad t = 1, 2, 3$$

$$\alpha \geq 0$$

$$y_i \in \{0, 1\}; \ i = 1, 2 \ .$$

Note that we have added the security of supply constraints to ensure the feasibility of the subsequent subproblem.

The optimal solution of this problem is shown in Tables B.12, B.13, B.14, and B.15.

Table B.12. Optimal solution status of every plant in every time period for Exercise 6.10: on-line (1) or off-line (0)

Period	1	2	3
C_1	0	1	0
C_2	1	1	1

Table B.13. Optimal production of every plant in every time period x_{it} for Exercise 6.10

Period (t)	1	2	3
x_{1t}	0	0.267	0
x_{2t}	0.8	2.333	0.2

Table B.14. Optimal height (angle) of every node in every time period δ_{it} for Exercise 6.10

Period (t)	1	2	3
δ_{1t}	−0.012	−0.032	−0.012
δ_{2t}	0.172	1.241	0.046
δ_{3t}	−0.159	−1.263	−0.058
δ_{4t}	−0.177	0.469	−0.031

Table B.15. Optimal sensitivities of every power balance equation in every time period for Exercise 6.10

Period (t)	1	2	3
Node 1	18.200	20.053	18.050
Node 2	18.200	18.583	18.050
Node 3	−18.200	−21.090	−18.050
Node 4	−18.200	−18.158	−18.050

The following GAMS program can be used to solve the problem:

```
$Title Exercise 6.10

file out/Exercise6.10.out/; put out;

Option mip=CPLEX;

Set
    lo production facilities /1,2/
    de consumption nodes /3,4/
    t time period /1*3/
    n line nodes /1*4/
    AL(n,n) dynamic set for the active lines
    IT iteration number /1*15/
    ITER(IT) dynamic set for activating cutting hyperplanes;

Alias(n,n1); Alias(t,t1); Alias(IT,IT1);

SCALARS
    error control error parameter /1/
    epsilon maximum tolerable error /1e-12/
    zup objective function upper bound /INF/
    zlo objective function lower bound /0/;

PARAMETERS
    B(n,n) susceptance of line ij
    emax(n,n) maximum capacity of the line
    xmax(lo) maximum production capacity/
    1       1.3
    2       2.5/
    xmin(lo) minimum production capacity/
    1       0.02
    2       0.02/
    f(lo) fixed cost for each production plant/
    1       20
    2       18/
    v(lo) variable cost for each production plant/
    1       0.1
    2       0.125/
    sup(lo) start up cost for each production plant/
```

```
          1       10
          2       17/
       u0(lo) initial status/
          1       0
          2       0/;

AL(n,n1)=no; AL('1','2')=yes; B('1','2')= 1.2; emax('1','2')=1.5;
AL('1','3')=yes; B('1','3')= 1.5; emax('1','3')=1.5;
AL('2','4')=yes; B('2','4')= 1.7; emax('2','4')=1.8;
AL('3','4')=yes; B('3','4')= 1.1; emax('3','4')=1.75;

Table d(de,t) demand data
              1       2       3
    3        0.2     2.5     0.1  4       0.6     0.1     0.1;

POSITIVE VARIABLES
    x(lo,t) actual production at a location and time
    c(lo,t)
    s(lo,t) start up cost;

BINARY VARIABLES uaux(lo,t);

VARIABLES
    e(n,n,t) energy sent from one location to another in time t
    delta(n,t) relative height of location
    cost total production cost
    u(lo,t)
    alpha;

PARAMETERS
    us(IT,lo,t) fixed values of the complicating variables
    zs(IT) objective function value for iteration IT
    lambda(IT,lo,t) dual variables associated with the complicating...
    ...variable;

Equations balance1,balance2,balance3,balance4 energy balances in...
nodes 1, 2, 3, and 4, respectively;
    balance1(t)..x('1',t)-e('1','2',t)-e('1','3',t)=e=0;
    balance2(t)..x('2',t)+e('1','2',t)-e('2','4',t)=e=0;
    balance3(t)..d('3',t)-e('1','3',t)+e('3','4',t)=e=0;
    balance4(t)..d('4',t)-e('2','4',t)-e('3','4',t)=e=0;

Equation translim transmission capacity limits;
    translim(n,n1,t)$(AL(n,n1))..e(n,n1,t)=l=emax(n,n1);

Equation edf transmitted commodity through lines;
    edf(n,n1,t)$(AL(n,n1))..e(n,n1,t)=e=B(n,n1)*sin(delta(n,t)...
    ...-delta(n1,t));

Equation prodlimax,prodlimin production capacity limits;
    prodlimax(lo,t)..x(lo,t)=l=u(lo,t)*xmax(lo);
    prodlimin(lo,t)..x(lo,t)=g=u(lo,t)*xmin(lo);

Equation copar cost coefficient definition;
    copar(lo,t)..c(lo,t)=e=f(lo)+v(lo)*x(lo,t);

Equation startup0a,startup0b,startupa,startupb start up cost
definition;
    startup0a(lo,'1')..s(lo,'1')$(u(lo,'1')-u0(lo)>0)=e=sup(lo);
    startup0b(lo,'1')$(u(lo,'1')-u0(lo) le 0)..s(lo,'1')=e=0;
    startupa(lo,t,t1)$(u(lo,t)-u(lo,t1)>0 and ORD(t1)+1=ORD(t))..
    s(lo,t)=e=sup(lo);
    startupb(lo,t,t1)$(u(lo,t)-u(lo,t1) le 0 and ORD(t1)+1=ORD(t))..
    s(lo,t)=e=0;

Equation costdf total production cost definition;
    costdf..cost=e=sum((lo,t),c(lo,t)*x(lo,t));
```

```
EQUATION cutting,secsupp,alphalo cutting planes;
    cutting(ITER)..alpha=g=zs(ITER)+sum((lo,t),lambda(ITER,lo,t)*...
    ...(uaux(lo,t)-us(ITER,lo,t)));
    secsupp(t)..sum(lo,uaux(lo,t)*xmax(lo))=g=sum(de,d(de,t));
    alphalo..alpha=g=0;

MODEL subproblem
/costdf,copar,balance1,balance2,balance3,balance4,...
...translim,edf,prodlimax,prodlimin/; MODEL master multiplier
updating /cutting,secsupp/;

lambda(IT,lo,t)=0; zs(IT)=0; us(IT,lo,t)=0; ITER(IT)=no;

loop(IT$(error>epsilon),
    put "    Iteration ",ORD(IT):3:0//;

    if(ORD(IT)=1,
        u.fx(lo,t)=1;
    else
        uaux.l(lo,t)=u.l(lo,t);
        SOLVE master using mip MINIMIZING alpha;
        put "Modelstat= ",master.modelstat,"; Solvestat= ",...
        ...master.solvestat/;
        put "alpha= ",alpha.l:12:6/;

        u.fx(lo,t)=uaux.l(lo,t);
        zlo=alpha.l;
        put "Optimal objective function lower bound= ",zlo:12:6/;
    );
    loop((lo,t),
        put "u(",lo.tl:2,",",t.tl:2,")= ",u.l(lo,t):12:6/;
    );

    SOLVE subproblem using nlp MINIMIZING cost;
    put "Modelstat= ",subproblem.modelstat,"; Solvestat= ",...
    ...subproblem.solvestat/;
    zs(IT)=cost.l;
    loop(lo,
        if(u.l(lo,'1')-u0(lo)>0,
            zs(IT)=zs(IT)+sup(lo);
        );
        loop((t,t1)$(u.l(lo,t)-u.l(lo,t1)>0 and ORD(t1)+1=ORD(t)),
            zs(IT)=zs(IT)+sup(lo);
        );
    );
    lambda(IT,lo,t)=u.m(lo,t);
    us(IT,lo,t)=u.l(lo,t);

    zup=zs(IT);
    put "cost= ",cost.l:12:6,"costTot= ",zs(IT):12:6,...
    ..."; Objective function upper bound= ",zup:12:6/;
    loop((lo,t),
        put "x(",lo.tl:2,",",t.tl:2,")= ",x.l(lo,t):12:6,...
        ..."; lambda= ",lambda(IT,lo,t):12:6/;
    );put /;

*       Updating error

    if(zlo ne 0,
        error=abs((zup-zlo)/zlo);
    else
        error=abs((zup-zlo));
    );
    put "Error= ",error:12:9//;
    ITER(IT)=yes;
);
```

B.7 Exercises from Chapter 7

Solution to Exercise 7.2. The solution of this problem can be obtained by means of the relaxation method stated in Sect. 7.1.1 using the following iterative process:

Step 0: Initialization. Initialize the iteration counter, $\nu = 1$, and let $r_1^{(\nu)} = g_1^{(0)}$ and $r_2^{(\nu)} = g_2^{(0)}$.

Step 1: Solve the master problem.

$$\underset{x_1, x_2}{\text{minimize}} \quad z = \left(\frac{x_1}{y_1}\right)^2 + \left(\frac{x_2}{y_2}\right)^2$$

subject to

$$g_1(\boldsymbol{x}, \boldsymbol{y}) = \frac{x_1 x_2}{y_1 y_2} \geq r_1^{(\nu)}$$

$$g_2(\boldsymbol{x}, \boldsymbol{y}) = \frac{x_2}{y_2} \sqrt{\frac{y_1}{x_1}} \geq r_2^{(\nu)}.$$

In this step the values of the $\boldsymbol{x}$ variables, $\boldsymbol{x}^{(\nu)}$, are obtained.

Step 2: Solve the subproblems. The following problems are solved for fixed values of the $\boldsymbol{x}$ variables, i.e., for $\boldsymbol{x} = \boldsymbol{x}^{(\nu)}$:

$$h_i^{(\nu)}(\boldsymbol{x}^{(\nu)}, \boldsymbol{y}) = \underset{u_1, u_2, v_1, v_2}{\text{minimum}} \sqrt{\sum_{j=1}^{2}\left(\frac{u_j - x_j^{(\nu)}}{x_j^{(\nu)} v_{x_j}}\right)^2 + \sum_{j=1}^{2}\left(\frac{v_j - y_j}{y_j v_{y_j}}\right)^2}$$

subject to

$$g_i(\boldsymbol{u}, \boldsymbol{v}) = 1$$

where

$$g_1(\boldsymbol{u}, \boldsymbol{v}) = \frac{u_1 u_2}{v_1 v_2} \quad \text{and} \quad g_2(\boldsymbol{u}, \boldsymbol{v}) = \frac{u_2}{v_2} \sqrt{\frac{v_1}{u_1}}.$$

The solution of this subproblems are $h_1^{(\nu)}$ and $h_2^{(\nu)}$, respectively.

Step 3: Check convergence. If the maximum relative error

$$\text{error}^{(\nu)} = \underset{\forall i}{\max} \left|\frac{h_i^{(\nu)} - h_i^{(\nu-1)}}{h_i^{(\nu)}}\right|$$

is smaller than the tolerance ε, stop the process. Otherwise go to Step 4.

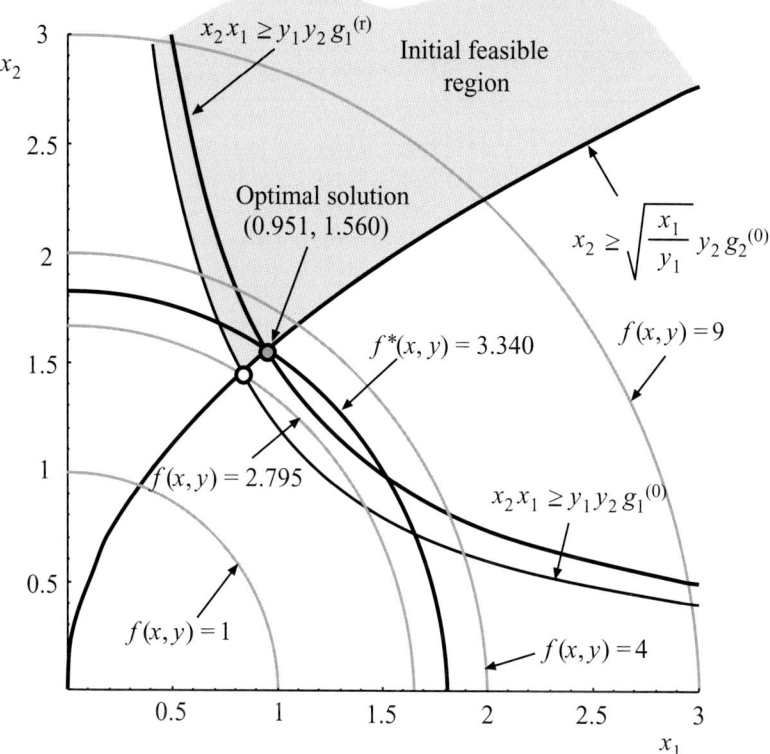

Fig. B.4. Illustration of the problem solution in Exercise 7.2

Step 4: Update values. Use the following rule to obtain $r^{(\nu)}$:

$$\begin{pmatrix} r_1^{(\nu+1)} \\ r_2^{(\nu+1)} \end{pmatrix} = \begin{pmatrix} r_1^{(\nu)} \\ r_2^{(\nu)} \end{pmatrix} + \rho \begin{pmatrix} h_1^{(0)} - h_1^{(\nu)} \\ h_2^{(0)} - h_2^{(\nu)} \end{pmatrix}.$$

Update the iteration counter $\nu \leftarrow \nu + 1$, and continue with Step 1.

In Fig. B.4 the graphical illustration of the problem solution is shown that is attained at the point $\boldsymbol{x}^* = (0.951, 1.560)^T$, with an objective function value $f(\mathbf{x}^*, \mathbf{y}^*) = 3.340$. Table B.16 illustrates the evolution of the iterative process showing the values of the variables and the error, corresponding to the following additional data:

$$\rho = 0.2, \quad \varepsilon = 0.00001 \ .$$

Alternatively, this problem can be solved using the cutting hyperplane method. To this end the following algorithm can be used:

Table B.16. Iterative process until the solution is obtained using the relaxation method

ν	$z^{(\nu)}$	$x_1^{(\nu)}$	$x_2^{(\nu)}$	$r_1^{(\nu)}$	$r_2^{(\nu)}$	$h_1^{(\nu)}$	$h_1^{(\nu)}$	Error$^{(\nu)}$
1	2.795	0.825	1.454	1.200	1.600	1.319	4.419	–
2	3.439	0.973	1.578	1.536	1.600	3.287	4.419	0.5988883
3	3.329	0.949	1.559	1.479	1.600	2.969	4.419	0.1070439
4	3.341	0.951	1.561	1.485	1.600	3.004	4.419	0.0114119
5	3.340	0.951	1.560	1.484	1.600	3.000	4.419	0.0013410
6	3.340	0.951	1.560	1.484	1.600	3.000	4.419	0.0001560
7	3.340	0.951	1.560	1.484	1.600	3.000	4.419	0.0000182
8	3.340	0.951	1.560	1.484	1.600	3.000	4.419	0.0000021

Step 0: Initialization. Initialize the iteration counter, $\nu = 1$.

Step 1: Solve the master problem.

$$\underset{x_1, x_2}{\text{minimize}} \quad \left(\frac{x_1}{y_1}\right)^2 + \left(\frac{x_2}{y_2}\right)^2$$

subject to

$$g_1(\boldsymbol{x}, \boldsymbol{y}) = \frac{x_1 x_2}{y_1 y_2} \geq g_1^{(0)}$$

$$g_2(\boldsymbol{x}, \boldsymbol{y}) = \frac{x_2}{y_2}\sqrt{\frac{y_1}{x_1}} \geq g_2^{(0)}$$

$$\boldsymbol{h}^{(s)} + \boldsymbol{\lambda}^{(s)T}(\boldsymbol{x} - \boldsymbol{x}^{(s)}) \geq \boldsymbol{h}_0; \quad s = 1, 2, \cdots, \nu - 1,$$

where the last equation constitutes an hyperplane reconstruction of the original constraint $\boldsymbol{h}(\boldsymbol{x}, \boldsymbol{y}) \geq \boldsymbol{h}_0$.

In this step the values of the $\boldsymbol{x}$ variables, $\boldsymbol{x}^{(\nu)}$, are obtained.

Step 2: Solve the subproblems. The following problems are solved for fixed values of the $\boldsymbol{x}$ variables, i.e., for $\boldsymbol{x} = \boldsymbol{x}^{(\nu)}$:

$$h_i^{(\nu)}(\boldsymbol{x}^{(\nu)}, \boldsymbol{y}) = \underset{u_1, u_2, v_1, v_2}{\text{minimum}} \sqrt{\sum_{j=1}^{2}\left(\frac{u_j - x_j}{x_j v_{x_j}}\right)^2 + \sum_{j=1}^{2}\left(\frac{v_j - y_j}{y_j v_{y_j}}\right)^2}$$

subject to

$$g_i(\boldsymbol{u}, \boldsymbol{v}) = 1$$
$$\boldsymbol{x} = \boldsymbol{x}^{(\nu)} : \boldsymbol{\lambda}_i^{(\nu)},$$

where
$$g_1(\boldsymbol{u},\boldsymbol{v}) = \frac{u_1 u_2}{v_1 v_2} \quad \text{and} \quad g_2(\boldsymbol{u},\boldsymbol{v}) = \frac{u_2}{v_2}\sqrt{\frac{v_1}{u_1}}\,.$$

The solution of this subproblems are $h_1^{(\nu)}$ and $h_2^{(\nu)}$, respectively.

Step 3: Check convergence. If the maximum relative error
$$\text{error}^{(\nu)} = \max_{\forall i} \left| \frac{h_i^{(\nu)} - h_i^{(\nu-1)}}{h_i^{(\nu)}} \right|$$
is smaller than the tolerance ε, stop the process. Otherwise, update the iteration counter $\nu \leftarrow \nu + 1$, and continue with Step 1.

The iterative procedure leads to the results shown in Table B.17 that provides the same information as Table B.16 using the alternative procedure. In this case the process converges after four iterations.

Table B.17. Iterative process until the solution is obtained using the cutting hyperplane method

ν	$z^{(\nu)}$	$x_1^{(\nu)}$	$x_2^{(\nu)}$	$r_1^{(\nu)}$	$r_2^{(\nu)}$	$h_1^{(\nu)}$	$h_1^{(\nu)}$	Error$^{(\nu)}$
1	2.795	0.825	1.454	1.200	1.600	1.319	4.419	–
2	3.333	0.950	1.559	1.481	1.600	2.980	4.419	0.5576132
3	3.340	0.951	1.560	1.484	1.600	3.000	4.419	0.0065065
4	3.340	0.951	1.560	1.484	1.600	3.000	4.419	0.0000009

Solution to Exercise 7.4. Consider the system of (7.136) relating forces and moments corresponding to the structure in Fig. 7.4a. If we add an additional diagonal piece as illustrated in Fig. 7.4b, the stiffness matrix of the structure has to be updated.

The rotation matrix (7.134) associated with this diagonal piece is
$$G = \begin{pmatrix} \frac{\sqrt{2}}{2} & \frac{\sqrt{2}}{2} & 0 \\ -\frac{\sqrt{2}}{2} & \frac{\sqrt{2}}{2} & 0 \\ 0 & 0 & 1 \end{pmatrix}, \tag{B.16}$$
whereas its stiffness matrix is obtained using (7.135):

$$K^{IV} = \begin{pmatrix} K_1^{IV} & | & K_2^{IV} \\ -- & + & -- \\ K_2^{IV^T} & | & K_3^{IV} \end{pmatrix}$$

$$= \begin{pmatrix} \frac{13}{2} & \frac{11}{2} & 3\sqrt{2} & -\frac{13}{2} & -\frac{11}{2} & -3\sqrt{2} \\ \frac{11}{2} & \frac{13}{2} & 3\sqrt{2} & -\frac{11}{2} & -\frac{13}{2} & -3\sqrt{2} \\ 3\sqrt{2} & 3\sqrt{2} & 4 & -3\sqrt{2} & -3\sqrt{2} & -2 \\ -\frac{13}{2} & -\frac{11}{2} & -3\sqrt{2} & \frac{13}{2} & \frac{11}{2} & 3\sqrt{2} \\ -\frac{11}{2} & -\frac{13}{2} & -3\sqrt{2} & \frac{11}{2} & \frac{13}{2} & 3\sqrt{2} \\ -3\sqrt{2} & -3\sqrt{2} & -2 & 3\sqrt{2} & 3\sqrt{2} & 4 \end{pmatrix}.$$

The stiffness matrix of the complete structure can be obtained from the stiffness matrices of all its pieces as shown in Fig. B.5.

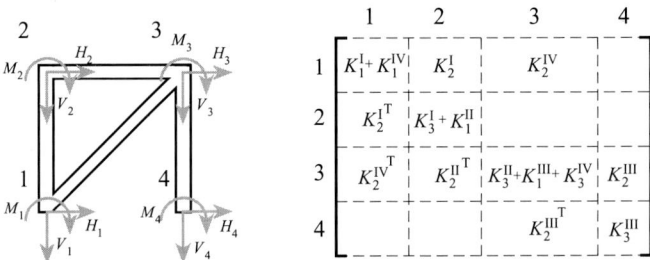

Fig. B.5. Stiffness matrix for the structure shown in Fig. 7.4b, and an illustration of how to build it

The final stiffness matrix relating forces and moments $H_i, V_i, M_i (i = 1,2,3,4)$ with displacements and rotations $h_i, v_i, \theta_i (i = 1,2,3,4)$ for $A_i = E_i = I_i = L_i = 1 (i = 1,2,3)$ for this new structure becomes

$$K = \begin{pmatrix}
\frac{37}{2} & \frac{11}{2}(6+3\sqrt{2}) & | & -12 & 0 & 6 & | & -\frac{13}{2} & -\frac{11}{2} & -3\sqrt{2} & | & 0 & 0 & 0 \\
\frac{11}{2} & \frac{15}{2} & 3\sqrt{2} & | & 0 & -1 & 0 & | & -\frac{11}{2} & -\frac{13}{2} & -3\sqrt{2} & | & 0 & 0 & 0 \\
(6+3\sqrt{2}) & 3\sqrt{2} & 8 & | & -6 & 0 & 2 & | & -3\sqrt{2} & -3\sqrt{2} & -2 & | & 0 & 0 & 0 \\
- & - & - & + & - & - & - & + & - & - & - & + & - & - & - \\
-12 & 0 & -6 & | & 13 & 0 & -6 & | & -1 & 0 & 0 & | & 0 & 0 & 0 \\
0 & -1 & 0 & | & 0 & 13 & 6 & | & 0 & -12 & 6 & | & 0 & 0 & 0 \\
6 & 0 & 2 & | & -6 & 6 & 8 & | & 0 & -6 & 2 & | & 0 & 0 & 0 \\
- & - & - & + & - & - & - & + & - & - & - & + & - & - & - \\
-\frac{13}{2} & -\frac{11}{2} & -3\sqrt{2} & | & -1 & 0 & 0 & | & \frac{39}{2} & \frac{11}{2} & (-6+3\sqrt{2}) & | & -12 & 0 & -6 \\
-\frac{11}{2} & -\frac{13}{2} & -3\sqrt{2} & | & 0 & -12 & -6 & | & \frac{11}{2} & \frac{39}{2} & (-6+3\sqrt{2}) & | & 0 & -1 & 0 \\
-3\sqrt{2} & -3\sqrt{2} & -2 & | & 0 & 6 & 2 & | & (-6+3\sqrt{2}) & (-6+3\sqrt{2}) & 12 & | & 6 & 0 & 2 \\
- & - & - & + & - & - & - & + & - & - & - & + & - & - & - \\
0 & 0 & 0 & | & 0 & 0 & 0 & | & -12 & 0 & 6 & | & 12 & 0 & 6 \\
0 & 0 & 0 & | & 0 & 0 & 0 & | & 0 & -1 & 0 & | & 0 & 1 & 0 \\
0 & 0 & 0 & | & 0 & 0 & 0 & | & -6 & 0 & 2 & | & 6 & 0 & 4
\end{pmatrix},$$
(B.17)

where the banded matrix structure is shown.

Note that the pattern of the initial structure shown in (7.136) is like the one in (7.127). Note also that after adding the new piece, the pattern of the new stiffness matrix, shown in (B.17) is also like that in (7.127), but with a larger width band, which does not justify the use of the proposed method.

Solution to Exercise 7.6. The solution of this problem can be obtained by means of the relaxation method stated in Sect. 7.1.1 using the following iterative scheme:

Step 0: Initialization. Initialize the iteration counter, $\nu = 1$, and let the actual safety factor and reliability index lower limit to their minimum values $F_{lo}^{(\nu)} = F_0$ and $\beta^{(\nu-1)} = \beta_0$.

Step 1: Solve the master problem. The construction cost minimization problem considering safety factor constraints is solved:

$$\underset{F_c,\, \tan \alpha_s}{\text{minimize}} \quad C_{co} = c_c v_c + c_a v_a$$

subject to

$$H = 1.8 H_s$$
$$T = 1.1 T_z$$
$$1/5 \leq \tan \alpha_s \leq 1/2$$
$$F_c = 2 + d$$
$$v_c = 10 d$$
$$v_a = \frac{1}{2}(D_{wl} + 2)\left(46 + D_{wl} + \frac{D_{wl} + 2}{\tan \alpha_s}\right)$$
$$\frac{R_u}{H} = A_u \left(1 - e^{B_u I_r}\right)$$
$$I_r = \frac{\tan \alpha_s}{\sqrt{H/L}}$$
$$\left(\frac{2\pi}{T}\right)^2 = g \frac{2\pi}{L} \tanh \frac{2\pi D_{wl}}{L}$$
$$F_c / R_u \geq F_{lo}^{(\nu)}.$$

In this step the values of the freeboard $F_c^{(\nu)}$ and seaside slope tangent $\tan \alpha_s^{(\nu)}$ are obtained. Note that the constraint related to the reliability index lower bound is not considered in this problem.

Step 2: Solve the subproblem. The reliability index associated with the new values of the design variables is calculated by means of the following problem:

$$\underset{H,\, T}{\text{minimize}} \quad \beta = \sqrt{z_1^2 + z_2^2}$$

subject to

$$\frac{R_u}{H} = A_u \left(1 - e^{B_u I_r}\right)$$

$$I_{\text{r}} = \frac{\tan\alpha_{\text{s}}^{(\nu)}}{\sqrt{H/L}}$$

$$\left(\frac{2\pi}{T}\right)^2 = g\frac{2\pi}{L}\tanh\frac{2\pi D_{\text{wl}}}{L}$$

$$\Phi(z_1) = 1 - e^{-2(H/H_{\text{s}})^2}$$

$$\Phi(z_2) = 1 - e^{-0.675(T/\bar{T})^4}$$

$$F_{\text{c}}^{(\nu)} = R_{\text{u}} .$$

The solution of this subproblem provides $\beta^{(\nu)}$.

Step 3: Check convergence. If the relative error

$$\text{error}^{(\nu)} = \left|\frac{\beta^{(\nu)} - \beta^{(\nu-1)}}{\beta^{(\nu)}}\right|$$

is smaller than the tolerance ε, stop the process, the optimal solution has been found. Otherwise, set $\nu \leftarrow \nu + 1$ and go to Step 4.

Step 4: Update values. Use the following rule to obtain $F_{\text{lo}}^{(\nu)}$:

$$F_{\text{lo}}^{(\nu)} = F_{\text{lo}}^{(\nu-1)} + \rho(\beta^{(0)} - \beta^{(\nu-1)}) .$$

If the resulting safety factor is lower than the minimum value ($F_{\text{lo}}^{(\nu)} < F_0$), then

$$F_{\text{lo}}^{(\nu)} = F_0 .$$

The optimal solution is

$$C_{\text{co}}^* = 6{,}512.2456, \quad F_{\text{c}}^* = 5.756, \quad \tan\alpha_s{}^* = 0.231 ,$$

which is attained after 12 iterations (see Table B.18) with a relaxation factor $\rho = 0.1$ and a relative error tolerance of $\varepsilon = 10^{-4}$. The probabilistic safety constraint is active $\beta^* = \beta_0 = 4.5$.

B.8 Exercises from Chapter 8

Solution to Exercise 8.2. Let (X_i, Y_i) be the coordinates of the random points, i.e.,

$$X_i \sim N(0,1), \quad Y_i \sim N(0,1); \quad i = 1, 2, \ldots, 20 ,$$

where $\sim$ indicates probability distribution.

Table B.18. Illustration of the iterative procedure in Exercise 7.6

Iterations	F_c	$\tan \alpha_s$	C_{co}	C_{in}	C_{to}	Error
1	5.344	0.259	5,991.6	1.247	3.526	0.276259
2	5.554	0.244	6,257.4	1.296	4.009	0.120514
3	5.657	0.237	6,387.2	1.321	4.256	0.057910
4	5.707	0.234	6,450.9	1.333	4.379	0.028197
5	5.732	0.232	6,482.2	1.339	4.440	0.013806
6	5.745	0.232	6,497.6	1.342	4.471	0.006776
7	5.751	0.231	6,505.1	1.343	4.486	0.003330
8	5.754	0.231	6,508.9	1.344	4.493	0.001637
9	5.755	0.231	6,510.7	1.344	4.496	0.000805
10	5.756	0.231	6,511.6	1.345	4.498	0.000396
11	5.756	0.231	6,512.0	1.345	4.499	0.000195
12	5.756	0.231	6,512.2	1.345	4.500	0.000096

Our problem can be stated as

$$\underset{r, x_0, y_0}{\text{minimize}} \quad z = r \tag{B.18}$$

subject to

$$(x_i - x_0)^2 + (y_i - y_0)^2 \leq r^2 \; : \; \mu_i; \quad i = 1, 2, \ldots, 20 \tag{B.19}$$

where $\mu_i (i = 1, 2, \ldots, 20)$ are the dual variables.

It is interesting to see that since a circle is defined by three points, apart from degenerate cases, only three points [those points defining the circle (see Fig. B.6)] lead to active constraints. Thus, the objective function is sensitive only to these three points. The same can be said for the sensitivities of $x_0, y_0,$ and r.

The sensitivities can be calculated analytically or numerically. For the first option, we first obtain the Lagrangian function

$$\mathcal{L}(x_0, y_0, r, \mu) = r + \sum_{i=1}^{20} \mu_i \left[(x_i - x_0)^2 + (y_i - y_0)^2 - r^2\right] \tag{B.20}$$

and, according to Theorem 8.2 the sensitivities of z^* to the data points are

$$\frac{\partial z^*}{\partial x_i} = 2\mu_i^*(x_i^* - x_0) \tag{B.21}$$

$$\frac{\partial z^*}{\partial y_i} = 2\mu_i^*(y_i^* - y_0) . \tag{B.22}$$

Let k, s, and t the indices of the data points with non-null values of μ_i, and call $\eta_k, \eta_s,$ and η_t to the corresponding μ values, then from (8.75) to (8.82) one gets

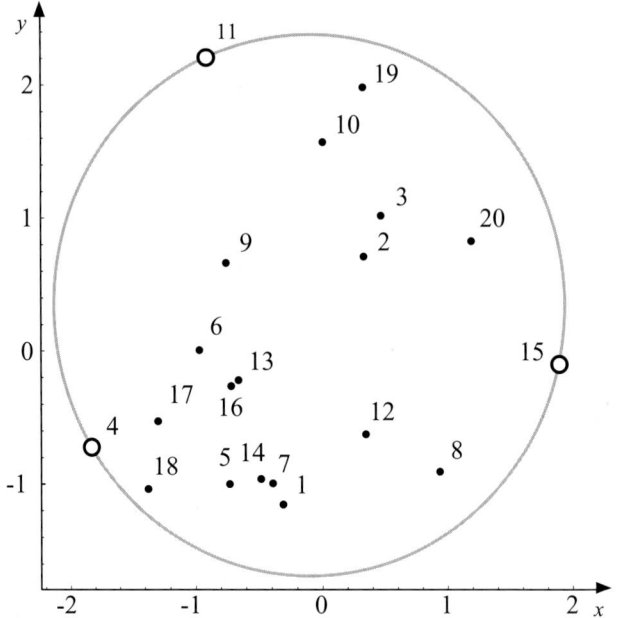

Fig. B.6. Illustration of the data points and the circle in Exercise 8.2. Note that the three data points 4, 11, and 15 define the circle

$$\boldsymbol{F_{xx}} = \left(\sum_{s=1}^{3} \eta_s\right)\begin{pmatrix} 2 & 0 & 0 \\ 0 & 2 & 0 \\ 0 & 0 & -2 \end{pmatrix}$$

$$\boldsymbol{F_{xa}} = \begin{pmatrix} -2\eta_1 & 0 & -2\eta_2 & 0 & -2\eta_3 & 0 \\ 0 & -2\eta_1 & 0 & -2\eta_2 & 0 & -2\eta_3 \\ 0 & 0 & 0 & 0 & 0 & 0 \end{pmatrix}$$

$$\boldsymbol{H_x} = \begin{pmatrix} -2(x_k - x_0) & -2(y_k - y_0) & -2r \\ -2(x_s - x_0) & -2(y_s - y_0) & -2r \\ -2(x_t - x_0) & -2(y_t - y_0) & -2r \end{pmatrix}$$

$$\boldsymbol{H_a} = \begin{pmatrix} 2(x_k - x_0) & 2(y_k - y_0) & 0 & 0 & 0 & 0 \\ 0 & 0 & 2(x_s - x_0) & 2(y_s - y_0) & 0 & 0 \\ 0 & 0 & 0 & 0 & 2(x_t - x_0) & 2(y_t - y_0) \end{pmatrix}$$

$$\boldsymbol{F_x} = (0 \ 0 \ 1)$$
$$\boldsymbol{F_a} = (0 \ 0 \ 0 \ 0 \ 0 \ 0),$$

and the sensitivities can be easily calculated from (8.98) to (8.100), i.e.,

$$\frac{\partial \boldsymbol{x}}{\partial \boldsymbol{a}} = -\boldsymbol{H_x^{-1}} \boldsymbol{H_a} \qquad (\text{B.23})$$

$$\frac{\partial \eta}{\partial \boldsymbol{a}} = (\boldsymbol{H_x^{-1}})^T \left[\boldsymbol{F_{xx}} \boldsymbol{H_x^{-1}} \boldsymbol{H_a} - \boldsymbol{F_{xa}}\right] \qquad (\text{B.24})$$

$$\frac{\partial z}{\partial a} = F_a - F_x H_x^{-1} H_a = F_a + \eta^T H_a. \qquad (B.25)$$

Alternatively to the analytical expressions, we can use the following program in GAMS to solve the problem and calculate the sensitivities by finite differences:

```
$ title Circle

file out/circle.out/;
put out;

SETS I number of points /1*20/;

ALIAS(I,I1);

SCALARS
    r0
    x00
    y00
    epsilon/0.001/;

PARAMETERS
    X(I) data points abscissas
    Y(I) data points ordinates
    sensrx(I),sensry(I),sensx0x(I),sensx0y(I),sensy0x(I),sensy0y(I)
    dual(I),dualx(I,I),dualy(I,I);

VARIABLES
    z objective function value
    x0 x coordinate of the center
    y0 x coordinate of the center
    r  radious of the center;

EQUATIONS
    zdef objective function value
    inside(I);

zdef..z=e=r;

inside(I)..sqr(X(I)-x0)+sqr(Y(I)-y0)=l=sqr(r);

MODEL circle/ALL/;

X(I)=normal(0,1);
Y(I)=normal(0,1);
r.l=10;
x0.l=0;
y0.l=0;

SOLVE circle USING nlp MINIMIZING z;
put " modelstat=",circle.modelstat," solvestat=",circle.solvestat/;
put "radio=",r.l:12:8," x0=",x0.l:12:8," y0=",y0.l:12:8/;

* sensitivities using finite differences

r0=r.l;
x00=x0.l;
y00=y0.l;
dual(I)=-inside.m(I);

loop(I1,
    X(I1)= X(I1)+epsilon;
    SOLVE circle USING nlp MINIMIZING r;
    sensrx(I1)=(r.l-r0)/epsilon;
    sensx0x(I1)=(x0.l-x00)/epsilon;
```

```
        sensy0x(I1)=(y0.1-y00)/epsilon;
        dualx(I1,I)=(-inside.m(I)-dual(I))/epsilon;
        X(I1)=X(I1)-epsilon;
);

loop(I1,
    Y(I1)= Y(I1)+epsilon;
    SOLVE circle USING nlp MINIMIZING z;
    sensry(I1)=(r.1-r0)/epsilon;
    sensx0y(I1)=(x0.1-x00)/epsilon;
    sensy0y(I1)=(y0.1-y00)/epsilon;
    dualy(I1,I)=(-inside.m(I)-dual(I))/epsilon;
    Y(I1)=Y(I1)-epsilon;
);

put "Data, dual variables and sensitivities"/;

loop(I,put I.tl:3," & " ,X(I):8:3," & ",Y(I):8:3," & ",...
...dual(I):8:3," & ", sensx0x(I):8:3," & ",sensy0x(I):8:3,...
..." & ",sensrx(I):8:3," & ",sensx0y(I):8:3," & ",...
...sensy0y(I):8:3," & ",sensry(I):8:3," \\"/;);
put "\\"/;

loop((I,I1),if( abs(dualx(I1,I)) > 0.0000001,put I.tl:3," & ",...
I1.tl:3," & ",dualx(I1,I):8:5," & ", dualy(I1,I):8:5," \\"/;);)
```

Table B.19 shows the data points, the dual variables, and the x_0, y_0, and r sensitivities. Note that all sensitivities are null, except those for data points 4, 11, and 15.

Table B.19. Data points, dual variables, and the x_0, y_0, and r sensitivities

i	x_i	y_i	μ_i	$\dfrac{\partial x_0}{\partial x_i}$	$\dfrac{\partial y_0}{\partial x_i}$	$\dfrac{\partial r}{\partial x_i}$	$\dfrac{\partial x_0}{\partial y_i}$	$\dfrac{\partial y_0}{\partial y_i}$	$\dfrac{\partial r}{\partial y_i}$
1	-0.313	-1.153	0.000	0.000	0.000	0.000	0.000	0.000	0.000
2	0.328	0.710	0.000	0.000	0.000	0.000	0.000	0.000	0.000
3	0.464	1.017	0.000	0.000	0.000	0.000	0.000	0.000	0.000
4	-1.830	-0.724	0.079	0.388	0.467	-0.274	0.240	0.289	-0.170
5	-0.732	-0.999	0.000	0.000	0.000	0.000	0.000	0.000	0.000
6	-0.972	0.006	0.000	0.000	0.000	0.000	0.000	0.000	0.000
7	-0.394	-0.995	0.000	0.000	0.000	0.000	0.000	0.000	0.000
8	0.935	-0.907	0.000	0.000	0.000	0.000	0.000	0.000	0.000
9	-0.759	0.663	0.000	0.000	0.000	0.000	0.000	0.000	0.000
10	0.000	1.571	0.000	0.000	0.000	0.000	0.000	0.000	0.000
11	-0.909	2.213	0.069	0.048	-0.290	-0.112	-0.111	0.670	0.258
12	0.344	-0.625	0.000	0.000	0.000	0.000	0.000	0.000	0.000
13	-0.662	-0.218	0.000	0.000	0.000	0.000	0.000	0.000	0.000
14	-0.486	-0.962	0.000	0.000	0.000	0.000	0.000	0.000	0.000
15	1.884	-0.108	0.097	0.563	-0.177	0.386	-0.129	0.040	-0.088
16	-0.721	-0.263	0.000	0.000	0.000	0.000	0.000	0.000	0.000
17	-1.299	-0.527	0.000	0.000	0.000	0.000	0.000	0.000	0.000
18	-1.375	-1.037	0.000	0.000	0.000	0.000	0.000	0.000	0.000
19	0.320	1.982	0.000	0.000	0.000	0.000	0.000	0.000	0.000
20	1.187	0.826	0.000	0.000	0.000	0.000	0.000	0.000	0.000

Table B.20. Sensitivities $\dfrac{\partial \eta_j}{\partial x_i}$ and $\dfrac{\partial \eta_j}{\partial y_i}$ of the dual variables η_j with respect to x_i and y_i

j	i	$\dfrac{\partial \eta_j}{\partial x_i}$	$\dfrac{\partial \eta_j}{\partial y_i}$
4	4	−0.02397	−0.00443
4	11	0.03649	−0.02972
4	15	−0.01254	0.03416
11	4	0.04959	−0.00070
11	11	−0.01842	0.02717
11	15	−0.03116	−0.02651
15	4	0.00750	0.02562
15	11	−0.00456	−0.02863
15	15	−0.00294	0.00302

Similarly, Table B.20 shows the sensitivities $\dfrac{\partial \eta_j}{\partial x_i}$ and $\dfrac{\partial \eta_j}{\partial y_i}$ of the dual variables η_j with respect to x_i and y_i.

The generalization to n dimensions of the problem (B.18)–(B.19) is

$$\underset{r,\, x_{01}, x_{02}, \ldots, x_{0n}}{\text{minimize}} \quad z = r \tag{B.26}$$

subject to

$$\sum_{j=1}^{n} (x_{ij} - x_{0j})^2 \leq r \;:\; \mu_i; \quad i = 1, 2, \ldots, m. \tag{B.27}$$

Solution to Exercise 8.4. The water supply problem can be stated as

$$\underset{x_1, x_2, \ldots, x_{15}}{\text{minimize}} \quad z = \sum_{i=1}^{15} c_i(x_i) = \sum_{i=1}^{15} (f_i + v_i x_i) \tag{B.28}$$

subject to the flow balance equations for all nodes (input amount of water equal to output amount of water including supplies and consumptions):

$$\begin{bmatrix} -1 & -1 & & & & & & & & & & & & & \\ 1 & & -1 & & & & & & & & & & & & \\ & 1 & & -1 & -1 & & & & & & & & & & \\ & & 1 & 1 & & -1 & & & & & & & & & \\ & & & & 1 & & -1 & & & & & & & & \\ & & & & & 1 & 1 & -1 & & & & & & & \\ & & & & & & & 1 & 1 & 1 & & & & & \\ & & & & & & & & & -1 & 1 & & & & \\ & & & & & & & & & -1 & & 1 & 1 & & \\ & & & & & & & & & & -1 & -1 & & 1 & \\ & & & & & & & & & & & & -1 & & 1 \\ & & & & & & & & & & & & & -1 & -1 \end{bmatrix} \begin{bmatrix} x_1 \\ x_2 \\ x_3 \\ x_4 \\ x_5 \\ x_6 \\ x_7 \\ x_8 \\ x_9 \\ x_{10} \\ x_{11} \\ x_{12} \\ x_{13} \\ x_{14} \\ x_{15} \end{bmatrix} = \begin{bmatrix} -q_1 \\ q_2 \\ q_3 \\ q_4 \\ q_5 \\ q_6 \\ q_7 \\ q_8 \\ q_9 \\ q_{10} \\ q_{11} \\ -q_{12} \end{bmatrix}$$

$$\tag{B.29}$$

where $c_i(x_i)$ is the cost associated with connection i and flow x_i, and
$$0 \leq x_i \leq r_i; \quad i = 1, 2, \ldots, 15, \tag{B.30}$$
which can be written as
$$\underset{x_1, x_2, \ldots, x_{15}}{\text{minimize}} \sum_{i=1}^{15} c_i(x_i) \tag{B.31}$$
subject to
$$\boldsymbol{Ax} = \boldsymbol{q} \tag{B.32}$$
$$\boldsymbol{x} \leq \boldsymbol{r} \tag{B.33}$$
$$-\boldsymbol{x} \leq \boldsymbol{0}, \tag{B.34}$$
where the matrices $\boldsymbol{A}, \boldsymbol{q}$, and $\boldsymbol{r}$ are those in (B.29) and (B.30).

Since all the r_i and q_j appear on the right- or left-hand sides of the constraints, we do not need a change of the statement of the problem, because Theorem 8.1 allows us to calculate the sensitivities. Thus, we solve the optimization problem (B.31)–(B.34) and use the dual variable values for calculating the sensitivities.

Alternatively, Theorem 8.2 can be used to solve the problem, as follows: The Lagrangian function is
$$\mathcal{L}(\boldsymbol{x}, \boldsymbol{q}, \boldsymbol{r}, \boldsymbol{\lambda}, \boldsymbol{\mu}^{(1)}, \boldsymbol{\mu}^{(2)}) = \sum_{i=1}^{15} c_i(x_i) + \boldsymbol{\lambda}^T(\boldsymbol{Ax} - \boldsymbol{q}) + \left(\boldsymbol{\mu}^{(1)}\right)^T(\boldsymbol{x} - \boldsymbol{r}) - \left(\boldsymbol{\mu}^{(2)}\right)^T \boldsymbol{x}. \tag{B.35}$$

Then, using Theorem 8.2, the sensitivities are
$$\frac{\partial z^*}{\partial q_i} = -\lambda_i^*; \quad i = 1, 2, \ldots, 12 \tag{B.36}$$
$$\frac{\partial z^*}{\partial r_i} = -\mu_i^*; \quad i = 1, 2, \ldots, 12. \tag{B.37}$$

Note that this solution coincides with that given by Theorem 8.1.

Solution to Exercise 8.6. To analyze the existence of partial derivatives we solve the system of inequalities (8.73)–(8.74), which in this case becomes

$$\left(\begin{array}{cc|ccc|cc|c} 0 & -2 & 0 & 0 & 0 & 0 & 0 & -1 \\ -2 & 0 & 0 & 0 & 0 & 1 & -1 & 0 \\ 0 & -2 & 0 & 0 & 0 & 1 & 0 & 0 \\ \hline 1 & 1 & -1 & 0 & 0 & 0 & 0 & 0 \\ -1 & 0 & 0 & 1 & 0 & 0 & 0 & 0 \end{array} \right) \begin{pmatrix} dx_1 \\ dx_2 \\ \hline da_1 \\ da_2 \\ da_3 \\ \hline d\mu_1 \\ d\mu_2 \\ \hline dz \end{pmatrix} = \boldsymbol{0}. \tag{B.38}$$

Note that the multiplier related to inactive constraint μ_3 is not considered in the study.

As we want to calculate all the sensitivities at once, we transform system (B.38) into (8.84) as follows:

$$\left(\begin{array}{cc|cc|c} 0 & -2 & 0 & 0 & -1 \\ \hline -2 & 0 & 1 & -1 & 0 \\ 0 & -2 & 1 & 0 & 0 \\ \hline 1 & 1 & 0 & 0 & 0 \\ -1 & 0 & 0 & 0 & 0 \end{array}\right) \left(\begin{array}{c} dx_1 \\ dx_2 \\ \hline d\mu_1 \\ d\mu_2 \\ \hline dz \end{array}\right) = \left(\begin{array}{ccc} 0 & 0 & 0 \\ \hline 0 & 0 & 0 \\ 0 & 0 & 0 \\ \hline -1 & 0 & 0 \\ 0 & 1 & 0 \end{array}\right) \left(\begin{array}{c} da_1 \\ da_2 \\ da_3 \end{array}\right). \quad (B.39)$$

Since matrix U is invertible, the problem is nondegenerate. The matrix with all the partial derivatives becomes

$$\left(\begin{array}{ccc} \dfrac{\partial x_1^*}{\partial a_1} & \dfrac{\partial x_1^*}{\partial a_2} & \dfrac{\partial x_1^*}{\partial a_3} \\ \dfrac{\partial x_2^*}{\partial a_1} & \dfrac{\partial x_2^*}{\partial a_2} & \dfrac{\partial x_2^*}{\partial a_3} \\ \hline \dfrac{\partial \mu_1^*}{\partial a_1} & \dfrac{\partial \mu_1^*}{\partial a_2} & \dfrac{\partial \mu_1^*}{\partial a_3} \\ \dfrac{\partial \mu_2^*}{\partial a_1} & \dfrac{\partial \mu_2^*}{\partial a_2} & \dfrac{\partial \mu_2^*}{\partial a_3} \\ \hline \dfrac{\partial z^*}{\partial a_1} & \dfrac{\partial z^*}{\partial a_2} & \dfrac{\partial z^*}{\partial a_3} \end{array}\right) = U^{-1} S = \left(\begin{array}{ccc} 0 & 1 & 0 \\ 1 & -1 & 0 \\ \hline 2 & -2 & 0 \\ 2 & -4 & 0 \\ \hline -2 & 2 & 0 \end{array}\right), \quad (B.40)$$

which gives all sensitivities. Note, for example, that the sensitivities with respect to parameter a_3 (third column in matrix B.40) are null, because a_3 only appears in the inactive constraint:

Solution to Exercise 8.8. In this case, (8.73)–(8.74) become

$$\left(\begin{array}{cc|cccccc|cc|cc|c} 6 & 4 & 9 & 4 & 0 & 0 & 0 & 0 & 0 & 0 & 0 & 0 & -1 \\ \hline 4 & 0 & 6 & 0 & 0 & -4 & 0 & 0 & 0 & 1 & -2 & 0 \\ 0 & 4 & 0 & 4 & 0 & 0 & 0 & 0 & 0 & 1 & 0 & 0 \\ \hline 1 & 1 & 0 & 0 & 0 & 3 & -1 & 0 & 0 & 0 & 0 & 0 \\ \hline -2 & 0 & 0 & 0 & 0 & 0 & 0 & -1 & 0 & 0 & 0 \end{array}\right) \left(\begin{array}{c} dx \\ dy \\ \hline da_1 \\ da_2 \\ da_3 \\ da_4 \\ da_5 \\ da_6 \\ da_7 \\ \hline d\lambda \\ \hline d\mu_2 \\ \hline dz \end{array}\right) = 0. \quad (B.41)$$

As we want to calculate all the sensitivities at once, we transform system (B.41) into (8.84) as follows:

$$\begin{pmatrix} 6 & 4 & | & 0 & | & 0 & | & -1 \\ -- & -+ & -+ & -+ & - \\ 4 & 0 & | & 1 & | & -2 & | & 0 \\ 0 & 4 & | & 1 & | & 0 & | & 0 \\ -- & -+ & -+ & -+ & - \\ 1 & 1 & | & 0 & | & 0 & | & 0 \\ -- & -+ & -+ & -+ & - \\ -2 & 0 & | & 0 & | & 0 & | & 0 \end{pmatrix} \begin{pmatrix} dx \\ dy \\ \overline{d\lambda} \\ \overline{d\mu_2} \\ \overline{dz} \end{pmatrix} = \begin{pmatrix} 9 & 4 & 0 & 0 & 0 & 0 & 0 \\ -- & -- & - & -- & - \\ 6 & 0 & 0 & -4 & 0 & 0 & 0 \\ 0 & 4 & 0 & 0 & 0 & 0 & 0 \\ -- & -- & - & -- & - \\ 0 & 0 & 0 & 3 & -1 & 0 & 0 \\ -- & -- & - & -- & - \\ 0 & 0 & 0 & 0 & 0 & 0 & -1 \end{pmatrix} \begin{pmatrix} da_1 \\ da_2 \\ da_3 \\ da_4 \\ da_5 \\ da_6 \\ da_7 \end{pmatrix}.$$

(B.42)

Since matrix U is invertible, the problem is nondegenerate. The matrix with all the partial derivatives becomes

$$\begin{pmatrix} \dfrac{\partial x^*}{\partial a_1} & \dfrac{\partial x^*}{\partial a_2} & \cdots & \dfrac{\partial x^*}{\partial a_7} \\ \dfrac{\partial y^*}{\partial a_1} & \dfrac{\partial y^*}{\partial a_2} & \cdots & \dfrac{\partial y^*}{\partial a_7} \\ -- & -- & & -- \\ \dfrac{\partial \lambda^*}{\partial a_1} & \dfrac{\partial \lambda^*}{\partial a_2} & \cdots & \dfrac{\partial \lambda^*}{\partial a_7} \\ -- & -- & & -- \\ \dfrac{\partial \mu_2^*}{\partial a_1} & \dfrac{\partial \mu_2^*}{\partial a_2} & \cdots & \dfrac{\partial \mu_2^*}{\partial a_7} \\ -- & -- & & -- \\ \dfrac{\partial z^*}{\partial a_1} & \dfrac{\partial z^*}{\partial a_2} & \cdots & \dfrac{\partial z^*}{\partial a_7} \end{pmatrix} = U^{-1} S = \begin{pmatrix} 0 & 0 & 0 & 0 & 0 & 0 & -1/2 \\ 0 & 0 & 0 & -3 & 1 & 0 & 1/2 \\ - & - & - & - & - & - & - \\ 0 & -4 & 0 & 12 & -4 & 0 & -2 \\ - & - & - & - & - & - & - \\ 3 & -2 & 0 & 4 & -2 & 0 & -2 \\ - & - & - & - & - & - & - \\ 9 & 4 & 0 & -12 & 4 & 0 & -1 \end{pmatrix}.$$

Note that the last row gives the sensitivities of z with respect to the $\boldsymbol{a}$-values, which are coincident with those obtained in Example 8.5. However, rows 1, 2, 3, and 4 give the sensitivities of x, y, λ, and μ_2 with respect to the same $\boldsymbol{a}$-values.

Solution to Exercise 8.10. The required GAMS program is

```
$title activeConstraints

file out/active.out/;
put out;

VARIABLES
    x
    y
    z objective function variable;

EQUATIONS
    zdef1 objective function 1
    zdef2 objective function 2
    zdef3 objective function 3
    c1 constraint 1
    c2 constraint 2
    c3 constraint 3
    c4 constraint 4;

zdef1..z=e=x*x+y*y;
zdef2..z=e=x+y;
zdef3..z=e=sqr(x-4)+sqr(y-2)-1;
```

```
c1..x*y=g=4;
c2..x+y=e=5;
c3..sqr(x-4)+sqr(y-2)=l=1;
c4..x*x+y*y=e=13;

* The three models are defined

MODEL A/zdef1,c1,c2,c3/;
MODEL B/zdef2,c1,c4,c3/;
MODEL C/zdef3,c1,c2,c4/;

* Model 1 is solved

SOLVE A USING nlp MINIMIZING z;
put "z=",z.l," modelstat=",A.modelstat," solvestat=",A.solvestat/;
put "x=",x.l:12:8," y=",y.l:12:8/;
put "mu1=",(-c1.m)," lambda=",(-c2.m)," mu2=",(-c3.m)//;

* Model 2 is solved

SOLVE B USING nlp MAXIMIZING z;
put "z=",z.l," modelstat=",B.modelstat," solvestat=",B.solvestat/;
put "x=",x.l:12:8," y=",y.l:12:8/;
put "mu1=",(-c1.m)," lambda=",(-c4.m)," mu2=",(-c3.m)//;

* Model 3 is solved

SOLVE C USING nlp MINIMIZING z;
put "z=",z.l," modelstat=",C.modelstat," solvestat=",C.solvestat/;
put "x=",x.l:12:8," y=",y.l:12:8/;
put "mu1=",(-c1.m)," lambda1=",(-c2.m)," lambda2=",(-c4.m)//
```

and the output file is

```
z=         13.00 modelstat=       2.00 solvestat=         1.00
x=  3.00000000 y=  2.00000000
mu1=        0.00 lambda=       -4.00 mu2=              1.00

z=          5.00 modelstat=       2.00 solvestat=         1.00
x=  3.00000000 y=  2.00000000
mu1=        0.00 lambda=       -0.25 mu2=             -0.25

z=          0.00 modelstat=       2.00 solvestat=         1.00
x=  3.00000000 y=  2.00000000
mu1=        0.00 lambda1=      -4.00 lambda2=           1.0
```

Solution to Exercise 8.12. The Karush–Kuhn–Tucker (KKT) conditions for this problem are

$$\begin{pmatrix} 2a_1 x_1 \\ 2x_2 \end{pmatrix} + \begin{pmatrix} x_2^2 \\ 2x_1 x_2 \end{pmatrix} \lambda + \begin{pmatrix} -1 \\ 0 \end{pmatrix} \mu = \begin{pmatrix} 0 \\ 0 \end{pmatrix}$$
$$x_1 x_2^2 - a_2 = 0$$
$$-x_1 + a_3 \leq 0 \quad\quad\quad (B.43)$$
$$\mu(-x_1 + a_3) = 0$$
$$\mu \geq 0,$$

and the corresponding solution, for the particular case $a_1 = a_3 = 1$ and $a_2 = 2$ is

$$x_1^* = 1, \ x_2^* = \sqrt{2}, \ \lambda^* = -1, \ \mu^* = 0, \ z^* = 3 \ . \quad\quad (B.44)$$

In Fig. B.7a the optimal solution and the feasible region of the problem (8.181)–(8.183) are shown. The graphical interpretation of the first equation

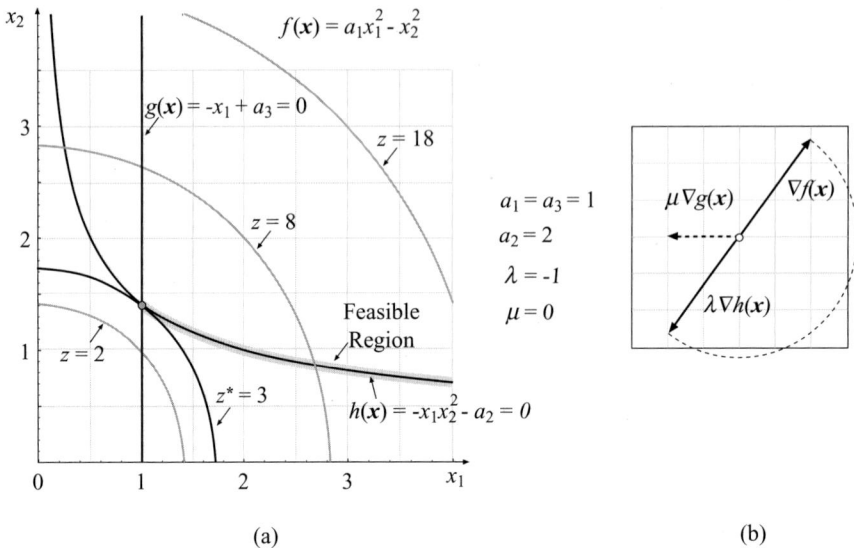

Fig. B.7. Illustration of: (**a**) the feasible region in the regular degenerate example, optimal values and (**b**) Karush–Kuhn–Tucker (KKT) multipliers

in (B.43) is shown in Fig. B.7b. Note that constraint $g(\boldsymbol{x})$ is not necessary for getting the optimal solution (B.44), this means that it could be removed and the same optimal solution would still remain.

A vector of changes

$$\delta\boldsymbol{p} = (dx_1, dx_2, da_1, da_2, da_3, d\lambda, d\mu, dz)^T$$

must satisfy the system (8.64)–(8.70), which for this example becomes

$$(2 \ \ 2\sqrt{2}) \begin{pmatrix} dx_1 \\ dx_2 \end{pmatrix} + (1 \ \ 0 \ \ 0) \begin{pmatrix} da_1 \\ da_2 \\ da_3 \end{pmatrix} - dz = 0 \quad \text{(B.45)}$$

$$\left(\begin{pmatrix} 2 & 0 \\ 0 & 2 \end{pmatrix} + \lambda \begin{pmatrix} 0 & 2\sqrt{2} \\ 2\sqrt{2} & 2 \end{pmatrix} + \mu \begin{pmatrix} 0 & 0 \\ 0 & 0 \end{pmatrix} \right) \begin{pmatrix} dx_1 \\ dx_2 \end{pmatrix}$$

$$+ \left(\begin{pmatrix} 2 & 0 & 0 \\ 0 & 0 & 0 \end{pmatrix} + \lambda \begin{pmatrix} 0 & 0 & 0 \\ 0 & 0 & 0 \end{pmatrix} + \mu \begin{pmatrix} 0 & 0 & 0 \\ 0 & 0 & 0 \end{pmatrix} \right) \begin{pmatrix} da_1 \\ da_2 \\ da_3 \end{pmatrix}$$

$$+ \begin{pmatrix} 2 \\ 2\sqrt{2} \end{pmatrix} d\lambda + \begin{pmatrix} -1 \\ 0 \end{pmatrix} d\mu = \begin{pmatrix} 0 \\ 0 \end{pmatrix} \quad \text{(B.46)}$$

$$(2 \ \ 2\sqrt{2}) \begin{pmatrix} dx_1 \\ dx_2 \end{pmatrix} + (0 \ \ -1 \ \ 0) \begin{pmatrix} da_1 \\ da_2 \\ da_3 \end{pmatrix} = (0) \quad \text{(B.47)}$$

$$(-1 \quad 0) \begin{pmatrix} dx_1 \\ dx_2 \end{pmatrix} + (0 \quad 0 \quad 1) \begin{pmatrix} da_1 \\ da_2 \\ da_3 \end{pmatrix} \leq (0) \quad \text{(B.48)}$$

$$d\mu \geq 0. \quad \text{(B.49)}$$

The M matrix in (8.73) particularized for this example is

$$M = \begin{bmatrix} 2 & 2\sqrt{2} & 1 & 0 & 0 & 0 & 0 & -1 \\ 2 & -2\sqrt{2} & 2 & 0 & 0 & 2 & -1 & 0 \\ -2\sqrt{2} & 0 & 0 & 0 & 0 & 2\sqrt{2} & 0 & 0 \\ 2 & 2\sqrt{2} & 0 & -1 & 0 & 0 & 0 & 0 \end{bmatrix}, \quad \text{(B.50)}$$

whereas the matrix N in (8.74) is

$$N = \begin{bmatrix} -1 & 0 & 0 & 0 & 1 & 0 & 0 & 0 \\ 0 & 0 & 0 & 0 & 0 & 0 & -1 & 0 \end{bmatrix}. \quad \text{(B.51)}$$

In this case, matrix U has no inverse because it is not a square matrix, the gradients of the constraints are linearly independent and one of the Lagrange multipliers is null, so we have a regular degenerate case. The system (8.73)–(8.74), using expressions (B.50) and (B.51), becomes

$$M\delta p = \begin{pmatrix} 2 & 2\sqrt{2} & 1 & 0 & 0 & 0 & 0 & -1 \\ 2 & -2\sqrt{2} & 2 & 0 & 0 & 2 & -1 & 0 \\ -2\sqrt{2} & 0 & 0 & 0 & 0 & 2\sqrt{2} & 0 & 0 \\ 2 & 2\sqrt{2} & 0 & -1 & 0 & 0 & 0 & 0 \end{pmatrix} \begin{pmatrix} dx_1 \\ dx_2 \\ da_1 \\ da_2 \\ da_3 \\ d\lambda \\ d\mu \\ dz \end{pmatrix} = \mathbf{0} \quad \text{(B.52)}$$

$$N\delta p = \begin{pmatrix} -1 & 0 & 0 & 0 & 1 & 0 & 0 & 0 \\ 0 & 0 & 0 & 0 & 0 & 0 & -1 & 0 \end{pmatrix} \begin{pmatrix} dx_1 \\ dx_2 \\ da_1 \\ da_2 \\ da_3 \\ d\lambda \\ d\mu \\ dz \end{pmatrix} \leq \mathbf{0}. \quad \text{(B.53)}$$

Note that we have not considered (8.70) yet. If we want (i) the inequality constraint to remain active the first equation in system (B.53) should be removed and included in system (B.52), whereas if we want (ii) the inequality constraint to be allowed to become inactive then the second equation in system (B.53) should be removed and included in system (B.52). The corresponding solutions are the cones

$$\begin{pmatrix} dx_1 \\ dx_2 \\ da_1 \\ da_2 \\ da_3 \\ d\lambda \\ d\mu \\ dz \end{pmatrix} = \rho_1 \begin{bmatrix} 1 \\ 0 \\ -2 \\ 2 \\ 1 \\ 1 \\ 0 \\ 0 \end{bmatrix} + \rho_2 \begin{bmatrix} 0 \\ \frac{1}{\sqrt{2}} \\ 1 \\ 2 \\ 0 \\ 0 \\ 0 \\ 3 \end{bmatrix} + \pi \begin{bmatrix} 0 \\ -\frac{1}{\sqrt{2}} \\ 2 \\ -2 \\ 0 \\ 0 \\ 6 \\ 0 \end{bmatrix} = \begin{bmatrix} \rho_1 \\ \frac{1}{\sqrt{2}}(\rho_2 - \pi) \\ -2\rho_1 + \rho_2 + 2\pi \\ 2(\rho_1 + \rho_2 - \pi) \\ \rho_1 \\ \rho_1 \\ 6\pi \\ 3\rho_2 \end{bmatrix} \quad (B.54)$$

and

$$\begin{pmatrix} dx_1 \\ dx_2 \\ da_1 \\ da_2 \\ da_3 \\ d\lambda \\ d\mu \\ dz \end{pmatrix} = \rho_1 \begin{bmatrix} 1 \\ 0 \\ -2 \\ 2 \\ 1 \\ 1 \\ 0 \\ 0 \end{bmatrix} + \rho_2 \begin{bmatrix} 0 \\ \frac{1}{\sqrt{2}} \\ 1 \\ 2 \\ 0 \\ 0 \\ 0 \\ 3 \end{bmatrix} + \pi \begin{bmatrix} 1 \\ 0 \\ -2 \\ 2 \\ 0 \\ 1 \\ 0 \\ 0 \end{bmatrix} = \begin{bmatrix} \rho_1 + \pi \\ \frac{1}{\sqrt{2}}\rho_2 \\ -2\rho_1 + \rho_2 - 2\pi \\ 2(\rho_1 + \rho_2 + \pi) \\ \rho_1 \\ \rho_1 + \pi \\ 0 \\ 3\rho_2 \end{bmatrix}, \quad (B.55)$$

respectively, where $\rho_1, \rho_2 \in \mathbb{R}$ and $\pi \in \mathbb{R}^+$, which give all feasible perturbations. Note, for example, that the component associated with $d\mu = (6\pi_1$ or $0)$ is always positive for (B.49) to hold, whereas the component related to the equality constraint $d\lambda = (\rho_1$ or $\rho_1 + \pi)$ can be positive or negative.

In order to study the existence of directional derivatives with respect to a_1 we use the directions $d\boldsymbol{a} = (1 \ 0 \ 0)^T$ and $d\boldsymbol{a} = (-1 \ 0 \ 0)^T$, and solve the two possible combinations of (8.84)–(8.85) that lead to

$$\begin{pmatrix} \frac{\partial x_1}{\partial a_1^+} \\ \frac{\partial x_2}{\partial a_1^+} \\ \frac{\partial \lambda}{\partial a_1^+} \\ \frac{\partial \mu}{\partial a_1^+} \\ \frac{\partial z}{\partial a_1^+} \end{pmatrix} = \begin{bmatrix} 0 \\ 0 \\ 0 \\ 2 \\ 1 \end{bmatrix}; \quad \begin{pmatrix} \frac{\partial x_1}{\partial a_1^-} \\ \frac{\partial x_2}{\partial a_1^-} \\ \frac{\partial \lambda}{\partial a_1^-} \\ \frac{\partial \mu}{\partial a_1^-} \\ \frac{\partial z}{\partial a_1^-} \end{pmatrix} = [\emptyset], \text{ and } \begin{pmatrix} \frac{\partial x_1}{\partial a_1^+} \\ \frac{\partial x_2}{\partial a_1^+} \\ \frac{\partial \lambda}{\partial a_1^+} \\ \frac{\partial \mu}{\partial a_1^+} \\ \frac{\partial z}{\partial a_1^+} \end{pmatrix} = [\emptyset]; \quad \begin{pmatrix} \frac{\partial x_1}{\partial a_1^-} \\ \frac{\partial x_2}{\partial a_1^-} \\ \frac{\partial \lambda}{\partial a_1^-} \\ \frac{\partial \mu}{\partial a_1^-} \\ \frac{\partial z}{\partial a_1^-} \end{pmatrix} = \begin{bmatrix} \frac{1}{3} \\ \frac{1}{3\sqrt{2}} \\ \frac{1}{3} \\ 0 \\ -1 \end{bmatrix},$$
(B.56)

respectively, where $[\emptyset]$ means that there is no solution, which implies that both directional derivatives exist (existence and uniqueness) but only the partial derivative of z with respect to a_1 exists $\dfrac{\partial z}{\partial a_1} = 1$. For the remaining variables the directional derivatives do not coincide in absolute value, therefore, the corresponding partial derivatives do not exist. Note that in the right-derivative the solution point remains the same but the Lagrange multiplier μ associated with the inequality constraint becomes different from zero as shown in

Fig. B.8a. For the left-derivative the solution point changes and the inequality constraint becomes inactive, note that the Lagrange multiplier associated with the equality constraint $h(\boldsymbol{x})$ changes but it is sufficient for getting a new optimal solution (see Fig. B.8b) whereas the one related to the inequality constraint remains equal to zero.

The directional derivatives with respect to a_2 are obtained using the directions $d\boldsymbol{a} = (0 \ \ 1 \ \ 0)^T$ and $d\boldsymbol{a} = (0 \ \ -1 \ \ 0)^T$, and solving the two possible combinations of (8.84)–(8.85) leading to

$$\begin{pmatrix} \frac{\partial x_1}{\partial a_2^+} \\ \frac{\partial x_2}{\partial a_2^+} \\ \frac{\partial \lambda}{\partial a_2^+} \\ \frac{\partial \mu}{\partial a_2^+} \\ \frac{\partial z}{\partial a_2^+} \end{pmatrix} = [\emptyset]; \quad \begin{pmatrix} \frac{\partial x_1}{\partial a_2^-} \\ \frac{\partial x_2}{\partial a_2^-} \\ \frac{\partial \lambda}{\partial a_2^-} \\ \frac{\partial \mu}{\partial a_2^-} \\ \frac{\partial z}{\partial a_2^-} \end{pmatrix} = \begin{bmatrix} 0 \\ -\frac{1}{2\sqrt{2}} \\ 0 \\ 1 \\ -1 \end{bmatrix}, \text{ and } \begin{pmatrix} \frac{\partial x_1}{\partial a_2^+} \\ \frac{\partial x_2}{\partial a_2^+} \\ \frac{\partial \lambda}{\partial a_2^+} \\ \frac{\partial \mu}{\partial a_2^+} \\ \frac{\partial z}{\partial a_2^+} \end{pmatrix} = \begin{bmatrix} \frac{1}{6} \\ \frac{1}{3\sqrt{2}} \\ \frac{1}{6} \\ 0 \\ 1 \end{bmatrix}, \quad \begin{pmatrix} \frac{\partial x_1}{\partial a_2^-} \\ \frac{\partial x_2}{\partial a_2^-} \\ \frac{\partial \lambda}{\partial a_2^-} \\ \frac{\partial \mu}{\partial a_2^-} \\ \frac{\partial z}{\partial a_2^-} \end{pmatrix} = [\emptyset],$$
(B.57)

respectively, which implies that both directional derivatives exist (existence and uniqueness) but only the partial derivative of z with respect to a_2 exists $\dfrac{\partial z}{\partial a_2} = 1$. For the remaining variables the partial derivatives do not exist. Note that in the right-derivative the solution point changes and the inequality constraint becomes inactive. The gradients of the objective and equality constraint remain with the same direction but different magnitude as shown in Fig. B.8c whereas for the left-derivative the solution point changes as well but the inequality constraint remains active with a Lagrange multiplier different from zero (see Fig. B.8d). Note that the inequality constraint forces the new solution point to move along its limit (see Fig. B.8d).

Analogously, the directional derivatives with respect to a_3 are obtained using the directions $d\boldsymbol{a} = (0 \ \ 0 \ \ 1)^T$ and $d\boldsymbol{a} = (0 \ \ 0 \ \ -1)^T$, and solving the two possible combinations of (8.84)–(8.85) leading to

$$\begin{pmatrix} \frac{\partial x_1}{\partial a_3^+} \\ \frac{\partial x_2}{\partial a_3^+} \\ \frac{\partial \lambda}{\partial a_3^+} \\ \frac{\partial \mu}{\partial a_3^+} \\ \frac{dz}{\partial a_3^+} \end{pmatrix} = \begin{bmatrix} 1 \\ -\frac{1}{\sqrt{2}} \\ 1 \\ 6 \\ 0 \end{bmatrix}; \quad \begin{pmatrix} \frac{\partial x_1}{\partial a_3^-} \\ \frac{\partial x_2}{\partial a_3^-} \\ \frac{\partial \lambda}{\partial a_3^-} \\ \frac{\partial \mu}{\partial a_3^-} \\ \frac{\partial z}{\partial a_3^-} \end{pmatrix} = [\emptyset], \text{ and } \begin{pmatrix} \frac{\partial x_1}{\partial a_3^+} \\ \frac{\partial x_2}{\partial a_3^+} \\ \frac{\partial \lambda}{\partial a_3^+} \\ \frac{\partial \mu}{\partial a_3^+} \\ \frac{\partial z}{\partial a_3^+} \end{pmatrix} = [\emptyset], \quad \begin{pmatrix} \frac{\partial x_1}{\partial a_3^-} \\ \frac{\partial x_2}{\partial a_3^-} \\ \frac{\partial \lambda}{\partial a_3^-} \\ \frac{\partial \mu}{\partial a_3^-} \\ \frac{\partial z}{\partial a_3^-} \end{pmatrix} = \begin{bmatrix} 0 \\ 0 \\ 0 \\ 0 \\ 0 \end{bmatrix},$$
(B.58)

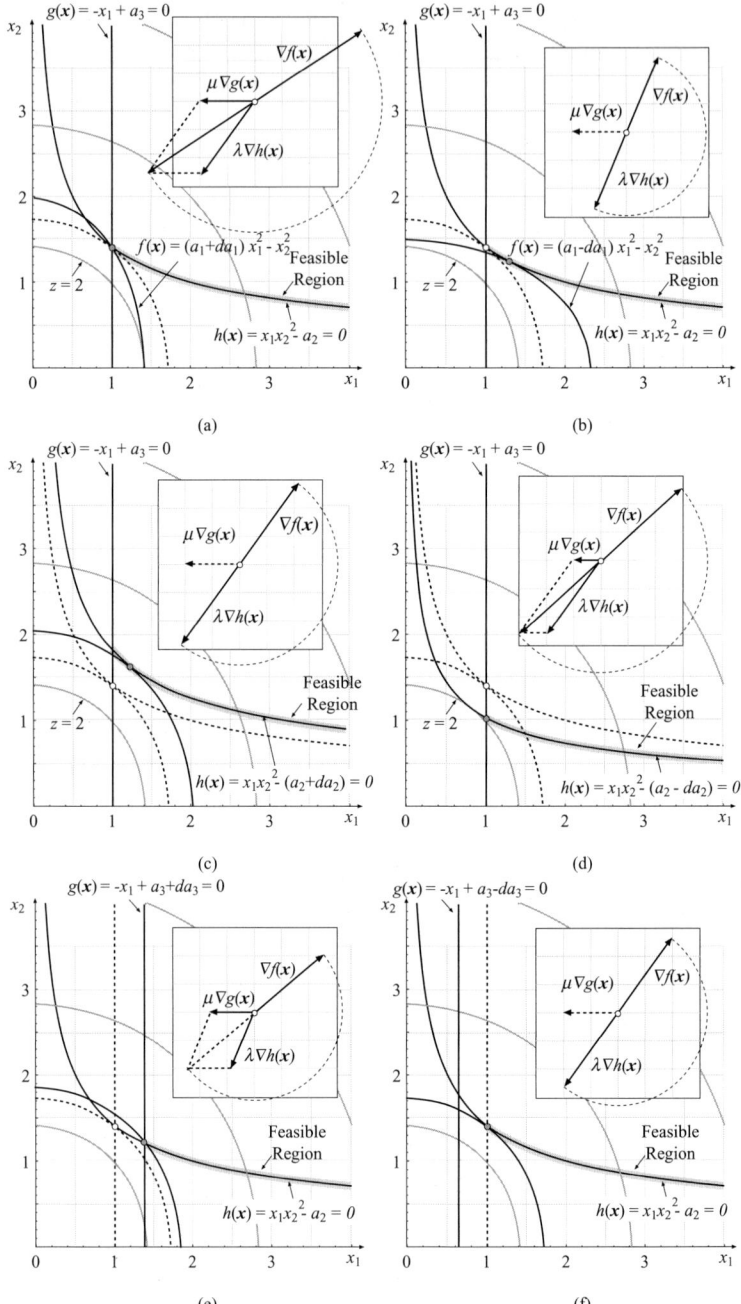

Fig. B.8. Illustration of the feasible regions, optimal values, and KKT multipliers of the modified problems due to changes in the parameters $\boldsymbol{a}$, for the regular degenerate case. (*Upper*) Positive and negative increments of a_1. (*Middle*) Positive and negative increments of a_2. (*Lower*) Positive and negative increments of a_3

respectively, which implies that both directional derivatives exist (existence and uniqueness) but only the partial derivative of z with respect to a_3 exists $\dfrac{\partial z}{\partial a_3} = 0$. For the remaining variables the partial derivatives do not exist. Note that in the right-derivative the solution point changes but the inequality constraint remains active with Lagrange multiplier different from cero. The inequality constraint forces the solution point to move to the right as shown in Fig. B.8e. For the left-derivative the solution point does not change and the inequality constraint becomes inactive. In Fig. B.8f it is shown how the feasible region moves left.

Solution to Exercise 8.14. The KKT conditions for this problem are

$$\begin{pmatrix} 2x_1 \\ 2x_2 \end{pmatrix} + \begin{pmatrix} -1 \\ 0 \end{pmatrix} \lambda + \begin{pmatrix} -1 & a_2 \\ -1 & -1 \end{pmatrix} \begin{pmatrix} \mu_1 \\ \mu_2 \end{pmatrix} = \begin{pmatrix} 0 \\ 0 \end{pmatrix}$$
$$-x_1 + a_1 = 0$$
$$-x_1 - x_2 + 2a_1 \le 0$$
$$a_2 x_1 - x_2 \le 0 \qquad (B.59)$$
$$\mu_1(-x_1 - x_2 + 2a_1) = 0$$
$$\mu_2(a_2 x_1 - x_2) = 0$$
$$\mu_1, \mu_2 \ge 0,$$

and the corresponding solution, for the particular case $a_1 = a_2 = 1$ are (see Fig. B.9a):

$$x_1 = x_2 = 1; \quad \mu_1 = \frac{4-\lambda}{2}; \quad \mu_2 = \frac{\lambda}{2}. \qquad (B.60)$$

Note that the dual problem has infinite solutions. Since the two inequality constraints are active, they will remain active or inactive in a neighborhood of the optimum depending on the values of the Lagrange multipliers. Then, a vector of changes

$$\delta p = (dx_1, dx_2, da_1, da_2, d\lambda, d\mu_1, d\mu_2, dz)^T$$

must satisfy the system (8.64)–(8.70), for all possible cases, the M matrix in (8.73) can be obtained from the following matrix:

$$M = \begin{bmatrix} 2 & 2 & 0 & 0 & 0 & 0 & 0 & -1 \\ 2 & 0 & 0 & \mu_2 & -1 & -1 & 1 & 0 \\ 0 & 2 & 0 & 0 & 0 & -1 & -1 & 0 \\ -1 & 0 & 1 & 0 & 0 & 0 & 0 & 0 \\ -1 & -1 & 2 & 0 & 0 & 0 & 0 & 0 \\ 1 & -1 & 0 & 1 & 0 & 0 & 0 & 0 \end{bmatrix}, \qquad (B.61)$$

by removing the rows corresponding to the null μ-multipliers, and the matrix N in (8.74) can be obtained from the matrix

522 B Exercise Solutions

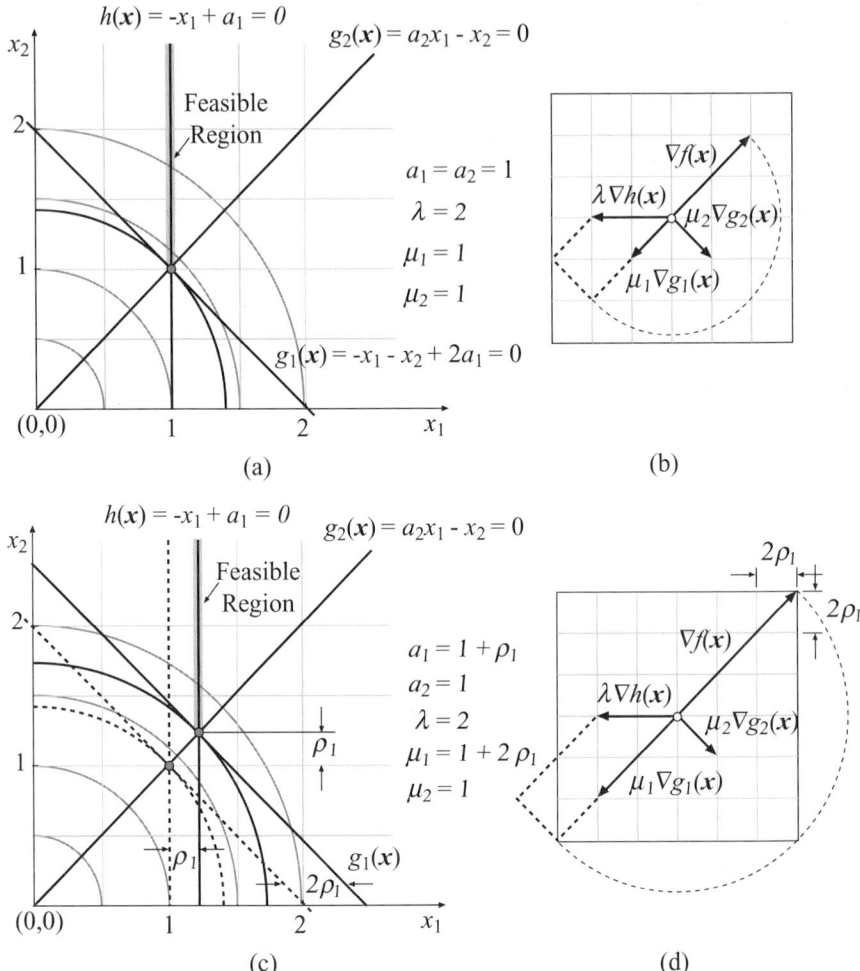

Fig. B.9. Illustration of the feasible regions in Case 1, optimal values and KKT multipliers of the initial and modified problems due to changes in the a_1 parameter. (*Upper*) Initial problem. (*Lower*) Positive increment of a_1

$$N = \begin{bmatrix} -1 & -1 & 2 & 0 & 0 & 0 & 0 & 0 \\ 1 & -1 & 0 & 1 & 0 & 0 & 0 & 0 \\ \hline 0 & 0 & 0 & 0 & 0 & -1 & 0 & 0 \\ 0 & 0 & 0 & 0 & 0 & 0 & -1 & 0 \end{bmatrix}, \qquad (B.62)$$

by removing the rows corresponding to the non-null μ-multipliers.

We analyze the only possible two different cases [see (B.60)]:

Case 1: $\mu_1, \mu_2 \neq 0$. For example $\lambda = 2$; $\mu_1 = 1$; $\mu_2 = 1$ [for a graphical interpretation of the first equation in system (B.59) see Fig. B.9b]. In this case, the matrix $\boldsymbol{U}$ is singular because the gradients of the active constraints are not linearly independent so we have a nonregular case. Since all μ-multipliers are non-null, the $\boldsymbol{N}$ matrix does not exist and the system (8.73)–(8.74), using expression (B.61), becomes

$$\boldsymbol{M}\delta\boldsymbol{p} = \begin{pmatrix} 2 & 2 & 0 & 0 & 0 & 0 & 0 & -1 \\ 2 & 0 & 0 & 1 & -1 & -1 & 1 & 0 \\ 0 & 2 & 0 & 0 & 0 & -1 & -1 & 0 \\ -1 & 0 & 1 & 0 & 0 & 0 & 0 & 0 \\ -1 & -1 & 2 & 0 & 0 & 0 & 0 & 0 \\ 1 & -1 & 0 & 1 & 0 & 0 & 0 & 0 \end{pmatrix} \begin{pmatrix} dx_1 \\ dx_2 \\ da_1 \\ da_2 \\ d\lambda \\ d\mu_1 \\ d\mu_2 \\ dz \end{pmatrix} = \boldsymbol{0}.$$

Note that in this example there is no need to consider (8.70) because the Lagrange multipliers are different from zero. The solution is the linear space

$$\begin{pmatrix} dx_1 \\ dx_2 \\ da_1 \\ da_2 \\ d\lambda \\ d\mu_1 \\ d\mu_2 \\ dz \end{pmatrix} = \rho_1 \begin{bmatrix} 1 \\ 1 \\ 1 \\ 0 \\ 0 \\ 2 \\ 0 \\ 4 \end{bmatrix} + \rho_2 \begin{bmatrix} 0 \\ 0 \\ 0 \\ 0 \\ 2 \\ -1 \\ 1 \\ 0 \end{bmatrix} = \begin{bmatrix} \rho_1 \\ \rho_1 \\ \rho_1 \\ 0 \\ 2\rho_2 \\ 2\rho_1 - \rho_2 \\ \rho_2 \\ 4\rho_1 \end{bmatrix}, \quad \rho_1, \rho_2 \in \mathbb{R}, \quad (B.63)$$

which gives all feasible perturbations. Note that the vector associated with ρ_2 corresponds to the feasible changes in the Lagrange multipliers owing to the linearly dependence on the constraint gradients (see Fig. B.10).
In Figs. B.9d the required changes in the Lagrange multipliers λ, μ_1, and μ_2 when a_1 is modified for the first equation in system (B.59) to hold, are shown.

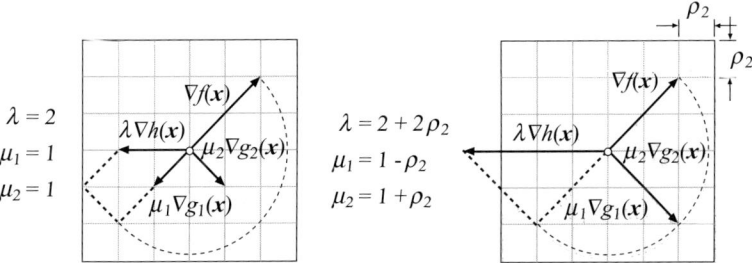

Fig. B.10. Case 1 (Nonregular point): Illustration of the feasible KKT multipliers changes due to changes in the ρ_2 parameter

In order to study the existence of partial derivatives with respect to a_1 we use the directions $d\boldsymbol{a} = (1\ \ 0)^T$ and $d\boldsymbol{a} = (-1\ \ 0)^T$, that imply [see (B.63)] $\rho_1 = 1$ and $\rho_1 = -1$, respectively, and

$$\begin{pmatrix} dx_1 \\ dx_2 \\ da_1 \\ da_2 \\ d\lambda \\ d\mu_1 \\ d\mu_2 \\ dz \end{pmatrix} = \begin{bmatrix} 1 \\ 1 \\ 1 \\ 0 \\ 0 \\ 2 \\ 0 \\ 4 \end{bmatrix} + \rho_2 \begin{bmatrix} 0 \\ 0 \\ 0 \\ 0 \\ 2 \\ -1 \\ 1 \\ 0 \end{bmatrix}, \quad \begin{pmatrix} dx_1 \\ dx_2 \\ da_1 \\ da_2 \\ d\lambda \\ d\mu_1 \\ d\mu_2 \\ dz \end{pmatrix} = \begin{bmatrix} -1 \\ -1 \\ -1 \\ 0 \\ 0 \\ -2 \\ 0 \\ -4 \end{bmatrix} + \rho_2 \begin{bmatrix} 0 \\ 0 \\ 0 \\ 0 \\ 2 \\ -1 \\ 1 \\ 0 \end{bmatrix}, \quad \text{(B.64)}$$

which implies that the following partial derivatives exist: $\dfrac{\partial x_1}{\partial a_1} = \dfrac{\partial x_2}{\partial a_1} = 1$, and $\dfrac{\partial z}{\partial a_1} = 4$ because they are unique (see Fig. B.9c). However, the partial derivatives $\dfrac{\partial \lambda}{\partial a_1}, \dfrac{\partial \mu_1}{\partial a_1}$, and $\dfrac{\partial \mu_2}{\partial a_1}$ do not exist, because the corresponding $d\lambda, d\mu_1, d\mu_2$ are not unique (they depend on the arbitrary real number ρ_2).

Alternatively, it is possible to consider the direction in which the desired partial derivative is looking for, $d\boldsymbol{a} = (1\ \ 0)^T$, and solve system (8.84)–(8.85) with $d\boldsymbol{a}$ and $-d\boldsymbol{a}$ leading to

$$\begin{pmatrix} \dfrac{\partial x_1}{\partial a_1^+} \\ \dfrac{\partial x_2}{\partial a_1^+} \\ \dfrac{\partial \lambda}{\partial a_1^+} \\ \dfrac{\partial \mu_1}{\partial a_1^+} \\ \dfrac{\partial \mu_2}{\partial a_1^+} \\ \dfrac{dz}{\partial a_1^+} \end{pmatrix} = \begin{bmatrix} 1 \\ 1 \\ 0 \\ 2 \\ 0 \\ 4 \end{bmatrix} + \rho_2 \begin{bmatrix} 0 \\ 0 \\ 2 \\ -1 \\ 1 \\ 0 \end{bmatrix}, \quad \begin{pmatrix} \dfrac{\partial x_1}{\partial a_1^-} \\ \dfrac{\partial x_2}{\partial a_1^-} \\ \dfrac{\partial \lambda}{\partial a_1^-} \\ \dfrac{\partial \mu_1}{\partial a_1^-} \\ \dfrac{\partial \mu_2}{\partial a_1^-} \\ \dfrac{dz}{\partial a_1^-} \end{pmatrix} = \begin{bmatrix} -1 \\ -1 \\ 0 \\ -2 \\ 0 \\ -4 \end{bmatrix} + \rho_2 \begin{bmatrix} 0 \\ 0 \\ 2 \\ -1 \\ 1 \\ 0 \end{bmatrix}.$$

(B.65)

As (8.85) does not exist in this case, this condition holds strictly and (B.65) provides the partial derivatives if the solution is unique. The partial derivatives obtained coincide with the ones obtained from (B.64), i.e., $\dfrac{\partial x_1}{\partial a_1} = \dfrac{\partial x_2}{\partial a_1} = 1$ and $\dfrac{\partial z}{\partial a_1} = 4$, whereas the partial derivatives $\dfrac{\partial \lambda}{\partial a_1}, \dfrac{\partial \mu_1}{\partial a_1}$, and $\dfrac{\partial \mu_2}{\partial a_1}$ do not exist, because they are not unique (they depend on the arbitrary real number ρ_2).

B.8 Exercises from Chapter 8

Since in system (B.63) $da_2 = 0$, the partial derivatives with respect to a_2 do not exist. The same result can be obtained considering the direction $d\boldsymbol{a} = (0\ \ 1)^T$ in (8.84) that has no solution, i.e., no derivative exists with respect to da_2.

Note that in this case the active constraints remain active (all μ multipliers are positive). This implies that the cone degenerates to a linear space.

Case 2: $\mu_1 = 0$, $\mu_2 \neq 0$. For example, $\lambda = 4$, $\mu_1 = 0$, $\mu_2 = 2$ (see Fig. B.11a or B.11b). In this case, the matrix $\boldsymbol{U}$ is singular because the gradients of the active constraints are not linearly independent, so we also have a nonregular case. The system (8.73)–(8.74), using expression (B.61) and (B.62), becomes

$$\mathbf{M}\delta\boldsymbol{p} = \begin{pmatrix} 2 & 2 & 0 & 0 & 0 & 0 & 0 & -1 \\ 2 & 0 & 0 & 2 & -1 & -1 & 1 & 0 \\ 0 & 2 & 0 & 0 & 0 & -1 & -1 & 0 \\ -1 & 0 & 1 & 0 & 0 & 0 & 0 & 0 \\ 1 & -1 & 0 & 1 & 0 & 0 & 0 & 0 \end{pmatrix} \begin{pmatrix} dx_1 \\ dx_2 \\ da_1 \\ da_2 \\ d\lambda \\ d\mu_1 \\ d\mu_2 \\ dz \end{pmatrix} = \mathbf{0} \quad \text{(B.66)}$$

$$\mathbf{N}\delta\boldsymbol{p} = \begin{pmatrix} -1 & -1 & 2 & 0 & 0 & 0 & 0 & 0 \\ 0 & 0 & 0 & 0 & 0 & -1 & 0 & 0 \end{pmatrix} \begin{pmatrix} dx_1 \\ dx_2 \\ da_1 \\ da_2 \\ d\lambda \\ d\mu_1 \\ d\mu_2 \\ dz \end{pmatrix} \leq \mathbf{0}. \quad \text{(B.67)}$$

Note that we have not considered (8.70) yet. If we want (i) the inequality constraint $g_1(\boldsymbol{x})$ to remain active the first equation in system (B.67) should be removed and included in system (B.66), whereas if we want (ii) the inequality constraint to be allowed to become inactive then the second equation in system (B.67) should be removed and included in system (B.66). The corresponding solutions are the cones

$$\begin{pmatrix} dx_1 \\ dx_2 \\ da_1 \\ da_2 \\ d\lambda \\ d\mu_1 \\ d\mu_2 \\ dz \end{pmatrix} = \rho \begin{bmatrix} 1 \\ 1 \\ 1 \\ 0 \\ 4 \\ 0 \\ 2 \\ 4 \end{bmatrix} + \pi \begin{bmatrix} 0 \\ 0 \\ 0 \\ 0 \\ -2 \\ 1 \\ -1 \\ 0 \end{bmatrix} = \begin{bmatrix} \rho \\ \rho \\ \rho \\ 0 \\ 4\rho - 2\pi \\ \pi \\ 2\rho - \pi \\ 4\rho \end{bmatrix}, \quad \text{(B.68)}$$

and

$$\begin{pmatrix} dx_1 \\ dx_2 \\ da_1 \\ da_2 \\ d\lambda \\ d\mu_1 \\ d\mu_2 \\ dz \end{pmatrix} = \rho \begin{bmatrix} 1 \\ 1 \\ 1 \\ 0 \\ 4 \\ 0 \\ 2 \\ 4 \end{bmatrix} + \pi \begin{bmatrix} -1 \\ 1 \\ -1 \\ 2 \\ 4 \\ 0 \\ 2 \\ 0 \end{bmatrix} = \begin{bmatrix} \rho - \pi \\ \rho + \pi \\ \rho - \pi \\ 2\pi \\ 4(\rho + \pi) \\ 0 \\ 2(\rho + \pi) \\ 4\rho \end{bmatrix}, \quad (B.69)$$

respectively, where $\rho \in \mathbb{R}$ and $\pi \in \mathbb{R}^+$. Analogously to the previous case, the vector associated with π for the first hypothesis corresponds to the feasible changes in the Lagrange multipliers owing to the linearly dependence on the constraint gradients but only positive increments are allowed because as $\mu_1 = 0$, a negative increment would imply a negative multiplier which does not hold system (B.59).

Figure B.11b shows a graphical interpretation of the first equation in system (B.59) particularized for this case. Note that constraint $g_1(\boldsymbol{x})$ is not necessary for getting the optimal solution (B.60), this means that it could be removed and the same optimal solution would still remain.

In order to study the existence of partial derivatives with respect to a_1 we use the directions $d\boldsymbol{a} = (1 \ 0)^T$ and $d\boldsymbol{a} = (-1 \ 0)^T$, that implies considering system (B.68), $\rho = 1$ and $\rho = -1$, respectively, leading to

$$\begin{pmatrix} dx_1 \\ dx_2 \\ da_1 \\ da_2 \\ d\lambda \\ d\mu_1 \\ d\mu_2 \\ dz \end{pmatrix} = \begin{bmatrix} 1 \\ 1 \\ 1 \\ 0 \\ 4 \\ 0 \\ 2 \\ 4 \end{bmatrix} + \pi \begin{bmatrix} 0 \\ 0 \\ 0 \\ 0 \\ -2 \\ 1 \\ -1 \\ 0 \end{bmatrix}, \quad \begin{pmatrix} dx_1 \\ dx_2 \\ da_1 \\ da_2 \\ d\lambda \\ d\mu_1 \\ d\mu_2 \\ dz \end{pmatrix} = \begin{bmatrix} -1 \\ -1 \\ -1 \\ 0 \\ -4 \\ 0 \\ -2 \\ -4 \end{bmatrix} + \pi \begin{bmatrix} 0 \\ 0 \\ 0 \\ 0 \\ -2 \\ 1 \\ -1 \\ 0 \end{bmatrix}, \quad (B.70)$$

and considering system (B.69), $\rho = 1, \pi = 0$, and $\rho = -1, \pi = 0$, respectively, leading to

$$\begin{pmatrix} dx_1 \\ dx_2 \\ da_1 \\ da_2 \\ d\lambda \\ d\mu_1 \\ d\mu_2 \\ dz \end{pmatrix} = \begin{bmatrix} 1 \\ 1 \\ 1 \\ 0 \\ 4 \\ 0 \\ 2 \\ 4 \end{bmatrix}, \quad \begin{pmatrix} dx_1 \\ dx_2 \\ da_1 \\ da_2 \\ d\lambda \\ d\mu_1 \\ d\mu_2 \\ dz \end{pmatrix} = \begin{bmatrix} -1 \\ -1 \\ -1 \\ 0 \\ -4 \\ 0 \\ -2 \\ -4 \end{bmatrix}, \quad (B.71)$$

respectively, which imply that the following partial derivatives exist: $\dfrac{\partial x_1}{\partial a_1} = \dfrac{\partial x_2}{\partial a_1} = 1$ and $\dfrac{\partial z}{\partial a_1} = 4$, because they are unique and have the same

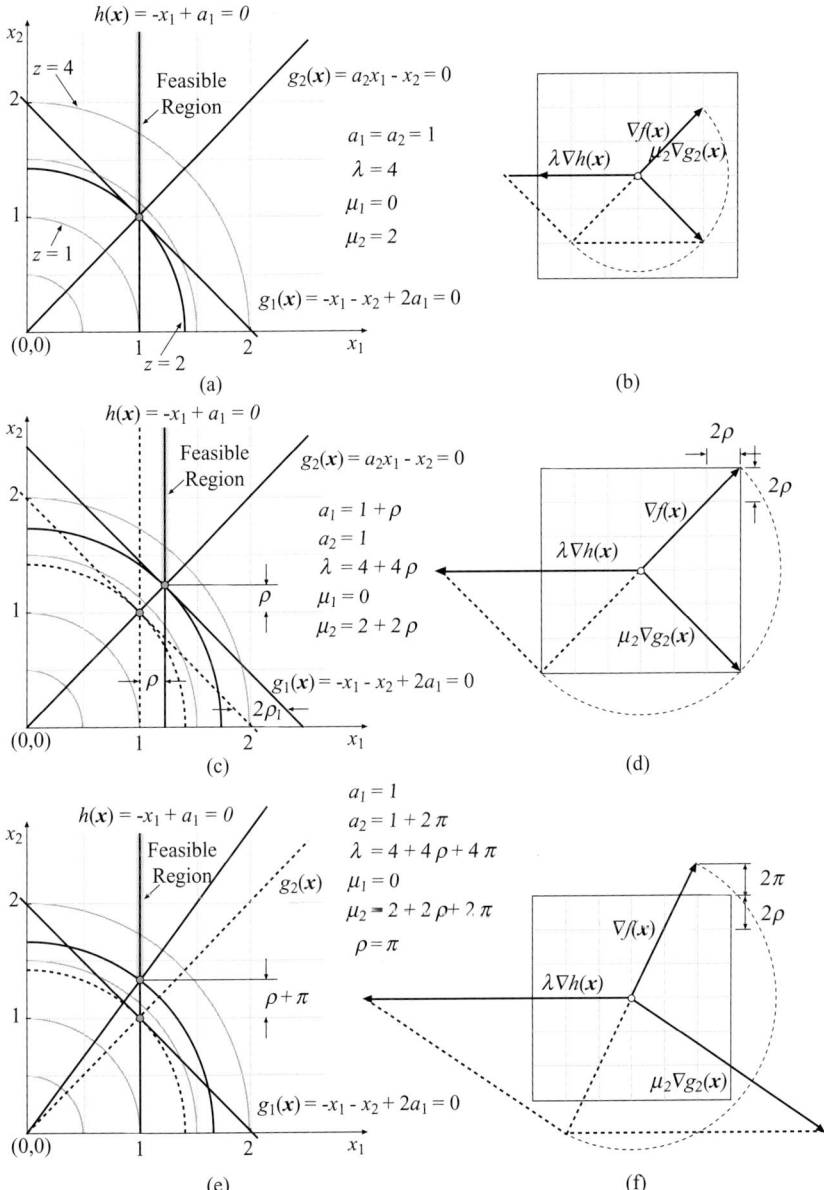

Fig. B.11. Illustration of Case 2 feasible regions, optimal values, and KKT multipliers of the initial and modified problems due to changes in the a_1 and a_2 parameters, respectively. (*Upper*) Initial problem. (*Middle*) Positive increment of a_1. (*Lower*) Positive increment of a_2

absolute value and different sign (see Fig. B.11c). However, the partial derivatives $\dfrac{\partial \lambda}{\partial a_1}$, $\dfrac{\partial \mu_1}{\partial a_1}$, and $\dfrac{\partial \mu_2}{\partial a_1}$ do not exist, because the corresponding $d\lambda$, $d\mu_1$, $d\mu_2$ are not unique [they depend on the arbitrary real number π in system (B.70)].

Alternatively, it is possible to consider the directions in which the desired directional derivatives are looked for, $d\boldsymbol{a} = (1 \quad 0)^T$ and $d\boldsymbol{a} = (-1 \quad 0)^T$, and solve the two possible combinations of system (8.84)–(8.85) leading to

$$\begin{pmatrix} \dfrac{\partial x_1}{\partial a_1^+} \\ \dfrac{\partial x_2}{\partial a_1^+} \\ \dfrac{\partial \lambda}{\partial a_1^+} \\ \dfrac{\partial \mu_1}{\partial a_1^+} \\ \dfrac{\partial \mu_2}{\partial a_1^+} \\ \dfrac{\partial z}{\partial a_1^+} \end{pmatrix} = \begin{bmatrix} 1 \\ 1 \\ 4 \\ 0 \\ 2 \\ 4 \end{bmatrix} + \pi \begin{bmatrix} 0 \\ 0 \\ -2 \\ 1 \\ -1 \\ 0 \end{bmatrix}, \quad \begin{pmatrix} \dfrac{\partial x_1}{\partial a_1^-} \\ \dfrac{\partial x_2}{\partial a_1^-} \\ \dfrac{\partial \lambda}{\partial a_1^-} \\ \dfrac{\partial \mu_1}{\partial a_1^-} \\ \dfrac{\partial \mu_2}{\partial a_1^-} \\ \dfrac{\partial z}{\partial a_1^-} \end{pmatrix} = \begin{bmatrix} -1 \\ -1 \\ -4 \\ 0 \\ -2 \\ -4 \end{bmatrix} + \pi \begin{bmatrix} 0 \\ 0 \\ -2 \\ 1 \\ -1 \\ 0 \end{bmatrix}, \quad \pi \in \mathbb{R}^+,$$

(B.72)

and

$$\begin{pmatrix} \dfrac{\partial x_1}{\partial a_1^+} \\ \dfrac{\partial x_2}{\partial a_1^+} \\ \dfrac{\partial \lambda}{\partial a_1^+} \\ \dfrac{\partial \mu_1}{\partial a_1^+} \\ \dfrac{\partial \mu_2}{\partial a_1^+} \\ \dfrac{\partial z}{\partial a_1^+} \end{pmatrix} = \begin{bmatrix} 1 \\ 1 \\ 4 \\ 0 \\ 2 \\ 4 \end{bmatrix}, \quad \begin{pmatrix} \dfrac{\partial x_1}{\partial a_1^-} \\ \dfrac{\partial x_2}{\partial a_1^-} \\ \dfrac{\partial \lambda}{\partial a_1^-} \\ \dfrac{\partial \mu_1}{\partial a_1^-} \\ \dfrac{\partial \mu_2}{\partial a_1^-} \\ \dfrac{\partial z}{\partial a_1^-} \end{pmatrix} = \begin{bmatrix} -1 \\ -1 \\ -4 \\ 0 \\ -2 \\ -4 \end{bmatrix},$$

(B.73)

where the same results as in system (B.70) and (B.71) are obtained.

In order to study the existence of partial derivatives with respect to a_2, we use the directions $d\boldsymbol{a} = (0 \quad 1)^T$ and $d\boldsymbol{a} = (0 \quad -1)^T$, which implies considering system (B.69), where $\rho = \pi = 1/2$ and as the value of π can just be positive, it is not possible to get $da_2 = -1$ from neither system (B.68) nor (B.69) and then no partial derivatives exist with respect to a_2. Therefore, as $\pi > 0$ only right-derivatives can exist:

$$\begin{pmatrix} dx_1 \\ dx_2 \\ da_1 \\ da_2 \\ d\lambda \\ d\mu_1 \\ d\mu_2 \\ dz \end{pmatrix} = \begin{bmatrix} 0 \\ 1 \\ 0 \\ 1 \\ 4 \\ 0 \\ 2 \\ 2 \end{bmatrix}, \qquad (B.74)$$

which implies: $\dfrac{\partial x_1}{\partial a_2^+} = 0$, $\dfrac{\partial x_2}{\partial a_2^+} = 1$, $\dfrac{\partial \lambda}{\partial a_2^+} = 4$, $\dfrac{\partial \mu_1}{\partial a_2^+} = 0$, $\dfrac{\partial \mu_2}{\partial a_2^+} = 2$, and $\dfrac{\partial z}{\partial a_2^+} = 2$ because they are unique (see Figs. B.11e and B.11f).

Alternatively, if we try to solve (8.84)–(8.85) using $d\bm{a} = (0 \;\; 1)^T$ and $d\bm{a} = (0 \;\; -1)^T$:

$$\begin{pmatrix} \frac{\partial x_1}{\partial a_2^+} \\ \frac{\partial x_2}{\partial a_2^+} \\ \frac{\partial \lambda}{\partial a_2^+} \\ \frac{\partial \mu_1}{\partial a_2^+} \\ \frac{\partial \mu_2}{\partial a_2^+} \\ \frac{\partial z}{\partial a_2^+} \end{pmatrix} = \pi \begin{bmatrix} 0 \\ 0 \\ -2 \\ 1 \\ -1 \\ 0 \end{bmatrix}, \; \begin{pmatrix} \frac{\partial x_1}{\partial a_2^-} \\ \frac{\partial x_2}{\partial a_2^-} \\ \frac{\partial \lambda}{\partial a_2^-} \\ \frac{\partial \mu_1}{\partial a_2^-} \\ \frac{\partial \mu_2}{\partial a_2^-} \\ \frac{dz}{\partial a_2^-} \end{pmatrix} = \pi \begin{bmatrix} 0 \\ 0 \\ -2 \\ 1 \\ -1 \\ 0 \end{bmatrix}, \; \begin{pmatrix} \frac{\partial x_1}{\partial a_2^+} \\ \frac{\partial x_2}{\partial a_2^+} \\ \frac{\partial \lambda}{\partial a_2^+} \\ \frac{\partial \mu_1}{\partial a_2^+} \\ \frac{\partial \mu_2}{\partial a_2^+} \\ \frac{\partial z}{\partial a_2^+} \end{pmatrix} = \begin{bmatrix} 0 \\ 1 \\ 4 \\ 0 \\ 2 \\ 2 \end{bmatrix}, \; \begin{pmatrix} \frac{\partial x_1}{\partial a_2^-} \\ \frac{\partial x_2}{\partial a_2^-} \\ \frac{\partial \lambda}{\partial a_2^-} \\ \frac{\partial \mu_1}{\partial a_2^-} \\ \frac{\partial \mu_2}{\partial a_2^-} \\ \frac{\partial z}{\partial a_2^-} \end{pmatrix} = [\emptyset],$$

(B.75)

where the same results as in system (B.74) are obtained.

In Figs. B.11d and B.11f the required changes in the Lagrange multipliers λ and μ_2 when a_1 and a_2 are modified, respectively, for the first equation in system (B.59) to hold, are shown.

Note that in this example constraint g_1 becomes inactive.

References

1. Bradley, S. P., Hax, A. C., and Magnanti, T. L. *Applied Mathematical Programming*. Addison-Wesley Publishing Company, Reading, MA, 1997.
2. Dantzig, G. B. *Linear Programming and Extensions*. Princeton University Press, Princeton, 1963.
3. Rockafellar, R. T. and Wets, R. J. B. Scenario and policy aggregation in optimization under uncertainty. *Mathematicals Operation Research 16* (1991), 119–147.
4. Escudero, L. F., de la Fuente, J. L., García, C., and Prieto, F. J. Hydropower generation management under uncertainty via scenario analysis and parallel computation. *IEEE Transactions on Power System 11*, 2 (1996), 683–689.
5. Bazaraa, M. S., Jarvis, J. J., and Sherali, H. D. *Linear Programming and Network Flows*, 2nd ed. John Wiley & Sons, New York, 1990.
6. Bloom, J. A. and Gallant, L. Modeling dispatch constraints in production cost simulations based on the equivalent load method. *IEEE Transactions on Power System 9*, 2 (1994), 598–611.
7. Edmonds, J. Submodular functions, matroids, and certain polyhedra. Combinatorial structure and their applications. In: *Proceedings of the Calgary International Conference* (New York, 1969), Gordon and Breach, pp. 69–87.
8. Nemhauser, G. L. and Wolsey, L. A. *Integer and Combinatorial Optimization*. John Wiley & Sons, New York, 1999.
9. Pérez-Ruiz, J. and Conejo, A. J. Multi-period probabilistic production cost model including dispatch constraints. *IEEE Transactions on Power System 15*, 2 (2000), 502–507.
10. Wallace, S. and Kall, P. *Stochastic Programming*. John Wiley & Sons, New York, 1995.
11. Higle, J. L. and Sen, S. *Stochastic Decomposition: A Statistical Method for Large Scale Stochastic Linear Programming*. Kluwer Academic Publisher: Dordrecht, The Netherlands, 1996.
12. Birge, J. R. and Louveaux, F. *Introduction to Stochastic Programming*. Springer-Verlag, New York, 1997.
13. Mínguez, R. *Seguridad, Fiabilidad y Análisis de Sensibilidad en Obras de Ingeniería Civil Mediante Técnicas de Optimización por Descomposición. Aplicaciones*. PhD thesis, University of Cantabria, Santander, Spain, 2003.

14. Mínguez, R., Castillo, E., and Hadi, A. S. Solving the inverse reliability problem using decomposition techniques. *Structural Safety, ASCE* **27** (2005), 1–23.
15. Castillo, E., Losada, M., Mínguez, R., Castillo, C., and Baquerizo, A. An optimal engineering design method that combines safety factors and failures probabilities: Application to rubble-mound breakwaters. *Journal of Waterways, Ports, Coastal and Ocean Engineering, ASCE* **130**, 2 (2004), 77–88.
16. Castillo, E., Mínguez, R., Ruíz-Terán, A., and Fernández-Canteli, A. Design and sensitivity analysis using the probability-safety-factor method. An application to retaining walls. *Structural Safety* **26** (2003), 159–179.
17. Castillo, E., Mínguez, R., Ruíz-Terán, A., and Fernández-Canteli, A. Design of a composite beam using the probability-safety-factor method. *International Journal for Numerical Methods in Engineering* **62** (2005), 1148–1182.
18. Castillo, C., Losada, M. A., Castillo, E., and Mínguez, R. Técnicas de optimización aplicadas al diseño de obras marítimas. In: *VII Jornadas de Ingeniería de Costas y Puertos*. Almería, España, 2003, pp. 27–30.
19. Castillo, C., Losada, M., Mínguez, R., and Castillo, E. Técnicas de optimización aplicadas al diseño de obras marítimas. In: *Procedimiento Metodológico Participativo para la Canalización, Recogida y Difusión de Estudios y Análisis Técnico-Científicos sobre los Documentos del Programa ROM, EROM 00* (2003), EROM, Puertos del Estado, Ministerio de Fomento.
20. Bazaraa, M. S., Sherali, H. D., and Shetty, C. M. *Nonlinear Programming, Theory and Algorithms*, 2nd ed. John Wiley & Sons, New York, 1993.
21. Castillo, E., Conejo, A., Pedregal, P., García, R., and Alguacil, N. *Building and Solving Mathematical Programming Models in Engineering and Science*. John Wiley & Sons, Inc., New York, 2001. Pure and Applied Mathematics: A Wiley-Interscience Series of Texts, Monographs and Tracts.
22. Chvátal, V. *Linear Programming*. W. H. Freeman and Company, New York, 1983.
23. Luenberger, D. G. *Linear and Nonlinear Programming*, 2nd ed. Addison-Wesley, Reading, MA, 1984.
24. Benders, J. F. Partitioning procedures for solving mixed-variables programming problems. *Numerische Mathematik 4* (1962), 238–252.
25. Geoffrion, A. M. Generalized Benders decomposition. *Jaurnal of Optimization Theory and Applications 10*, 4 (1972), 237–260.
26. Lasdon, L. S. *Optimization Theory for Large Systems*. MacMillan, New York, 1970.
27. Floudas, C. A. *Nonlinear and Mixed-Integer Optimization. Fundamentals and Applications*. Oxford University Press, New York, 1995.
28. Golub, G. B. and Van Loan, C. F. *Matrix Computations*, 3rd ed. The Johns Hopkins University Press, USA, 1996.
29. Walras, L. *Elements of Pure Economics, or, The Theory of Royal Wealth*. American Economic Association and the Royal Economy Society by Irwin, R. D., Homewood, 1874. Translated into english by William Jaffé from: "Élements d'Économie Politique Pure; ou la Théorie de la Richesse Sociale" in 1954.
30. Bertsekas, D. P., Lauer, G. S., Sandell, N. R., and Posbergh, T. A. Optimal short-term scheduling of large-scale power systems. *IEEE Transactions on Automatic Control 28*, 1 (1983), 1–11.
31. Everett, H. Generalized Lagrange multiplier method for solving problems of optimum allocation of resources. *Operations Researchs* **11** (1963), 399–417.

32. Zhuang, F. and Galiana, F. D. Toward a more rigorous and practical unit commitment by lagrangian relaxation. *IEEE Transactions on Power Systems 3*, 2 (1988), 763–773.
33. Polyak, B. T. *Introduction to Optimization.* Optimization Software, Inc., New York, 1987.
34. Redondo, N. J. and Conejo, A. J. Short-term hydro-thermal coordination by Lagrangian relaxation: Solution of the dual problem. *IEEE Transactions on Power Systems 14*, 1 (1999), 89–95.
35. Bertsekas, D. P. *Nonlinear Programming.* Athena Scientific, Belmont, MA, 1995.
36. Wu, Y., Debs, A. S., and Marsten, R. E. A direct nonlinear predictor-corrector primaldual interior point algorithm for optimal power flow. *IEEE Transactions on Power Systems 9*, 2 (1994), 876–883.
37. Cohen, G. Auxiliary problem principle and decomposition of optimization problems. *Journal of Optimization Theory and Applications 32*, 3 (1980), 277–305.
38. Bertsekas, D. *Constrained Optimization and Lagrange Multiplier Methods.* Academic Press, New York, 1982.
39. Conejo, A. J., Nogales, F. J., and Prieto, F. J. A decomposition procedure based on approximate newton directions. *Mathematical Programming 93*, 3 (2002), 495–515.
40. Saad, Y. *Iterative Methods for Sparse Linear Systems.* PWS Publishing, New York, 1996.
41. Brooke, A., Kendrick, D., and Meeraus, A. *Release 2.25 GAMS A User's Guide.* South San Francisco, 1992.
42. Bixby, R. E. Solving real-world linear programs: A decade and more of progress. *Operations Research 50*, 50 (2002), 3–15.
43. Barnhart, C., Johnson, E., Nemhauser, G. L., Savelsbergh, M., and Vance, P. Branch-and-price: Column generation for solving huge integer programs. *Operations Researh 46*, 3 (1998), 316–329.
44. Geoffrion, A. M. and Graves, G. W. Multicommodity distribution system-design by Benders decomposition. *Management Science 20*, 5 (1974), 822–844.
45. Kelley, J. E. The cutting-plane method for solving convex programs. *Journal of the SIAM 8*, 4 (1960), 703–712.
46. Pereira, M. V. F. and Pinto, L. M. V. G. Multistage stochastic optimization applied to energy planning. *Mathematical Programming 52*, 2 (1991), 359–375.
47. Haffner, S., Monticelli, A., García, A., Mantovani, J., and Romero, R. Branch and bound algorithm for transmission system expansion planning using a transportation model. *IEEE Proceedings-Generations Transmission and Distribution.* (2000), 149–156.
48. Romero, R. and Monticelli, A. A hierarchical decomposition approach for transmission network expansion planning. *IEEE Transactions on Power Systems 9*, 1 (1994), 373–380.
49. Bloom, J. A. Solving an electricity generating capacity expansion planning problem by generalized Benders decomposition. *Operations Research 31*, 1 (1983), 84–100.
50. Bloom, J. A., Caramanis, M., and Charny, L. Long-range generation planning using generalized Benders decomposition Implementation and experience. *Operations Research 32*, 2 (1983), 290–313.

51. Castillo, E., Conejo, A., Mínguez, R., and Castillo, C. An alternative approach for addressing the failure probability-safety factor method with sensitivity analysis. *Reliability Engineering and System Safety 82* (2003), 207–216.
52. Dempe, S. *Foundations of Bilevel Programming*. Kluwer Academic Publishers: Dordrecht, The Netherlands, 2002.
53. Bard, J. F. *Practical Bilevel Optimization: Algorithms and Applications*. Kluwer Academic Publishers: Dordrecht, The Netherlands, 1998.
54. Arroyo, J. M. and Galiana, F. D. On the solution of the bilevel programming formulation of the terrorist threat problem. *IEEE Transactions on Power Systems 20*, 2 (2005), 789–797.
55. Castillo, E., Conejo, A., Castillo, C., Mínguez, R., and Ortigosa, D. A perturbation approach to sensitivity analysis in nonlinear programming. *Journal of Optimization Theory and Applications 128*, 1 (2006).
56. Castillo, E., Cobo, A., Jubete, F., and Pruneda, R. E. *Orthogonal Sets and Polar Methods in Linear Algebra: Applications to Matrix Calculations, Systems of Equations and Inequalities, and Linear Programming*. John Wiley & Sons, New York, 1999.
57. Castillo, E., Jubete, F., Pruneda, E., and Solares, C. Obtaining simultaneous solutions of linear subsystems of equations and inequalities. *Linear Algebra and its Applications*, 346 (2002), 131–154.
58. Castillo, E., Conejo, A., Mínguez, R., and Castillo, C. A closed formula for local sensitivity analysis in mathematical programming. *Engineering Optimization* (2005) (in press).
59. Padberg, M. *Linear Optimization and Extensions*. Springer, Berlin, Germany, 1995.
60. Castillo, E. and Jubete, F. The γ-algorithm and some applications. *International Journal of Mathematical Education in Science and Technology 35*, 3 (2004), 369–389.
61. Press, W. H., Teukolsky, S. A., Vetterling, W. T., and Flannery, B. P. *Numerical Recipes in C*. Cambridge University Press, New York, 1992.
62. Murty, K. G. *Linear Programming*. John Wiley & Sons, New York, 1983.
63. Rosenblatt, M. Remarks on a multivariate transformation. *Annals Mathematical Statistics 23*, 3 (1952), 470–472.
64. Carpentier, J. Optimal power flows (survey). *Electric Power & Energy Systems 1*, 1 (1979), 142–154.
65. Stott, B., Alsac, O., and Monticelli, A. Security analysis and optimization. In: *Proceedings of the IEEE, Special Issues on Computers in Power Systems* (1987), Vol. 75, pp. 1623–1644.
66. Alsac, O., Bright, J., Prais, M., and Stott, B. Further development in LP-based optimal power flow. *IEEE Transactions on Power Systems 5*, 3 (1990), 697–711.
67. Merlin, A. and Sandrin, P. A new method for unit commitment at electricité de france. *IEEE Transactions on Power Apparatus and Systems PAS 102* (1983), 1218–1225.
68. Ferreira, L. A. F. M., Anderson, T., Imparato, C. F., Miller, T. E., Pang, C. K., Svoboda, A., and Vojdani, A. F. Short-term resource scheduling in multi-area hydrothermal power system. *Electric Power & Energy Systems 11*, 3 (1989), 200–212.
69. Wang, S. J., Shahidehpour, S. M., Kirschen, D. S. Mokhtari, S., and Irisarri, G. D. Short-term generation scheduling with transmission and environmental

constraints using an augmented lagrangian relaxation. *IEEE Transactions on Power Systems 10*, 3 (1995), 1294–1301.
70. Sheblé, G. B. and Fahd, G. N. Unit commitment literature synopsis. *IEEE Transactions on Power Systems 9*, 1 (1994), 128–135.
71. Schaw, J. A direct method for security-constrained unit commitment. *IEEE Transactions on Power Systems 10*, 3 (1995), 1329–1342.
72. Ruzic, S. and Rajakovic, N. A new approach for solving extended unit commitment problem. *IEEE Transactions on Power Systems 6*, 1 (1991), 269–277.
73. Batut, J., and Renaud, A. Daily generation scheduling optimization with transmission constraints: A new class of problems. *IEEE Transactions on Power Systems 7*, 3 (1992), 982–989.
74. Alguacil, N. and Conejo, A. J. Multi-period optimal power flow using Benders decomposition. *IEEE Transactions on Power Systems 15*, 1 (2000), 196–201.
75. Balériaux, H., Jamoulle, E., and Linard de Gertechin, F. Simulation de l'explotation d'un parc de machines thermiques de production d'lectricit coupl des stations de ponpage. *Revue E. Societé Royale Belge des Electriciens 7*, 3 (1967), 225–245.
76. Bloom, J. A. and Charny, L. Long range generation planning with limited energy and storage plants. part I: Production costing. *IEEE Transactions on Power Apparatus and Systems PAS 102*, 9 (1983), 2861–2870.
77. Booth, R. R. Power system simulation model based on probability analysis. *IEEE Transactions on Power Systems PAS 91*, 1 (1972), 62–69.
78. Rakic, M. V. and Marcovic, Z. M. Short term operation and power exchange planning of hydro-thermal power systems. *IEEE Transactions on Power Systems 9*, 1 (1994), 359–365.
79. Yan, H., Luh, P. B., Guan, X., and Rogan, P. M. Scheduling of hydrothermal power systems. *IEEE Transactions on Power Systems 8*, 3 (1993), 1135–1365.
80. Wood, A. J. and Wollenberg, B. F. *Power Generation, Operation and Control*, 2nd ed. John Wiley & Sons, New York, 1996.
81. Jimónez, N. and Conejo, A. J. Short-term hydro-thermal coordination by lagrangian relaxation: Solution of the dual problem. *IEEE Transactions on Power Systems 14*, 1 (1999), 89–95.
82. Nogales, J., Conejo, A. J., and Prieto, F. J. A decomposition methodology applied to the multi-area optimal power flow problem. *Annals of Operations Research 120*, 1–4 (2003), 99–116.
83. Fiacco, A. V. Introduction to sensitivity and stability analysis in nonlinear programming. Academic Press, New York, 1983.

Index

Active constraints 147
Algorithm
   augmented Lagrangian decomposition 207
   augmented Lagrangian relaxation 207
   Benders decomposition 116, 225
     GAMS code 403
     example 118
     for MILP 245
   coordinate descent decomposition 285
   cutting plane 278
   Dantzig-Wolfe decomposition 77, 83
     GAMS code 397
   Lagrangian relaxation 194
   optimality condition decomposition 216, 389
   outer linearization 258, 259
     convergence 264
   relaxation method 272
   updating safety factor bounds 355
All sensitivities at once 321
Alternative formulation 93
Analysis of structures 288
Applications 62, 69, 93, 99, 158, 245, 251, 264, 303, 349
Augmented Lagrangian 187, 205
   decomposition 205
Augmented Lagrangian decomposition 187, 207
   algorithm 207
Augmented Lagrangian relaxation 222
   algorithm 207
   multiplier updating 208
   penalty parameter updating 208
   separability 208

Banded matrix structure 287, 291
Benders decomposition 30, 111, 223, 369
   algorithm
     for MILP 245
   algorithm 61, 114, 116, 225, 251, 264
     example 118
     for MINLP 251
   application 368
   bounds 116
   convergence 250, 257, 369
   GAMS code 403
   MINLP 251
   mixed-integer linear programming 245
   scheme 245
Benders description 111
Bi-level problems 47, 52
Bilevel decomposition 271
Bilevel problems 353
Bilevel programming 280
Binary variables 245
Bounds 87, 116
   Benders decomposition 116
   Dantzig-Wolfe 87
   updating safety factors
     method 1 355
Bridge crane design 361

Index

Bundle method 199

Capacity expansion planning 4, 32, 33, 53, 54, 57
  revisited 2 53, 57
Civil 62
Classical design 46
Coal and gas procurement 28
Communication net 158
Complementary slackness condition 143
Complicating constraints 5, 7, 8, 12, 18, 40, 67, 68, 72, 74, 99, 109, 110, 136, 187, 188, 214, 216, 234, 243, 251, 271, 377
  linear programming 8
  mixed-integer programming 55
  nonlinear case 257
  nonlinear programming 39
  problem structure 70
Complicating constraints and variables 136
Complicating constraints that prevent a distributed solution 67
Complicating constraints that prevent an efficient solution 69
Complicating variables 6–8, 30, 35, 37, 38, 54, 57, 59, 67, 109–111, 135, 223, 224, 239, 243, 245, 251, 264
  dual problem 110
  linear programming 28, 107
  mixed-integer programming 57
  nonlinear case 251
  nonlinear programming 53
  prevent a distributed solution 107
  preventing a straightforward solution 108
  problem structure 110
Conductance 43
Convergence 250, 257, 264
Convergence properties 216–218, 369
Convex 164, 173, 244, 251
Convex combination 74, 99, 378, 379
  constraint 77
  linear 74, 98
Convex linear combination 93
Convexity 6, 164
Coordinate descent decomposition 285
  algorithm 285

Cutting hyperplane 257, 279, 359, 414
  example 278
Cutting plane
  algorithm 278
Cutting plane method 197
Cutting planes 359

Dantzig-Wolfe 18, 21
  bounds 87
Dantzig-Wolfe decomposition 233, 243
  algorithm 77, 83, 99, 235, 239, 397
  example 79
  example revisited 89
  GAMS code 79
  geometric interpretation 83
  master problem 237
  method 77
  procedure 136
  technique 233, 374
Decentralized 5, 8, 67, 181, 218, 385, 386
Decision variables 3
Decomposable structure 3, 7
Decomposed 5, 12, 41, 44, 54, 59, 72, 98, 189, 214, 219, 220, 243, 384, 388, 428, 434, 455
Decomposition 73, 188
  Benders 61, 111, 223
  Dantzig-Wolfe 18, 21, 77
  in linear programming 67, 107
  mixed-integer programming 243
Decomposition in nonlinear programming 187
Decomposition Structure 213
Decomposition Techniques
  other 271
Decomposition techniques 7, 8, 18, 28, 30, 41, 62, 67, 176, 217, 385
Degenerate inequality constraint 148
design 4
Directional and partial derivatives 318
Directional derivatives 320
Distributed 67, 107, 111, 136, 192, 214, 216–218, 243
Dual 155
  example 174
  problem 152
Dual feasibility condition 143
Dual function 161, 188

Index 539

Dual infeasibility   195
Dual price   157
Dual problem   149, 188
  in standard form   150
  obtaining   151
Dual variable   6, 74, 77, 115, 141, 175, 181, 214, 224, 246, 252, 280, 303, 304, 307–310, 339, 359, 371, 382, 389
Duality   141
  in linear programming   149
  in nonlinear programming   161
  theorems   154
Duality and separability   176
Duality for convex problems
  theorem   173
Duality gap   141, 171, 172, 189, 244
Duality theory   305

Energy production   14, 18, 22, 24, 43, 62, 374
Energy production model   23
Equality constraints   142
Equilibrium problems   282
Example
  cutting hyperplane   278
  Dantzig-Wolfe decomposition   79
  Dantzig-Wolfe revisited   89
  energy production model   23
  linear programming   167

Facet   27, 28, 375, 377, 380
Failure mode   46
Feasibility cuts   253
Feasibility region   3, 27, 94, 199, 202–204, 257
Feasible region   70, 83, 112, 142, 145, 155, 171, 304
  boundaries   113
Feasible solution   142
Flow application   292

General method for obtaining all sensitivities   315
Gradient of the dual function   164

Hessian   164, 167
Hydroelectricity   62
Hydrothermal coordination   381

  problem formulation   383
  solution approach   384

Industrial   62, 361
Inequality constraints   142
Infeasibility   128
Integer variable   245

Karush, Kuhn, and Tucker   142
Karush–Kuhn–Tucker
  conditions   142
  first order conditions   143
Karush–Kuhn–Tucker conditions   190, 305, 306, 311
  optimality   181
Karush–Kuhn–Tucker first- and second-order optimality conditions   142
Karush–Kuhn–Tucker second-order necessary conditions   149
Kuhn–Tucker multipliers   143

Lagrangian   161
Lagrangian relaxation   187, 210, 215, 233, 239, 369, 381
  Dantzig-Wolfe   233
  algorithm   194
  decomposition   188
  in LP   234
Linear programming   4
  complicating constraints   8, 67
  complicating variables   28, 107
  dual   167
  dual problem   151
  duality   149
  mixed-integer   4, 243, 244, 264
  primal   167
  problem   7, 9, 70, 109, 136, 197, 202, 233, 309, 377
  sensitivities   309
Lower bound   11, 46, 87, 116, 170, 189, 203, 205, 225, 235, 244, 251, 276, 352, 385

Master problem   69, 74, 79, 87, 88, 93, 99, 114, 115, 141, 198, 238, 245, 280, 360, 369, 371–373, 378
  alternative formulation   93
  dual   236

issues 88
Mathematical programming 3
Matrix analysis of structures 288
Mill problem 155
Mixed design 47
Mixed-integer
  linear programming 244
  nonlinear programming 251
Mixed-integer linear programming
  Benders decomposition 245
Mixed-integer programming 55, 57
  decomposition 243
Modern design 46
Modified Lagrangian relaxation 211
Multiarea electricity network 42
Multiarea optimal power flow 385
  problem formulation 387
  solution approach 389
Multiplier updating 195, 208

Network constrained unit commitment 368
Network constraint
  problem formulation 370
  solution approach 371
Newton algorithm 147
Nonanticipativity constraints 16, 17, 19
Nonlinear programming 4
  complicating constraints 39
  complicating variables 53
  decomposition 187
  duality 141, 161
  method 495
  mixed-integer 4, 243, 251, 493
  problem 7, 142, 312
  sensitivity 312

Objective function 3, 4, 6, 40, 46, 57, 73, 76, 87, 115, 116, 128, 141, 142, 145, 189, 195, 207, 225, 235, 245, 253, 258, 286, 303, 306, 308, 310, 315, 350, 369, 379, 385
Obtaining
  the dual from a primal in standard form 150
Obtaining the dual problem 151
Only equality constraints 145
Only inequality constraints 145

operation problems 4
Optimality condition decomposition 187, 210, 385
  algorithm 216, 389
  example 218
optimization 3
Other decomposition techniques 271
Outer linearization 57
  algorithm 258, 259
  convergence 264
  example 260

Partial derivatives 320
Particular cases 321
Partitions 70
Penalty parameter updating 208
Polymatroid 27
Prevent a distributed solution 107
Preventing a straightforward solution 108
Primal 155
Primal and dual decomposability 109
Primal and dual optimality 171
Primal feasibility condition 143
Primal infeasibility 244
Primal variable 6
Probabilistic design 47, 51, 351, 356, 365, 410, 414
Problem with decomposable structure 73
Production costing 374
  problem formulation 376
  solution approach 377
Production scheduling 39–41

Reduced cost 75, 77, 375, 376, 378, 380
Regular degenerate case 328
Regular nondegenerate case 326
Regular point 148, 308, 310, 339
Regularity conditions 4, 273, 274, 304
Relaxation method 272
  algorithm 272
  wall design 276
Relaxed problem 73–76, 78, 79, 88
Reliability 228
  analysis 349
  bounds 48
  constraints 45, 46
  estimation 52

index   47, 52, 276
indices   349, 353
Reliability-based optimization   48
Reliability-based Optimization of a
     Rubblemound Breakwater   228
River basin operation   19, 21
Rubblemound breakwater   48
   GAMS code   407

Safety factor   47, 48, 276, 277, 299, 349,
     351, 353, 367
Same active constraints   323
Scenario tree   15
Scenarios   13
Second-order
   necessary conditions   149
   sufficient conditions   149
Sensitivities   155, 305
   active constraints   339
   dependence on a common parameter
     312
   dual variables   175
   linear programming   309
   objective function   308, 310
     general formula   310
Sensitivity   157, 175, 181, 224, 308, 311,
     315
   analysis   7, 303, 304, 310, 410, 414
   local   303
   nonlinear programming   312
Sensitivity analysis   6
Sensitivity in Regression models   389
Separability   208
Set of all feasible perturbations   317
Shadow price   157
Software   397, 403, 407, 410
Special case   145
Standard form
   dual problem   150
Stochastic   12
Stochastic hydro scheduling   12, 18
Stopping criteria   204
structural properties   4
Subdifferential   164–166
Subgradient   164, 166, 195, 196, 211,
     222, 239, 381
   example   165
Subgradient and subdifferential   164
Subgradients   164
Subproblem   72, 73, 76, 78, 93, 97, 110,
     114, 116, 141, 181, 189, 196, 209,
     214, 224, 243, 253, 280, 286, 360,
     369, 372, 379, 388
   infeasibility   128, 253
   reduced   215
Subproblems   5
Sufficient conditions   309
Susceptance   43
Symmetry of the duality relation   150

Theorem
   dual in standard form   150
   duality   154
   duality for convex problems   173
   local duality   189
   primal and dual decomposability
     109
   sensitivities   175, 308
   sensitivities of active constraints
     339
   symmetry of the duality relationship
     150
   weak duality   154, 170
Transnational soda company   8
Trust region method   202

Uncertainty   13, 62
Unconstrained problems   145
Unit commitment   55, 369
Updating safety factor bounds
   algorithm   355
Upper bound   11, 47, 70, 87, 89, 116,
     128, 203, 204, 224, 236, 238, 245,
     276, 353, 355, 385

Variable
   dual   6, 74, 77, 115, 141, 175, 181,
     214, 224, 246, 252, 280, 303, 304,
     307–310, 339, 359, 371, 382, 389
   primal   6
Vector of mismatches   164

Wall design   45, 349
Wall problem
   GAMS code   410
Water supply system   5, 6, 36, 60
   revisited   60
Weak duality
   theorem   170
Weak duality lemma   154
Weighting problem   74, 76, 77